Exploring Genomics, Proteomics and Bioinformatics

Exploring Genomics, Proteomics and Bioinformatics

Edited by **Charles Malkoff**

SYRAWOOD
PUBLISHING HOUSE
New York

Published by Syrawood Publishing House,
750 Third Avenue, 9th Floor,
New York, NY 10017, USA
www.syrawoodpublishinghouse.com

Exploring Genomics, Proteomics and Bioinformatics
Edited by Charles Malkoff

International Standard Book Number: 978-1-68286-253-7 (Hardback)

Printed in China.

Contents

Preface

This book on genomics, proteomics and bioinformatics aims to provide in-depth knowledge in understanding molecular cloning of DNA, functions and structure of proteins and genomes, and biological data. It strives to provide a fair idea about this discipline and to help develop a better understanding of the latest advances and computational methods within this field. The topics included in this book on protein profiling, genome analysis, gene expression, genetic interactions, properties and interactions of proteins are of utmost significance and bound to provide incredible insights to readers. It will prove to be immensely beneficial to students and researchers engaged in this field.

All of the data presented henceforth, was collaborated in the wake of recent advancements in the field. The aim of this book is to present the diversified developments from across the globe in a comprehensible manner. The opinions expressed in each chapter belong solely to the contributing authors. Their interpretations of the topics are the integral part of this book, which I have carefully compiled for a better understanding of the readers.

At the end, I would like to thank all those who dedicated their time and efforts for the successful completion of this book. I also wish to convey my gratitude towards my friends and family who supported me at every step.

Editor

Molecular marker analysis to differentiate a clonal selection of Centennial Seedless grapevine

Manisha P. Shinde, Anuradha Upadhyay*, Lalitkumar B. Aher and G. S. Karibasappa

National Research Centre for Grapes, Solapur Road, Pune -412307, Maharashtra, India.

Microsatellite and amplified fragment length polymorphism (AFLP) markers were used to differentiate Manjari Naveen, a clonal selection of Centennial Seedless variety of grape. Twenty one (21) microsatellite primers could not detect variation between parent variety and its clone. AFLP analysis with 40 primer combinations generated 1093 markers. Three AFLP markers were polymorphic; two markers were present only in Centennial Seedless, whereas one was present only in Manjari Naveen. These markers will be useful for establishing genetic identity, variety registration and protection of breeder's right.

Key words: *Vitis vinifera*, AFLP, microsatellites, clonal selection.

INTRODUCTION

Grapevine (*Vitis vinifera* L) is one of the most important fruit crops in India, presently cultivated over an area of about 110 000 ha, and mainly used for table purposes. Development of varieties suitable for different grape growing regions of the country and requiring minimum inputs in term of agrochemicals and labor is the focus of grape improvement. Besides conventional methods of breeding, clonal selection is an important method of cultivar improvement in grapes. Clones with phenotypic variations are observed due to spontaneous somatic mutation in the regenerative cells. The clones with considerable difference in traits are considered as different cultivars, the most important example being the clones of Pinot viz. Pinot Noir, Pinot Blanc, Pinot Gris, each one of which is recognized as a different cultivar. A clonal selection of grape variety Centennial Seedless was identified at National Research Centre for Grapes in 2002 with a balanced canopy, self-thinned bunches and ripening early by about 20 to 25 days at Pune as compared to Thompson Seedless, the predominant table variety. This selection was further evaluated for several years in multi-locational trials and was released as Manjari Naveen in 2008 for commercial cultivation in India, for table and raisin purpose.

It is important to characterize this selection in order to establish its genetic identity for variety registration and protection. Molecular markers are the most efficient tools for genetic characterization and variety identification. While microsatellite markers have proven to be effective for varietal identification, amplified fragment length polymorphism (AFLP) markers which scan larger proportion of the genome have been more useful for the identification of grape somatic mutants and clonal variants (Blaich et al., 2007; Upadhyay et al., 2011). The aim of this study was therefore, to identify a molecular marker to distinguish Manjari Naveen from its parent Centennial Seedless.

MATERIALS AND METHODS

Six years old vines of *Vitis vinifera* L. var. Centennial Seedless and Manjari Naveen grown at research farms of National Research Centre for Grapes, Pune (latitude 18.5236° N, longitude 73.8478° E) were used for the study. Young fresh leaf samples were collected and DNA was extracted using the DNeasy plant mini kit (Qiagen, Hilden, Germany), following the supplier's instructions. The DNA concentration was determined on nanophotometer (IMPLEN) and quality was tested by agarose gel electrophoresis. Twenty one (21) microsatellite primers used in this study were VMC4f8, VMC7b1, VMC7f2, VMC7g3, VMC8g3.2, VrZAG62, VrZAG79, VVIB23, VVIB63, VVII52, VVIP17A, VVIP22, VVIP25B,

*Corresponding author. E-mail: anu_upadhyay@yahoo.com.

Table 1. Details of AFLP analysis of Centennial Seedless and Manjari Naveen.

Primer pair record	No.
Total number of selective primer pairs used	40
Number of primer pairs resulting amplification	37
Total number of AFLP markers obtained	1093
Average number of markers per primer pair	29
Number of polymorphic markers	3

VVIP31, VVIP44, VVIV15B, VVIV16, VVIV33, VVMD21 and VVMD27. The microsatellite analysis was performed as follows. The polymerase chain reaction (PCR) reaction mixture (10 µL) contained 20 ng genomic DNA;1X PCR buffer;100 µM of each dNTP; 2.5 mM $MgCl_2$; 0.66 µM forward primer labeled with FAM, VIC or NED, 0.66 µM reverse primer and 1.0 U of Taq-polymerase (Bangalore Genei Pvt. Ltd., India). The temperature profile consisted of the following steps: 10 min at 94°C followed by 35 cycles for 1 min at 94°C, 1 min at 54 or 56°C and 1 min at 72°C and a final extension for 10 min at 72°C. PCR products were analyzed on ABI 3130 genetic analyzer using 36 cm capillary filled with POP7 polymer. GeneMapper Version 4.0 (Applied Biosystems, USA) was used to determine the peak size using local Southern method and allele call.

AFLP analysis was performed using AFLP analysis system I (Invitrogen Life Technologies, USA) following manufacturer's instructions with some modifications as reported earlier in Upadhyay et al. (2011). In brief, 500 ng DNA was used to restrict digestion with *Eco*RI-*Mse*I, followed by adaptor ligation and pre amplification as per the manufacturer's instructions. However, selective amplification fluorescently labeled *Eco*RI primers (Applied biosystems, USA) were used in combination with *Mse*I primers. A total of 40 primer combinations were studied. After amplification, 1 µL of amplified product was added to 15 µL formamide containing 0.1 µL ROX internal size standard and denatured at 94°C for 5 min and analyzed on 3130 genetic analyzer (Applied Biosystems, USA) using a 36 cm capillary filled with POP7 polymer. GeneMapper Version 4.0 (Applied Biosystems) was used for data scoring, following the instructions given in the AFLP system analysis getting started guide (Applied Biosystems, USA). All primer combinations that produced polymorphic AFLP markers were repeated to confirm the results.

RESULTS AND DISCUSSION

Twenty one (21) microsatellite primers detected 32 alleles in Centennial Seedless and Manjari Naveen. These two genotypes were heterozygous at 11 loci and two alleles were detected at these 11 loci. However the allele profiles of Centennial Seedless and its clone Manjari Naveen were identical at all the 21 loci. Thus these primers failed to detect any variation between these two genotypes. The probability of identical genotypes for these primers has been estimated to be 7.5×10^{-23} in a separate set of 47 grape genotypes in our laboratory. Thus identical allele profile at 21 loci confirmed the varietal origin of Manjari Naveen. There are contrasting reports on the ability of microsatellite primers to differentiate the grapevine clones.

Microsatellites either fail to differentiate the clones or show limited genetic variability among clones of different varieties. Attempts have also been made to develop highly variable microsatellite loci for clones of a variety (Regner et al., 2006) but the use of microsatellite markers for differentiating the clones of a variety remains limited.

AFLP analysis of Centennial Seedless and its clone Manjari Naveen (Table 1) produced three reproducible AFLP markers that differentiated the mutant clone from its parental genotype. Out of 40 *Eco*RI:*Mse*I AFLP primer pairs used for the selective amplification, 37 primer pairs amplified both the genotypes. A total of 1093 markers were obtained with 37 primer pairs. The number of markers detected by each primer pair ranged from 12 to 55 with an average of 29 markers per primer combination. Thirty four (34) primer pairs resulted in identical AFLP fragment pattern among two genotypes. However, variation between Centennial Seedless and Manjari Naveen was detected by three primer pairs. Primers E-AGG:M-CAC and E-ACC:M-CAG gave AFLP markers of 438 bp and 280 bp size respectively, which were detected only in Centennial Seedless (Figure 1). Another AFLP marker of 408 bp size obtained with primer pair E-AAG:M-CAG was present in Manjari Naveen but absent in Centennial Seedless (Figure 2).

AFLP marker technique has good resolving power, which makes it suitable system for the detection of somatic mutations responsible for clonal variation. High resolution power is due to high polymorphic index and ability to scan wider genome area. The efficiency of AFLP can further be enhanced by using large number of primers and using different restriction enzymes to target the variations arising due to point mutations in restriction sites for different enzymes. We have used a large number of primers in this study and were able to differentiate clone Manjari Naveen from its parent Centennial Seedless though; the percentage of polymorphic bands was little. Several works have reported successful use of AFLP to distinguish clonal selections of a variety (Scott et al., 2000; Upadhyay et al., 2011).

AFLPs have also been used for clonal differentiation of cultivars like Pinot Noir (Blaich et al., 2007), Primitivo (Fanizza et al., 2005), Riesling (Anhalt et al., 2011) and several others. On the other hand, there are reports of inability of AFLP markers to differentiate the clones of a variety (Fanizza et al., 2003). Thus AFLP analysis was successfully used to differentiate Manjari Naveen and Centennial Seedless. The available markers will be useful for identification and variety protection.

ACKNOWLEDGEMENT

The authors acknowledge the financial assistance from the Department of Biotechnology, Government of India, New Delhi.

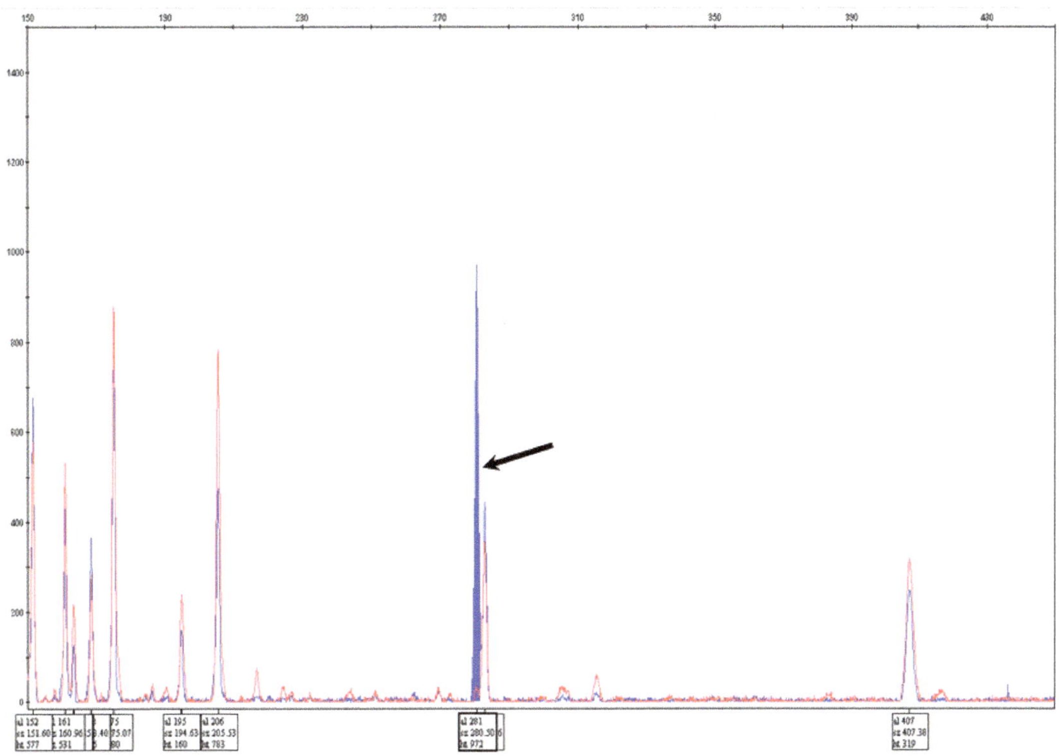

Figure 1. Polymorphic AFLP marker present in Centennial Seedless.

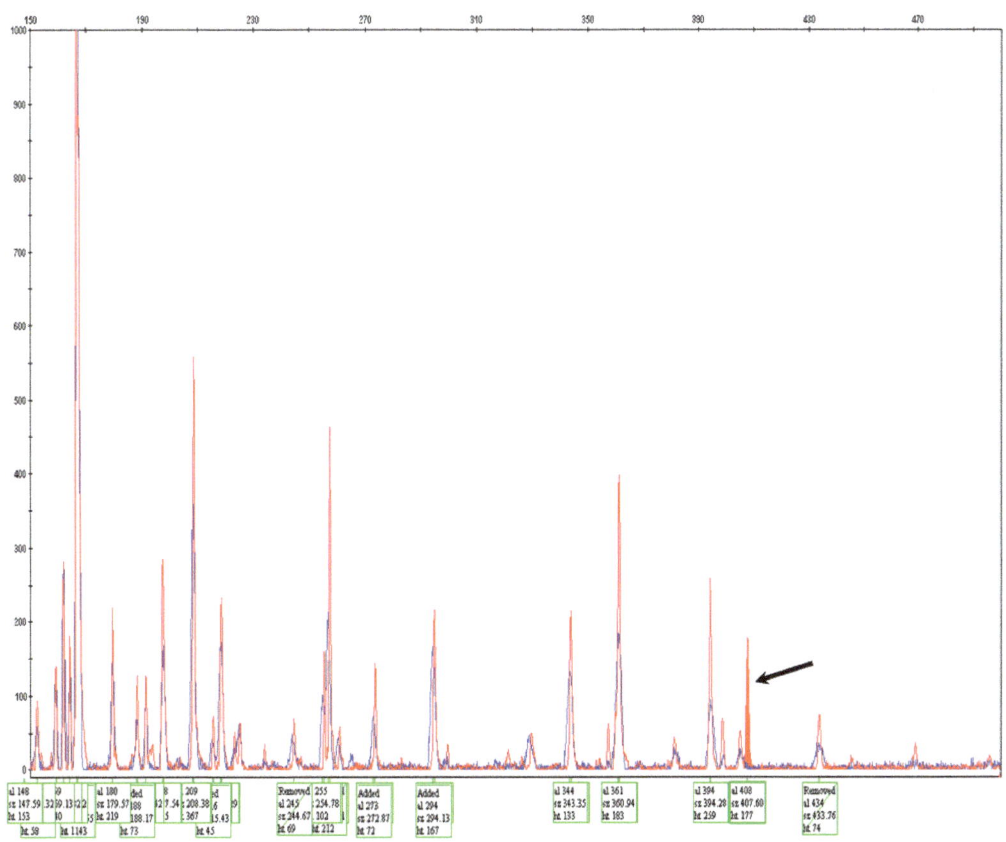

Figure 2. Polymorphic AFLP marker present in Manjri Naveen.

REFERENCES

Anhalt UCM, Martínez SC, Rühl E, Forneck A (2011). Dynamic grapevine clones - an AFLP-marker study of the *Vitis vinifera* cultivar Riesling comprising 86 clones. Tree Genet. Genomes 7:739-746.

Blaich R, Konradi J, Rühl E, Forneck A (2007). Assessing genetic variation among Pinot Noir (*Vitis vinifera* L.) clones with AFLP markers. Am. J. Enol. Viticult. 58:526-529.

Fanizza G, Chaabane R, Ricciardi L, Resta P (2003). Analysis of a spontaneous mutant and selected clones of cv. Italia (*Vitis vinifera*) by AFLP markers. Vitis 42:27-30.

Fanizza G, Lamaj F, Resta P, Ricciardi L, Savino V (2005). Grapevine cvs Primitivo, Zinfadel and Crljenak kastelanski: molecular analysis by AFLP. Vitis 44:147-148.

Regner F, Hack R, Santiago, JL (2006). Highly variable *Vitis* microsatellite loci for the identification of Pinot Noir clones. Vitis 45:85-91.

Scott K, Ablett E, Lee L, Henry R (2000). AFLP markers distinguishing an early mutant of Flame Seedless grape. Euphytica 113:245-249.

Upadhyay A, Aher LB, Karibasappa GS (2011). Detection of variation among clonal selections of grapevine (*Vitis vinifera* L.) cv. Kishmish Chernyi by AFLP analysis. J. Hortic. Sci. Biotechnol. 86:230-234.

Isolation of four hydrocarbon effluent-degrading *Bacillaceae* species and evaluation of their ability to grow under high-temperature or high-salinity conditions

Saïd Ben Hamed[1,2]*, Abderrazak Maaroufi[1], Abdeljelil Ghram[1], Ben Ali Gam Zouhaier[2] and Marc Labat[2]

[1]Laboratory of Microbiology, Bioprocess Group, Institute Pasteur of Tunisia (IPT), Bp 74, 13 place Pasteur, Belvédère-1002 Tunis, Tunisia.
[2]Laboratoire de Microbiologie IRD, Biotechnologie Microbienne des Environnements Chauds, UMR D180, IRD ESIL MIO, Aix-Marseille Université, 163 avenue de Luminy, 13288 Marseille cedex 9, France.

Four bacterial strains belonging to the family *Bacillaceae* were isolated from a polluted site and tested for their efficiency in degrading a refinery effluent highly polluted with hydrocarbons. Among 13 strains isolated, four were selected for their efficiency. Each of these four strains demonstrated a strong ability to grow as a single strain on a hydrocarbon effluent (HCE) as sole carbon source. In batch assays using clarified wastewater as diluent, the strains achieved high-percentage decreases in total hydrocarbon concentration within 18 days: 98% with *Bacillus licheniformis* STK08, 87% with *Geobacillus stearothermophilus* STM04, 80% with *Lysinibacillus sphaericus* STZ75 and 72% with *Bacillus firmus* STS84. The decreases were greater during the first three days of treatment, with 73, 66, 39 and 47% recorded for *Bacillus licheniformis* STK08, *G. stearothermophilus* STM04, *L. sphaericus* STZ75 and *B. firmus* STS84, respectively. Growth assays run under different conditions showed that *B. licheniformis* STK08 and *G. stearothermophilus* STM04 were able to grow at salinities of up to 120 g/L and at 55°C. Potential biosurfactant production tested using two methods namely modified drop collapse (MDC) and blue agar plate (BAP) demonstrated that the four *Bacillaceae* species are biosurfactant producers.

Key words: Hydrocarbons, biodegradation, pure culture, high salinity, high temperature.

INTRODUCTION

Petroleum oil is a major driver of worldwide economic activity, but it has also created major pollution problems due to repeated accidental spills during petroleum component transport and storage and during refining process. Many physical and chemical methods can remove toxic petroleum components, such as incineration, chlorination, ozonation and combustion, but they unfortunately require elaborate facilities and substantial amounts of additional fuels (Mrayyan and Battikhi, 2005). Moreover, these processes, especially the chemical ones, produce intermediates that may present toxicity levels similar to the original substances (Tomei et al., 2003). Thus, some studies tend to look for other process which seems to be more efficient and less costly. In fact, several studies have demonstrated that biological approaches are better suited to the degradation of these pollutants due to their low cost, their ability to convert contaminants to harmless end-products (Rahman et al., 2002) and the proven high sensitivity of the technologies employed. Efficient treatments hinge on precise knowledge of the polluting load, the physical and chemical properties of the pollutant, the environmental conditions and the specific competencies of the micro-

*Corresponding author. E-mail: benhamed_med@yahoo.fr.

organisms involved (Rahman et al., 2002; Hui et al., 2007; Das and Chandran, 2011).

Studies of biodegradation processes have demonstrated that various species belonging to different genera such as *Achromobacter, Alcaligenes, Azospirillum, Bacillus, Cycloclasticus, Micrococcus, Pseudalteromonas, Pseudomonas, Sphingomonas, Staphylococcus, Vibrio* and *Xanthomonas* (Hedlund et al., 2001; Zhuang et al., 2003; Viñas et al., 2005; Alonso-Gutiérrez et al., 2009) are able to use one or more polycyclic aromatic hydrocarbons (PAHs) as sole carbon source. These genera are capable of degrading crude oil in pure and mixed cultures with natural or constructed bacterial consortia. However, the degradation rate usually depends on the composition of the crude oil used and its bio-accessibility. Also, compounds such as saturates, aromatics and polar compounds present in different crude oil samples are degraded to different degrees by the same organisms (Tzarkova and Groudeva, 2000).

Hydrocarbon degradation has been widely reported in laboratory-scale batch studies. Indeed, Chavan and Mukherji (2008) demonstrated that green algae and cyanobacteria can remove 98.99% of total petroleum hydrocarbon. Likewise, using bacterial consortium isolated from oil contaminated site, Malik and Ahmed (2011) revealed percentages of hydrocarbon removal of 94.64 and 93.75 of aliphatic and aromatics hydrocarbons, respectively. Conclusions of all biodegradation studies show that bacterial efficiency is usually related to the acclimation period and growth conditions, mainly temperature and salinity.

In fact, hydrocarbon biodegradation in high salt medium was slow, because high NaCl in medium disrupts cell membrane, denaturates some proteins such as enzymes, or changes osmotic forces; these factors could be lethal for microorganisms (Minai-Tehrani et al., 2006). Nevertheless, temperature improves hydrocarbon degradation and enhances its bio-accessibility for bacteria (Margesin and Schinner, 2001). Regarding the influence of environmental conditions on the hydrocarbon degradation rate, it seems interesting to seek for efficient bacterial strains, which could be explored for the bioremediation of hydrocarbon-polluted sites in terrestrial or marine environments, especially under high temperature or high salinity conditions. Within this framework, four species belonging to the family *Bacillaceae* were isolated from sites highly polluted with hydrocarbons. The isolates were tested to evaluate their ability to grow at different temperatures and salt concentrations, to produce biosurfactant and to degrade a refinery effluent in pure culture efficiently.

MATERIALS AND METHODS

Sampling site and characterization of the hydrocarbon effluent

Samples of a hydrocarbon effluent (HCE) were collected from a rejection mud pit run by the Tunisian Refining Industries Company (STIR) located at Bizerte in Northern Tunisia. The samples collected from the STIR were transferred to pre-sterilized glass bottles and analyzed by measuring Total Hydrocarbon Concentration (THC), Chemical Oxygen Demand (COD), Biochemical Oxygen Demand (BOD_5), pH and salinity (Ben Hamed et al., 2010).

Isolation of hydrocarbon-degrading bacteria

Samples from highly polluted wastewater were collected and serially diluted and plated onto basal medium amended with 1% (v/v) HCE as the sole carbon source. The plates were incubated at 30°C for three days. Individual colonies were streaked successively onto the same medium (basal medium with 1% (v/v) HCE) for isolation and purification (Auffret et al., 2009).

The basal medium used for strain isolation contained per liter: 0.5 g KH_2PO_4, 1.5 g K_2HPO_4, 1.0 g NH_4Cl, 0.5 g $MgSO_4$. $7H_2O$, 5.0 g NaCl and 1.0 ml of trace metal solution. The trace metal solution contained, per liter: 1.5 g nitrilotriacetic acid, 5.0 g $MnSO_4.2H_2O$, 0.01 g $FeSO_4$, 0.1 g $CaCl_2.2H_2O$, 0.1 g $ZnSO_4.7H_2O$, 0.01 g $CuSO_4.5H_2O$, 0.1 g $CoCl_2$, 0.01 g $Al_2(SO_4)_3.16H_2O$, 0.01 g H_3BO_4, 0.01 g $Na_2MoO_4.2H_2O$, 18 g/L agar (Van Hamme and Ward, 2001). The pH of the medium was adjusted to 7.5, and incubation was run at 37°C for 48 h.

Morphological and biochemical characterization of isolated bacteria

Strains were identified using morphological and biochemical characteristics (colony size and shape, pigmentation, cell shape, cell arrangement, cell motility, Gram staining, oxidase and catalase tests). The main characteristics of each strain were checked using an API 50CH system gallery and its associated software (APILAB 3.3.3, Biomérieux Corporation) (Hedi et al., 2009).

DNA extraction and 16S rRNA gene amplification and sequencing

DNA was extracted from bacterial cultures using a Wizard Genomic DNA Purification Kit from Promega. The 16S rRNA gene of the isolate strain was amplified by adding 0.5 µL of extracted DNA to a thermocycler microtube containing 10 µL of taq polymerase q Biogène, 0.5 µL of 50 nM Fd1 and Rd1 primers and 38.5 µL of sterilized distilled water. The taq polymerase q Biogène contained the taq buffer, $MgCl_2$ $6H_2O$, dNTPs and the taq polymerase. The universal primers Fd1 and Rd1 (Fd1, 5'-AGAGTTTGATCCTGGCTCAG-3' and Rd1, 5'-AAGGAGGTGATCCAGCC-3') were used to obtain a PCR product of almost 1.5 kb corresponding to base positions 8-1542 based on *Escherichia coli* numbering of the 16S rRNA gene (Winker and Woese, 1991). The sample was placed in a hybrid thermal reactor thermocycler (BIOMetra, Leusden, The Netherlands), denatured for 1 min at 95°C and subjected to 30 cycles for 20 s at 95°C, 30 s at 55°C, and 1 min and 30 s at 72°C. This was followed by a final elongation step lasting for 5 min at 72°C. The PCR products were analyzed on 1% (w/v) agarose gels and sent to GATC Company (GATC Biotech Ltd., Germany) for sequencing. Multiple alignment analyses with CLUSTAL X (Thompson et al., 1997) were run, and all alignment gaps were treated as missing data. The obtained sequences then underwent a BLAST search (version 2.2.23 available on line at Http://blast.ncbi.nlm.nih.gov/blast.cgi) to determine the closest related species registered in the GenBank database (Figure 1).

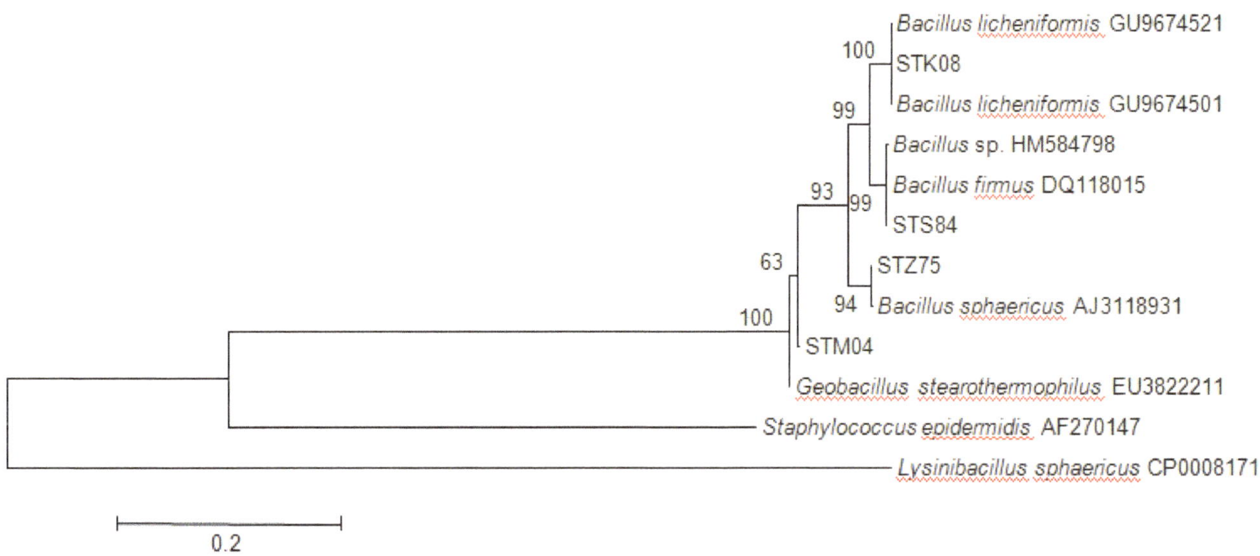

Figure 1. 16rRNA gene sequence-based dendrogram obtained by neighbour-joining method, showing the phylogenetic positions of STK08, STS84, STZ75 and STM04 *Bacillaceae* isolates. Bootstrap confidence levels greater than 50% are indicated at the internodes.

Biosurfactant production screening using the modified drop collapse method (MDC)

To prepare the assay, plates were rinsed successively with hot water, 75% ethanol, distilled water and dried with air. After preparation, plates were equilibrated and coated with a thin layer of crude oil. The preparation was left for 24 h to ensure a uniform oil coating. Bacterial suspensions of all isolated strains were prepared and OD (600 nm) was adjusted to 0.8 for each strain. A volume of 0.5 µl of each bacterial suspension was transferred on the thin oil layer.

The shape of the drop was inspected after 1 min; if the drop remained beaded, the result was scored as negative (-). If the drop collapsed, the result was scored as positive (+) (Bodour and Miller-Maier, 1998).

Biosurfactant production screening using the blue agar plate method (BAP)

This method shows an ionic biosurfactant producing strains by color reaction. Mineral salts agar medium (MSA) (Siegmund and Wagner, 1991) supplemented with carbon sources (glycerol: 2%), cetyltrimethylammonium bromide (CTAB: 0.5 mg/ml), methylene blue (MB: 0.2 mg/ml) were prepared. Plates was streaked with bacteria of interest and incubated at 30°C for 24 h. A dark blue halo around the culture was considered as positive for biosurfactant production. Sodium dodecyl sulfate (SDS: 1 mg/l) and sterile distilled water were used respectively as positive and negative controls.

Bacterial growth conditions

To check the potential use of these strains for biodegradation in biotechnological processes under extreme conditions, strain performances were tested under halophilic and thermophilic conditions. Bacterial growth assays were carried out on liquid basal medium with 1% (v/v) of HCE as sole carbon source, with 2, 60 or 120 g/L of salt, a temperature of 30 or 55°C, or simultaneous combinations of these different conditions of temperature and salinity (Table 1).

The increase in biomass was quantified by measuring the OD at a wavelength of 600 nm. A reading of OD_{600} is frequently used in studies characterizing the physiology of hydrocarbon utilization (Ciric et al., 2010).

Pre-culture of the isolated strains for biodegradation study

Each of the four isolated strains was cultivated in a 1 L flask bottle containing 800 ml of sterile clarified wastewater (CWW) as diluent, low COD (120 mg/l), and trace hydrocarbons (THC <10 mg/l). Each flask bottle was inoculated with 20% (v/v) of a single strain in pure culture, and the HCE was added at 1% (v/v). The initial COD and initial THC of the hydrocarbon effluent were high, at 312 and 71 g/l, respectively, while the COD and THC concentrations of the inoculum were low, at 0.625 and 0.15 g/L, respectively. Thus the inoculum represented about 5% of total COD whereas HCE represented about 95% of total COD of the culture medium in each assay. Consequently, the HCE degradation run in these 1 L flask bottles were tested as main rather than sole carbon source. After incubation at 30°C for 48 h in a reverse shaker (130 rpm), biomasses of each single strain of these pre-grown cultures (20% v/v) were used to inoculate each bioreactor.

Bioreactor design and biodegradation conditions

In-bioreactor biodegradation assays were conducted in a batch system at 30°C and in 2 g/L of NaCl, in view of the bacterial growth conditions previously obtained (Table 1). As done earlier, CWW was used as diluent, with HCE as main carbon source. The use of CWW as culture medium diluent in these in-bioreactor biodegradation assays was justified by the potential exploitation of this by-product combination (CWW and HCE) in industrial processes, as this would avoid the need for amendment of costly nutrients, essentially nitrogen and phosphorus organic or inorganic compounds. The HCE was added to obtain a final THC of 750 mg/l. Salinity was adjusted to 2 g/l and pH was adjusted to 7.2. Four bioreactors were trialed. Two one-liter flask bottles of each pre-grown single strain were necessary to inoculate each 10-L

Table 1. Bacterial growth under different conditions of temperature and salinity and biosurfactant production

Strain	T: 30°C; NaCl: 2 (g/L)	T: 30°C; NaCl: 60 (g/L)	T: 30°C; NaCl: 120 (g/L)	T: 55°C; NaCl: 2 (g/L)	T: 55°C; NaCl: 60 (g/L)	T: 55°C; NaCl: 120 (g/L)	Biosurfactant production (using MDC and BAP)
Bacillus licheniformis STK08	+++	+++	++	++	++	+	+
Bacillus stearothermophilus STM04	+++	+++	++	++	++	+	+
Bacillus sphaericus STZ75	+++	+	-	+	-	-	+
Bacillus firmus STS84	+++	+	+	-	-	-	+

Growth at different conditions of temperature and salinity is considered: - if $OD_{600nm} \leq 0.1$; + for $0.1 \leq OD_{600nm} \leq 0.4$; ++ for $0.4 \leq OD_{600nm} \leq 0.6$; +++ for $OD_{600nm} \geq 0.7$.

bioreactor filled up to 8 L (final volume) and fed with a 5 L min^{-1} airflow. Each bioreactor was inoculated with 20% (v/v) of a single strain in pure culture, and the HCE was added at 1% (v/v). As seen earlier, for each essay, the effluent (HCE) represented the main percentage of COD and THC in the bioreactor (93.4 and 95%, respectively). The inoculum represented a minor percentage of COD and THC (3.7 and 4.0%, respectively), while CWW represented less than 3% of total COD and less than 1% of total THC of each in-bioreactor batch assay.

The initial THC concentration of each bioreactor batch assay was adjusted to 750 mg/l (CWW + HCE + inoculum) giving an initial COD mix concentration of about 3300 mg/l. Infeeds and outfeeds of each bioreactor were equipped with a 0.22 µm filter to maintain sterile conditions. Homogenization was ensured with a bubble diffuser and a magnetic stirrer. Biodegradation assays were carried out at ambient temperature (30°C). A negative in-bioreactor control assay with similar COD (2680 mg/l) but without inoculation was run in parallel with inoculated assays at the same final adjusted volume of 8 L.

Assessment of hydrocarbon degradation in batch assays

Each assay was analyzed periodically (at three-day intervals). HCE degradation was monitored by measuring THC and COD coupled with total suspended solids (TSS). Viable cell count was made on agar plates by decimal dilution, spreading 0.1 ml of samples taken from each bioreactor (final volume 8 L). Inoculated agar plates contained basal medium supplemented with HCE as sole carbon source (1% v/v).

COD and BOD_5 were measured by standard methods (McGraw, 1985). In the initial effluent sample, and in samples collected periodically from the bioreactor, THC was analyzed on a gas chromatography (GC) system equipped with a flame ionization detector (FID) according to French standard NF EN ISO 9377-2. Each sample was acidified (pH = 2) and centrifuged at 10000 rpm for 15 min with hexane (5:1). After a separation phase, the organic phase was cleaned with Florisil® to remove the most polar substances.

After adding the internal standard (IS) (5-α-androstane, 2 mg/l), 2 µl of each extracted sample was injected under constant flow and analyzed by GC (Perkin Elmer-Auto system XL) coupled with FID and a capillary column (length 30 m; ID 0.53 mm; thickness 3 µm). The carrier gas was helium (He) (12.5 psi, flow rate 35 ml min^{-1}). A control assay under the same conditions but without hydrocarbon was also run in parallel. The oven temperature program was as follows: initial temperature (50°C) with a 1 min hold, ramped up at 10°C min^{-1} up to 300°C with a 25 min hold. Peak areas of hydrocarbon components were quantified using Perkin Elmer software.

Statistical analysis

Statistical analysis was performed using the SPSS 10.0 software. Data underwent a one-way ANOVA test, and means were compared using Duncan's multiple range tests at 5% significance level.

RESULTS AND DISCUSSION

The analysis of the initial HCE demonstrated an important concentration of hydrocarbon (THC 28.8 $\times 10^3$ mg/l), DCO (96 × 10^3 mg/l) and BOD_5 (37.4 × 10^3 mg/l). These analyses showed also that the HCE is alkaline (pH: 12) and holding an important salt concentration (NaCl concentration: 90 g/l).

Assays realized on mineral agar plates supplemented with HCE as a sole carbon source allowed the isolation of 13 bacterial strains. Among these isolates, four showed a strong ability We first studied the morphological and biochemical characteristics of these isolated bacterial strains using the API 50CH gallery system (Biomérieux Corporation). The four most efficient strains were all identified as rod-shaped, Gram-positive, oxidase-positive and motile. The API system evaluated these four strains as affiliated to *Bacillus* genus, and tentatively identified them as *Bacillus licheniformis* STK08, *Bacillus stearothermophilus* STM04, *Bacillus sphaericus* STZ75 and *Bacillus firmus* STS84 (Table 1). Phylogenetic affiliation with DNA extraction and 16S rRNA gene amplification and sequencing gave similar affiliations for the four strains but the genera names of two strains were transferred as new combinations. *B. sphaericus* was transferred to *Lysinibacillus sphaericus* comb. nov. (strain STZ75) (Nazina et al., 2001) and *B. stearothermophilus* was transferred to *Geobacillus stearothermophilus* (strain STM04) (Ahmed et al.,

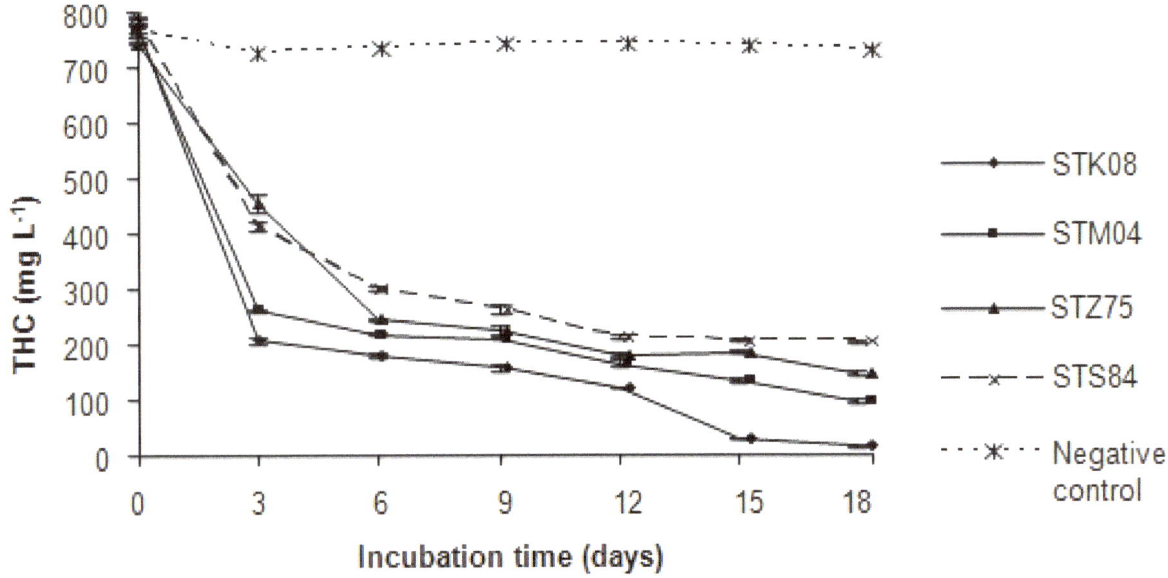

Figure 2. Total hydrocarbon concentration (THC) variation during the treatment time. The data are mean values (n = 3). The bars indicate the SDs (standard deviations). Some bars are smaller than the symbols. STK08: *Bacillus licheniformis;* STM04*: Geobacillus stearothermophilus*; STZ75: *Lysinibacillus sphaericus*; STS84*: Bacillus firmus.*

2007). These new names were not taken into consideration when the API system was utilized.

Bacterial growth under high salt concentrations and high temperature

The performances of these four strains at high salinity and high temperature were studied on basal medium supplemented with 1% (v/v) of HCE as sole carbon source (no CWW).

Bacterial growth results (Table 1) demonstrated that *B. licheniformis* STK08 and *G. stearothermophilus* STM04 strains could tolerate up to 120 g/L of NaCl and a temperature up to 55°C, but bacterial growth proved more efficient at 60 g/L of NaCl at 55°C. The *L. sphaericus* STZ75 strain was able to grow at 60 g/L of NaCl up to 30°C, and at high temperature (55°C) under low salinity (2 g/L NaCl), but was sensitive to high salinity combined with high temperature. The *B. firmus* STS84 strain did not grow at high temperature even with only 2 g/L of NaCl, but grew at 30°C and salt concentrations reaching 120 g/L.

Given that all four strains were capable of strong growth in both solid and liquid media at 30°C and 2 g/L of NaCl with hydrocarbon as sole carbon source, we tested the performance of each of these strains separately via bioreactor batch assays under these conditions.

Hydrocarbon biodegradation in bioreactor batch assays

THC variation in batch assays during the treatment period indicated that hydrocarbon concentration in each bioreactor inoculated with the four strains decreased within the 18 days of incubation (Figure 2). High-percentage decreases in THC were reached in 18 days, that is, 98% with *B. licheniformis* STK08, 87% with *G. stearothermophilus* STM04, 80% with *L. sphaericus* STZ75 and 72% with *B. firmus* STS84 (Figure 2). This decrease was greater during the first three days of treatment, where THC removal was 73, 66, 39 and 47% with *B. licheniformis* STK08, *G. stearothermophilus* STM04, *L. sphaericus* STZ75 and *B. firmus* STS84, respectively. The THC values of the non-inoculated bioreactor remained fairly constant (Figure 2).

In addition, these four strains were all able to produce biosurfactants following the two methods used (MDC and BAP) (Table 1). COD also showed a decrease starting from the first day of treatment, and recorded COD removals were 78, 75, 78 and 82% (data not shown) with *B. licheniformis* STK08, *B. firmus* STS84, *G. stearothermophilus* STM04 and *L. sphaericus* STZ75, respectively. The main phase of both THC and COD removal by all four strains was achieved in the first three days of treatment. These percentages also demonstrated a high level of HCE biodegradability by each of these four strains performed separately.

In these bioreactor batch assays, the highest THC removal (98%) was recorded with *B. licheniformis* STK08. This strain showed a second marked decrease in THC between days 12 and 15 (Figure 2). Similarly, the colony forming unit (CFU) variation of this strain demonstrated that peak bacterial proliferation was also recorded within these first three days (3.2×10^9 CFU mL^{-1}), decreasing progressively thereafter. The TSS value recorded with

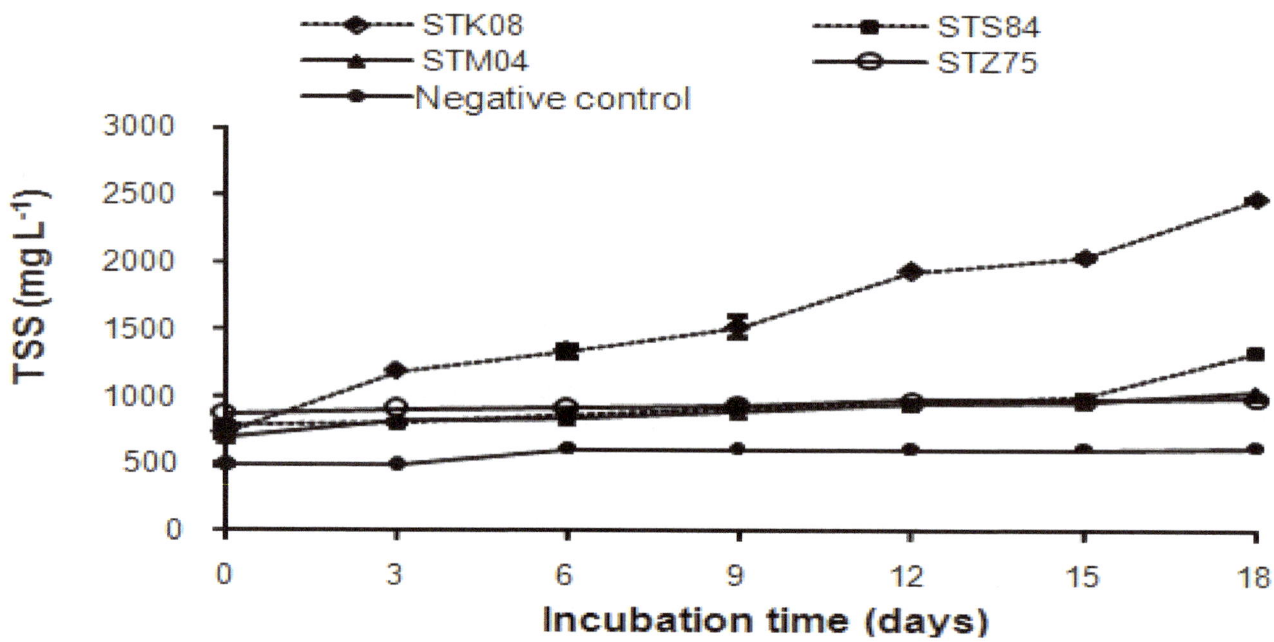

Figure 3. Variation in total suspended solids (TSS) variation during the treatment. The data are mean values (*n* = 3). The bars indicate the SDs (standard deviations). Some bars are smaller than the symbols. STK08*: Bacillus licheniformis;* STM04*: Geobacillus stearothermophilus;* STZ75*: Lysinibacillus sphaericus;* STS84*: Bacillus firmus.*

this strain, which was 735 mg/l on day 1, reached 1190 mg/l after three days and then doubled to 2476 mg/l between days 3 and 18 of treatment (Figure 3).

In the batch assay inoculated with *B. stearothermophilus* STM04, THC value decreased sharply (66%) during the first three days of treatment, then slowed progressively, reaching a final biodegradation rate of 87% after 18 days of treatment. Unlike strain STK08, strain STM04 showed a moderate increase in TSS during the treatment period, from 694 mg/l at day 1 to 955 mg/l at day 18 of treatment.

Figure 2 shows that THC decrease was weaker in assays carried out with *L. sphaericus* STZ75 and *B. firmus* STS84 strains, reaching about 145 and 203 mg/l, respectively, at day 18. Nevertheless, TSS values remained almost constant with these two strains, even after 18 days of treatment. Like strain STM04, strains STS84 and STZ75 showed a moderate increase in TSS, from 790 mg/l STS84 and 882 mg/l for STZ75 at day 1 to 1332 mg/l for STS84 and 990 mg/l for STZ75 at day 18 (Figure 3).

The different results under these conditions (30°C and 2 g/L NaCl) confirmed those found during isolation steps in agar plates, demonstrating that all four selected strains were able to degrade hydrocarbons contained in the HCE. The strong decrease in THC recorded mainly in the first days of treatment demonstrated that all these four strains were well acclimated to this HCE pollutant.

Time-course plots of THC variation revealed that the four strains selected in this work were able to degrade the HCE used as a main carbon source. Although all strains performed well on agar plates, there was clear variability between biodegradation rates of different strains in pure culture cultivated in liquid medium.

After 18 days of treatment, the highest biodegradation rate was recorded by *B. licheniformis* STK08 (98%) followed by *G. stearothermophilus* STM04 (87%) then *L. sphaericus* STZ75 (80%) and *B. firmus* STS84 (73%). The highest biodegradation rate for all the strains was essentially recorded in the first three days of treatment, as these highly adapted strains enabled fast biodegradation of the HCE. This fast decrease in both COD and THC, starting from day 1, disagrees with an efficient strain acclimation period in pre-culture. By contrast, Lotfabad and Gray (2002) showed that the presence of some PAH, such as phenanthrene, introduced a lag phase of 4.5 to 5 days in the degradation of other aromatic compounds.

The high THC removal recorded with our strain of *B. licheniformis* STK08 confirmed results obtained by Bayoumi et al. (2009) who report that *B. licheniformis* APIS473 has a high capacity to completely degrade anthracene and to reduce concentrations of some PAH compounds including acenaphtylene, acenaphtene, benzo(a)anthracene, benzo(b)fluoranthene, benzo(g,h,i)perylene and indeno(1,2,3-c,d)pyrene. Our study is the first to demonstrate that this species grow efficiently at high salt concentrations (up to 120 g/L), at high temperature (55°C) and under both conditions simultaneously.

As reported by Farinazleen et al. (2004), species *G. stearothermophilus* tolerates high temperatures (60°C) and is able to degrade 80 to 89% of crude oil (5 g/L), but this study proves the capacity of *G. stearothermophilus* strain STM04 to also grow under high salinity (120 g/L) simultaneously with high temperature (55°C). Our results showed strong bacterial growth with *B. licheniformis* STK08 and *G. stearothermophilus* STM04 up to 55°C. High temperature is often a limiting factor for strains, but also has potentially positive effects since high temperatures decrease the viscosity of hydrocarbons and so increase their solubility and accessibility and favor their metabolism (Joseph and Colwell, 1990). Nevertheless, Sorkhoh et al. (1993) demonstrated that *B. stearothermophilus* grew best on pentadecane (C15), hexadecane (C16), heptadecane (C17) but it is less efficient for other hydrocarbon compounds with shorter and longer chains such as *n*-Alkanes, *n*-alkenes and aromatic hydrocarbons.

Although high salinity is known to decrease hydrocarbon degradation ability (Minai-Tehrani et al., 2006; Mille et al., 1991; Riis et al., 2003) by decreasing the number and types of substrates utilizable by microorganisms (Oren, 1988; Oren et al., 1992; Riis et al., 2003), both *B. licheniformis* STK08 and *G. stearothermohilus* STM04 demonstrated a capacity to grow under high salt concentrations (120 g/L). Similarly, Obuekwe et al. (2005) reported that microorganisms cultured under stress conditions expended more energy on maintenance requirements. Therefore, provision of extra energy sources such as supplemental crude oil in the rich medium not only potentially satisfied the increased maintenance needs but also explained the increased biomass development observed.

B. firmus STS84 and *L. sphaericus* STZ75 grown on agar plates with HCE as sole carbon source appeared to show less competent biodegradation than when grown in liquid medium. The moderate biodegradation rate recorded at the end of the treatment period (73 and 80%, respectively) prompted us to conclude that these two strains were unable to degrade complex compounds present in the effluent. Mancera-Lopez et al. (2007) reported that a strain of *B. firmus* B05PR isolated from highly-polluted sites (60,000 and 500,000 mg of total petroleum hydrocarbon per kg of dry soil) was able to remove 90% of aliphatic hydrocarbons but only 80% of total hydrocarbons. These two species were previously known for their ability to degrade some hydrocarbon components and to grow at 60°C (Mohamed et al., 2006). In our study, the more moderate performance of *B. firmus* STS84 and *L. sphaericus* STZ75 compared with *G. stearothermophilus* STM04 and *B. licheniformis* STK08 might be due to difficulties in transporting a fraction of the hydrocarbon pollutant into cells. Haner et al. (1994) reported that a major fraction of dissolved organic carbon could not be directly transported into cells because of their high molecular weights or lack of transport systems.

Using the two qualitative methods (MDC and BAP), results demonstrated that the four *Bacillaceae* strains selected in this study were biosurfactant producers. This can explicate the THC degradation performance obtained with these strains in pure culture. Knowing that biodegradation depends strongly on hydrocarbon emulsion, the use of biosurfactant producer strains in bioremediation technology seems to offer more potential than chemical surfactant, due to their structural diversity, biodegradability and biocompatibility relative to synthetic surfactant (Abalos et al., 2004).

In conclusion, this study isolated four *Bacillaceae* strains that all demonstrated a high capacity to degrade heterogeneous HCE derived from a hydrocarbonic refining process. The high THC removal obtained in batch assays with these four strains indicated that clarified wastewater could be used as HCE diluent in the biotechnological process to degrade the main hydrocarbon compounds. In addition, the ability of these four strains, particularly *B. licheniformis* STK08 and *G. stearothermophilus* STM04 to grow at high temperature (55°C) and high salt concentrations (120 g/L) simultaneously and to produce biosurfactant, supports the use of these *Bacillaceae* strains for the bioremediation of hydrocarbon-polluted sites in terrestrial or marine environments, especially under high temperature or high salinity conditions.

ACKNOWLEDGEMENTS

The authors thank the Tunisian Ministry of Scientific Research, Technology and Competency Development for the financial support, the Chemical Analysis Laboratory at the Tunisian Society of Refining Industry (STIR), Bizerte, Tunisia. Special thanks to the laboratory of the Office National des Assainissements (ONAS), Bizerte, Tunisia, for their technical support.

REFERENCES

Abalos A, Vinas M, Sabaté J, Manresa MA, Solanas AM (2004). Enhanced biodegradation of Casablanca crude oil by a microbial consortium in presence of a rhamnolipid produced by *Pseudomonas aeruginosa* AT10. Biodegradation 15:249-260.

Ahmed I, Yokota A, Yamazoe A, Fujiwara T (2007). Proposal of *Lysinibacillus boronitolerans* gen. nov., sp. nov., and transfer of *Bacillus fusiformis* to *Lysinibacillus fusiformis* comb. nov. and *Bacillus sphaericus* to *Lysinibacillus sphaericus* comb. nov. Int. J. Syst. Evol. Microbiol. 57:1117-1125.

Alonso-Gutiérrez J, Lekunberri I, Teira E, Gasol JM, Figueras A, Novoa B (2009). Bacterioplankton composition of the coastal upwelling system of 'Ría de Vigo' NW Spain. FEMS. Microbiol. Ecol. 70:493-505.

Auffret M, Labbe D, Thouand G, Greer CW, Fayolle-Guichard F (2009). Degradation of a mixture of hydrocarbons, gasoline, and diesel oil additives by *Rhodococcus aetherivorans* and *Rhodococcus wratislaviensis*. Appl. Environ. Microbiol 75:7774-7782.

Bayoumi RA (2009). Bacterial bioremediation of polycyclic aromatic hydrocarbons in heavy oil contaminated soil. J. Appl. Sci. Res. 5:197-211.

Ben Hamed S, Rezgui R, Halleb A, Ghram A, Oueslati R, Labat M, Maaroufi A (2010). Efficiency of refinery sludge biodegradation using municipal wastewater and activated sludge and effect of hydrocarbon concentration on culturable bacteria community. Ann. Microbiol. 60:747-755.

Bodour AA, Miller-Maier R (1998). Application of a modified drop collapse technique for surfactant quantitation and screening of biosurfactant producing microorganisms. J. Microb. Methods 32:273-280.

Chavan A, Mukherji S (2008). Treatment of hydrocarbon-rich wastewater using oil degrading bacteria and phototrophic microorganisms in rotating biological contactor: effect of N:P ratio. J. Hazard Mater. 15:63-72.

Ciric L, Philp JC, Whiteley AS (2010). Hydrocarbon utilization within a diesel-degrading bacterial consortium. FEMS Microbiol. Lett. 303:116-122.

Das N, Chandran P (2011). Degradation of petroleum hydrocarbon contaminants: An Overview. Biotechnol. Res. Int. 2011:1-13.

Farinazleen MG, Abdul Rahman RN, Salleh AB, Basri M (2004). Biodegradation of hydrocarbons in soil by microbial consortium. Int. Biodeterior. Biodegrad. 54:61-67.

Haner A, Mason CA, Hamero G (1994). Death and lysis during aerobic thermophilic sludge treatment characterization of recalcitrant products. Water Res. 28:863-869.

Hedi A, Sadfi N, Fardeau ML, Rebib H, Cayol JL, Ollivier B, Boudabous A (2009). Studies on the biodiversity of halophilic microorganisms isolated from El-Djerid Salt Lake (Tunisia) under aerobic conditions. Int. J. Microbiol. pp. 1-17.

Hedlund BP, Staley JT (2001). *Vibrio cyclotrophicus* sp. nov., a polycyclic aromatic hydrocarbon (PAH)-degrading marine bacterium. Int. J. Syst. Evol. Microbiol. 51:61-66.

Hui LI, Zhang Y, Kravchenko I, Hui X, Zhang C (2007). Dynamic changes in microbial activity and community structure during biodegradation of petroleum compounds : A laboratory experiment. J. Environ. Sci-China. 19:1003-1013.

Joseph GL, Colwell RR (1990). Microbial degradation of hydrocarbons in the environment. Microbiol. Rev. 9:305-315.

Lotfabad S, Gray M (2002). Kinetics of biodegradation of mixtures of polycyclic aromatic hydrocarbons. Appl. Microbiol. Biotechnol. 60:361-366.

Malik ZA, Ahmed S (2012). Degradation of petroleum hydrocarbons by oil field isolated bacterial consortium. Afr. J. Biotechnol. 11:650-658.

Mancera-Lopez ME, Rodriguez-Casasola MT, Rios-Leal E, Esparza-Garcia F, Chavez-Goómez B, Rodriguez-Vazquez R, Barrera-Cortés J (2007). Fungi and bacteria isolated from two highly polluted soils for hydrocarbon degradation. Acta. Chim. Slov. 54:201-209.

Margesin R, Schinner F (2001). Biodegradation and bioremediation of hydrocarbons in extreme environments. Appl. Microbiol. Biotechnol. 56:650-663.

Mc Graw Hill (1985). Standard methods for the examination of water and wastewater. 16th ed., American Public Health Association (APHA). New York, USA.

Mille G, Almallah M, Bianchi M, Van Wambeke F, Bertrand JC (1991). Effect of salinity on petroleum biodegradation. Fresen. J. Anal. Chem. 339:788-791.

Minai-Tehrani D, Herfatmanesh A, Azari-dekhordi F, Minooi S (2006). Effect of salinity on biodegradation of aliphatic fractions of crude oil in soil. Pak. J. Biol. Sci. 9:1531-1535.

Mohamed ME, Al-Dousary M, Hamzah RY, Fuchs G (2006). Isolation and characterization of indigenous thermophilic bacteria active in natural attenuation of bio-hazardous petrochemical pollutants. Int. Biodeterior. Biodegrad. 58:213-223.

Mrayyan B, Battikhi MN (2005). Biodegradation of total organic carbon (TOC) in Jordanian petroleum sludge. J. Hazard. Mater. 120:127-134.

Nazina TN, Tourova TP, Poltaraus AB, Novikova EV, Grigoryan AA, Ivanova AE, Lysenko AM, Petrunyaka VV, Osipov GA, Belyaev SS, Ivanov MV (2001). Taxonomic study of aerobic thermophilic bacilli: descriptions of *Geobacillus subterraneus* gen. nov., sp. nov. and *Geobacillus uzenensis* sp. nov. from petroleum reservoirs and transfer of *Bacillus stearothermophilus*, *Bacillus thermocatenulatus*, *Bacillus thermoleovorans*, *Bacillus kaustophilus*, *Bacillus thermoglucosidasius* and *Bacillus thermodenitrificans* to *Geobacillus* as the new combinations *G. stearothermophilus*, *G. thermocatenulatus*, *G. thermoleovorans*, *G. kaustophilus*, *G. thermoglucosidasius* and *G. thermodenitrificans*. Int. J. Syst. Evol. Microbiol. 51:433-446.

Obuekwe CO, Badrudeen AM, Al-Saleh E, Mulder JL (2005). Growth and hydrocarbon degradation by three desert fungi under conditions of simultaneous temperature and salt stress. Int. Biodeterior. Biodegrad. 56:197-205.

Oren A (1988). Anaerobic degradation of organic compounds at high salt concentration. Antoine van Leeuwenhoek 54:267-277.

Oren A, Gurevich P, Azachi M, Henis Y (1992). Microbial degradation of pollutants at high salt concentrations. Biodegradation 3:387-398.

Rahman KS, Rahman T, Lakshmanaperumalsamy J, Banat P (2002). Towards efficient crude oil degradation by a mixed bacterial consortium. Bioresour. Technol. 85:257-261.

Riis V, Kleinsteuber S, Babel W (2003). Influence of high salinities on the degradation of diesel fuel by bacterial consortia. Can. J. Microbiol. 49:713-721.

Siegmund I, Wagner F (1991). New method for detecting rhamnolipids excreted by *Pseudomonas* spp. during growth on mineral agar. Biotechnol. Tech. 5:265-268.

Sorkhoh NA, Ibrahim AS, Ghannoum MA, Radwan SS (1993). High-temperature hydrocarbon degradation by *Bacillus stearothermophilus* from oil-polluted Kuwaiti desert. Appl. Microbiol. Biotechnol. 39:123-126.

Thompson JD, Gibson TJ, Plewniak F, Jeanmougin F, Higgins DG (1997). The CLUSTALX windows interface: flexible strategies for multiple sequence alignment aided by quality analysis tools. Nucleic Acids Res. 24:4876-4882.

Tomei MC, Annesini MC, Luberti R, Cento G, Senia A (2003). Kinetics of 4-nitrophenol biodegradation in a sequencing batch reactor. Water Res. 37:3803-3814.

Tzarkova EK, Groudeva VI (2000). Bioremediation of petroleum polluted soils by adsorptive immobilized *Corynebacterium* sp. RB-96. Biotechnol. Biotechnol. Equip. 14:72-75.

Van Hamme JD, Ward OP (2001). Volatile hydrocarbon biodegradation by a mixed-bacterial culture during growth on crude oil. J. Ind. Microbiol. Biotechnol. 26:356-362.

Viñas M, Sabaté J, Espuny MJ, Solanas A M (2005). Bacterial Community Dynamics and Polycyclic Aromatic Hydrocarbon Degradation during Bioremediation of Heavily Creosote-Contaminated Soil. Appl. Environ. Microbiol. 71:7008-7018.

Winker S, Woese CR (1991). A definition of the domains Archaea, Bacteria and Eucarya in terms of small subunit ribosomal rRNA characteristics. Syst. Appl. Microbiol. 13:161-165.

Zhuang WQ, Tay JH, Maszenan AM, Kumholz LR, Tay, STL (2003). Importance of Gram positive naphtalène-degrading bacteria in oil-contaminated tropical marine sediments. Lett. Appl. Microbiol. 36:251-257.

Genomic DNA extraction from sapwood of *Pinus roxburghii* for polymerase chain reaction studies

Anita Rawat*, Santan Barthwal and H. S. Ginwal

Division of Genetics and Tree Propagation, Forest Research Institute, P.O.I.P.E. Kaulagarh Road, Dehradun 248195, Uttarakhand, India.

A method for extraction of genomic DNA from sapwood tissues of mature tall trees of *Pinus roxburghii*, where collection of needle tissues is extremely difficult has been standardized. The extracted DNA was comparable to that obtained from the needle tissue in terms of yield and purity. The yield of extracted DNA ranged from 6.98 to 19.668 µg / 100 mg tissue and A_{260} / A_{280} ratio ranged from 1.70 to 1.87. The polymerase chain reaction (PCR) amplification of the DNA extracted from sapwood tissue using random amplification of polymorphic DNA (RAPD), inter simple sequence repeat (ISSR) and simple sequence repeat (SSR) markers was similar to that of DNA extracted from the needle tissues.

Key words: *Pinus roxburghii,* DNA extraction, sapwood, random amplification of polymorphic DNA (RAPD), inter simple sequence repeat (ISSR), simple sequence repeat (SSR).

INTRODUCTION

Pinus roxburghii Sargent (Chir pine or long needle pine), with a long rotation of 120 years is a precious timber-resin resource, native to the outer range and principal valleys of the Himalayas from Afghanistan to Arunachal Pradesh in northeastern India between 450 and 2300 m altitude. It covers very large areas as pure forests and also with other species, particularly at its upper and lower limits. The species provides a variety of goods for the society and the resin industry of the country. Tree improvement and conservation programs have been initiated with the use of DNA marker techniques to study the population genetic structure and genetic diversity in this species. However, extraction of genomic DNA is the first essential step for molecular diversity/similarity analysis. Generally, the DNA is extracted from the young tissues in their active growth for obtaining high quality and yield. Young leaves are the most preferred starting material. DNA from young needles can be extracted following a combination of the protocols given by Doyle and Doyle (1990) and Stange et al. (1998). However, in case of *P. roxburghii* where the trees are very tall (~30 m) and situated in steep hilly terrain, obtaining foliage tissues is highly laborious and time consuming. Therefore, an alternative method of extraction was developed using sapwood peeled from the trunk of trees at approachable height.

*Corresponding author. E-mail: anitasrawat@gmail.com.

MATERIALS AND METHODS

The cetyl-trimethylammonium bromide (CTAB) extraction buffer consisted of 100 mM Tris-HCl (pH-8.0), 20 mM ethylenediaminetetraacetic acid (EDTA) (pH-8.0), 1.4 mM NaCl, 3% CTAB, 1% polyvinylpyrollidone (PVP) and 0.2% β-mercaptoethanol. The bark of the tree was obtained using bark shaver and transported to the laboratory wrapped in zip pouch inside ice box. The sapwood (developing xylem near cambium and bark) was separated and cut into fine pieces. One gram fine-shredded tissue was ground in liquid nitrogen and incubated in 5 ml extraction buffer for 2 h at 65°C in water bath. The mixture was extracted with equal volume of chloroform : isoamyl alcohol (24:1). Two phases were separated by centrifuging at 5000 rpm for 20 min using refrigerated centrifuge (Hermile model: Z233 MK2) at 4°C. An equal volume of cold isopropanol was added to the supernatant, mixed well and incubated overnight at -20°C. DNA was pelleted by centrifuging at 5000 rpm for 20 min at 4°C. The pellet was washed with 1 ml of 76% ethanol for 45 min by gentle inverting and then pelleted by centrifuging at 5000 rpm for 5 min. The pellet was washed again

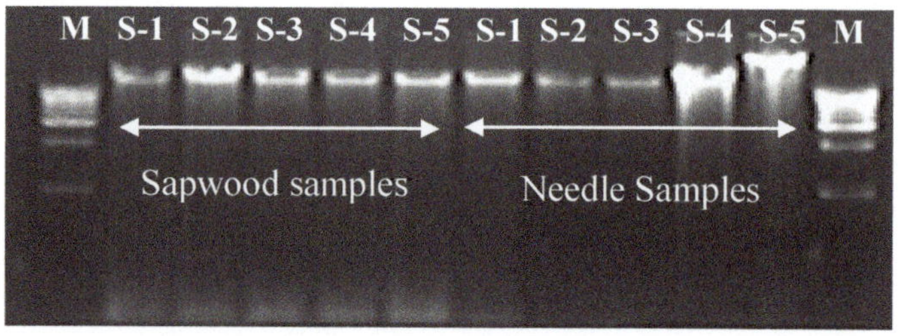

Figure 1. Genomic DNA extracted from sapwood and needles of five individuals of *P. roxburghii* on 1% agarose gel. M: 1 kb DNA ladder.

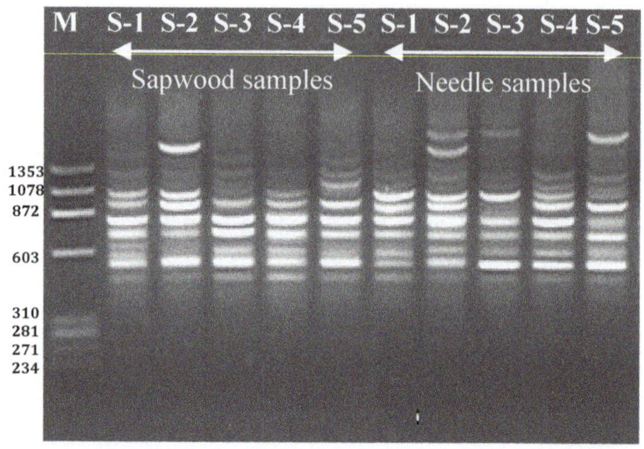

Figure 2. RAPD profile of sapwood and needle samples of five genotypes of *P. roxburghii* generated by the primer M-182. Sequence: 5'GTT CTC GTG T 3', M: ф X 174 DNA/ *Hae* III Digest.

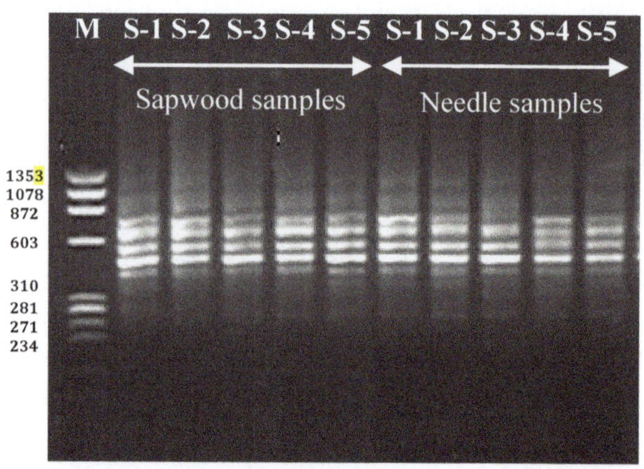

Figure 3. ISSR profile of sapwood and needle samples of 5 genotypes of *P. roxburghii* generated by the primer UBC-809. Sequence: 5'AGAGAGAGAGAGAGAGG-3', M: ф X 174 DNA/ *Hae* III Digest.

with 500 µl of 70% alcohol by centrifuging at 5000 rpm for 15 min. After vacuum drying for 10 min, the pellet was re-dissolved in 50 µl of Tris-EDTA (TE) buffer. The quantitative analysis was done based on A_{260}/A_{280} spectral absorbance ratio using Biophotometer (Eppendorf). The size, purity and integrity of the DNA was also confirmed through electrophoresis on 1% agarose gel (Figure 1) containing ethidium bromide (0.5 µg/ml) at 80 V for 1 h. Finally, the extracted DNA was subjected to PCR amplification using random amplification of polymorphic DNA (RAPD), inter simple sequence repeat (ISSR) and simple sequence repeat (SSR) primers and compared with DNA extracted from the young needle tissues.

RAPD analysis

The extracted DNA was subjected to RAPD analysis using primer M-182, 5'GTT CTC GTG T 3' (Mosseler et al., 1992). PCR for RAPD marker amplification was performed using 10 ng genomic DNA as a template. PCR was performed on a thermal cycler (Biorad, Mycycler) incorporating 10 ng genomic DNA to a 25 µl reaction mix containing 1X *Taq* buffer, 3 mM $MgCl_2$, 0.2 mM each of dNTPs, 0.4 µM primer and 1 unit of *Taq* polymerase (Applied Biosystems). The PCR amplification parameters consisted of: one cycle at 94°C for 2 min; 41 cycles each at 94°C for 45 s; 37°C for 1 min and 72°C for 1 min followed by a final elongation at 72°C for 10 min. Amplification products were visualized on 1.5% agarose gel containing ethidium bromide (0.5 µg/ml) run at 100 V for 3 h (Figure 2).

ISSR analysis

ISSR analysis was carried out in a 20 µl volume of reaction mixture containing 1X *Taq* polymerase buffer, 1.75 mM $MgCl_2$, 0.2 mM each of dNTPs, 0.4 µM primer, 1 unit of *Taq* polymerase (Applied Biosystems) and 10 ng template DNA. ISSR primer (University of British Columbia, UBC-809 Sequence: 5'AGAGAGAGAGAGAGAGG-3' was used for the amplification (Hong et al., 2007). The PCR amplification was programmed for: one cycle at 94°C for 5 min; 40 cycles each at 94°C for 30 s, 57°C for 30 s and 72°C for 1 min followed by a final extension at 72°C for 10 min. Amplification products were visualized on 2.0% agarose gel (Figure 3).

SSR analysis

For SSR analysis, primer Pt 71936 was used (forward sequence: 5' TTCATTGGAAATACACTAGCCC 3' and reverse sequence: 5'AAAACCGTACATGATTCCC 3') (Vendramin et al., 1996). PCR amplification was carried out in a 15 µl reaction mixture containing 1X *Taq* polymerase buffer, 2.5 mM $MgCl_2$, 0.2 mM each of dNTPs, 0.2 µM of each primer, 1 unit of *Taq* polymerase (Applied Biosystems) and 15 ng template DNA. The PCR amplification program consisted of: one cycle at 95°C for 5 min; 35 cycles each at 94°C for 1 min, 55°C for 1 min and 72°C for 1 min followed by a final extension at 72°C for 8 min. Amplification products were visualized on 3% metaphor-agarose gel (Figure 4).

RESULTS AND DISCUSSION

The developed protocol yielded DNA that was comparable to the quality of DNA extracted from the needle tissues as evident from the A_{260}/A_{280} ratios and the yield of the DNA from needle and sapwood tissues (Table 1). The value of 1.8 indicates a high purity of the extracted DNA while values below and above 1.8

Table 1. Comparison of the yield and quality of DNA extracted from sapwood and needle samples of *P. roxburghii*.

S/N	Sample	A_{260}/A_{280} (ratio)		DNA yield (µg/100 mg tissue)	
		Sapwood	Needle	Sapwood	Needle
1	S-1	1.87	1.86	6.98	22.632
2	S-2	1.71	1.72	18.108	12.792
3	S-3	1.70	1.72	12.398	14.016
4	S-4	1.82	1.72	18.108	46.232
5	S-5	1.74	1.68	19.668	31.204

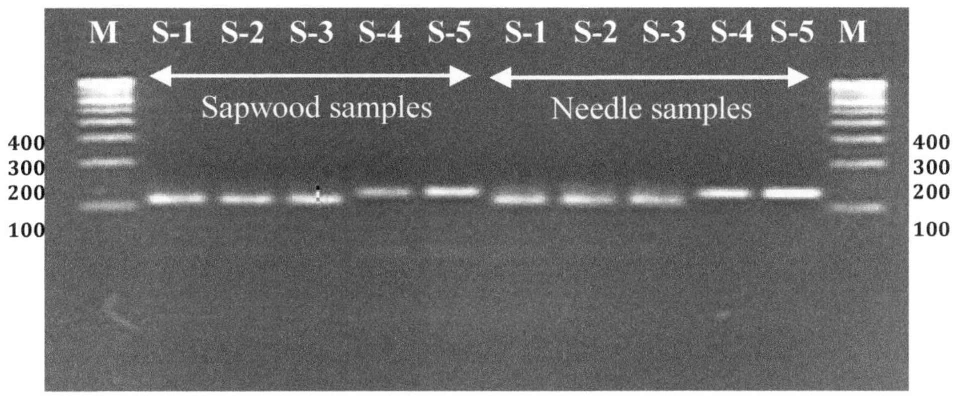

Figure 4. SSR profile of sapwood and needle samples of 5 genotypes of *P. roxburghii* using SSR primer Pt71936. Forward: 5' TTCATTGGAAATACACTAGCCC 3'; Reverse: 5' AAAACCGTACATGAGATTCCC 3'. M: 100 bp DNA ladder.

denotes contaminations of proteins and RNAs, respectively. Samples S1 and S4 however yielded lesser quantity of DNA than that expected, as the extraction is difficult from the sclerenchymatous tissues. The quality of the DNA was also evident from the agarose gel (Figure 1). The results of RAPD (Figure 2), ISSR (Figure 3) and SSR (Figure 4) marker analyses were almost similar for DNA extracted from the sapwood and needles of the same genotype. There are other published reports of DNA isolation from wood of *Cyclobalanopsis* species (Ohyama et al., 2001), Oak (Deguilloux et al., 2002) and species of Dipterocarpaceae (Rachmayanti et al., 2006, 2009). The differences in genomic DNA extracted from bark and wood of different zones in *Robinia* trees were studied using RAPD-PCR (De Filippis and Magel, 1998). In sapwood and heartwood, the random amplification of polymorphic DNA and reverse transcription polymerase chain reaction of RNA was studied (Magel et al., 2002). DNA from processed wood and herbarium samples of *Gonystylus bancanus* was reported (Asif and Cannon, 2005) for identification of endangered tropical timber species. The protocol reported for DNA isolation from *G. bancanus* (Asif and Cannon, 2005) with modifications worked well for *P. roxburghii* in our study. The DNA extracted from sapwood was equally good like that obtained from green needle tissue in terms of both purity and yield. The quality of DNA was also found adequate for RAPD, ISSR and SSR analysis. The results indicate the usefulness of the protocol for the molecular marker investigations of *P. roxburghii*. Also, the method of DNA extraction from sapwood of *P. roxburghii* is extremely useful for molecular studies involving mature tall trees, where collection of foliage tissue is difficult and labour intensive and often professional tree climber is hired for such work in the field.

REFERENCES

Asif MJ, Cannon CH (2005). DNA Extraction From Processed Wood: A case study for the identification of an endangered timber species (*Gonystylus bancanus*). Plant Mol. Biol. Rep. 23:185-192.

De Filippis L, Magel E (1998). Differences in genomic DNA extracted from bark and from wood of different zones in *Robinia* trees using RAPD-PCR. Trees Struc. Funct. 12:377-384.

Deguilloux MF, Pemongel MH, Petit RJ (2002). Novel perspectives in wood certification and forensics: dry wood as a source of DNA. Proceedings of the Royal Society B: Biological Sciences. 269:1039-1046.

Doyle JJ, Doyle JL (1990). A rapid total DNA preparation procedure for fresh plant tissue. Focus 12:13-15.

Hong YP, Kwon HY, Kim IS (2007). ISSR markers revealed inconsistent phylogeographic patterns among populations of Japanese red pines in Korea. Silvae Genet. 43:167-176.

Magel E, Hauch S, De Filippis LF (2002). Random amplification of polymorphic DNA and reverse transcription polymerase chain reaction of RNA in studies of sapwood and heartwood. In: Wood formation in trees, Chaffey, N., (Ed.). Harwood Academic Publishing, Malaysia.

Mosseler A, Egger KN, Hughes GA (1992). Low levels of genetic diversity in Red pine confirmed by random amplified polymorphic DNA markers. Can. J. For. Res. 22:1332-1337.

Ohyama M, Baba K, Itoh T (2001). Wood identification of Japanese *Cyclobalanopsis* species (Fagaceae) based on DNA polymorphism of the intergenic spacer between trnT and trnL 5'exon. J. Wood Sci. 47:81-869.

Rachmayanti Y, Leinemann L, Gailing O, Finkeldey R (2006). Extraction, amplification and characterization of wood DNA from Dipterocarpaceae. Plant Mol. Biol. Rep. 24:45-55.

Rachmayanti Y, Leinemann L, Gailing O, Finkeldey R (2009). DNA from processed and unprocessed wood: Factors influencing the isolation success. Forensic Science International: Genetics 3:185-192.

Stange C, Prehn D, Johnson PA (1998). Isolation of *Pinus radiata* genomic DNA suitable for RAPD analysis. Plant Mol. Biol. Rep. 16:1-8.

Vendramin GG, Lelli L, Rossi P, Morgante M (1996). A set of primers for the amplification of 20 chloroplast microsatellites in *Pinaceae*. Mol. Ecol. 5:595-598.

Occurrences of pathogenic *Vibrio parahaemolyticus* from Vellar estuary and shrimp ponds

K. M. Alagappan[1,2] , B. Deivasigamani[2] and S. Balamurugan[2]

[1]Centre of advanced study in Marine Biology, Faculty of Marine Sciences, Annamalai University, Parangipettai -608502, Tamil Nadu South India.
[2]Department of Biochemistry and Biotechnology, Pondicherry University, Puducherry – 605014, India.

***Vibrio parahaemolyticus* is the predominant seafood pathogen associated with human gastroenteritis. Samples were collected from Vellar estuary, shrimp ponds and shrimp for characterization of *V. parahaemolyticus*. A total of 26 blue green centre (BG) *Vibrio* strains were isolated and characterized through biochemical tests, *toxR* gene and 16S rRNA gene sequencing. Based on pathogenic characteristics, six strains were confirmed as pathogenic *V. parahaemolyticus*. This report implies that preventative measure must be taken before consumption of fish and shrimp.**

Key words: Estuary, shrimp pond, *Vibrio parahaemolyticus*, protease, haemolysis, *toxR* gene.

INTRODUCTION

Vibrio harveyi and *Vibrio parahaemolyticus*, are opportunistic pathogens that cause symptomatic infections when the shrimp are stressed. *V. parahaemolyticus* is the dominant species in shrimp affected by red disease and tail necrosis (Jayasree et al., 2006). Most of the pathogenic *Vibrio* sp. has the virulence characteristic like haemolytic activity and proteolytic activity. Potentially, pathogenic *V. parahaemolyticus* strains could be differentiated from the related nonpathogenic marine vibrios based on haemolytic and non haemolytic activity. Moreover, extracellular products secreted by the *V. parahaemolyticus* were found to be highly toxic to tiger prawn (Sudheesh and Xu, 2001). However, gastroenteritis caused by *V. parahaemolyticus* has been associated with foods prepared with seaweeds (Mahmud et al., 2007). In 2006, an outbreak of *V. parahaemolyticus* infections resulted in 177 cases and was linked to the consumption of contaminated raw shellfish including oysters (Balter et al., 2006).

Identification and detection of total and pathogenic *V. parahaemolyticus* by polymerase chain reaction (PCR) have been extensively studied. A toxin regulatory gene (*toxR*) sequence specific to *V. parahaemolyticus*, which is present in all the strains irrespective of their ability to produce thermostable direct hemolysin (tdh) or tdh related hemolysin (trh), has been applied for definitive identification of the bacterial isolates by several workers (Jacksic et al., 2002; Kim et al., 1999). Vellar estuary is situated in Parangipettai, Southeast coast of India. This estuary is the water hub for ~150 shrimp ponds and it is used as the nursery ground for fish fry and fingerlings. Studying of pathogenic *V. parahaemolyticus* is vital for seafood consumption, therefore, this work was scrutinized to enumerate *Vibrio* sp. from Vellar estuary, shrimp ponds, healthy shrimp tissues and infected shrimp tissues; characterization of isolated *Vibrio* strains for haemolytic activity and protease activity and identification of *V. parahaemolyticus* through specific *toxR* gene and partial sequencing of 16S rRNA gene.

MATERIALS AND METHODS

Sample collection and enumeration of *Vibrio* sp.

Samples (water, W and sediment, S) were collected from five different stations (A to E) in Vellar estuary (Figure 1). All the samples were collected during monsoon season which usually occur in the month of October to first part of December 2009. Healthy and infected shrimp (*P. monodon*), water, and sediment samples were collected

*Corresponding author. E-mail: alagumicro@yahoo.co.in.

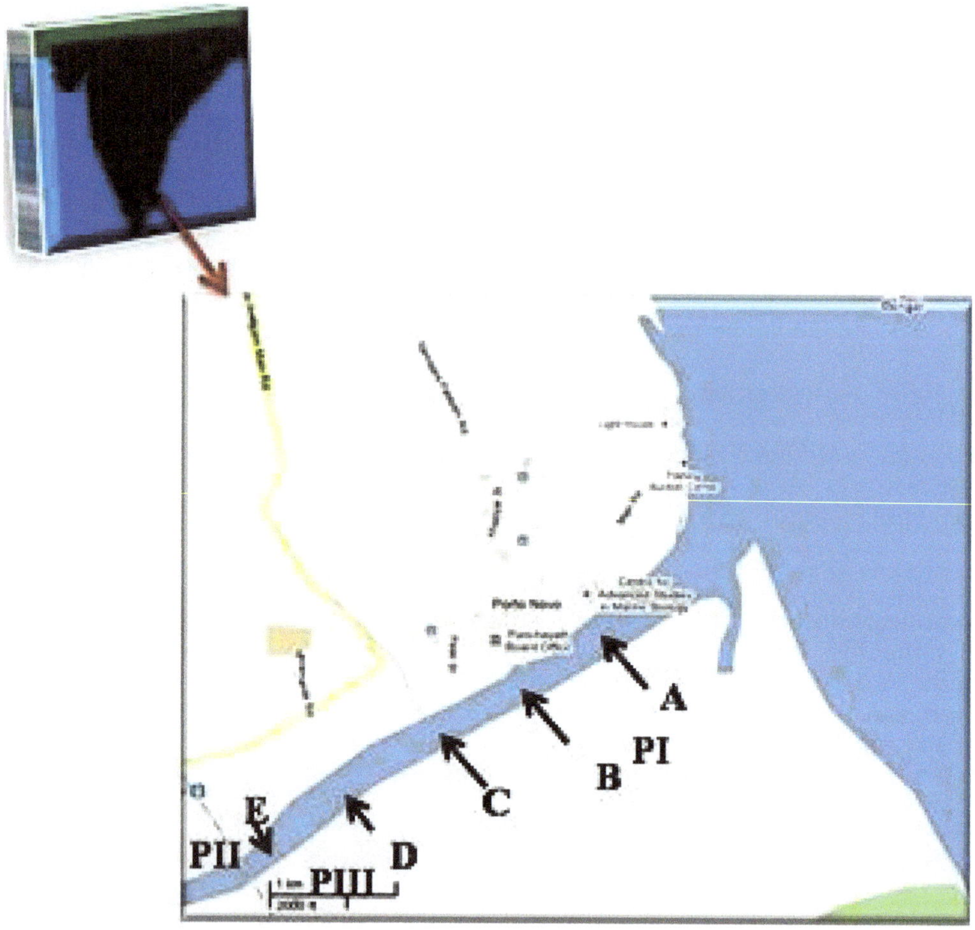

Figure 1. Sample collection stations in Vellar estuary (A to E) and shrimp ponds (PI to PIII).

separately from three shrimp culturing ponds after 70 days of culture, Parangipettai, Tamil Nadu, India and brought to the laboratory with sterile containers. However, heavy mortality in shrimp was not observed while collecting samples. Samples from water and sediment of three ponds (P) were designated as PW-I, PS-I, PW-II, PS-II, PW-III and PS-III and samples from shrimp surface (SS) and infected shrimp tissues (IST). All the samples were collected in sterile glass vials and brought to the laboratory under cold conditions for the isolation of *Vibrio* sp. and *V. parahaemolyticus*. One gram of sediment samples was added to 100 ml of sterile 50% sea water and mixed well. The filtrate was serially diluted and plated on thiosulphate citrate bile salt sucrose (TCBS) agar medium. Similarly, 1 ml of water was serially diluted and plated on TCBS agar medium. Animal tissue (1 g wet weight) were ground, homogenized on a glass homogenizer with a small amount of 1% peptone water with 3% NaCl, and covered with a fourfold amount of peptone water and simultaneously was aseptically pipette into TCBS agar. For isolation of *V. parahaemolyticus*, previously used methodology was followed (Alagappan et al., 2010)

Haemolytic activity and protease activity for BG strains

The Kanagawa phenomenon (KP) or tdh phenotype detection was carried out on Wagatsuma agar as described by Elliot et al. (1992). Protease enzyme activity for isolated BG strains was characterized on skim milk agar (Himedia, Mumbai). All the strains were streaked on the skim agar plates and incubated at 28°C for 24 h. Proteolytic bacteria exhibiting zones of clearance on skim milk agar was confirmed as positive.

Bacterial DNA isolation

The biochemically identified isolate was streaked on nutrient agar plates and the colonies were inoculated in the nutrient broth supplemented with NaCl (2% w/v), and incubated at 37°C under shaking (120 rpm) for 16 to 18 h. The broth cultures were centrifuged (10 000 rpm, 4°C, 10 min) to obtain the pellet and DNA was isolated by followed the method of Sambrook et al. (1989).

Amplification of *toxR* gene

The PCR targeting *toxR* gene of *V. parahaemolyticus* was performed for isolated BG strains based on the method of Kim et al. (1999) using a thermocycler (Thermalcycler, Genei, Bangalore, India) and PCR amplification was done by using PCR mastermix (Medox^{TM}, Chennai, India). Primers like toxR forward (5´-GTCTTCTGACGCAATCGTTG-3´) and reverse (5´-ATACGAGTGGTTGCTTGCTGTCATG-3´) were purchased from Bioserve Biotechnologies (India) Pvt Ltd, Hyderabad, India. Each cycle consisted of three step reactions that is, initial denaturation (95°C, 1 min) followed by 35 cycles of denaturation (94°C, 30 s), annealing temperature was calculated with T_m value of primer

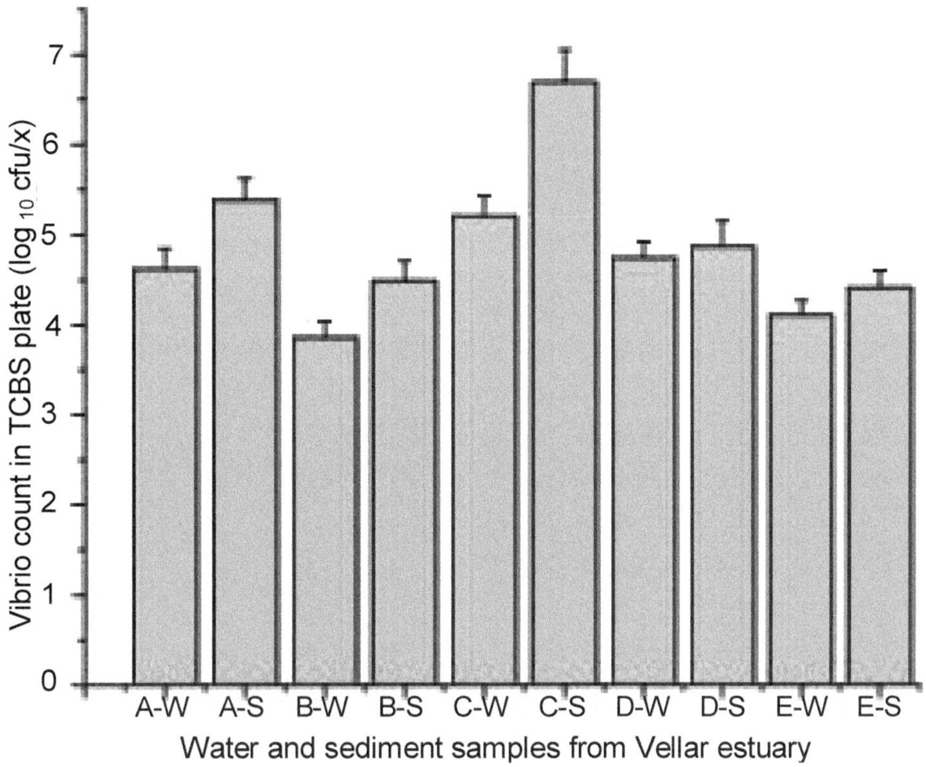

Figure 2. Total *Vibrio* spp. count in TCBS plate of water and sediment samples collected from five different stations in Vellar estuary. Each bar represents the mean value from three determinations with the standard deviation (mean ± SD). X = ml or g.

(51°C, 1 min) and extension (72°C, 1 min 30 s) followed by final extension (72°C, 10 min). The PCR products were resolved on agarose (1.5% w/v) gel electrophoresis. The gel was stained with ethidium bromide (0.5 mg/ml) and visualized under a ultraviolet (UV) transilluminator (Alpha Imager, Innotech Corporation, USA). Genei^{TM} 100 bp DNA Ladder (Genei, Bangalore, India) was used as a molecular size marker.

16S rRNA gene amplification and phylogenetic tree construction

After *toxR* gene PCR reactions for BG strains, two toxR positive and negative strains were confirmed as *V. parahaemolyticus* or other species, respectively through 16S rRNA gene sequence (Kumaran et al., 2010). The amplified PCR products were purified using a Medox PCR product purification kit (Medox^{TM}, Chennai, India). All the PCR 16S rRNA gene products were sequenced by using ABI 3130 (four capillary) or 3730XI (96 capillary) electrophoresis instruments (VIMTA LABS Pvt. Limited, Hyderabad, India). Four bacterial sequences BG24-1446 nt, BG6-1394 nt, BG17-1243 nt and BG12-1441 nt were performed. The data were analyzed using applied biosystem DNA editing and assembly software and sequence comparisons were obtained using the Micro Seq Software. Sequence similarity search was made for the four bacterial 16S rRNA gene sequences by applying their sequence to blast search of the NCBI, then phylogenetic tree was constructed for four strains with Gram positive bacteria *Rhodococcus* sp. SH15 (HM590053) and Gram negative bacteria *Pseudomonas* sp. KUMS3 (FJ596148) served as out group.

RESULTS AND DISCUSSION

For isolation of *Vibrio* spp. and *V. parahaemolyticus*, samples from five different stations in Vellar estuary, three shrimp ponds and healthy and infected animal were collected and characterized. During the monsoon season, samples were collected which was highly influenced by freshwater and salinity was also low compared to other seasons. This condition may induce the invasion of pathogenic forms into estuary and shrimp culturing ponds due to rainfall and inflow of fresh water with sewage contamination. Seasonal changes affect the culturing shrimp in ponds and diseases were more during monsoon and post monsoon period (Muralidhar et al., 2010).

When compared with all the estuary and shrimp pond samples, *Vibrio* sp. count was observed as slightly higher in sediments than in waters (Figures 2 and 3). High *Vibrio* count was observed in station C water and sediment samples when compared with other stations and minimum was observed in stations B and E. Station C waters are connected with sewages and highly contaminated with fecal matters. Eja et al. (2008) also reported that the estuary was constantly faecally polluted, coupled with high rates of infection of shellfish by *V. parahaemolyticus*, *V. cholerae* non-01 and *V. alginolyticus*. Moreover, plankton composition plays an important and independent

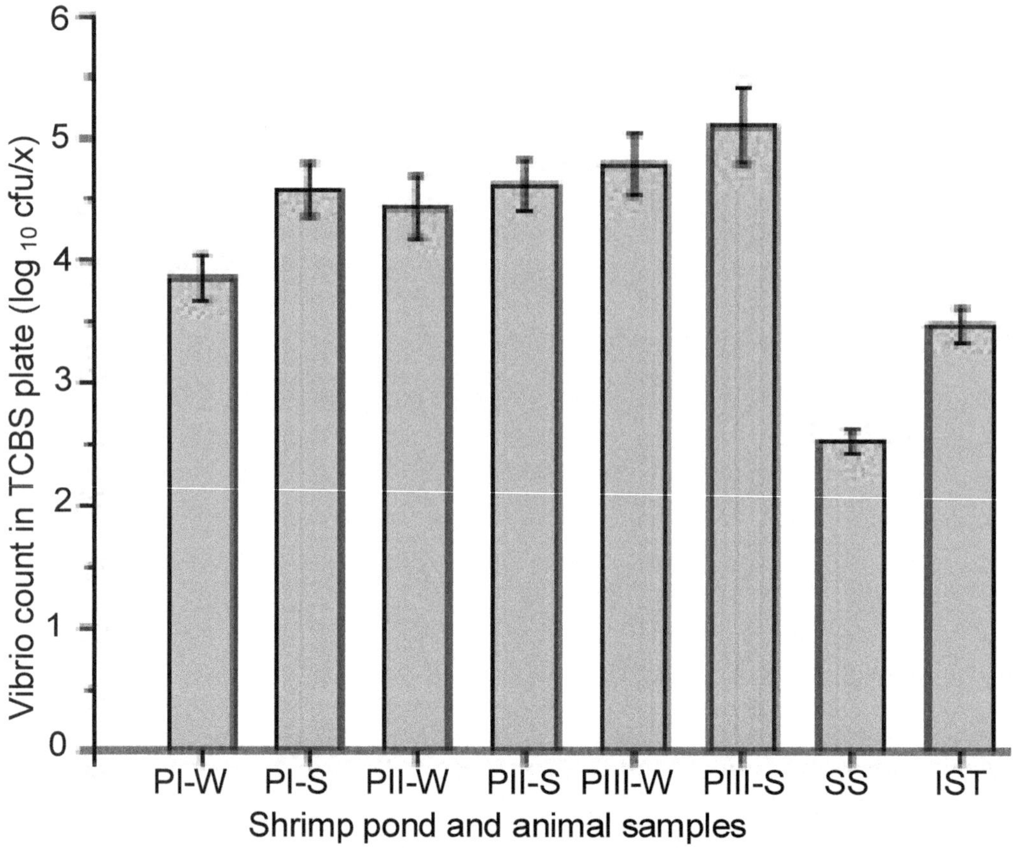

Figure 3. Total *Vibrio* count in TCBS plate of three different shrimp ponds and animal samples. Each bar represents the mean value from three determinations with the standard deviation (mean ± SD). In log_{10} cfu/x: X for ml in PI, II, III (W); for g in PI, II, III (S) and IST; for ml in one shrimp (15 g weight) surface (ss) was completely swabbed and dipped into 10 ml alkaline peptone water.

role as a driver of the total culturable *Vibrio* community in natural estuarine systems (Turner et al., 2009). Shrimp pond III sediment and water samples showed high count of *Vibrio* when compared with other ponds. A total of 26 BG strains were isolated and ~70% of the strains obeyed all the biochemical tests for *V. parahaemolyticus*.

Pathogenic characteristics like Kanagawa phenomenon, protease and toxR were used as validating factors of *V. parahaemolyticus*. Totally 17 protease and 13 β –haemolyic strains were characterized. In that four strains were isolated from infected shrimp with 2 β -haemolysis and 3 protease production. Moreover, different *Vibrio* extracellular products (ECP) have been identified and proposed as putative virulence factors that are pathogenic to shrimp (Harris and Owens, 1999). Ten (10) BG strains showed positive for both protease and β –haemolysis and considered to be pathogens. However, Sudheesh and Xu (2001) reported that extracellular products of *V. parahaemolyticus* were toxic to *Penaeus orientalis* and it was found to be a poor producer of hemolysins. Environmental strains of *V. parahaemolyticus* are typically not human pathogens; therefore, it is important to have data which states that the occurrence of virulent strains like *V. parahaemolyticus* and *V. cholerae* infection through consumption of cultured shrimp. Due to the marine environmental conditions, identification of marine vibrios is difficult. However, BG colonies also produced by *V. vulnificus* could not easily identify the causative agent whether *V. parahaemolyticus* or *V. vulnificus*. Moreover, *V. vulnificus* is also an important *Vibrio* spe-cies which can cause wound infections and septicemia with a high mortality rate and also associated with disease outbreaks, either with ingestion of contaminated seafood or infectious wounds by contaminated sea water (Nascimento et al., 2001).

Therefore, molecular techniques like PCR amplification for species specific gene have given higher importance than biochemical tests. All strains of *V. parahaemolyticus* harbor the *toxR* gene and it has been suggested that PCR amplifying the *toxR* gene could be used for detection of *V. parahaemolyticus* (Kim et al., 1999).

Some authors suggested that tdh and trh are used to characterize its identification and pathogenicity associated with human disease (DePaola et al., 2003). Moreover, virulence factor genes called tdh and trh are usually found in strains isolated from clinical specimens and less

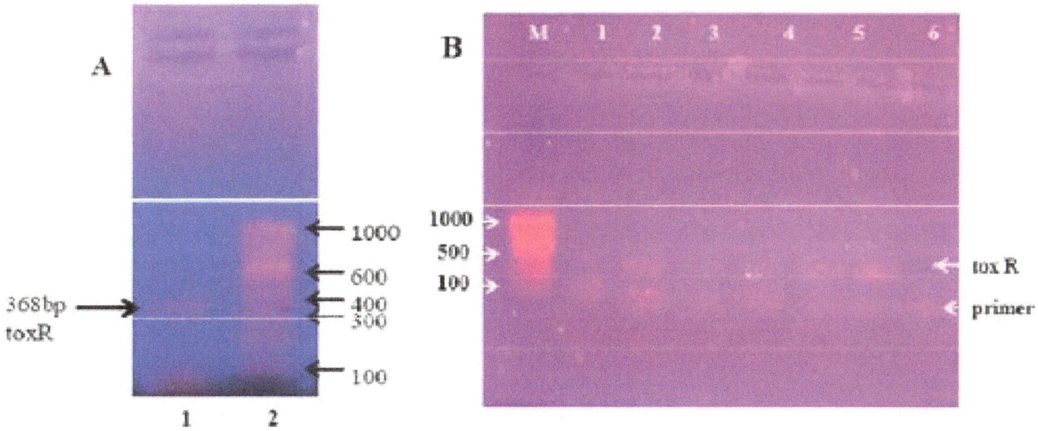

Figure 4. PCR product of amplified *toxR* gene from BG strains. **A.** Agarose gel electrophoresis of PCR amplified products for *toxR* gene (368 bp). Lane 1, *toxR* gene of BG24 strain; Lane 2, DNA ladder (1000 - 100 bp). **B.** Lane M, DNA ladder (1000 - 100 bp). Lanes 1 to 6, BG 1, 3, 6, 8, 15 and 19 strains toxR gene products.

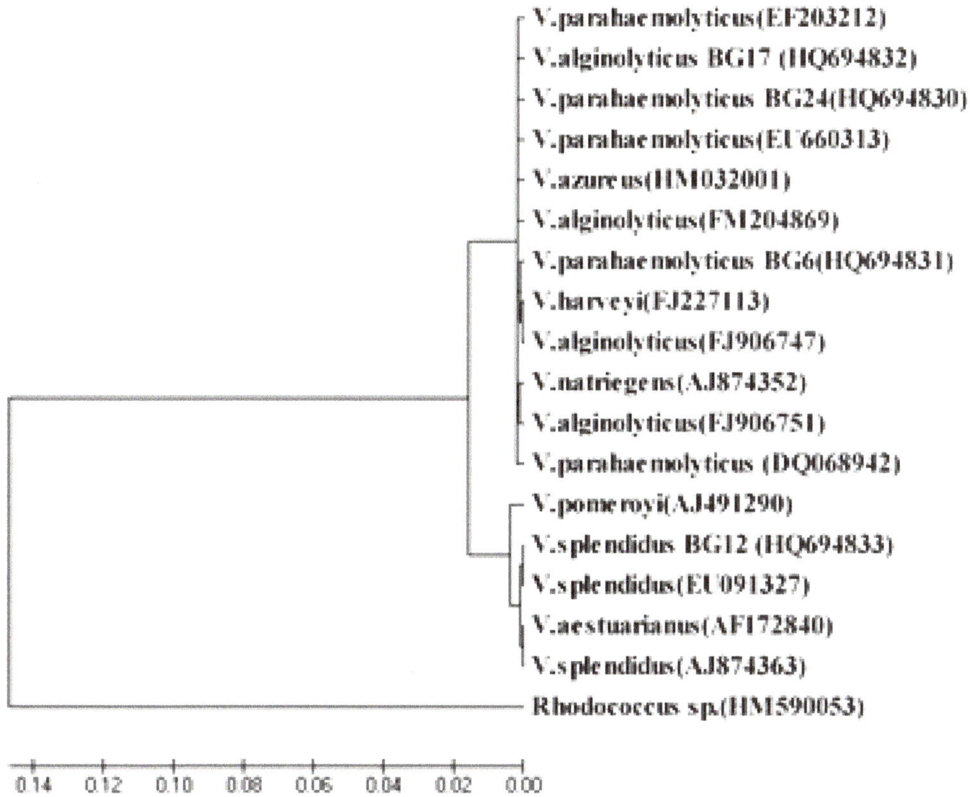

Figure 5. Phylogenetic tree constructed through Kimura 2-parameter model using the neighborhood- joining method. Bar scale 0.02. *Pseudomonas* sp. KUMS3 and *Rhodococcus* sp. SH15 were used as an out group. Queried sequence: BG24, *V. parahaemolyticus* (HQ694830); BG6, *V. parahaemolyticus* (HQ694831); BG17, *V. alginolyticus* (HQ694832); BG12, *V. splendidus* (HQ694833).

than 5% of environmental isolates producing thermostatble direct hemolysin (Wong et al., 2000). However, Johnson et al. (2009) isolated *V. parahaemolyticus* containing tdh and trh from Mississippi coastal environments and they found that there was no source-specific sequestering in oysters, water or sediment.

From 26 BG strains, 11 (BG: 1, 3, 6, 8, 15, 16, 18, 19, 23, 24 and 25) were found to be positive for *toxR* gene and identified as *V. parahemolyticus* (Figures 4A and B). Similarly, Chakraborty and Surendran (2008) also confirmed that 24% of their 72 *V. parahaemolyticus* cultures showed positive for 368-bp *toxR* gene fragment. In that of the 11 BG strains, six (BG1, BG3, BG6, BG15, BG19 and BG24) produced positive for protease activity, β –haemolysis and *toxR* gene which were confirmed as highly pathogenic to shrimp and also humans. Biochemical results for identification of *V. parahaemolyticus* from clinical samples may produce 100% good results but in marine strains different types of sugars utilization occurred. Therefore, biochemical identification of marine strains may be a suitable method but however, the results may value up to genus level and not to species level. Therefore, two positive and two negative toxR *gene* strains were sequenced for 16S rRNA gene to confirm its species level. Similarity between these strains and previously identified strains were compared with NCBI Blast and the 16S rRNA sequenced four strains matching maximum percentage were observed. After blasting, maximum percentage similarity (98%) were observed and two toxR positive strains were belongs to *V. parahaemolyticus* (BG24 and BG6) and negative strains were *V. alginolyticus* (BG17) and *V. splendidus* (BG12) (Figure 5). From the results blue green centered colonies were also belongs to other *Vibrio* species other than that of *V. vulnificus*. Moreover, identification of toxR gene in marine samples is the assured technique to confirm marine *V. parahaemolyticus*. Moreover, six BG strains (BG1, BG3, BG6, BG15, BG19 and BG24) were confirmed as pathogenic *V. parahaemolyticus* based on three virulence characteristics.

TCBS plating and biochemical tests are not correct tools to identify pathogenic marine *V. parahaemolytcius*. When compared with terrestrial bacterial strains, marine bacterial strains may differ with biochemical characteristics. Moreover, pathogenic *V. parahaemolyticus* isolated from estuary, shrimp pond and shrimp may pose human health risk when consumed. During favorable conditions, pathogenic strains may cause infection in estuarine fishes and culturing shrimp in shrimp ponds.

ACKNOWLEDGEMENT

We thank our Dean Prof. T. Balasubramanian, Annamalai University for his kind cooperation and encouragement during our research. First author want to thank University Grand Commission, India for his fellowship.

REFERENCES

Alagappan KM, Deivasigamani B, Somasundaram ST, Kumaran S (2010). Occurrence of *Vibrio parahaemolyticus* and Its Specific Phages from Shrimp Ponds in East Coast of India. Curr. Microbiol. 61: 235-240.

Balter S, Hanson H, Kornstein L, Lee L, Reddy V, Sahl S, Stavinsky F, Fage M, Johnson G, Bancroft J, Keene W, Koepsell J, Williams M, MacDonald K, Napolilli N, Hofmann J, Bopp C, Lynch M, Moore K, Painter J, Puhr N, Yu P (2006). *Vibrio parahaemolyticus* infections associated with consumption of raw shellfish - three states. Journal of the American Medical Association. 296: 2309-2310.

Chakraborty RD, Surendran PK (2008). Occurrence and distribution of virulent strains of *Vibrio parahaemolyticus* in seafoods marketed from Cochin (India). World. J. Microbiol. Biotechnol. 24: 1929-1935.

DePaola A, Ulaszek J, Kaysner CA, Tenge BJ, Nordstrom JL, Wells J, Puhr N, Gendel SM (2003). Molecular, serological, and virulence characteristics of *Vibrio parahaemolyticus* isolated from environmental, food, and clinical sources in North America and Asia. Appl. Environ. Microbiol. 69(7): 3999-4005.

Eja ME, Abriba C, Etok CA, Ikpeme EM, Arikpo GE, Enyi-Idoh KH, Ofor UA (2008). Seasonal Occurrence of Vibrios in Water and Shellfish Obtained from the Great Kwa River Estuary, Calabar, Nigeria. Bull. Environ. Contam. Toxicol. 81: 245-248.

Elliot EL, Kaysner CA, Tamplin ML (1992). Media and reagents. In: USFDA bacteriological analytical manual, (Eds.) 7th ed. AOAC International, Arlington, Va. p. 508.

Harris LJ, Owens L (1999). Production of exotoxins by two luminous *Vibrio harveyi* strains known to be primary pathogens of *Penaeus monodon* larvae. Dis. Aquat. Org. 38: 11-22.

Jacksic S, Suncica U, Petrak T, Bazulic D, Lada GK (2002). Occurrence of *Vibrio* spp. In: Sea fish, shrimps and bivalve mollusks harvested from Adriatic sea. Food Control. 13: 491-493.

Jayasree L, Janakiram P, Madhavi R (2006). Characterization of *Vibrio* spp. associated with diseased shrimp from culture ponds of Andhra Pradesh (India). J. World Aquac. Soc. 37: 523-532.

Johnson CN, Flowers AR, Young VC, Gonzalez-Escalona N, DePaola A, Noriea NF. 3rd., Grimes DJ (2009). Genetic Relatedness Among tdh+ and trh+ *Vibrio parahaemolyticus* Cultured from Gulf of Mexico Oysters (*Crassostrea virginica*) and Surrounding Water and Sediment. Microb. Ecol. 57: 437-443.

Kim YB, Okuda J, Matsumoto C, Takahashi N, Hashimoto S, Nishibuchi M (1999). Identification of *Vibrio parahaemolyticus* strains at the species level by PCR targeted to the toxR gene. J. Clin. Microbiol. 37: 1173-1177.

Kumaran S, Deivasigamani B, Alagappan KM, Sakthivel M, Guru Prasad S (2010). Isolation and characterization of *Pseudomonas* sp. KUMS3 from Asian sea bass (*Lates calcarifer*) with fin rot. W. J. Microbiol. Biotechnol. 26: 359-363.

Mahmud ZH, Neogi SB, Kassu A, Wada T, Islam MS, Nair GB, Ota F (2007). Seaweeds as a reservoir for diverse *Vibrio parahaemolyticus* populations in Japan. Int. J. Food. Microbiol. 118: 92-96.

Muralidhar M, Kumarana M, Muniyandia B, Aberyb NW, Umeshc NR, De Silvab SS, Jumnongsongd S (2010). Perception of climate change impacts and adaptation of shrimp farming in India: Farmer focus group discussion and stakeholder workshop Report, 2nd ed. Network of Aquaculture Centers in Asia-Pacific. p. 75.

Nascimento SMM, Vieira RHSF, Theophilo GND, Rodrigues DP, Vieira GHF (2001). *Vibrio vulnificus* as a health hazard for shrimp consumers. Revista do Instituto de Medicina Tropical de Sao Paulo. 43: 263- 266.

Sambrook J, Fritsch EF, Maniatis T (1989). Molecular Cloning: A Laboratory Manual, 2nd ed. Cold Spring Harbor. Laboratory Press, NY.

Sudheesh PS, Xu HS (2001). Pathogenicity of *V. parahaemolyticus* in tiger prawn *Penaeus monodon* Fabricius: possible role of extracellular proteases. Aquaculture. 196: 37-46.

Turner JW, Good B, Cole D, Erin KL (2009). Plankton composition and environmental factors contribute to Vibrio seasonality. The ISME Journal. 3: 1082-1092.

Wong HC, Liu SH, Ku LW, Lee IY, Wang TK, Lee YS, Lee CL, Kuo LP, Shih DYC (2000). Characterization of *Vibrio parahaemolyticus* isolates obtained from foodborne illness outbreaks during 1992 through 1995 in Taiwan. J. Food. Prot. 63: 900-906.

5

Optimization of chemiluminescent detection of mitochondrial RNA-protein interaction by nonradioactive mobility shift

Ahmed M. El-Shehawi[1,2] and Mona M. Elseehy[2]

[1]Department of Biotechnology, Faculty of Science, Taif University, Taif 21974, Saudi Arabia.
[2]Department of Genetics, Faculty of Agriculture, Alexandria University, Alexandria, Egypt.

RNA mobility shift is one among many procedures used to study RNA-protein interaction. Yet, there are some limitations for the radioactive RNA mobility shift including; 1) the risk of using radiolabeled nucleotides, 2) the long time to get the results; this could range from days to weeks, and 3) its high cost as compared to nonradioactive techniques. In this study, we optimized a nonradioactive procedure using dig-11-UTP nucleotide and chemiluminescent detection for mitochondrial RNA-protein interaction. The optimizations include the quality limiting steps such as using non-specific competitors of RNA probe, UV-cross linking time, electrotransfer to membranes, application with various protein extracts, and the examination for false positive RNA-protein complexes using proteinase K digestion. The results show that the optimizations carried out in the study significantly enhanced the quality of the results obtained with this procedure.

Key words: RNA mobility shift, mitochondria, wheat, RNA-protein interaction, chemiluminescent detection, nonradioactive.

INTRODUCTION

Genetic information stored in DNA is converted into biological data in protein sequences through the two complex processes: transcription and translation. RNA binding proteins play an important role in these two and other related biological processes such as splicing and RNA editing (Mulligan et al., 1992; Lavigueur et al., 1993; Suzuki et al., 2010). The interface of RNA-binding proteins is a wide stage with many interconnected players that organize and control the final image of gene expression. RNA mobility shift assay is a vital method for studying RNA-protein interaction which is used in studying the interplay between RNA sequences and protein factors to: 1) modify the structure of RNA molecules (Li et al., 1999; Ficzycz and Ovsenek, 2002); 2) regulate gene expression at the posttranscriptional and translational levels (Le et al., 1997; Mochow-Grundy et al., 2001; Wei et al., 2001; Randall et al., 2002); 3) determine the essential *cis acting* sequences in RNA molecules (Ruth et al., 1999; Berkowitz et al., 1993; Dixon et al., 2000; Totary-Jain et al., 2005; Crooks et al., 2007); 4) study the structure/function relationship of secondary structure domains of RNA molecules (Blyn, 1996; Le et al., 1997; Shen et al., 2001); 5) study the minimum essential parts for RNA-protein binding (Berkowitz et al., 1993; Graff et al., 1998; Jonas et al., 2008) and 6) study the role of 3' and 5' untranslated regions of mRNA (Chang et al., 1994; Le et al., 1997; Thekkumkara and Linas, 2003; Totary-Jain et al., 2005; Jonas, 2008). The technique is widely applied in various array of living organisms including animal and animal cell cultures (Ficzycz and Ovsenek, 2002; Totary-Jain et al., 2005; Crooks et al., 2007; Suzuki et al., 2010), plant cells (Shen et al., 2001), bacteria (Wei et al., 2001; Jonas et al., 2008) and viruses (Graff et al., 1998; Kim et al., 1998; Li et al., 1999). Mobility shift assay is one of the predominant procedures that are used in analysis of RNA-protein interaction (Thomson et al., 1999) because of many reasons including its higher sensitivity and lake of reliable reproducible alternatives.

Nonradioactive methods have been developed to replace the radioactive ones. They have some advantages over the radioactive labeling methods. They have shorter exposure time which means getting faster results of an experiment. In addition, they save time and money because of the long half life of labeling molecules and they offer homogenous labeling with minimum number of probe preparation. More importantly, they are safer because they do not include exposure to radioactive molecules. Nonradioactive labeling using Digoxigenin-11-UTP along with chemiluminescent detection is among nonradioactive methods which were introduced to replace the radioactive procedures (Vaquero et al., 1998).

Although, the method has all advantages of nonradioactive methods, yet there has not been enough optimization for the protocol to enhance its efficiency. In this report, we have made some optimizations for the key steps which determine the quality of the results obtained by this protocol using wheat mitochondrial protein extracts and *orf256-coxI* RNA.

MATERIALS AND METHODS

Mitochondrial proteins

Mitochondrial protein extracts of four wheat lines [cytoplasmic male sterile (CMS); *Triticum aestivum* (Ta); fertility restored (FR1); *Triticum temopheevi* (Tt)] were prepared according to a published method (Song and Hedgcoth, 1994).

Electrotransfer of proteins to PVDF membrane

Proteins were blotted to a PVDF membrane (Immobilon-P, Millipore) using a BioRad semi-dry electrobloter for 1 h at 2 mA/ cm^2 of PVDF membrane using the manufacturer instructions and buffers (Millipore, USA; El-Shehawi et al., 2003).

RNA probe

RNA probe was prepared using *in vitro* transcription from the mitochondrial gene *orf256-coxI* (Rathburn and Hedgcoth, 1991) using *in vitro* transcription kit according to the manufacturer instructions (Epecentre, USA) with replacement of the ^{32}P-UTP with DIG-11-UTP. The ratio of UTP : DIG-11-UTP in the *in vitro* transcription reaction was 0.65:0.35 as recommended by the manufacturer (Roche). The RNA probe was precipitated after transcription according to the manufacturer's instructions (Roche).

RNA mobility shift

Mobility shift assays were done by modifications of a nonradioactive method (Berger et al., 1993). Mitochondrial protein extracts (20 µg), were incubated in a 20 µL reaction containing 40 U RNase inhibitor in mobility shift assay buffer (10 mM Tris-HCl, pH 7.5; 50 mM NaCl; 1 mM MgCl; 0.5 mM EDTA; 4% glycerol; 0.5 mM DTT) for 5 min at room temperature. Yeast RNA (4 µg), was added (Kafert et al., 1998) and the reaction was incubated for 10 min. RNA probe was added and incubation was continued for 30 min at room temperature. The reaction mixture was UV cross-linked for 10 min at 254 nm at 5 cm from the UV source (Zehner et al., 1997). Unprotected RNA probe was removed by the addition of 1 µL of RNase cocktail (Ambion, USA) and incubation for 15 min at 37°C. Following SDS-PAGE (Laemmli, 1970), separated components were electrotransfered to PVDF membrane. The membrane was heated for 1 h at 80°C to fix RNA to the membrane.

The blot was blocked by three changes of 30 min each of blocking buffer (10 mM Tris-HCl, pH 7.5; 150 mM NaCl; 10% glycerol; 2.5% Nonidet P-40; 0.1 mM DTT; 5% casein) and rinsed in detection buffer (50 mM Tris-HCl, pH 7.5; 50 mM NaCl; 0.25% casein, USB). The blot was incubated in anti-digoxigenin-alkaline phosphatase conjugate (Boehringer Mannheim) solution, diluted (1:10,000) in detection buffer for 30 min with gentle shaking at room temperature. The blot was washed four times for 10 min each time with gentle shaking in detection buffer and once in assay buffer (0.1 M diethanolamine, pH 10; 1 mM $MgCl_2$; 0.02% sodium azide) (Tropix). CDP-star was applied (2.5 µL of 1:100 diluted CDP-Star in assay buffer per cm^2 of membrane). The blot was exposed to X-ray film for 30 min to 1 h.

RESULTS AND DISCUSSION

Nonspecific competitor

Yeast RNA and heparin are usually used as nonspecific competitors in mobility shift assays. Yeast RNA is added before the probe to block nonspecific sites, whereas heparin is added after binding the probe to its sites to remove the nonspecific binding. Various concentrations of yeast RNA as well as heparin were tested with only Ta extract (Figure 1). Heparin seems to give faint diffused bands at lower concentrations (2.5 and 5 µg/µL). These bands became undetectable at 10 µg/µL. The same bands detected using yeast RNA was sharper and more intensed (Figure 1), therefore yeast RNA serves as a better nonspecific competitor than heparin.

UV cross-linking time

Time of UV cross-linking is a sensitive factor for stabilizing the RNA probe to its target protein factor especially if binding is not stable on native gels. The exposure time of RNA-protein complexes to the UV is a limiting factor. Overexposure to UV could lead to cross-linking of the probe molecules that can be detected as bands giving misleading results. Underexposure to the UV could result in missing some proteins which may be important.

UV cross-linking times ranging from 2 to 15 min were tried. The intensity of RNA-protein bands increased with

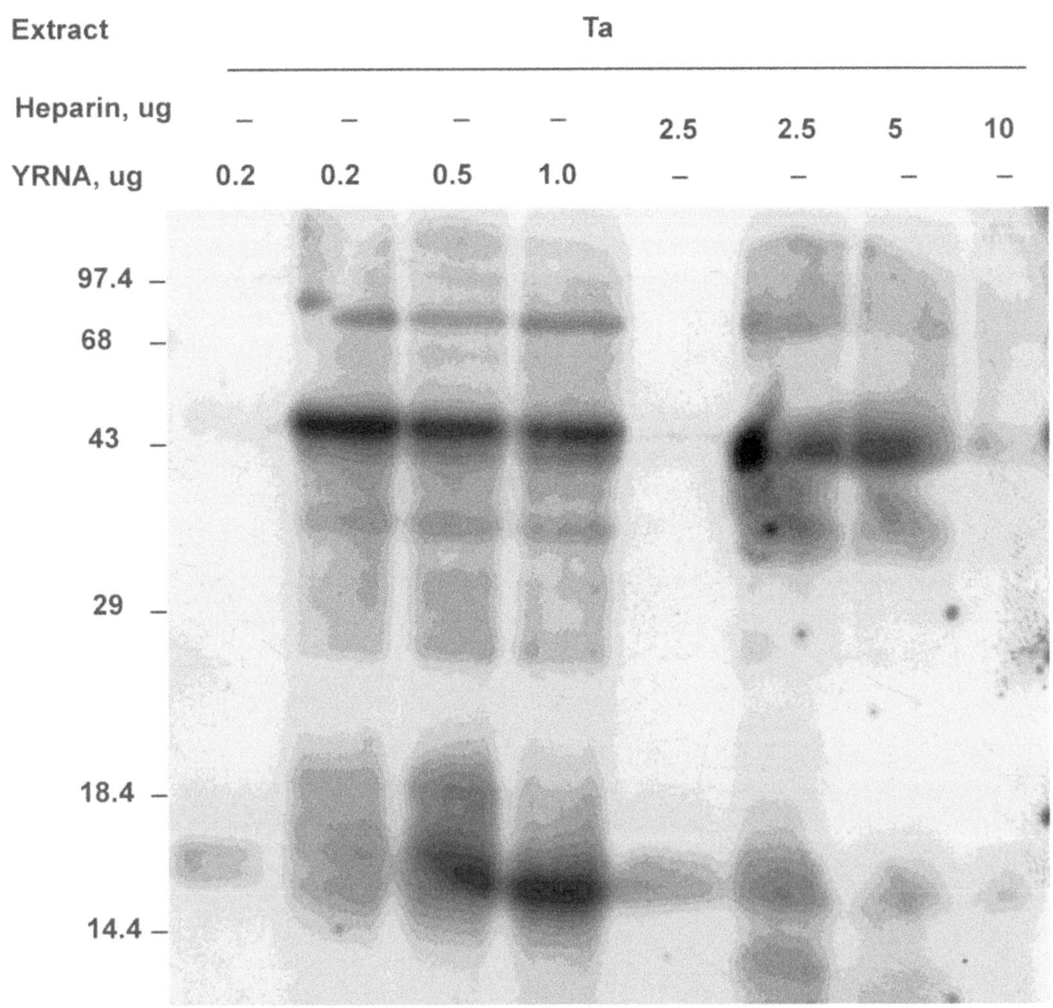

Figure 1. Effect of non-specific competitors on the formation of mitochondrial RNA-protein complexes. Mitochondrial protein extract (20 µg) was incubated with DIG-labeled *orf256-coxI* RNA probe in the presence of increased concentration of yeast RNA or heparin. For yeast RNA, 0.2 (lane 1, 2), 0.5, (lane 3) and 1.0 (lane 4) µg/µL were used, whereas, 2.5 (lane 5, 6), 5 (lane 7) and 10 µg/µL (lane 8) of heparin were used. Lane 1 and 5 do not contain any protein extract and served as negative controls.

increasing time of UV cross-linking (Figure 2A). It appeared that 10 min of UV cross-linking is an optimum time under these conditions since it showed a reasonable intensity of the 46 kD band and minimal intensity of the 16 kD bands which may be due to UV cross-linking of probe fragments.

Electrotransfer to PVDF membrane

Electrotransfer is an essential determinant of the nonradioactive mobility shift because if it is not done appropriately, could lead to missing of the RNA-protein complexes. During electrotransfer, double PVDF membranes were used to test if some proteins leak through the first membrane. No bands were detected on the second membrane indicating that there is no protein leakage through the first membrane (Figure 2B).

Also, the gel was stained after electrotransfer to make sure there was no significant amount of proteins retained in the gel. It seems that there was no RNA-protein complexes transferred to the second membrane. Staining the gel also indicated that most of the proteins were transferred to the PVDF membrane.

Using of different protein extracts

To confirm that this procedure can be used with different sources of mitochondrial proteins, another three wheat mitochondrial extracts (CMS, FR1, Tt) were tested against the *orf256-coxI* probe under the same conditions. Five protein bands were detected between 200 and 68 kD, some

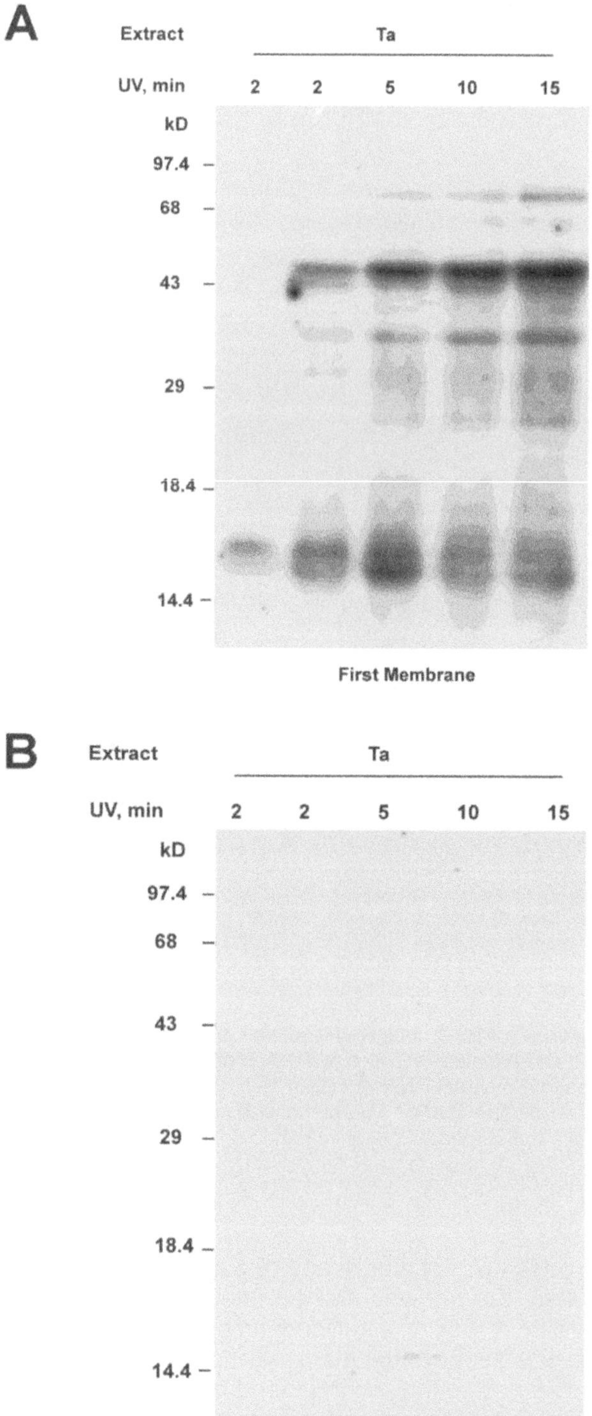

Figure 2. Effect of UV cross-linking and electrotransfer time on the formation (a) of and transfer (b) of mitochondrial RNA-protein complexes. Ta mitochondrial protein extracts (20 μg) was incubated with *orf256-coxI* RNA probe in the presence of 0.2 μg/μL yeast RNA. After RNA-Protein binding, samples were subjected to 254 nm UV light for 2 (lane 1, 2), 5 (lane 3), 10 (lane 4) or 15 (lane 5) min on ice at 5 cm from the UV source. Double PVDF membrane was used during electrotransfer (A, B) to test if some proteins leaked through the first membrane. The rest of the protocol is done as stated in the text.

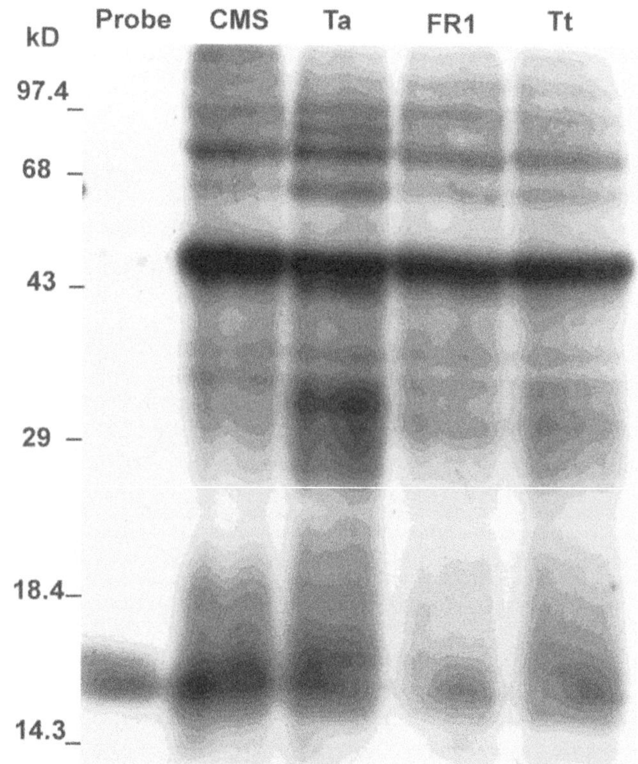

Figure 3. Optimization of RNA-protein interaction with various mitochondrial protein extracts. Different mitochondrial protein extracts were incubated with *orf256-coxI* RNA probe under the above conditions. Lane 2, CMS; lane 3, Ta; lane 4, FR1; lane 5, Tt. Lane 1 is negative control since it does not contain any protein extract.

of which differ in intensity, but all are detected in all extracts (Figure 3). One heavy band of about 46 kD and two bands of about 35 kD were also detected in all extracts.

There was one band at about 16 kD that was detected in all extracts and was present in the free probe lane as well. This band might represent UV cross-linked fragments of the RNA probe. Giving similar bands with various mitochondrial protein extracts confirm that this procedure is reproducible with different protein sources (Figure 3).

Proteinase K treatment

Since the bands detected on the mobility shift chemilumnigram may be detected from either real RNA-protein complexes or cross-linked fragments of the RNA probe, to distinguish between both possibilities, Proteinase K was used to digest proteins after UV cross-linking and before RNase treatment.

Proteinase K treatment abolished all bands detected under these conditions (Figure 4). Results obtained from this study provide an evidence that these optimizations contribute significantly in enhancing the quality of nonradioactive mobility shift assay to study mitochondrial

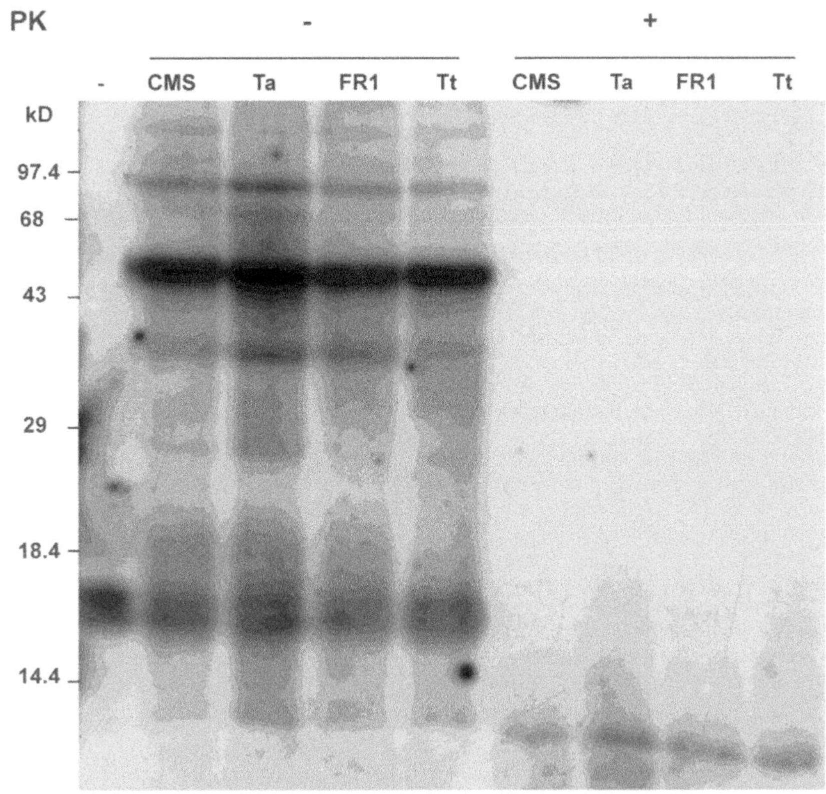

Figure 4. Differentiation of real RNA-protein complexes using proteinase K digestion. Mitochondrial proteins from various wheat lines (Figure 3 legend) were used in mobility shift reactions as explained in the text. Lanes 6, 7, 8 and 9 show identical reactions except that the reaction mixes were treated with Proteinase K (PK) for 10 min at 50°C. Proteinase K treatment was done after UV cross-linking and before RNase treatment.

plant RNA-protein interaction.

REFERENCES

Berger RM, Duncan R, Berman B (1993). Nonradioactive gel mobility shift assay using chemiluminescent detection. Biotechniques 15:650-651.

Berkowitz RD, Luban J, Goff SP (1993). Specific binding of human immunodeficiency virus type 1 gag polyprotein and nucleocapsid protein to viral RNAs detected by RNA mobility shift assays. J. Virol. 67:7190-7200.

Blyn LB, Swiderek KM, Richards O, Stahl DC, Semler BL, Ehrenfeld E (1996). Poly(rC) binding protein 2 binds to stem-loop IV of the poliovirus RNA 5' noncoding region: identification by automated liquid chromatography-tandem mass spectrometry. Proc. Natl. Acad. Sci. U.S.A. 93:11115-11120.

Chang DY, Nelson B, Bilyeu T, Hsu K, Darlington GJ, Maraia RJ (1994). A human Alu RNA-binding protein whose expression is associated with accumulation of small cytoplasmic Alu RNA. Mol. Cell Biol. 14:3949-3959.

Crooks DR, Ghosh MC, Braun-Sommargren M, Rouault TA, Smith DR (2007). Manganese targets m-aconitase and activates iron regulatory protein 2 in AF5. J. Neurosci. Res. 85:1797-1809.

Dixon DA, Kaplan CD, McIntyre TM, Zimmerman GA, Prescott SM (2000). Post-transcriptional control of cyclooxygenase-2 gene expression. The role of the 3'-untranslated region. J. Biol. Chem. 275:11750-11757.

El-Shehawi AM, Elseehy MM, Hedgcoth C (2003). orf256-coxI RNA binds to wheat mitochondrial proteins. Arab J. Biotechnol. 6:29-38.

Ficzycz A, Ovsenek N (2002). The Yin Yang 1 transcription factor associates with ribonucleoprotein (mRNP) complexes in the cytoplasm of Xenopus oocytes. J. Biol. Chem. 277:8382-8387.

Graff J, Cha J, Blyn LB, Ehrenfeld E (1998). Interaction of poly(rC) binding protein 2 with the 5' noncoding region of hepatitis A virus RNA and its effects on translation. J. Virol. 72:9668-9675.

Jonas K, Edwards AN, Simm R, Romeo T, Romling U, Melefors O (2008) The RNA binding protein CsrA controls cyclic di-GMP metabolism by directly regulating the expression of GGDEF proteins. Mol. Microbiol. 70:236-257

Kafert S, Winzen R, Loos A, Bollig F, Resch K, Holtmann H. (1998). Protein binding of the mRNAs for the 55 kDa tumor necrosis factor receptor and the glucose transporter 1: sequence homology and competition for cellular proteins. FEBS Lett. 421:2-6.

Kim YM, Kim JH, Kim KH, Hwang YS, Paik YK (1998). Overexpression and characterization of the cDNA encoding the coat protein of cucumber mosaic virus (Strain ABI) isolated in Korea. Mol. Cells. 8:416-423.

Laemmli UK (1970). Cleavage of structural proteins during the assembly of the head of bacteriophage T4. Nature 227:680-685.

Lavigueur A, La Branch H, Kornblihtt AR, Chabot B (1993). A splicing enhancer in the human fibronectin alternate ED1 exon interacts with SR proteins and stimulates U2 snRNP binding. Genes Dev. 7:2405-2417.

Le H, Tanguay RL, Balasta ML, Wei CC, Browning KS, Metz AM, Goss DJ, Gallie DR (1997). Translation initiation factors eIF-iso4G and eIF-4B interact with the poly(A)-binding protein and increase its RNA binding activity. J. Biol. Chem. 272:16247-16255.

Li B, Wyman TE, Moudgil T, Marracci GH, Ju CF, Machida CA (1999).

Nucleocytoplasmic export of type D simian retrovirus genomic RNA: identification of important genetic subregions and interacting cellular proteins. Virology 264:37-54.

Mochow-Grundy M, Dermody TS (2001). The reovirus S4 gene 3' nontranslated region contains a translational operator sequence. J. Virol. 75:6517-6526.

Mulligan GJ, Guo W, Wormsley S, Helfman DM (1992). Polypyrimidine tract binding protein interacts with sequences involved in alternative splicing of beta-tropomyosin pre-mRNA. J. Biol. Chem. 267:25480-25487.

Randall JD, Cahill CM, Eder PS, Huang X, Gunshin H, Leiter L, McPhee J, Sarang SS, Utsuki T, Greig NH, Lahiri DK, Tanzi RE, Bush AI, Giordano T, Gullans SR (2002). An iron-responsive element type II in the 5'-untranslated region of the Alzheimer's amyloid precursor protein transcript. J. Biol. Chem. 277:45518-5528.

Rathburn H, Hedgcoth C (1991). A chimeric open reading frame in the 5' flanking region of coxI mitochondrial DNA from cytoplasmic male-sterile wheat. Plant Mol. Biol. 16:909-912.

Ruth JH, Esnault S, Jarzembowski JA, Malter JS (1999). Calcium ionophore upregulation of AUUUA-specific binding protein activity is contemporaneous with granulocyte macrophage colony-stimulating factor messenger RNA stabilization in AML14.3D10 cells. Am. J. Respir. Cell Mol. Biol. 21:621-628.

Shen Y, Danon A, Christopher DA (2001). RNA binding-proteins interact specifically with the Arabidopsis chloroplast psbA mRNA 5' untranslated region in a redox-dependent manner. Plant Cell Physiol. 42:1071-1078.

Song J, Hedgcoth C (1994). A chimeric gene (*orf256*) is expressed as protein only in cytoplasmic male-sterile lines of wheat. Plant Mol. Biol. 26:535-539.

Suzuki MG, Imanishi S, Dohmae N, Asanuma M, Matsumoto S (2010). Identification of a male-specific RNA binding protein that regulates sex-specific splicing of Bmdsx by increasing RNA binding activity of BmPSI. Mol. Cell Biol. 30(24):5776-5786.

Thekkumkara TJ, Linas SL (2003). Evidence for involvement of 3'-untranslated region in determining angiotensin II receptor coupling specificity to G-protein. Biochem. J. 370:631-639.

Thomson AM, Rogers JT, Walker CE, Staton JM, Leedman PJ (1999). Optimized RNA gel-shift and UV cross-linking assays for characterization of cytoplasmic RNA-protein interactions. Biotechniques 27:1032-1040.

Totary-Jain H, Naveh-Many T, Riahi Y, Kaiser N, Eckel J, Sasson S (2005). Calreticulin destabilizes glucose transporter-1 mRNA in vascular endothelial and smooth muscle cells under high-glucose conditions. Circ. Res. 97:1001-1008.

Vaquero C, Liao Y, Fischer R (1998). Nonradioactive UV cross-linking assayfor the study of protein-RNA binding. Biotechniques 24:68-71.

Wei BL, Brun-Zinkernagel AM, Simecka JW, Pruss BM, Babitzke P, Romeo T (2001). Positive regulation of motility and flhDC expression by the RNA-binding protein CsrA of *Escherichia coli*. Mol. Microbiol. 40:245-256.

Zehner ZE, Shepherd RK, Gabryszuk J, Fu T, Al-Ali M, Holmes WM (1997). RNA-binding interactions within the 3' untranslated region of vimentin mRNA. Nucleic Acids Res. 25:3362-3370.

Genetic transformation of *cry1EC* gene into cotton (*Gossypium hirsutum* L.) for resistance against *Spodoptera litura*

R. Pushpa, T. S. Raveenderan, S. Rajeswari, P. Amalabalu and D. Punitha

Centre for Plant Breeding and Genetics, Tamil Nadu Agricultural University, Coimbatore-641 003, India.

Cotton is the chief fibre crop of global importance. It plays a significant role in the national economy. Cotton crop is vulnerable to a number of insect species, especially to the larvae of lepidopteron pests. 60% insecticides sprayed on cotton are meant to control the damage caused by bollworm complex. Transgenic technology has become a popular option for the development of bollworm resistant cotton. *Spodoptera litura* is one of the notorious emerging pests in cotton. Introduction of *cry1EC* gene into commercial variety have very significant importance. This research work was carried out to transform chimeric δ-endotoxin *Cry1EC* into cotton. The tobacco cutworm (*Spodoptera litura*) is a polyphagous foliage insect which is susceptible to the chimeric δ-endotoxin *Cry1EC*. Six month-old highly friable embryogenic calli derived from cotyledonary explants of Coker 310 were used for transformation using *Agrobacterium* tumefaciens strain LBA4404 harboring plasmid pBI101.1 carrying the marker gene neomycin phosphotransferase II (*npt II*) and a synthetic *cry 1EC* gene under a constitutive 35 S promoter. *Agrobacterium* treated calli were selected on MS medium containing, 50 mg/L kanamycin, 500 mg/L cefotaxime, 30 g/L maltose and 0.4% phytagel. Embryos developed on kanamycin resistant calli were maintained on the same medium till somatic embryos matured. The cotyledonary stage embryos (3 to 5 mm size) were germinated on MS basal slat with 0.1 mg/L GA3 + 1.0 mg/L IAA, 30 g/L sucrose and solidified with 0.4% phytagel. The regenerated putative transgenic plants were hardened and transferred to the transgenic green house. Transgenic plants were confirmed by polymerase chain reaction (PCR) amplification of 800 bp npt II fragment, and 578 bp amplification of *cry1Ec* gene. Transgenic plant with single copy insertion of *cry1EC* was selected in T_0 by southern blot hybridization. Insect bioassay using *Spodoptera litura* larvae of first instar stages on T_0 plants showed 70% mortality. Not much data has been published on the toxicity of the endotoxins to *S. litura*, which is a common pest in warm and humid climates. Efforts are aimed at recovering more efficient transgenic plants through efficient transformation system and developing high resistant transgenic cotton against *S. litura* and will paved a way for promising future in cotton production.

Key words: *Agrobacterium* transformation, *cry 1EC*, molecular analysis, insect bioassay.

INTRODUCTION

Cotton (*Gossypium* spp.) is an important source of natural fibre, and plays a significant role in the Indian economy. *Spodoptera litura* is a polyphagous insect pest affecting various economically important crops and in case of severe infestation has caused significant loss to cotton crop in india. It primarily feed on foliage but also feed on fruiting parts of cotton plant (Eveleens et al., 1973). Larval feeding on cotton leaves can cause devastating losses in yield. The insect pest management in cotton is mainly dominated by the use of broad-spectrum insecticides, which disrupt the beneficial insect fauna too thus leading to pest resurgence and outbreaks of secondary pests, as well as risk to human health and environment (Yousefi, 2000). It is estimated that 60% of the insecticides sprayed on cotton are meant to control the damage caused by bollworm complex (Chaudhary and Loroia, 2001). These pests have developed high level of resistance due to repeated application of insecticides leading to heavy expenditure, crop failures and vicious cycle of debt for farmers. Development of new genetic material possessing resistance to insects, especially for bollworms has not made the required impact in achieving complete resistance due to lack of genetic variation for resistance and ploidy differences limit the utilization of wild species too.

One option to reduce the insecticide use on cotton is the exploitation of transgenic *Bt* cotton as a component of integrated pest management (Gore et al., 2001). Bt cotton, the first transgenic non-food crop, has provided a specific, safe and effective tool for the control of lepidopterous pests (Wu and Guo, 2005). Transgenic cotton cultivars expressing *Cry1Ac* have been released for commercial cultivation; worldwide this endotoxin is moderately toxic to *Helicoverpa zea* but poorly effective *against Spodoptera exigua* and *S. frugiperda* (Perlak et al., 2001). The genus *Spodoptera* contains five primary species (*litura, littoralis, exigua, frugiperda* and *exempta*), that occur worldwide. These are the pests of cotton, tobacco, castor, groundnut, a number of grain legumes and vegetable crop. The primary natural δ-endotoxin reported to show significant toxicity to *S. exigua* (Sanchis et al., 1988; Visser et al., 1990) and *S. littoralis* (Hone´e et al., 1991) is *Cry1C.*

Plants expressing *Cry1Ca* at a high level were reported to cause mortality and effectively give protection against *Spodoptera* sp. (Strizhov et al., 1996; Mazier et al., 1997). However, transgenic Bt cotton with *Cry1Ac* proved not to be effective against *Spodoptera* spp. (Yu et al., 2004). A number of studies have shown that *Spodoptera* sp is not susceptible to Bt cotton containing Cry1Ac toxin so, there is a chance that this pest may become the major and alarming pest in Bt cotton field. As transgenic crops may perform differently under different ecological conditions so, there is a need to determine the efficacy of transgene against *Spodoptera*

Currently, cotton varieties containing Bt do not offer acceptable control of *Spodoptera.* A synthetic *cry1EC* gene has been designed by NBRI specifically against *S. litura* by replacing amino acids 530 to 587 in a poorly active Cry1Ea protein with a homologous 70 amino acid stretch of Cry1Ca in domain III (Singh et al., 2004).

Expression of the synthetic *cry1EC* gene in transgenic tobacco and cotton gave significant levels of resistance against *S. litura* (Singh et al., 2004). This elite Bt Cotton carrying *cry1EC* will give broader insect resistance coverage, particularly against *Spodoptera* (tobacco caterpillar) and expected to provide season long uniform protection of the target pests throughout the entire field. The present study reports introduction of *cry1EC* gene into cotton plants through *agrobacterium* mediated transformation and efficient regenerating of plants *via* somatic embryogenesis

MATERIALS AND METHODS

Bacterial strain and vector

The *Agrobacterium* strain, LBA4404 harbouring plasmid pBI101.1 and carrying synthetic *cry1EC* (Source: NBRI; Figure 1) was used in this study. The plasmid contains (i) the *cry1EC* gene, which encodes an insecticidal protein *Cry1EC*, (ii) 35S promoter from cauliflower mosaic virus to drive the expression of *cry1EC* gene in all parts of the plants (iii) *nptII* gene, the plant selectable marker, which encodes the enzyme, neomycin phosphotranferase II (NPTII).

Plant material

Acid-delinted cotton seeds of cotton varieties, Coker 310 were obtained from the germplasm collections of the Department of Cotton, Centre for Plant Breeding and Genetics, Tamil Nadu Agricultural University, Coimbatore. Seeds were surface sterilized with 70% ethanol for 2 min and then washed three times with sterile distilled water. They were again surface-sterilized with 0.1% mercuric chloride for 10 min followed by three washes with sterile distilled water. The seeds were germinated on half strength MS (Murashige and Skoog, 1962) medium supplemented with 1% (w/v) sucrose and 10 g/L agar. The pH of the medium was adjusted to 5.7 to 5.8 (by using 0.1 N KOH or 0.1 N HCl) prior to autoclaving the media for 15 min at 121°C and 15 psi pressure.

Callus induction

Explants from five to seven day-old seedlings were used for callus induction. Both hypocotyl (4 to 6 mm) and cotyledonary leaf (16 mm^2) sections were plated onto callus induction medium MS basal salt, supplement with 0.1 mg/L of 2,4-D and 0.5 mg/L of kinetin, 30 g/L maltose and solidified with 10 g/L agar. Vigorously growing loose and light yellow callus was transferred to a fresh somatic embryo induction medium containing MS salts with 1.9 g/L KNO_3, 30 g/L sucrose and solidified with 0.4 % phytagel (Sigma) (pH 5 to 8). Embryogenic calli with high proliferation activity were obtained after about three or four rounds of subculture with an interval of 2 weeks. Six month-old highly friable embryogenic calli derived from cotyledonary explants of Coker 310 were used for transformation.

Transformation and regeneration

Preculture

Six month-old highly friable embryogenic calli were precultured on somatic embryogenic medium containing (MS salts, 1.9 g/L KNO_3,

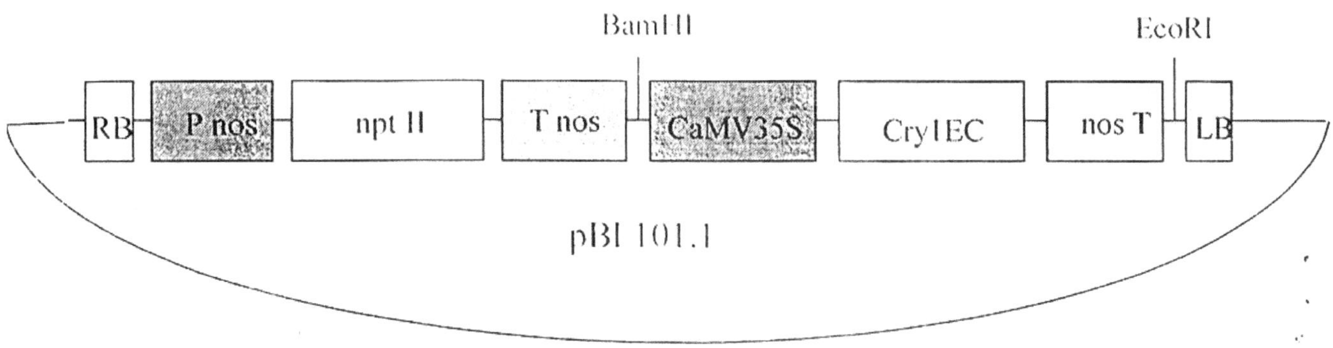

Figure 1. Linear map of T-DNA of pBI 101.1 carrying *cry1Ec* gene.

30 g/L maltose and 0.4 % phytagel) for seven days prior to cocultivation.

Cocultivation

A loopful of *Agrobacterium* was inoculated in 3 ml of AB medium (Chilton et al., 1974) containing kanamycin (100 mg/L) and grown at 28°C in a rotary shaker at 175 rpm. From the overnight culture, 0.5 ml was transferred to 30 ml AB liquid medium with kanamycin (100 mg/L) and was allowed to grow for 12 h to 0.5 to 0.6 OD (optical density) at 600 nm. The culture was centrifuged at 4000 rpm for 10 min and the pellet was resuspended in 20 ml of AAM medium (Hiei et al., 1994) containing 100 μM acetosyringone. After preculture, the embryogenic calli were treated with the bacterial suspension by shaking vigorously for 30 min. The bacterial suspension was then removed using pipette and the calli were blot dried on sterile tissue paper towels. The nearly dry calli were placed on Whatman No.1 filter paper which was overlaid on somatic embryogenic medium containing 100 μM acetosyringone (cocultivation medium). Cocultivation of calli was carried out under the dark for three days.

Selection and regeneration

After cocultivation, the calli were blot dried on sterile tissue paper towels and transferred to Somatic embryogenic medium containing, 50 mg/L kanamycin, 500 mg/L cefotaxime, 30 g/L maltose and 0.4% phytagel. After two months, embryos developed on kanamycin resistant calli were maintained on the same medium till somatic embryos matured. The cotyledonary stage embryos (3 to 5 mm size) were germinated on plant regeneration medium (MS basal slat) with 0.1 mg/L GA_3 + 1.0 mg/L IAA, 30 g/L sucrose and solidified with 0.4% phytagel. The regenerated putative transgenic plantlets attained adequate growth by producing 4 to 6 leaves with sufficient root system; they were removed from the bottles and washed in running tap water carefully to remove the media adhering to roots. They were subsequently transplanted in pots containing sterilized soil. The plants were covered with polyethylene bags and kept in the culture room. After 15 days, the polyethylene bags were removed and well established plants were transferred to transgenic greenhouse.

Polymerase chain reaction (PCR) and Southern blot analysis

Total DNA was extracted from leaves of putative transgenic plants and non transgenic wild type plants following the method described by Permingeat et al. (1998). In the extraction buffer, 0.5 M glucose was added to avoid browning by polyphenolic compounds and improve the quality of DNA samples for further molecular analysis.

PCR was performed as described by Surekha et al. (2005) with a few modifications. Reactions were performed in a final volume of 25 μl and the mixture contained 50 to 100 ng of genomic DNA, 2.5 μl of 10X PCR buffer, 200 nmM of each of dNTPs, 70 ng of forward and reverse primers and two units of Taq DNA polymerase. Amplification was performed in a MYCycler, (Biorad, USA). The primer used for amplification for *cry1EC* gene was Forward primer 5' CCA GAG AAC GAG ATC TTG GAC 3', and Reverse primer 5' AGT ATT GTA CCA TCT AAC AGC GTA 3' and for amplifying *nptII* gene, the sequence was 5' ATG ATT GAA CAA GAT GGA TTG CACG 3', 5' TCA GAA GAA CTC GTC AAG AAG GC 3'. PCR amplification condition consist of pre-incubation at 94°C for 5 min leading to 30 cycles of melting at 94°C for 1 min, annealing at 55°C for 1 min, and synthesis at 72°C for 1 min followed by an extension at 72°C for 15 min. After amplification, 10 μl of the product was used for electrophoretic analysis on 0.8 % agarose gel. For southern analysis, 10 μg of genomic DNA extracted from putative transformants, untransformed control plant and 10 ng of plasmid DNA were digested with the *Eco*RI and *Hind*III enzyme, electrophoresed on 1% agarose gel and transferred onto a nylon membrane and probed with *cry1EC* gene. All standard procedures were followed as described by Sambrook et al. (1989).

Insect bioassay

Detached leaf feeding tests were done on T_0 plants and non-transformed (control) plants for insect resistance using 1st instar larvae of the pest *Spodeptera litura*. Leaves collected from the transgenic plants, were washed thoroughly using double distilled water to remove any fungicide, pesticide or dirt, blot dried, and were placed in sterile Petri dishes on a wet filter paper. Two replications were carried out for each transgenic plant individually. Five larvae were released into each Petri dish and observations were made every 24 h regarding the mortality, and growth of the larvae. After 3 days, insect mortality % age was calculated as given below:

$$\text{Mortality (\%)} = \frac{\text{Number of larvae dead}}{\text{Total Number of larvae released}} \times 100$$

Analysis of T_1 plants

To test the functional expression of the *nptII* gene in the T_0 progeny a germination test performed. Seeds collected from T_0 plants and untransformed plants were germinated on the medium containing 75 mg/L kanamycine. PCR analysis was carried out in the germinated seedling to check the presence of transgene.

RESULTS AND DISCUSSION

Transformation, selection and regeneration

Genetic transformation in cotton has so far resulted in the development of commercially successful insect and herbicide resistant cultivars (Perlak et al., 1990; Rajasekaran et al., 1996 and Keller et al., 1997). *Agrobacterium*-mediated transformation followed by regeneration *via* somatic embryogenesis remains the method of choice (Wilkins et al., 2004). This method is, by far, the most efficient means for generating transgenic upland cotton in most of the laboratories (Leelavathi et al., 2004; Light et al., 2005; Jin et al., 2006). It is a multi-step process involving labour intensive work over a 10 to 12 month period starting from cocultivation of *Agrobacterium* with cotyledonary or hypocotyl explants, followed by production and maintenance of several hundreds of calli derived from independent transformation events on a selection medium, induction of embryogenesis in each callus line and development and germination of somatic embryos into normal plants. The above mentioned hurdles have been overcome partially by the use of embryogenic calli with a defined embryogenic potential as the explants source for transformation. The embryogenic calli offers a large population of cell competent for the *Agrobacterium* transformation, resulting in a high frequency of transformation; such a high frequency of transformation is accompanied by the shortening of the culture period to 3 to 5 months as compared to the conventional transformation method (Leelavathi et al., 2004; Wu et al., 2005; Jin et al., 2006). Based on this background, *Agrobacterium* transformation method was followed to transform the Coker genotypes in the present study. Six month-old highly friable embryogenic calli derived from cotyledonary explants were used for *Agrobacterium* mediated transformation.

The aminoglycoside, kanamycin has been extensively used as a selective antibiotic in transformation experiments, mainly because several plant transformation vectors include neomycin phosphotransferase II (*nptII*) gene as a selectable marker. Only transformed cells can grow in the presence of kanamycin. Barring a few reports, most of the workers on cotton have used kanamycin as a selection agent (Wu et al., 2005; Jin et al., 2005; Jin et al., 2006). In the present study, the minimum lethal concentration of kanamycin on cotyledon, hypocotyls and embryogenic calli explants was 50 mg/L. Higher level of kanamycin (75 and 100 mg/L) killed most of the explants. Therefore, a concentration of 50 mg/L kanamycin was used to select transformed cells or tissues in this investigation. Zhao et al. (2006) described the development of method for the recovery of genetically transformed cotton *via Agrobacterium* mediated transformation without using any antibiotics, while most workers used kanamycin as antibiotic (Tohidfar et al., 2005). During the first selection on SEM3K50 medium, the cocultivated calli become dark brown or black. Fresh growth was observed from the kanamycin resistant calli in the third selection while non transformed calli did not produce new calli and dried off. The newly formed calli were selected and subcultured. The globular somatic embryos were formed on the kanamycin resistant calli after one month of culture. The embryos developed on kanamycin resistant calli matured and germinated normally, while the non-transformed embryos became brown and subsequently died. In the present study, out of 30 embryogenic calli explants cultured in the first experiment, 10 plants were regenerated at a regeneration frequency of 33.3 % while in the second experiment, out of the 42 embryogenic calli cocultivated, five plants were regenerated at a regeneration frequency of 11.9% and in subsequent experiment, the regeneration frequency was 14.1 and 5.0 and the transformation efficiency of 6.6, 7.1 and 2.5% was recorded in the first, second and third experiment respectively. The number of calli co-cultivated and number of putative transgenic plants regenerated are summarized in the Table 1 and Figure 2.

Molecular confirmation and insect bioassay

Transformation with different *cry* genes developed for resistance against a variety of insect pests has been reported earlier in other crops. Chickpea expressing *cry1A(c)* show resistance to pod borer (*Helicoverpa armigera*) (Sanyal et al. 2005). Pigeonpea expressing *cry1EC* show resistance to tobacco caterpillar (*Spodoptera litura*) (Surekha et al. 2005). Transformation with different *cry1EC*genes has been reported earlier in cotton (Kumar et al., 2009). Putative transgenic plants were subjected to various molecular analysis techniques to determine the presence of gene of interest in the putative transgenic plants and its expression. 50 plants were shifted to sterile soil in which only 10 plants survived till maturity. Their DNA samples were used in PCR to identify their transgenic nature. The results indicate that, eight plants were PCR positive and showed amplification of the expected 800 bp of the *nptII* gene with *nptII* specific primers (Figure 3) and 578 bp of *cry1EC* gene (Figure 4). No amplified product was detected from non transformed plant and the transformation efficiency measured as % of confirmed transgenic plant out of total number of plants raised and it was found to be 80% in T_0 generation.

Transgenic plant showing highest larval mortality rate has been selected for southern blotting. In Southern blot hybridization experiment incubation of the membrane in

Table 1. *Agrobacterium* mediated transformation in Coker 310.

Experiment number	Number of calli cocultivated	Frequency of explant survival (%)	Frequency of somatic embryogenesis (%)	Number of plants regenerated	Regeneration frequency (%)	Number of plants positive for PCR (*cry1Ec*)	Transformation frequency (%)
1	30	50.0	26.6	10	33.3	2	6.6
2	42	59.5	33.3	5	11.9	3	7.1
3	120	41.6	28.3	17	14.1	3	2.5
4	118	46.6	38.1	6	5.0	-	-
5	32	40.6	28.1	12	37.5	-	-
6	46	54.3	32.6	-	-	-	-
7	89	22.4	31.4	-	-	-	-
Mean	-	45.0	31.2	-	-	-	-

Figure 2. *Agrobacterium* mediated transformation at Coker 310 using embryogenic calli as explants. **a,** Embryogenic calli after cocultivation. **b,** Fresh growth of embryogenic calli on selection medium. **c,** Globular stage embryos development on SEMK25. **d,** Maturation and germination of somatic embryos on SEMK25. **e,** regenerated plant. **f,** Fertile transgenic plants with flowering and boll setting.

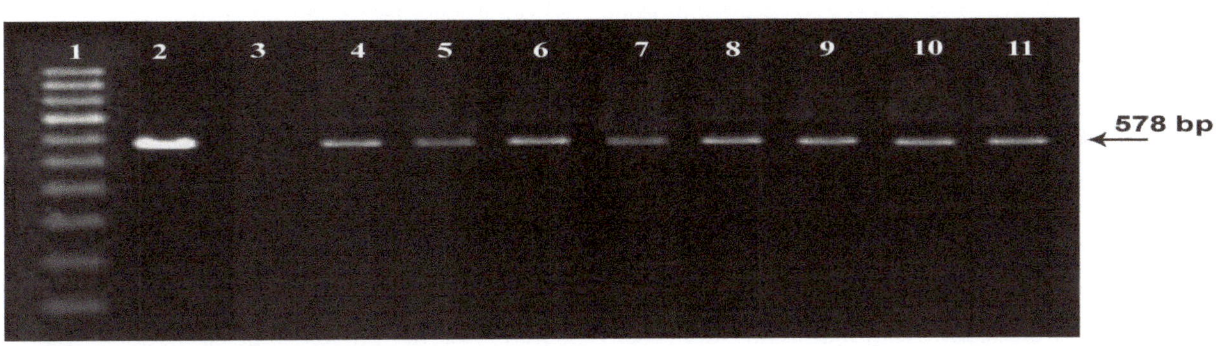

Figure 3. PCR amplification for *cry1EC* gene in transgenic cotton plants. Lane 1, 100 bp ladder; lane 2, plasmid DNA (positive control); lane 3, non transgenic control plant- Coker 310; lane 4 -11, putative transgenic plants.

solution containing radiolabelled *cry1EC* specific probe resulted in hybridization signal in transgenic plants while such signal could not detected in non-transformed control plants. The hybridization signal could be detected in plasmid

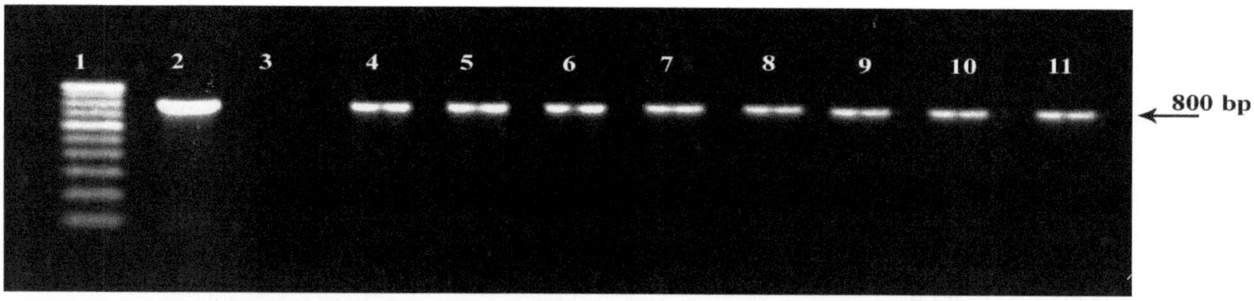

Figure 4. PCR amplification for *nptII* gene in transgenic cotton plants. Lane 1,100 bp ladder; lane 2, plasmid DNA (positive control); lane 3, non transgenic control plant- Coker 310; lane 4 -11, putative transgenic plants.

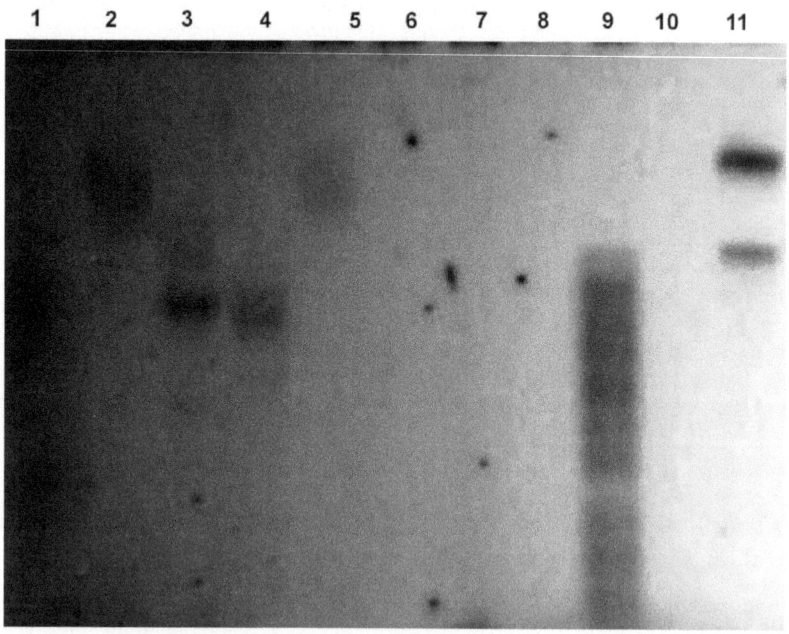

Figure 5. Southern blotting. Lane 1, MW (Lamda Hind III); lane 2, uncut genomic DNA of transgenic plant; lane 3, genomic DNA of transgenic plant digested with EcoR1; lane 4, genomic DNA of transgenic plant digested with Hind III; lane 5, uncut genomic DNA of transgenic plant, lane 6, blank, lane 7, genomic DNA of non transgenic control plant digested with Hind III; lane 8, blank; lane 9, plasmid control digested with Hind III; lane 10, blank; lane 11, uncut plasmid control.

control as well. In the present study, PCR and southern blot hybridization analysis confirmed the stable integration of plasmid DNA into plant genome (Figure 5).

The transforment possessed a single copy insertion. The toxicity of T_0 plants expressing *cry 1EC* protein was tested using feeding bioassays with first instar larvae of *S. litura*. PCR confirmed positive plants with healthy plant morphology were subjected to feeding. Mortality and loss in weight/retardation in growth were recorded for assessing the effect of the protein on the larvae. All the experiments were conducted along with a non-transformed cotton plant for comparison. Bioassay on detached leaves showed significant variation from the control wild-type plants (Table 2). The highest mortality of the larvae found in the transgenic plants was 70 %. Singh et al. (2004), Surekha et al. (2005) and Tiwari et al. (2008) recorded 100% mortality. All the larvae survived on the control plant.

The larvae survived on transgenic plant showed a greater reduction in size and loss of weight. Sujatha et al. (2009) observed varying degrees of larval mortality and slow growth in larvae fed on transgenic leaf tissue of castor. The growth of surviving larvae was also found inhibited. This could be because of lower level of expression of the synthetic gene. It is possible that effective plants could be recovered with the develop-

Table 2. Insecticidal activity of transgenic plants containing *cry1EC* gene to *Spodeptera litura*.

Transgenic plant	% mortality (mean ± SD)
1	70 ± 0.2[a]
2	50 ± 0.2[abc]
3	70 ± 0.2[a]
4	60 ± 0.2[ab]
5	50 ± 0.2[abc]
6	30 ± 0.2[c]
7	40 ± 0.2b[c]
8	30 ± 0.2[c]
Non transgenic –control plant	0 ± 0.2[d]

Insect bioassays were performed with *S. litura* first instar larvae. The readings were taken after 3 days of feeding on the leaves of transgenic T_0 plants.

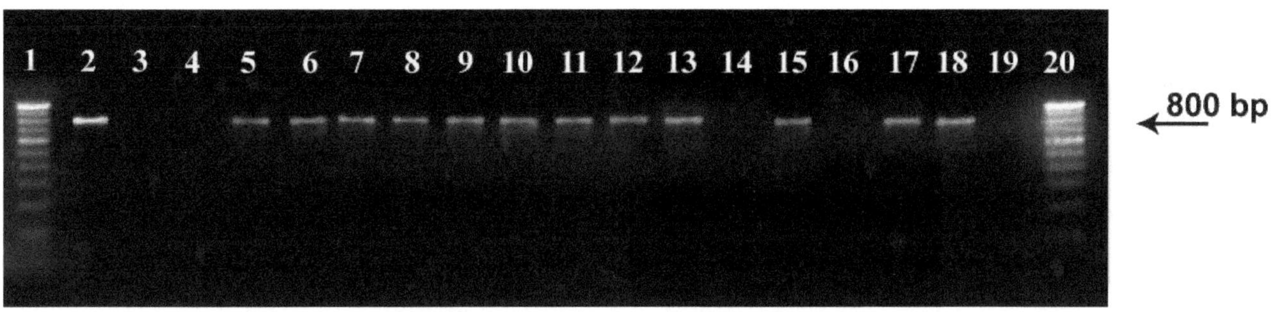

Figure 6. PCR amplification for *cry1EC* gene in T_1 transgenic cotton plants. Lane 1, 100 bp ladder; lane 2, plasmid DNA (positive control), lane 3, non transgenic control plant- Coker 310; lane 4-20, putative T_1 transgenic plants.

ment of more number of independent transgenic plants.

Transgenic inheritance analysis

The selected T_0 plant was fertile and set seeds, but the boll size was much reduced and number of boll was low when compared to non-transformed control plant. T_1 seeds were obtained from the glasshouse-grown T_0 transgenic plants. The inheritance of the introduced genes in the T_1 generation was studied using germination test. A total of 16 T_0 seeds were used for germination, in which 12 seeds were germinated. None of the seeds from non-transformed plants germinated on medium containing 75 mg/L kanamycin. For transformed plants tested, the ratio of resistance to sensitive was 3:1; the expected ratio for single dominant gene. The seedlings were scored for analyzing the segregation of *nptII* reporter gene in the progeny. PCR result also confirmed 3:1 ratio, out of 16 T_1 plants, and 12 plants were found to be positive (Figure 6).

Thus, the development of transgenic cotton against *S. litura* and the use of a hybrid protein with *Cry1E* backbone in a transgenic crop are useful developments. Not much data has been published on the toxicity of the δ-endotoxins to *S. litura*, which is a common pest in warm and humid climates. Efforts are aimed at recovering more efficient transgenic plants through efficient transformation system and developing high resistant transgenic cotton against *S.litura* will paved a way for promising future in cotton production.

REFERENCES

Chaudhary B, Loroia G (2001). Technological developments and cotton production in India and China. Curr. Sci. 80(8):8-25.

Chilton MD, Currier TC, Farrand SK, Bendich AJ, Gordon MP, Nester EW (1974). *Agrobacterium tumefaciens* DNA and PS8 bacteriophage DNA not detected in crown gall tumours. Proc. Natl. Acad. Sci. USA. 71:3672-3676.

De Maagd RA, Weemen-Hendriks M, Stiekema W, Bosch D (2000) *Bacillus thuringiensis* delta-endotoxin Cry1C domain III can function as a specificity determinant for *Spodoptera exigua* in different, but not all, Cry1-Cry1C hybrids. Appl. Environ. Microbiol. 66:1559-1563.

Eveleens KG, Van Den Bosch R, Dehler LE(1973). Secondary outbreak induction of beet armyworm by experimental insecticide applications in cotton in California. Environ. Ent. 2:497-503.

Gore J, Leonard BR, Damczyk JJ (2001). Bollworm (Lepidoptera: Noctuidae) survival on 'Bollgard' and 'Bollgard II' cotton flower buds and flower components. J. Econ. Entomol. 94:1445-1451.

Hiei Y, Ohta, S, Komari T, Kumashiro T (1994). Efficient transformation of rice (*Oryza sativa* L.) mediated by *Agrobacterium* and sequence analysis of the boundaries of the T-DNA. Plant J. 6:271-282.

Hone´e G, Convents D, Rie JV, Jansens S, Peferoen M ,Visser B (1991) The C-terminal domain of the toxic fragment of Bacillus thuringiensis crystal protein determines receptor binding. Mol Microbiol. 5:2799-2806.

Jin S, Zhang X, Liang S, Nie, Yichun, Guo X, Huang C (2005). Factors affecting transformation efficiency of embryogenic callus of Upland cotton (*Gossypium hirsutum*) with *Agrobacterium tumefaciens*. Plant Cell Tiss. Org. Cult. 81:229-237.

Jin S, Zhang X, Nie Y, Guo X, Liang S , Zhu, H (2006). Identification of novel elite genotypes for *in vitro* culture and genetic transformation of cotton. Biol. plant. 50 (4):519-524.

Kumar M, Shukla AK, Singh H, Tuli R(2009). Development of insect resistant transgenic cotton lines expressing *cry1EC* gene from an insect bite and wound inducible promoter. J. Biotechnol. 140(3-4):143-148.

Lambert B, Buysse L, Decock C, Jansens S, Piens C, Saey B (1996). A *Bacillus thuringiensis* insecticidal crystal protein with a high activity against members of the family noctuidae. Appl. Environ. Microbiol. 62:80-86.

Leelavathi S, Sunnichan VG, Kumria R, Vijaykanth, GP, Bhatnagar, RK, Reddy VS (2004). A simple and rapid *Agrobacterium*-mediated transformation protocol for cotton (*Gossypium hirsutum*): Embryogenic calli as a source to generate large numbers of transgenic plants. Plant Cell Rep. 22:465-470.

Light GG, Mahan JR, Roxas VP , Allen RD (2005). Transgenic cotton (*Gossypium hirsutum* L.) seedlings expressing a tobacco glutathione S-transferase fail to provide improved stress tolerance. Planta. 222:346-354.

Mazier M, Chaufaux J, Sanchis V, Lereclus D, Giband M and Tourneur J (1997) .The crylC gene from Bacillus thuringiensis provides protection against *Spodoptera littoralis* in young transgenic plants. Plant Sci. 127:179-190.

Murashige T, Skoog F (1962). A revised medium for rapid growth and bioassays with tobacco tissue culture. Physiol. Plant. 15:474-497.

Perlak FJ, Oppenhuizen M, Gustafson K, Voth R, Sivasupramaniam S, Heering D (2001) Development and commercial use of Bollgard cotton in the USA – early promises versus today's reality. Plant J. 27:489-501.

Permingeat HR, Romagnoli VM, Vallejos RH (1998). A simple method for isolating high yield and quality DNA from cotton (*Gossypium hirsutum* L.) leaves. Plant Mol. Biol. Rep.11:122-127.

Rajasekaran K, Grula JW, Hudspeth RL, Pofelis S, Anderson DM (1996). Herbicide-resistant Acala and Coker cottons transformed with a native gene encoding mutant forms of acetohydroxy acid synthase. Mol. Breed. 2:307-319.

Sambrook J, Fritsch EF, Maniatis T (1989). Molecular cloning. A laboratory manual. II edition. Cold Spring Harbor Laboratory Press, Cold Spring Harbor, New York, USA.

Sanchis V, Lereclus D, Menou G, Chaufaux J, Lecadet MM (1988). Multiplicity of δ- endotoxin genes with different insecticidal specificities in *Bacillus thuringiensis* aizawai. Mol. Microbiol. pp. 393-404.

Sanyal I, Singh AK, Kaushik M, Amla DV (2005). *Agrobacterium* mediated transformation of chickpea (*Cicer arietinum* L.) with *Bacillus thuringiensis* cry1Ac gene for resistance against pod borer insect *Helicoverpa armigera*. Plant Sci. 168:1135-1146.

Singh PK, Kumar M, Chaturvedi CP, Yadav D, Tuli R (2004). Development of a hybrid δ-endotoxin and its expression in tobacco and cotton for control of a olyphagous pest *Spodoptera litura*. Transgenic Res. 13:397-410.

Strizhov N, Keller M, Mathur J, Koncz-Kalman Z, Bosch D, Prudovsky E (1996). A synthetic crylC gene, encoding a *Bacillus thuringiensis* δ-endotoxin, confers Spodoptera resistance in alfalfa and tobacco. Proc. Natl. Acad. Sci. USA 93:15012-15017.

Sujatha M, Lakshminarayana M, Tarakeswari M, Singh PK, Tuli R (2009). Expression of the cry1EC gene in castor (*Ricinus communis* L.) confers field resistance to tobacco caterpillar (*Spodoptera litura* Fabr) and castor semilooper (*Achoea janata* L.). Plant Cell Rep. 28:935-946.

Surekha CH, Beena MR, Arundhati A, singh PK, Tuli R, Dutta gupta A, Kirti, PB (2005). *Agrobacterium* mediated genetic transformation of pigeon pea (*Cajanus cajan (L.)* Millsp.) using embryonal segments and development of transgenic plants for resistance against *Spodoptera*. Plant Sci. 169:1074-1080.

Tiwari S, Mishra D K, Singh A, Singh PK, Tuli R (2008). Expression of a synthetic *cry1EC* gene for resistance against *Spodoptera litura* in transgenic peanut *(Arachis hypogaea L.)*. Plant Cell Reports 27(6):1017-1025.

Tohidfar M, Mohammadi M, Ghareyazie B (2005). *Agrobacterium*-mediated transformation of cotton (*Gossypium hirsutum*) using a heterogous bean chitinae gene. Plant Cell Tiss. Organ Cult. 83:83-96.

Visser B, Munsterman E, Stoker A, Dirkse WG (1990). A novel *Bacillus thuringiensis* gene encoding a *Spodoptera exigua* – specific crystal protein. J. Bacteriol. 172:6783-6788.

Wilkins TA, Mishra R, Trolinder NL (2004). *Agrobacterium*-mediated transformation and regeneration of cotton. Food Agric. Environ. 2:179-187.

Wu J, Zhang X, Nie Y, Luo X (2005). High-efficiency transformation of *Gossypium hirsutum* embryogenic calli mediated by *Agrobacterium tumefaciens* and regeneration of insect-resistant plants. Plant Breed. 124:142-146.

Yousefi (2000). Agrochemical in South Africa. Available: http://ww.occuphealth.fi/e/info/anl/199/agro03.htm.

Yu Y, Kang X, Lu Y, Liang J, Wang H, Wu J, Yang Y (2004). Effects of the transgenic Bt cotton on the increase in population of *Spodoptera litura* Fabricius. Jiangsu J. Agric. Sci. 20:169-172.

Zhao FY, Li YF, Xu P (2006). *Agrobacterium* – mediated transformation of cotton (*Gossypium hirsutum* L. cv. Zhongmian 35) using glyphosate as a selectable marker. Biotechnol. Lett. 28:1199-1207.

7

Assessment of the protein quality of twenty nine grain amaranth (*Amaranthus* spp. L.) accessions using amino acid analysis and one-dimensional electrophoresis

Akin-Idowu, Pamela Eloho[1,2] , Odunola, Oyeronke Adunni[2], Gbadegesin, Michael Adedapo[2], Oke, Abiola[3] and Orkpeh, Uterdzua[3]

[1]Biotechnology Unit, National Horticultural Research Institute, P.M.B. 5432 Idi-Ishin, Ibadan, Oyo State, Nigeria.
[2]Cancer Research and Molecular Biology Laboratories, Department of Biochemistry, University of Ibadan, Oyo State, Nigeria.
[3]Vegetable Program, National Horticultural Research Institute, P.M.B. 5432 Idi-Ishin, Ibadan, Oyo State, Nigeria.

Protein deficiency in diets adversely affects growth and development. Novel source of high quality protein and its utilization is essential in improving the nutritive status of the vulnerable groups. Total protein content and protein fractions of 29 amaranth accessions and a soybean cultivar used as reference were determined. The amino acid composition of ten representative accessions of amaranth was also determined. Total protein content ranged from 11.77 to 19.01 g/100 g. One-dimensional gel electrophoretic separations revealed albumin, globulin and glutelin as the major protein fractions; prolamin was not detected. All accessions had similar seed protein electrophoretic profile, ranging from 6.5 to 66 kDa. The glutelin fraction of all amaranth accessions shared similar electrophoretic bands with the soybean cultivar at the 4 to 14, 24 to 36 and 65 to 66 kDa regions. All amaranth accessions contained a good balance of the nine essential amino acids. The sum of essential amino acids ranged from 31.22 to 44.88 g/100 g and 60.87 g/100 g total protein in amaranth and soybean, respectively; limited only in tryptophan and leucine for amaranth, and methionine for soybean. Amaranth is a good source of high quality protein and may serve as a nutritive substitute for some cereals in functional foods.

Key words: Amaranthus, amino acid, gel electrophoresis, protein quality, protein fractions.

INTRODUCTION

Proteins constitute an important group of biomacromolecules that are involved in physiological functions (Wright, 1987). Protein malnutrition is still a major challenge in developing countries (UNICEF, 2009; Black et al., 2008). Therefore, it is essential to broaden the food base by utilization of underutilized crops with promising nutritive potential as recognized by National Academy of Science (NAS) (1984). Natural vegetable proteins are useful materials owing to their safeness, high biocompatibility, nutritional value and low cost. Finding new vegetable proteins rich in essential amino acids is important for the food and pharmaceutical industries (Paredes-Lopez et al., 1988). Most cereals are deficient in some amino acids, for instance, corn is deficient in lysine and tryptophan, rice in lysine and threonine (Deshpande et al., 1955; Bressani and Garcia-Vela, 1990) and wheat in lysine (Howe et al., 1965). Amaranth (*Amaranthus*) belongs to a nutritious class of pseudo-cereals and it has been identified as a very promising food crop because of its exceptional nutritive value as judged by its protein and lipid content, as well as for its essential amino acid composition that has relatively high lysine content (Teutonico and Knorr, 1985). There are three main species of amaranth with desirable agronomic traits namely, *Amaranthus caudatus, Amaranthus cruentus* and *Amaranthus hypochondriacus*. The protein content of *A. cruentus, A. caudatus* and *A. hypochondriacus* are 13.2 to 18.2 g/100 g, 17.6 to 18.4 g/100 g and 17.9 g/100 g, respectively (Gorinstein et al., 1998).

The lysine contents of amaranth species are high ranging from 3.2 to 6.4 g/100 g when compared with those found in most cereals, 2.2 to 4.5 g/100 g. The sulphur containing amino acids (2.6 to 5.5 g/100 g) is higher than that of the most important legumes (1.4 g/100 g) such as pea, beans and soybeans (Gorinstein et al., 1998; Juan et al., 2007). In addition, amaranths are also good sources of minerals and vitamins and they contain larger amounts of these nutrients than most of the common cereals and legumes (Muyonga et al., 2008).

The major storage proteins in most cereal grains such as wheat, barley, rye, maize and sorghum are the alcohol-water soluble prolamins (Gorinstein et al., 1991a; Kreis et al., 1985). Oats and rice have globulins and glutelins, respectively, as major proteins but small amounts of prolamins (Yamagata et al., 1982). In amaranth, the major storage proteins are the globulins and glutelins, while the amount of alcohol-soluble prolamins is low as in oats and rice (Gorinstein et al., 1991a).

Protein content and amino acid composition depend on genotype and growing conditions (Gorinstein et al., 2002). Also, the storage protein composition of soybean has been reported to be influenced by plant nutrient availability (Krishnan et al., 2005). Most investigations on amaranth species have focused on storage protein fractions. However, assessment of the protein quality of *Amaranthus* species based on total protein content, protein fractions and amino acid composition, with reference to soybean an excellent source of high-quality protein (Zarkadas et al., 2007b) and amino acid requirement for human according to the FAO/WHO (1991) and the USFDA (1993) compositional data is scarce.

Thus, this study was carried out to determine the levels of total protein and individual amino acids in 29 grain amaranth accessions grown under field conditions at the National Horticultural Research Institute (NIHORT), Ibadan, Nigeria. Amaranth accessions were also screened for differences in genetic variability in storage protein subunits by 1-DE gel electrophoresis and were compared with the soybean cultivar used as reference.

Findings from this study would assist plant breeders and nutritionists in their selection of high protein quality amaranth accessions.

MATERIALS AND METHODS

Twenty nine accessions of *Amaranthus* belonging to five species: *A. caudatus, A. cruentus, Amaranthus hybrid, A. hypochondriacus* and *Amaranthus hybridus* were used in this study (Table 1). Twenty seven of the accessions were obtained from the USDA-ARS North Central Regional Plant Introduction Station (NCRPIS) in Ames, USA; and two were obtained from the National Horticultural Research Institute (NIHORT) germplasm, Ibadan, Nigeria. The twenty nine accessions were planted in the experimental field of NIHORT in 2010 in three replicates in a randomized complete block design and harvested at maturity.

Sample preparation

Whole mature seeds of each accession planted in 3 replicates were pooled together and ground in a mill. The resulting flour was defatted with n-hexane at a flour/hexane ratio of 1:10 (w/v) prior to protein extraction and fractionation (Barba de la Rosa et al., 1992).

Protein extraction

Proteins were extracted stepwise according to solubility in different solvents following the method of Landry and Moureaux (1980). Fractionation sequences were performed first in distilled water (albumin), followed by 0.5 M NaCl (globulin), 55% 2-propanol (prolamin-like), 55% 2-propanol with 0.6% 2-mercaptoethanol (prolamin), sodium borate buffer (pH 10) with 0.6% 2-mercaptoethanol and 0.5 M NaCl (glutelin-like), borate buffer (pH 10) with 0.6% 2-mercaptoethanol and 0.5% sodium dodecyl sulfate (glutelin).

Protein determination

Nitrogen content was determined by the micro-Kjeldahl method (Hach, 1990) and the protein content was calculated using the nitrogen/protein conversion factor of 6.25.

One dimensional electrophoresis (1-DE)

Sodium dodecyl sulfate polyacrylamide gel electrophoresis (SDS-PAGE) was performed according to the procedure of Laemmli (1970). Protein extracts were combined and diluted in a sample buffer that contained 60 mM Tris-HCl (pH 6.8), 2% w/v SDS, 3.33% v/v β-mercaptoethanol, 10% glycerol and 0.05% of bromophenol blue. The samples were heated at 98°C for 5 min before loading onto a vertical slab gel (Mini-PROTEAN II Electrophoresis Cell) using a 12% gradient separating (acrylamide/bis acrylamide) gel and a 4% (w/v) stacking gel containing 0.1% SDS. Low range molecular weight marker (6.5 to 66 kDa) obtained from Sigma Chemical Co was used for the estimation of the protein subunits. Molecular weights standard are: bovine serum albumin (66 kDa), ovalbumin (45 kDa), phosphate dehydrogenase (36 kDa), trypsinogen (24 kDa), trypsin inhibitor (20 kDa) and aprotinin (6.5 kDa). Electrophoresis was done at 125 V for about 2 h. At the end of the run, the gels were stained with Coomasie Brilliant Blue R250 in methanol/water/acetic acid (4:5:1 v/v/v) and destained with methanol/water/acetic acid (4:5:1 v/v/v) over night.

Table 1. Species and accessions of 29 grain amaranth used in this study, their pedigree and total protein contents (g/100 g).

Species	Accession Number[a]	Accession code	Origin	Plant name	Total protein content (g/100 g)
Amaranthus caudatus	1	P1 490458	Bolivia	LSK 38	14.12±0.26
	2	PI 511679	Argentina	RRC 551	17.30±1.12
	3	PI 553073	United States, New Jersey	LOVE-LIES-BLESSING	11.77±0.26
	4	P1 642741	Bolivia	Oscar Blanco	13.83±0.15
Amaranthus cruentus	5	PI 477913	Mexico	RRC 1011	16.10±0.20
	6	PI 511719	Guatemala	Niqua, alegria, chang	15.04±0.59
	7	PI 515959	United states, Montana	Montana-3	15.39±0.11
	8	PI 538319	United states, Pennsylvania	K266	16.03±0.09
	9	PI 590992	China	TIBET	18.42±0.17
	10	P1 604666	United states, Pennsylvania	RRC 1027	14.45±0.13
	11	PI 641047	Nigeria, Oyo	CEN/IB/97/AMA008	15.71±0.06
	12	PI 641045	Nigeria, Oyo	CEN/IB/97/AMA005	15.36±0.08
Amaranthus hybrid	13	PI 538325	United states, Pennsylvania	K593	16.57±0.08
	14	PI 538326	United States, Pennsylvania	D70-1	13.63±0.09
	15	PI 538327	United States, Pennsylvania	DI36-1	17.75±0.09
	16	Ames 1974	Nigeria	RRC 18C	16.46±0.08
	17	Ames 2256	Nigeria	SP 12C	17.96±0.62
	18	Ames 5644	Nigeria	RRC 1044	12.81±0.30
	19	Ames 5647	Nigeria	RRC 1047	15.27±0.82
Amaranthus hypochondriacus	20	PI 337611	Uganda	P 373	13.04±0.20
	21	PI 511731	Mexico	RRC 646	12.76±0.28
	22	PI 558499	United States, Nebraska	PLAINSMAN	12.82±0.04
	23	PI 590991	China, Shanxi	ZHEN PING	14.01±0.35
	24	PI 615696	India, Himachal Pradesh	Amapurna	15.16±0.12
	25	P1 619250	United States, Pennsylvania	K 116	14.51±0.51
	26	PI 633596	Nepal	Jumla	12.72±0.56
	27	Ames 1972	Nigeria	RRC 18A	15.53±0.43
Amaranthus hybridus	28	NH 84/444-4	Nigeria	NH Purple	19.01±0.21
	29	NAC 3	Nigeria	NH Green	16.77±0.73

Amino acid analysis

Only 10 out of the 29 amaranth accessions were used for this analysis; and they included two accessions from each of the five amaranth species. A soybean cultivar (TGX 1448-2E) obtained from International Institute of Tropical Agriculture (IITA) germplasm, was also included in this study and used for comparison as a known high quality protein.

Amino acid analysis was carried out at AltaBioscience Laboratory, University of Birmingham, Edgbaston, UK, according to the methods of Spackman et al. (1958) and Barrett (1985). Amino acid score (AAS) is the concentration of the limiting amino acid in the food protein, which is expressed according to the method of Young and Steinke (1992) and Zarkadas et al. (2007b) as the proportion or percentage of the concentration of the same amino acid in a standard or reference pattern such as for the diet of a 2-5-year-old child.

$$\text{AAS} = \frac{\text{AA content (mg/g of protein) of food protein}}{\text{AA content of FAO/WHO (1991) pattern for a 2-5-year old child}}$$

Statistical analysis

Analysis of variance was done on the protein and amino acid data using Statistical Analysis System (SAS) (2003). Duncan multiple range test was used to determine significant differences among the amaranth accessions. P values less than 0.05 were considered statistically significant.

RESULTS AND DISCUSSION

The total protein content of the twenty nine amaranth accessions evaluated ranged from 11.77 to 19.01 g/100 g and differed significantly ($p<0.05$) among the accessions (Table 1). *A. hybridus* (accession 28) had the highest protein content (19.01 g/100 g), followed by *A. cruentus* (accession 9) having 18.42 g/100 g and *A. hybrid* (accession 17) having 17.96 g/100 g. The total protein content of most of the amaranth accessions evaluated in this study were higher than values obtained for wheat (13.5 to 14.5 g/100 g), maize (10.6 to 13.8 g/100 g), barley (10 to 14.9 g/100 g) and oats (12.4 to 12.9 g/100 g) by Gorinstein et al. (2002); but lower than 32 to 38 g/100 g obtained for soybean by Zarkadas et al. (2007a, b). Total protein contents (11.77 to 19.01 g/100 g) of the 29 amaranth accessions are similar to 13 to 18 g/100 g obtained for some amaranth species by Dagmar et al. (2012). Bressani and Garcia-Vela (1990) also reported the total protein content of *A. caudatus, A. hypochondriacus and A. cruentus* to be 14.5 to 14.6, 14.7 to 15.9 and 15.3 to 16.1 g/100 g, respectively. Similar results were obtained for the total protein content of *A. caudatus* (16.6 g/100 g) by Gorinstein et al. (1998, 1999), *A. caudatus* (11.0 g/100 g) by Repo-Carrasco-Valencia et al. (2010) and *A. hypochondriacus* (14.2 to 15 g/100 g) by Czerwinski et al. (2004).

Significant variation ($p<0.05$) in the amino acid composition was observed among the ten selected amaranth accessions evaluated (Table 2). The amaranth accessions evaluated were high in glutamic acid, aspartic acid, arginine, glycine and lysine. Similar trend was observed in the amino acid composition of the soybean cultivar (TGX 1448-2E) used for comparison in this study. Zarkadas et al. (2007a, b) reported high levels of glutamic acid, aspartic acid, arginine, lysine, leucine and glycine in some soybean cultivars evaluated. Most of the amaranth accessions evaluated in this study were observed to have higher glycine, methionine and cysteine contents than the soybean cultivar used as reference, while non-essential amino acids were higher in soybean than in amaranth. The lysine contents: 7.89 g/100 g protein of *A. hypochondriacus* (accession 27), 7.88 g/100 g protein of *A. hybridus* (accession 28) and 7.77 g/100 g protein of *A. caudatus* (accession 2) are higher than the lysine contents (7.0 g/100 g protein) obtained in hen's whole egg (FAO/WHO, 1991); 3.2 to 6.4 g/100 g protein obtained in amaranth (Gorinstein et al., 1991b) and 2.2 to 4.5 g/100 g protein found among the most common cereals (Gorinstein et al., 1998). The sulphur containing amino acids (methionine + cysteine) contents of amaranth accessions in this study ranged from 1.77 to 2.74 g/100 g protein with some accessions having higher values than 2.15 g/100 g protein obtained for the soybean cultivar (TGX 1448-2E). The methionine + cysteine contents of the amaranth accessions evaluated were also higher than 1.4 g/100 g protein obtained for important legumes such as pea and beans (Gorinstein et al., 1998). All amaranth accessions evaluated had higher histidine, lysine and threonine contents than values obtained for hen's whole egg (FAO/WHO, 1991).

The essential amino acid EAA profiles and protein ratings of the 10 accessions of amaranth evaluated are compared with reference amino acid patterns for humans (requirement for a 2-5-year old preschool child) as recommended by FAO/WHO (1991) and USFDA (1993) (Table 3). All amaranth accessions evaluated contained all the nine essential amino acids (EAA_9) with values higher than the EAA requirement for a 2-5-year old preschool child. Amaranth accessions were limited only in tryptophan, which ranged from 0.72 to 0.91 g/100 g per protein and is comparable to 1.1 g/100 g per protein recommended for a 2-5-year old preschool child (FAO/WHO, 1991).

The sum of essential amino acids of amaranth in this study ranged from 31.22 to 44.88 g/100 g per protein and is lower than 60.87 g/100 g per protein obtained for the soybean cultivar (TGX 1448-2E) used for comparison. Results of this study are similar to earlier results obtained in grain amaranth (47.6 g/100 g) and soybean (60.3 g/100 g) by Gorinstein et al. (2002). Amaranth and soybean have higher amounts of total essential amino acids than 33.9 g/100 g reference protein pattern value of FAO/WHO (1991) for the diet of a 2- to 5-year-old child, and close to 51.2 g/100 g given for hen's whole egg. These results indicate that grain amaranth provide a good balance of total essential amino acids.

From the calculated amino acid score, the limiting amino acid in amaranth is tryptophan and leucine while in soybean, it is methionine and tryptophan (Table 4). Result of this study is not in agreement with result of Bressani et al. (1987) in which threonine is the limiting amino acid in amaranth protein. Earlier assessment of protein quality of soybean reported methionine as the limiting amino acid (Zarkadas et al., 2007a), this is similar to result of the soybean cultivar (TGX 1448-2E) used as reference in this study.

Table 2. Comparison of the amino acid (AA) composition and total protein content (g/100 g of total protein; mean ± SD) of ten selected grain amaranth accessions with a soybean cultivar (TGX 1448-2E).

AA	2	3	9	10	17	18	21	23	27	28	Soybean (TGX 1448-2E)	Mean AA	LSD	CV
Aspartic acid	11.20 ± 0.20	8.60 ± 0.60	10.60 ± 0.52	10.80 ± 0.60	10.80 ± 0.36	9.47 ± 0.10	8.87 ± 0.10	11.40 ± 0.20	11.50 ± 0.40	11.00 ± 0.50	20.57 ± 0.45	13.22	0.69	3.09
Threonine	4.82 ± 0.02	3.50 ± 0.44	4.53 ± 0.03	4.58 ± 0.03	4.50 ± 0.30	4.22 ± 0.04	3.86 ± 0.06	4.78 ± 0.03	5.11 ± 0.03	4.79 ± 0.03	6.70 ± 0.30	5.28	0.31	3.52
Serine	7.37 ± 0.02	5.79 ± 0.02	7.11 ± 0.02	7.39 ± 0.10	7.22 ± 0.20	6.20 ± 0.20	5.70 ± 0.60	7.61 ± 0.01	8.14 ± 0.04	7.48 ± 0.06	8.40 ± 0.26	7.89	0.37	2.77
Glutamic acid	22.90 ± 0.40	14.10 ± 0.17	19.90 ± 0.90	22.40 ± 0.20	21.70 ± 0.30	18.40 ± 0.40	16.00 ± 0.57	23.20 ± 0.10	23.70 ± 0.20	21.50 ± 0.50	29.10 ± 0.26	23.82	0.72	1.78
Proline	4.95 ± 0.03	3.42 ± 0.02	4.64 ± 0.04	5.02 ± 0.03	5.05 ± 0.02	4.14 ± 0.03	3.39 ± 0.02	5.22 ± 0.11	5.23 ± 0.10	4.87 ± 0.07	7.60 ± 0.20	5.56	0.14	1.46
Glycine	8.82 ± 0.02	6.96 ± 0.17	8.03 ± 0.03	8.81 ± 0.01	8.67 ± 0.07	7.44 ± 0.04	6.97 ± 0.02	9.11 ± 0.10	9.04 ± 0.03	8.24 ± 0.02	6.10 ± 0.00	8.58	0.11	0.79
Alanine	4.68 ± 0.08	3.30 ± 0.30	4.30 ± 0.20	4.34 ± 0.01	4.37 ± 0.04	4.06 ± 0.03	3.59 ± 0.20	4.71 ± 0.01	4.70 ± 0.03	4.54 ± 0.04	6.45 ± 0.30	5.04	0.27	3.11
Cysteine	2.00 ± 0.20	1.13 ± 0.03	1.57 ± 0.02	1.88 ± 0.08	1.57 ± 0.03	1.32 ± 0.02	0.98 ± 0.01	1.54 ± 0.04	1.55 ± 0.03	1.47 ± 0.03	1.50 ± 0.07	1.64	0.12	4.41
Valine	5.15 ± 0.04	3.64 ± 0.03	4.69 ± 0.06	4.86 ± 0.05	4.98 ± 0.03	4.61 ± 0.01	4.19 ± 0.01	5.35 ± 0.05	5.38 ± 0.05	5.16 ± 0.02	7.30 ± 0.60	5.69	0.31	3.24
Methionine	2.74 ± 0.04	1.80 ± 0.20	2.35 ± 0.04	2.44 ± 0.04	2.32 ± 0.02	2.49 ± 0.02	1.77 ± 0.10	2.56 ± 0.10	2.45 ± 0.04	2.45 ± 0.02	2.15 ± 0.20	2.52	0.17	3.93
Isoleucine	4.75 ± 0.05	3.31 ± 0.02	4.37 ± 0.02	4.48 ± 0.02	4.65 ± 0.04	4.19 ± 0.02	3.94 ± 0.02	4.89 ± 0.02	4.98 ± 0.03	4.75 ± 0.10	7.25 ± 0.50	5.35	0.26	2.92
Leucine	7.53 ± 0.03	5.33 ± 0.03	6.94 ± 0.04	7.10 ± 0.01	7.22 ± 0.20	6.55 ± 0.15	6.00 ± 0.87	7.65 ± 0.02	7.74 ± 0.03	7.38 ± 0.03	12.25 ± 0.40	8.54	0.5	3.49
Tyrosine	3.12 ± 0.10	2.12 ± 0.01	2.82 ± 0.02	3.26 ± 0.05	2.88 ± 0.02	2.44 ± 0.03	2.27 ± 0.02	2.92 ± 0.02	3.06 ± 0.06	3.05 ± 0.05	4.95 ± 0.40	3.44	0.22	3.73
Phenylalanine	5.71 ± 0.02	4.04 ± 0.04	5.01 ± 0.01	5.39 ± 0.05	5.31 ± 0.00	4.77 ± 0.10	4.43 ± 0.02	5.66 ± 0.06	5.74 ± 0.04	5.44 ± 0.04	8.20 ± 0.20	6.17	0.13	1.21
Histidine	4.49 ± 0.10	3.07 ± 0.02	4.22 ± 0.10	4.48 ± 0.03	4.45 ± 0.03	3.53 ± 0.03	3.36 ± 0.06	4.53 ± 0.03	4.70 ± 0.20	4.30 ± 0.30	5.29 ± 0.11	4.70	0.21	2.65
Lysine	7.77 ± 0.02	5.65 ± 0.08	7.01 ± 0.01	7.42 ± 0.03	7.30 ± 0.30	6.84 ± 0.03	6.06 ± 0.06	7.88 ± 0.08	7.89 ± 0.07	7.55 ± 0.05	10.45 ± 0.70	8.39	0.40	2.80
Arginine	11.5 ± 0.50	6.70 ± 0.50	9.94 ± 0.03	11.30 ± 0.00	11.20 ± 0.40	9.13 ± 0.02	8.09 ± 0.05	11.40 ± 0.20	11.80 ± 0.70	11.40 ± 0.05	12.15 ± 0.20	11.52	0.57	2.91
Tryptophan	0.91 ± 0.02	0.88 ± 0.04	0.84 ± 0.04	0.91 ± 0.07	0.79 ± 0.04	0.72 ± 0.08	0.78 ± 0.08	0.91 ± 0.03	0.89 ± 0.04	0.81 ± 0.08	1.28 ± 0.26	0.88	0.16	1.09
Total protein (g/100 g dry matter)	17.30	11.77	18.42	14.45	17.96	12.81	12.76	14.01	15.53	19.01	32.63	15.40	0.73	2.65

Mean values and standard deviation (±) for three replicates.

Both amaranth and soybean could supply preschool child and adult requirements of histidine, isoleucine, leucine, lysine, methionine, cysteine, phenylalanine, tyrosine, threonine and valine.

The sodium dodecyl sulfate polyacrylamide gel electrophoretic separation (Figures 1 to 3) revealed that albumin, globulin and glutelin are the major storage proteins in amaranth. Albumins, globulins and glutelins showed similar patterns for homologous protein fractions isolated from different species. Earlier reports of Silva-Sanchez et al. (2008), Czerwinski et al. (2004) and Gorinstein et al. (1998) showed that albumin and globulin are the major fractions of amaranth seed proteins. Prolamin was not detected in all the 29 grain amaranth accessions evaluated, this is not in agreement with results of Gorinstein et al. (1998) who reported the presence of prolamin, though in very low amount which may be due to the fact that it is probably derived from the perisperm of the amaranth seeds. Konishi et al. (1985) and Gorinstein et al. (1991) reported that the amount of alcohol soluble proteins in amaranth was as low as in oats and rice. Albumins are a protein fraction with polypeptides of very heterogenous sizes, with low molecular weight components being the most abundant as observed in the 20 and 24 kDa regions (Figure 1). Electrophoretic pattern of albumin fractions of all the 29 accessions was similar; all subunits migrated between 6.5 and 66 kDa and had characteristic bands at 20 and 24 kDa, this was similar to the soybean cultivar used for comparison. Accessions 5, 6, 7 (*A. cruentus*) and 14 and 15 (*A. hybrid*) differed from others mainly in the region of 22 kDa. Globulin fractions had distinct bands at 14 to 36 kDa and just above the 45 kDa and shared similar bands with soybean cultivar used for reference in the regions of 20 to 36 kDa and just above the 45 kDa (Figure 2). Result of this study is similar to result of

Table 3. Comparison of the essential amino acid (EAA) scores of ten selected grain amaranth accessions and one soybean cultivar with hen's whole egg, and the FAO/WHO EAA Requirements of a 2-5-year-old child. The values are in mg of amino acid/g of total protein.

EAA	EAA[a] requirements for a preschool child (2-5 year old)	2[b]	3	9	10	17	18	21	23	27	28	Soybean (TGX 1448-2E)	Egg[a]
Histidine	19	44.9	30.7	42.2	44.8	44.5	35.3	33.6	45.3	47.0	43.0	52.9	22
Isoleucine	28	47.5	33.1	43.7	44.8	46.5	41.9	39.4	48.9	49.8	47.5	72.5	54
Leucine	66	75.3	53.3	69.4	71.0	72.2	65.5	60.0	76.5	77.4	73.8	122.5	86
Lysine	58	77.7	56.5	70.1	74.2	73.0	68.4	60.6	78.8	78.9	75.5	104.5	70
Methionine + cysteine	25	47.4	29.3	39.2	43.2	38.9	38.1	18.68	41.0	40.0	39.2	36.5	57
Phenylalanine + tyrosine	63	88.3	61.6	78.3	86.5	81.9	72.1	67.0	85.8	88.0	84.9	131.5	93
Threonine	34	48.2	35.0	45.3	45.8	45.0	42.2	38.6	47.8	51.1	47.9	67.0	47
Tryptophan	11	9.1	8.8	8.4	9.07	7.9	7.2	7.8	9.1	8.9	8.07	12.8	17
Valine	35	51.5	36.4	46.9	48.6	49.8	46.1	41.9	53.5	53.8	51.6	73.0	66
mg/g Total protein EAA9	339	438.7	312.2	399.6	416.6	415.2	379.2	343.9	442.1	448.8	426.3	608.7	512
Total protein (%) EAA9	33.9	43.9	31.2	40.0	41.7	41.5	37.9	34.4	44.2	44.9	42.6	60.9	51.2
Percent amino acid score[c]		73	73	73	73	73	73	73	73	73	73	91	97

[a] Data from FAO/WHO (1991). [b]Calculation of protein ratings of the ten amaranth accessions and a soybean cultivar (TGX 1448-2E) was carried out by comparison of the amino acid composition of hen's whole egg with that of the reference pattern established by FAO/WHO (1991) for a preschool child (2-5- year-old). [c] True protein digestibility values were taken from the USFDA (Federal Register, Appendix B (1993) and FAO (1973) scoring pattern.

Table 4. Amino acid scores of ten selected accessions of grain amaranth, one soybean cultivar (TGX 1448-2E) and their limiting amino acid.

EAA	EAA Requirement	2	3	9	10	17	18	21	23	27	28	30 (Soybean)
Histidine	19	2.36	1.61	2.22	2.36	2.34	1.86	1.77	2.38	2.47	2.26	2.78
Isoleucine	28	1.70	1.18	1.56	1.60	1.66	1.50	1.41	1.75	1.79	1.70	2.59
Leucine	66	1.14	0.81	1.05	1.08	1.09	0.99	0.91	1.16	1.17	1.12	1.85
Lysine	58	1.34	0.97	1.21	1.28	1.25	1.18	1.04	1.36	1.36	1.30	1.80
Methionine + cysteine	25	1.89	1.17	1.57	1.73	1.56	1.52	0.75	1.64	1.60	1.57	1.46
Phenylalanine + tyrosine	63	1.40	0.98	1.24	1.37	1.30	1.14	1.06	1.36	1.40	1.35	2.09
Threonine	34	1.42	1.03	1.33	1.35	1.32	1.24	1.13	1.41	1.5	1.41	1.97
Tryptophan	11	0.83	0.80	0.77	0.82	0.72	0.65	0.71	0.83	0.81	0.73	1.16
Valine	35	1.47	1.04	1.34	1.39	1.42	1.32	1.20	1.53	1.54	1.47	2.08
Limiting amino acid		Leu, Tryp	Leu, Tryp	Leu, Tryp	Leu, Tryp	Leu, Tryp	Leu, Tryp	Leu, Tryp	Leu, Tryp	Leu, Tryp	Leu, Tryp	Met, Tryp

Leu, Leucine; Try, tryptophan; Met, methionine.

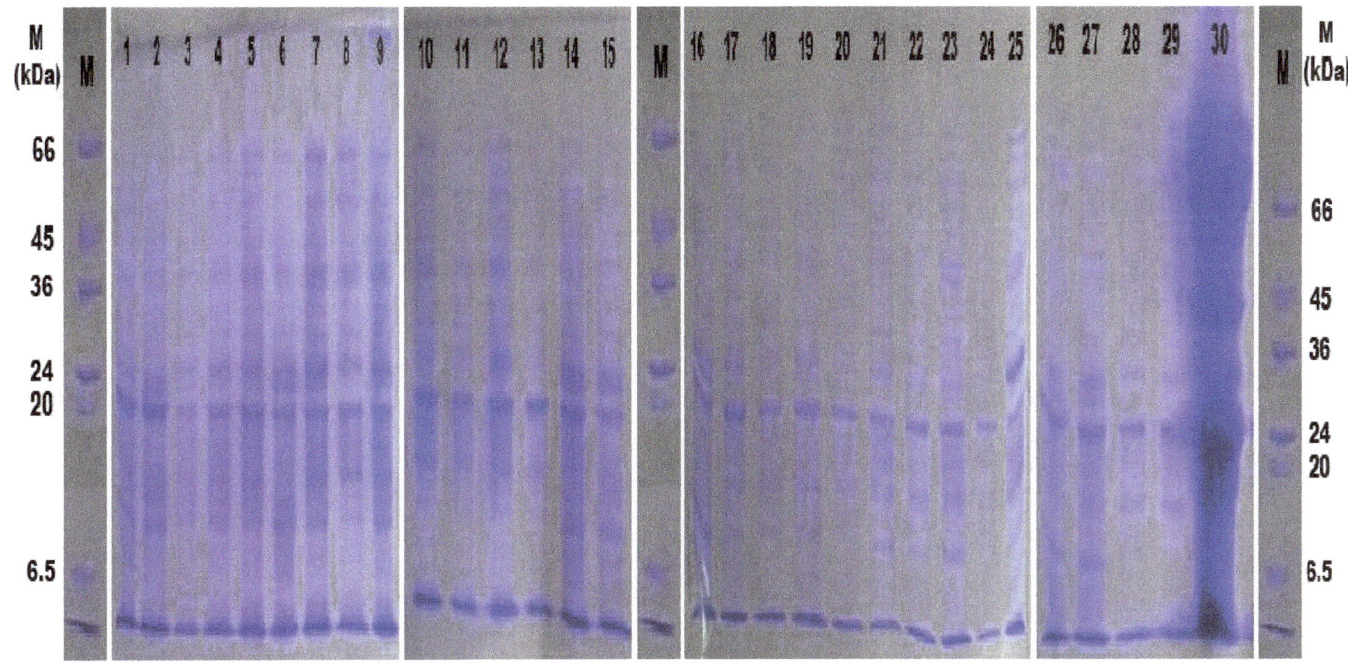

Figure 1. SDS-PAGE of albumin fractions of 29 accessions of *Amaranth* seeds. Lanes 1 - 4 *A. caudatus;* 5 - 12 *A. cruentus*; M-marker; 13 - 19 *A. hybrid*; 20 - 27 *A hypochondriacus*; 28 - 29, *A. hybridus*; 30, soybean (TGX 1448-2E).

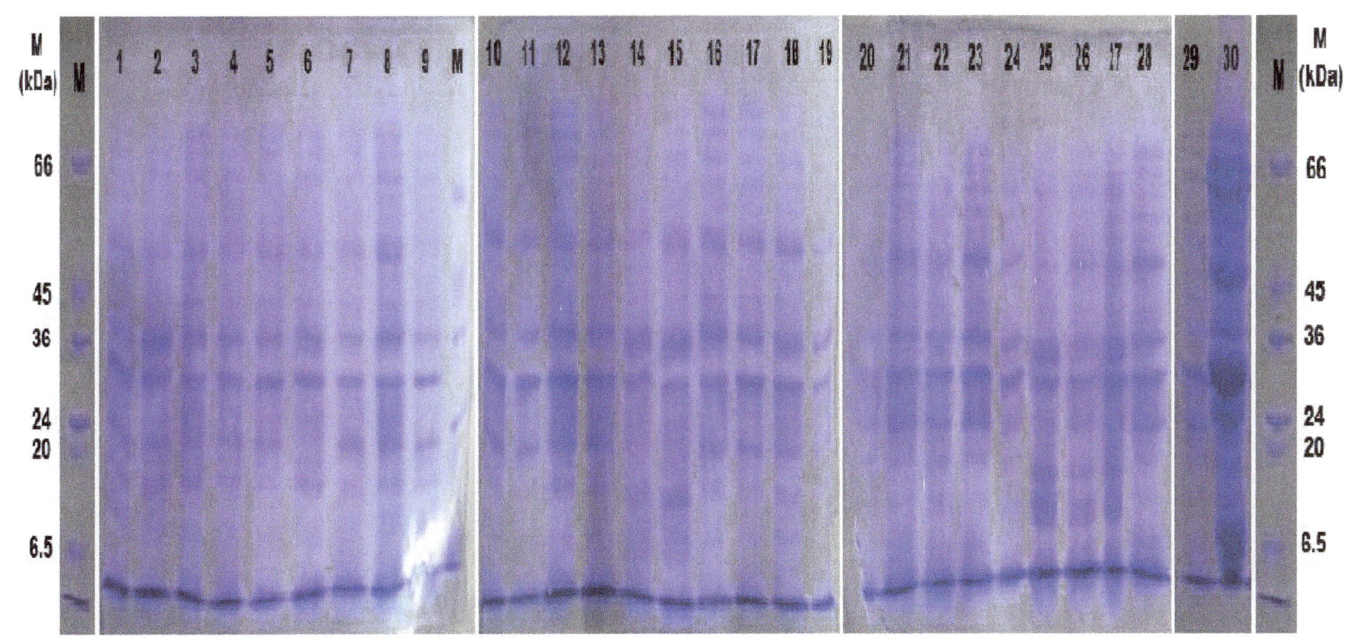

Figure 2. SDS-PAGE of globulin fractions of 29 accessions of *Amaranth* seeds. Lanes 1 - 4 *A. caudatus;* 5 - 12 *A. cruentus*; M-marker; 13 - 19 *A. hybrid*; 20 - 27 *A hypochondriacus*; 28 and 29, *A. hybridus*; 30 soybean (TGX 1448-2E).

Gorinstein et al. (1991b) who reported a major globulin band at 14 to 18 kDa of some amaranth species, but this is not in agreement with the result of Barba de la Rosa et al. (1992) who reported that all globulins showed a major band at 38 kDa. The amaranth glutelin fractions had three main non-separated subunits in the region of 4 to 14 kDa, 24 to 34 kDa and 65 to 66 kDa (Figure 3). These protein subunits were similar in all the 29 grain amaranth

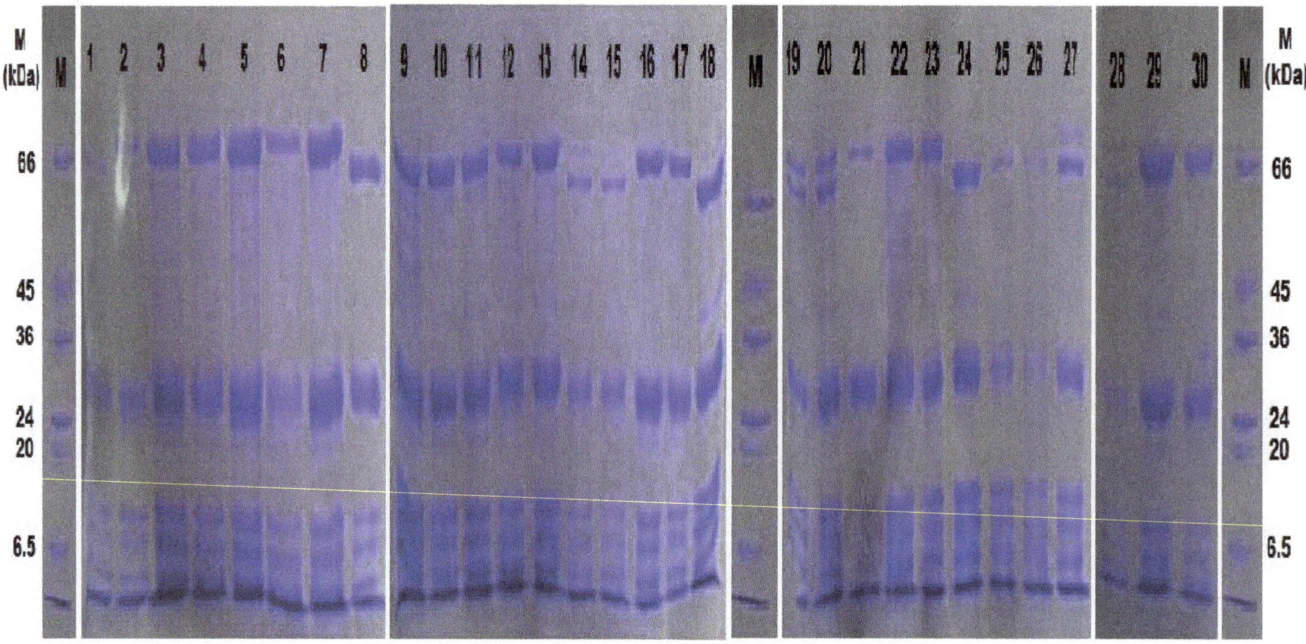

Figure 3. SDS-PAGE of glutelin fractions from 29 accessions of *Amaranth* seeds. Lanes 1 - 4 *A. caudatus;* 5 - 12 *A. cruentus*; 13 - 19 *A. hybrid*; M-marker; 20 - 27 *A. hypochondriacus*; 28 - 29, *A. hybridus*; 30 Soybean (TGX 1448-2E).

accessions. There were outstanding similarities between the electrophoretic patterns of all amaranth accessions evaluated and the soybean cultivar used for comparison. The abundant components in the glutelin fraction between 4 and 14 kDa, 30 kDa and 65 and 66 kDa of all 29 amaranth accessions evaluated in this study are similar to the results of Gorinstein et al. (2002) for *A. caudatus,* maize and rice. It will therefore be possible to use *Amaranthus* species in blends with other cereals for food nutrients.

Conclusion

Amaranth is a valuable food resource, containing a higher level of protein than cereals like maize, wheat, sorghum, rye and oat. It may therefore play an important role in human diets. Lysine and methionine which are limiting in most cereals are present in substantial high amount in amaranth. Most of the amaranth accessions had higher amounts of essential amino acids than the requirements of a 2-5-year old child and comparable amounts as whole egg protein. All accessions were high in non-essential amino acids such as glutamic acid, histidine, aspartic acid and glycine, limited only in tryptophan and leucine. Albumin, globulin and glutelin are the major fractions found in grain amaranth. Prolamin was not detected and so it is not a major fraction in amaranth. A close similarity between the glutelin fraction of amaranth and soybean was observed in the regions of 4 to 14 kDa, 30 kDa and 65 to 66 kDa. Amaranth could be a nutritive substitute for cereals and improve value in different diets in developing countries where protein malnutrition is still a major challenge.

ACKNOWLEDGEMENT

The authors are grateful to the USDA-ARS North Central Regional Plant Introduction Station (NCRPIS) in Ames, USA for providing the amaranth seeds, as well as Lava Kumar of the Virology Unit, IITA, Ibadan, Nigeria for providing the facilities to carry out this investigation.

REFERENCES

Barba de la Rosa AP, Gueguen J, Paredes-Lopez O, Viroben G (1992). Fractionation procedures, electrophoretic characterization and amino acid composition of amaranth seed proteins. J. Agric. Food Chem. 40:931-936.

Barrett GC (1985). Chemistry and Biochemistry of the Amino acids. Chapman and Hall, pp. 414-425.

Black RE, Allen LH, Bhutta ZA, Caulfield LE, De Onis M, Ezzati M (2008). Maternal and child undernutrition: global and regional exposures and health consequences. Lancet 371:243-260.

Bressani R, Garcia Vela LA (1990). Protein fractions in amaranth grain and their chemical characterization. J. Agric. Food Chem. 38:1206-1209.

Bressani R, Gonzalez JM, Zuniga J, Breuner M, Elias LG (1987). Yield, selected chemical composition and nutritive value of 14 selections of amaranth grain representing four species. J. Sci. Food Agric. 38:347-356.

Czerwinski J, Bartnikowska E, Leontowicz H, Lange E, Leontowicz M, Katrich E, Trakhtenberg S, Gorinstein S (2004). Oat (*Avena sativa* L.) and amaranth (*Amaranthus hypochondriacus*) meals positively affect

plasma lipid profile in rats fed cholesterol-containing diets. J. Nutr. Biochem. 15:622-629.
Dagmar J, Petra HČ, Mária D (2012). Characterisation of the Amaranth Genetic Resources in the Czech Gene Bank, Genetic Diversity in Plants, Prof. Mahmut Caliskan (Ed.), ISBN: 978-953-51-0185-7, InTech, Available at: http://www.intechopen.com/books/genetic-diversity-inplants/. Accessed August 12, 2012.
Deshpande PD, Harper AE, Quiros-Perez F, Elvehjem CA (1955). Further observations on the improvement of polished rice with protein and amino acid supplements. J. Nutr. 57:415-428.
FAO/WHO (1991). Expert Consultation. Protein Quality Evaluation, FAO/WHO Nutrition Meetings, Report Series 51. Rome, Italy: Food and Agriculture Organization/World Health Organization.
Gorinstein S, Arnao de Nue I, Arruda P (1991a). Alcohol-soluble and total proteins from amaranth seeds and their comparison with other cereals. J. Agric. Food Chem. 39:848-850.
Gorinstein S, Jaramillo NO, Medina OJ, Rodrigues WA, Tosello GA, Parades-Lopez O (1999). Evaluation of some cereals, plants and tubers through protein composition. J. Protein Chem. 18:687-693.
Gorinstein S, Mashe R, Greene LJ, Arruda P (1991b). Evaluation of four *Amaranthus* species through protein electrophoretic patterns and their amino acid composition. J. Agric. Food Chem. 39:851-854.
Gorinstein S, Pawelzik E, Delgado-Licon E, Haruenkit R, Weisz M, Trakhtenberg S (2002). Characterization of pseudocereal and cereal proteins by protein and amino acid analyses. J. Sci. Food Agric. 82:886-891.
Gorinstein S, Zemser M, Fliess A, Shnitman I, Parades-Lopez O, Yamamoto k, Kobayashi S, Taniguchi H (1998). Computational Analysis of the Amino Acid Residue Sequences of Amaranth and some other Proteins. Biosci. Biotech. Biochem. 62(10):1845-1851.
Hach (1990). System for food, feed and beverages analysis procedure. HACH Company, Loveland, Colorado, USA.
Howe EE, Jansen GR, Gilfillan EW (1965). Amino acid supplementation of cereal grains as related to the world food supply. Am. J. Clin. Nutr. 16:315-320.
Juan R, Pastor J, Alaiz M, Vioque J (2007). Electrophoretic characterization of *Amaranthus* L. seed proteins and its systematic implication. Bot. J. Linnean Soc. 155:57-63.
Konishi Y, Fumita Y, Ikeda K, Okuno K, Fuwa H (1985). Isolation and characterization of globulin from seeds of *Amaranthus hypochondriacus*. Agric. Biol. Chem. 49(5):1453-1459.
Kreis M, Shewry PR, Forde BG, Forde J, Miflin BJ (1985). Structure and evolution of seed storage proteins and their genes with particular reference to those of wheat, barley and rye. In Oxford Surveys of Plant Molecular and Cell Biology; Miflin BJ, Ed. Oxford University Press, Oxford, U.K. pp. 253-317.
Krishnan HB, Bennett JO, Kim WS, Krishnan AH, Mawhinney TP (2005). Nitrogen lowers the sulphur amino acid content of soybean [*Glycine max* (L.) Merr.] by regulating the accumulation of Bowman-Birk protease inhibitor. J. Agric. Food Chem. 53:6347-6354.
Laemmli UK (1970). Cleavage of structural proteins during the assembly of the head of bacteriophage. Nature 227:680-685.
Landry J, Moureaux T (1980). Distribution and Amino acid composition of protein groups located in different histological plants of Maize grain. J. Agric. Food Chem. 28: 1186-1191.
Muyonga JH, Nabakabya D, Nakimbugwe DN, Masinde D (2008). Efforts to promote amaranth production and consumption in Uganda to fight malnutrition. Chapter 8 from Using Food Science and Technology to Improve Nutrition and Promote National Development, Robertson, GL & Lupien, JR (Eds), International Union of Food Science & Technology.
National Academy of Science (NAS) (1984). Amaranth: Modern Prospects for an Ancient Crop. National Academy Press: Washington, D.C.
Paredes-Lopez O, Guzman-Maldonado H, Ordorica-Falomia C (1988). Food proteins from emerging seed sources. In New and Developing Sources of Food Proteins. Edited by Hudson BJF, Chapman and Hall (1988). Pp. 240-279.
Repo-Carrasco-Valencia RAM, Encina CR, Binaghi MJ, Greco CB, Ronayne de Ferrer PA (2010). Effects of roasting and boiling of quinoa, Kiwicha and Kaniwa on composition and availability of minerals *in vitro*. J. Sci Food Agric. 90: 2068-2073.
Silva-Sanchez C, Barba De La Rosa AP, Leon-Galvan MF, De Lumen BO, De Leon-Rodriguez A, Gonzalez De Mejia E (2008). Bioactive Peptides in Amaranth (*Amaranthus hypochondriacus*) Seed. J. Agric. Food Chem. 56:1233-1240.
Spackman DH, Stein WH, Moore S (1958). Automatic recording apparatus for use in the chromatography of amino acids. Anal. Biochem. 30:1190-1206.
Statistical Analysis System (SAS) (2003). Version 9.1. SAS Institute Inc., Cary, NC, USA.
Teutonico RA, Knorr D (1985). Amaranth: composition, properties and applications of a rediscovered food crop. Food Technol. 39(4):49-61.
UNICEF (2009). Tracking Progress on Child and Maternal Nutrition. New York.
USFDA (1993). Department of Health and Human Services (HHS), Food and Drug Administration (FDA), 21 CFR Parts 1 and 101. Food labeling: mandatory status of nutrition labeling and nutrient content revision, format for nutrition label. Federal Register 58:2079-2195.
Wright DJ (1987). The seed globulins in Developments in Food Proteins. Edited by Hudson BJF, Elsevier Applied Science, Chapter 3, New York, pp. 299-335.
Yamagata H, Sugimoto T, Tanaka K, Kasai Z (1982). Biosynthesis of storage proteins in developing rice seeds. Plant Physiol. 70:1094-1100.
Young VR, Steinke FH (1992). Protein and amino acid requirements in relation to dietary food protein needs. In Steinke FH, Waggle DH, Volgarev MN (Eds). New protein foods in human health: nutrition, prevention and therapy. Boca Raton, FL; CRC Press, pp. 9-31.
Zarkadas CG, Gagnon C, Poysa V, Gleddie S, Khanizadeh S, Cober ER, Guillemette RJD (2007a). Assessment of the protein quality of fourteen soybean [*Glycine max* (L.) Merr.] cultivars using amino acid analysis and two-dimensional electrophoresis. Food Res. Int. 40:129-146.
Zarkadas CG, Gagnon C, Poysa V, Khanizadeh S, Cober ER, Chang V, Gleddie S (2007b). Protein quality and identification of the storage protein subunits of tofu and null soybean genotypes, using amino acid analysis, one- and two-dimensional gel electrophoresis, and tandem mass spectrometry. Food Res. Int. 40:111-128.

8

Variation of gene effects of six agronomic traits with water regimes in durum wheat

F. Bnejdi[1]*, M. Saadoun[2], N. Rassaa[3] and M. El Gazzah[1]

[1]Laboratoire de Génétique et Biométrie, Département de Biologie, Faculté des Sciences de Tunis, Université Tunis El Manar 2092, Tunisia.
[2]Laboratoire de Protection des Végétaux, Institut National de la Recherche Agronomique de Tunisie, Tunisia.
[3]Laboratoire de physiologie des céréales, Ecole supérieure d'agriculture du Kef. Tunisie.

This study was carried out to determine the relative importance of additive, dominance and epistatic effects of six agronomic traits evaluated in two crosses under irrigated and rainfed conditions. Separate generations mean analyses revealed that gene effects were dependent upon water regime. Under irrigated regime, only additive and dominance effects were implicated in inheritance of all traits, except plant height in the two crosses and grains per spike in the cross Razzek × Chili. However, under rainfed condition, additive, dominance and epistasis effects were induced in the genetic control of these traits. Thus, epistasis effect was water regime depended and seems to play a significant role in the genetic adaptability of populations. The variation of inheritance and the presence of genetic interactions complicate the procedure of breeding. Under irrigated regime, selection would be simpler because only additive and dominance effects were implicated in the genetic control of all traits. Consequently, the genetic mechanism would not be stable when plants are grown in environment with lower water availability. However, for the stability of suitable cultivars, selection under rainfed condition (lower water availability) was suggested. The moderate to high narrow-sense heritability estimates for the majority of traits indicate that selection would be efficient.

Key words: Durum wheat, epistatic effect, genetic adaptability, heritability, water regime.

INTRODUCTION

Durum wheat (*Triticum durum* Desf) is one of the leading cereal crops, in many countries in the world (Maniee et al., 2009; Kahrizi et al., 2010a, b; Mohammadi et al., 2010). In the Mediterranean basin, durum wheat is grown mostly in rainfall areas with different environmental conditions (Edmeades et al., 1989; Loss and Siddique, 1994; Dunkeloh and Jacobeit, 2003). Water deficit is one of the main abiotic stresses that affect production and quality of wheat cultivated in arid and semi-arid regions (Ehdaie and Shakiba, 1996; Plaut et al., 2004; Passioura, 2007; Talebi et al., 2009). In Tunisia, durum covers more than half of the cereal areas (Bnejdi and El Gazzah, 2008), where many regions were not exploited due to the wide fluctuation of precipitations and absence of tolerant varieties. To improve crop productivity, it is necessary to understand the mechanism of plant responses to drought stress with the ultimate goal of improving crop performance in the vast areas of the world where rainfall is limiting or unreliable (Huang et al., 2008). Therefore, selection and breeding for drought tolerance has been the main challenge of wheat breeders throughout the last 50 years (Lopez et al., 2003). The choice of an efficient breeding procedure depends to a large extent on knowledge of the genetic system controlling the character to be selected. Generations mean analyses provide information on the relative importance of average effects of the genes (additive effects, dominance deviations and effects due to non allelic genetic interactions) in determining genotypic values of the individuals and,

*Corresponding author. E-mail: fethibnejdi@yahoo.fr.

Table 1. Means of six morphologic traits (± SE) and narrow-sense heritability of parents and offspring populations from two crosses under rainfed condition.

Population	PH	NHP	SL	SPS	GPS	100-GW
Karim 80 × Inrat 69						
P_2(20)y	96.8±1.4^{e}	5.4±0.6^{a}	6.0±0.1^{a}	19.8±0.6^{ab}	49.8±2.3^{a}	7.62±0.7^{a}
BC_1P_2(50)	105.5±2.4^{c}	5.2±0.8^{a}	3.9±0.1^{b}	18.17±0.4^{b}	42.0±2.4^{b}	7.05±2.5^{a}
F_1(20)	101.0±2.5^{d}	3.3±0.2^{c}	6.5±0.1^{a}	17.4±1.0^{b}	29.1±1.4^{c}	4.93±1.4^{bc}
F_2(100)	89.2±1.7^{f}	5.3±0.2^{a}	5.3±0.1^{ab}	18.1±0.2^{b}	48.5±1.3^{a}	5.86±2.3^{b}
BC_1P_1(50)	117.8±2.6^{a}	5.7±0.4^{a}	2.4±0.1^{c}	21.31±0.5^{a}	51.5±1.5^{a}	5.4±1.2^{b}
P_1(20)	111.1±0.9^{b}	4.8±0.4^{b}	5.2±0.2^{b}	21.5±0.4^{a}	49.7±2.2^{a}	4.02±0.6^{c}
h^2_n	0.36	0.40	0.57	0.41	0.35	0.60
Razzek × Chili						
P_2(20)	93.8±1.0^{a}	9.4±0.7^{a}	7.2±0.1^{a}	17.8±0.7^{b}	37.2±2.6^{b}	5.34±0.8^{a}
BC_1P_2(50)	90.2±0.9^{a}	5.6±0.4^{b}	6.97±0.10^{a}	15.2±0.3^{c}	38.3±1.6^{b}	4.75±0.2^{a}
F_1(20)	100.0±1.9^{a}	8.0±0.9^{a}	5.7±0.1^{b}	16.4±1.2	37.1±1.6^{b}	3.88±0.9^{bc}
F_2(100)	79.2±0.9^{c}	5.1±0.3^{b}	5.57±0.1^{b}	14.1±0.3^{c}	37.1±1.1^{b}	4.01±1.8^{b}
BC_1P_1(50)	84.2±0.9^{b}	5.3±0.4^{b}	5.96±0.1^{b}	20.2±0.7^{a}	41.4±1.3^{a}	4.09±0.9^{b}
P_1(20)	80.7±0.9^{bc}	7.7±0.9^{a}	5.81±0.1^{b}	19.4±0.3^{a}	41.3±2.4^{a}	4.99±0.8^{a}
h^2_n	0.44	0.46	0.45	0.41	0.71	0.37

For each cross, means within a column bearing different letters differ (P<0.05). [Y]Number of plants evaluated in each generation. PH, Plant height; NHP, number of heads per plant; SL, spike length; SPS, spikelets per spike; GW, grain weight.

consequently, mean genotypic values of families and generations (Viana, 2000).

This study was carried out to determine the relative importance of additive, dominance and epistatic effects of six agronomic traits on durum wheat in two crosses under irrigated and rainfed conditions.

MATERIALS AND METHODS

The study was carried out at Sidi Thabet site under irrigated and rainfed condition (400 ml/year), during the growing season 2009-2010. The experiment was laid out in randomized complete block design with two replications. Irrigated plots were watered at different growing stages (sowing, tillering, flowering and grain filing). Non-irrigated plots were grown under rainfed condition. Sowing was done in the middle of November in all experiments. Four Tunisian durum wheat genotypes were selected on the basis of their differential yield. Karim 80 and Razzek were known by their excellent productivity (P_1). Inrat 69 and Chili were known by their excellent grain protein content (P_2). Two crosses were made as follows: Karim 80 × Chili and Razzek × Inrat 69. F_1 were self-pollinated to produce F_2. Backcrosses of F_1 to each parent were noted as BC_1P_1 and BC_1P_2, respectively.

Experiments with 12 populations including parental lines (P_1 and P_2), F_1, F_2 and backcrosses (BC_1P_1 and BC_1P_2) were grown under irrigated and rainfed conditions in a randomised complete block design with two replications. From each plant, the following data were recorded: plant height (PH) in cm, number of heads per plant (NHP), spike length (SL) in cm, spikelets per spike (SS), number of grains per spike (NGS) and 100-grain weight (100-GW) in g. The number of evaluated plants, equal for each replication was significantly higher in segregating populations (Tables 1 and 2). Prior to analysis, transforming the data by log, square root, arc-sine and arc-sine of square root had no effect on data distribution or in removing epistatic effects. Analyses of variance by population and water regime using SAS software version 6 (SAS Institute, 1990) indicated that the replication and generation × replication effects were not significant. Therefore, the generations mean analyses were made without adjusting the data for replication.

Gene effects

Weighted least squares regression analyses were used to solve for mid-parent (m), pooled additive (d), pooled dominance (h) and pooled digenic epistatic (i, l and j) genetic effects, following the models and assumptions described in Mather and Jinks (1982). A simple additive-dominance genetic model containing only the m, d and h effects was tested first, using the joint scaling test described in Rowe and Alexander (1980). Adequacy of the genetic model was assessed using a chi-square goodness-of-fit statistic derived from deviations from this model. If statistically significant at $p < 0.05$, the genetic models containing digenic epistatic effects were then tested until the Chi-square statistic was non-significant.

Heritability

Homogeneity of variances of non-segregation generations was tested by using Bartlett's test (Bartlett, 1937), and when the variances were heterogeneous, the environmental variance (VE) was replaced by an adequate number of separate parameters and pooled to produce a single environmental variance. Additive, dominance and environmental variance components were estimated using the maximum likelihood method with the observed variance of the six basic generations being used as the initial weights ($df/(2\times S^2)$) until the chi-squared test value reached a minimum (Lynch and Walsh, 1998). Narrow-sense heritability was calculated as follows: $h^2n = VA/(VA + VD + VE)$; where VA is the

Table 2. Means of six morphologic traits (±SE) assessed in parents and offspring populations from two crosses under irrigated regime.

Population	PH	NHP	SL	SPS	GPS	100-GW
Karim 80 × Inrat 69						
$P_2(20)^Y$	102.0±1.0^b	8.0±0.5^a	6.75±0.1^a	21.00±0.3^a	66.50±2.1^a	6.00±0.1^a
$BC_1P_2(50)$	117.0±1.6^a	7.0±0.3^a	6.43±0.1^b	22.00±0.3^a	54.0±1.6^b	3.90±0.1^c
$F_1(20)$	109.0±1.8^a	7.0±0.1^a	7.13±0.2^a	20.00±0.7^{ab}	62.00±5.0^a	6.50±0.1^a
$F_2(100)$	99.0±1.3^b	4.45±0.2^c	6.80±0.1^a	17.00±0.1^b	45.50±0.9^c	5.30±0.1^b
BC_1P_1 (50)	121.0±1.1^a	5.0±0.2^{bc}	6.30±0.1^b	24.00±0.2^a	53.50±1.1^b	2.40±0.1^d
$P_1(20)$	112.5±0.6^{ab}	5.45±0.4^b	6.28±0.1^b	23.00±0.4^a	56.50±1.8^b	5.20±0.1^b
h^2_n	0.41	0.42	0.40	0.49	0.26	0.41
Razzek × Chili						
$P_2(20)$	97.30±0.9^a	10.0±0.6^a	7.46±0.1^a	18.90±0.2^a	53.00±2.2^a	5.23±0.1^a
$BC_1P_2(50)$	90.23±0.9^b	8.0±0.4^b	7.23±0.1^a	18.00±0.1^a	50.00±1.6^a	5.00±0.1^a
$F_1(20)$	100.0±1.8^a	10.4±0.5^a	6.00±0.1^b	17.00±0.5^{ab}	40.00±2.6^b	5.23±0.1^a
$F_2(100)$	79.23±0.8^c	5.0±0.78^d	7.23±0.1^a	16.89±0.2^{ab}	39.09±1.0^b	4.22±0.1^b
$BC_1P_1(50)$	84.23±0.9^c	7.0±0.3^{bc}	7.00±0.1^{ab}	18.00±0.1^{ab}	42.23±1.6^b	4.89±0.1^a
$P_1(20)$	80.70±0.9^c	7.2±0.9^c	6.93±0.1^{ab}	19.00±0.2^a	44.23±2.6^b	4.63±0.2^{ab}
h^2_n	0.21	0.55	0.59	0.41	0.52	0.42

For each cross, means within a column bearing different letters differ (P<0.05). ^{Y}Number of plants evaluated in each generation. PH, Plant height; NHP, number of heads per plant; SL, spike length; SPS, spikelets per spike; GW, grain weight.

additive genetic component of variance; VD is the dominance genetic component of variance and VE is the environmental variance (Kearsey and Pooni, 1996).

RESULTS

The mean values and their standard errors for the analyzed traits of the two crosses at irrigated and rainfed condition are presented in Tables 1 and 2. For the four parents, large differences were observed between the majority of traits assessed under irrigated and rainfed conditions. For the majority of genotypes, means of the six traits at irrigated regime were higher than at rainfed condition. Concerning each agronomic trait, the response of genotypes at each condition differed. Under irrigated and rainfed conditions, the highest PH value was observed for Inrat 69 and the lowest value for Karim 80. Highest NHP was observed in Karim 80 and Razzek under irrigated regime. SL varied from 2.4 to 7.46 and the highest value was obtained for Chili under rainfall regime. SPS, GPS and 100-GW varied between genotypes and water regime.

Results of separate generations mean analyses are given in Tables 3 and 4. Under irrigated regime and for the two crosses, the three-parameter model was sufficient to explain the variation in generation means for all traits, except PH in the two crosses and GPS in the cross Razzek × Chili. Therefore, under rainfed condition, the additive-dominance model failed to explain variation in generation means for all traits and digenic epistatic model was applied and adequately revealed. The additive effect was significant with a negative sign in all cases. The dominance effect was negative in the majority of cases and was significant only in three cases. For all traits, narrow-sense heritability varied between crosses and water regime and ranged from 21 and 71%.

DISCUSSION

Significant differences were found among the generation means for all the traits which revealed the presence of genetic diversity between cultivars. Generations mean analysis indicated that the mode of gene effects implicated in the control of the majority of traits was depended upon water regime. Our results reveal the implication of epistasis in inheritance of the six agronomic traits under rainfed condition. In contrast, under irrigated regime, epistasis was not implicated in the genetic control of all traits, except PH in the two crosses and GPS in the cross Razzek × Chili. Recent studies suggest that epistatic effects are implicated in inheritance of quantitative traits in many species, for examples: wheat and Fusarium head blight (Waldron et al., 2008), sesame and powdery mildew (Rao et al., 2011), common bean and Fusarium root rot (Mukankusi et al., 2011) and maize and common smut (Namayandeh et al., 2011). Result of generation mean analysis revealed that inheritance of all traits, except PH was site dependent. In favourable environment, the plant implores only additive and

Table 3. Estimates of additive, dominance and epistatic effects (±SE) for six agronomic traits in two durum wheat crosses evaluated at irrigated regime.

Parameter	PH	NHP	SL	SPS	GPS	100-GW
Karim 80 × Inrat 69						
m	28.17±1.02*	4.09±0.03*	6.32±0.02*	21.02±0.06*	59.54±0.30*	4.33±0.01*
d	5.09±0.13*	0.58 ± 0.04*	0.0 ±0.01*	-0.84±0.05*	3.87 ± 0.23*	0.03±0.01*
h	202.48± 2.5*	2.67 ± 0.05*	0.56±0.04*	-4.68±0.12*	21.44 ± 0.62*	1.05±0.02*
i	79.13±1.17*	-	-	-	-	-
l	121.6 ±1.77*	-	-	-	-	-
j	-	-	-	-	-	-
P	0.34	0.54	0.62	0.38	0.44	0.25
Razzek × Chili						
m	1049.61±0.70*	6.94±0.09*	7.56±0.01*	18.83±0.07*	34.22±0.60*	4.44±0.02*
h	-8.30±0.14*	-0.12±0.07*	-0.22±0.01*	0.04±0.06	-4.04±0.38*	-0.1±0.02*
h	45.92±1.83*	1.30±0.16*	-1.35±0.02*	-2.37±0.15*	10.67±1.08*	0.64±0.03*
i	32.23±0.68*	-	-	-	16.49±0.66*	-
l	-2.92±1.39*	-	-	-	-	-
J	28.60±0.56*	-	-	-	23.62±1.13*	-
P	-	0.61	0.59	0.31	0.22	0.36

m, d, h, i, l and j = mean, additive, dominance, additive × additive, dominance × dominance and additive × dominance genetic effects, respectively. P = probability of adequateness of the model; *$p<0.05$ as compared to zero.
PH, Plant height; NHP, number of heads per plant; SL, spike length; SPS, spikelets per spike; GW, grain weight.

Table 4. Estimates of additive, dominance and epistatic effects (±SE) for six agronomic traits in two durum wheat crosses evaluated at rainfed condition.

Parameter	PH	NHP	SL	SPS	GPS	100-GW
Karim 80 × Inrat 69						
m	783.49±1.62*	4.32±0.31*	756.16±0.10*	135.46±0.27*	364.77±0.35*	7.29±0.68*
d	17.15±0.18*	-0.33±0.07*	-0.4±0.01*	0.85±0.08*	-0.05±0.35*	-1.80±0.10*
h	213.35±4.39*	4.93±0.91*	-27.9 ±0.24*	12.73±0.80*	-5.35±3.60*	-2.21±0.90*
i	89.8 ±1.61*	0.79±0.29*	-8.6 ±0.10*	6.56±0.25*	-7.0±1.26*	-1.45±0.69*
l	-126.5±2.99*	-5.95 ± 0.60*	20.2±0.14*	-9.42±0.67*	-22.3±2.38*	-
j	-38.9 ±1.34*	-	3.8±0.06*	-7.98±0.28*	-18.9±1.25*	5.88±0.90*
P	-	0.27	-	-	-	0.45
Razzek × Chili						
m	58.0±0.54*	19.46±0.29*	5.01±0.09*	135.50±0.27*	27.46±1.05*	379±1.05*
d	-6.62±0.11*	-0.85±0.13*	-0.19±0.01*	0.85 ± 0.08*	-0.65±0.27*	2.05±0.40*
h	43.18±0.85*	-9.05±0.83*	5.53±0.26*	12.85 ± 0.80*	28.90±2.85*	26.55±2.87*
i	-	1.4±0.26*	1.98±0.09*	6.6±0.26*	12.00±0.97*	11±0.97*
l	29.33±0.55*	9.9±0.65*	-4.84±0.17*	-9.5 ± 0.67*	-19.27±1.94**	-17.7±1.95*
J	24.89±0.45*	2.3±0.33*	-	-7.9 ± 0.28*	-	-10.3±1.1
P	0.56	-	0.91	-	0.31	-

m, d, h, i, l and j = Mean, additive, dominance, additive × additive, dominance × dominance and additive × dominance genetic effects, respectively. P = probability of adequateness of the model; *$p<0.05$ as compared to zero. PH, Plant height; NHP, number of heads per plant; SL, spike length; SPS, spikelets per spike; GW, grain weight.

dominance effects while in unfavourable environment, all mode of gene effects are induced. Epistasis seems to play a significant role in the genetic adaptability of cultivars when cultivated under water deficiency. Similar results were reported for resistance to *Septoria tritici* in durum wheat (Bnejdi et al., 2011) and resistance to

Phytophthora nicotianae (Bnejdi et al., 2010, Ceballos et al., 1998) found that epistasis was more pronounced in non-acid than in acid solution for grain yield in maize. Genetic interactions complicate the procedure of fixation of desirable genes in the suitable varieties. Selection under irrigated regime will be efficient and simple to fix the additive genetic effect; however, selection under rainfed condition (lower water availability) could be complicated but more stable. As regards the presence of additive and non-additive gene effects in inheritance of the six traits in the two crosses, the recurrent selection followed by pedigree breeding or a selective diallel mating system may be useful in improving the six agronomic traits in wheat. Under irrigated and rainfed conditions, for the majority of traits, the moderate to high narrow-sense heritability indicated a higher involvement of genetic effects in the expression of these traits and selection is expected to be efficient.

In conclusion, inheritance of the majority of traits varied from simple to complicated depending upon water regime. So, for the cultivars designed for exploitation in a wide range of environmental conditions, selection under unfavourable environment (lower water availability) was recommended for their stability.

REFERENCES

Bartlett MS (1937). Some examples of statistical methods of research in agriculture and applied biology. J. Roy. Stat. Soc. 4:137-183.

Bnejdi F, El Gazzah M (2008). Inheritance of resistance to yellowberry in durum wheat. Euphytica 163:225-230.

Bnejdi F, Saadoun M, Allagui MB, Colin H, El Gazzah M (2010). Relationship between epistasis and aggressiveness in resistance of pepper (*Capsicum annuum* L.) to *Phytophthora nicotianae*. Genet. Mol. Biol. 33:279-284.

Bnejdi F, Saadoun M, El Gazzah M (2011). Genetic adaptability of the inheritance of the resistance to different levels of aggressiveness of *Septoria tritici* isolates in durum wheat. Crop Prot. 30:1280-1284.

Ceballos H, Pandey S, Narro L, Perez-Velazquez JC (1998). Additive, dominant, and epistatic effects for maize grain yield in acid and non-acid soils. Theory Appl. Genet. 96:662-668.

Dunkeloh A, Jacobeit J (2003). Circulation dynamics of Mediterranean precipitation variability 1948-1998. Int. J. Climatol. 23:1843-1866.

Edmeades GO, Bolanos J, Lafitte HR, Rajaram S, Pfeiffer WH, Fischer RA (1989). Traditional approaches to breeding for drought resistance in cereals. In: Baker, FWG (ed) Drought Resistance in Cereals. ICSU Press C.A.B International: pp. 27-52.

Ehdaie B, Shakiba MR (1996). Relationship of internode-specific weight and water-soluble carbohydrates in wheat. Cereal Res. Comm. 24:61-67.

Huang RH, Gu L, Chen JL, Huang G (2008). Recent progresses in studies of the temporal-spatial variations of the East Asian monsoon system and their impacts on climate anomalies in China. Chinese J. Atmos. Sci. 32:691-719.

Kahrizi D, Cheghamirza K, Kakaei M, Mohammadi R, Ebadi A (2010a). Heritability and genetic gain of some morpho-physiological variables of durum wheat (*Triticum turgidum* var. durum.). Afr. J. Biotechnol. 9(30):4687-4691.

Kahrizi D, Maniee M, Mohammadi R, Cheghamirza K (2010b). Estimation of genetic parameters related to morpho-agronomic traits of durum wheat (*Triticum turgidum* var. durum). Biharean Biol. 4(2):93-97.

Kearsey MJ, Pooni HS (1996). The genetical analysis of quantitative traits. 1st edition. Chapman and Hall, London p. 381.

Lopez CG, Banowetz GM, Peterson CJ, Kronstad WE (2003). Dehydrin expression and drought tolerance in seven wheat cultivars. Crop Sci. 43:577-582.

Loss SP, Siddique KHM (1994). Morphological and physiological traits associated with wheat yield increases in Mediterranean environments. Adv. Agron. 52:229-276.

Lynch M, Walsh B (1998). Genetics and Analysis of Quantitative Traits, Sinauer Associates, Inc, Sunderland. p. 980.

Maniee M, Kahrizi D, Mohammadi R (2009). Genetic Variability of Some Morpho-physiological Traits in Durum Wheat (*Triticum turgidum* var. durum) . J. Appl. Sci. 9(7):1383-1387.

Mather K, Jinks JL (1982). Biometrical Genetics. London: Chapman and Hall Ltd, London, p. 396.

Mohammadi R, Armion M, Kahrizi D, Amri A (2010). Efficiency of screening techniques for evaluating durum wheat genotypes under mild drought conditions. Int. J. Plant Prod. 4(1):11-24.

Mukankusi C, Derera J, Melis R (2011). Genetic analysis of resistance to fusarium root rot in common bean. Euphytica, 182: 11-23.

Namayandeh A, Choukan R, Mohahhadi SA, Kamelmanesh MM, Ghasemi S, Hervan EM (2011). Genetic analysis of resistance to common smut in maize (*Zea mays* L.) using triple test cross. Afr. J. Agric. Res. 6:630-635.

Passioura JB (2007). The drought environment: physical, biological and agricultural perspectives. J. Exp. Bot. 58:113-117.

Plaut Z, Butow BJ, Blumenthal CS, Wrigley CW (2004). Transport of dry matter into developing wheat kernels and its contribution to grain yield under post-anthesis water deficit and elevated temperature. Field Crop. Res. 86:185-198.

Rao PVR, Anuradha G, Prada MJ, Shankar VG, Reddy KR, Reddy NPE, Siddiq EA (2011). Inheritance of powdery mildew tolerance in sesame. Arch. Phyto. Plant Prot. 1:1-9.

Rowe KE, Alexander WL (1980). Computations for estimating the genetic parameter in joint-scaling tests. Crop Sci. 20:109-110.

SAS Institute (1990) SAS/STAT user's guide. Version 6, 4th edn. SAS Institute, Cary.

Talebi R, Fayaz F, Naji AM (2009). Effective selection criteria for assessing drought stress tolerance in durum wheat (*Triticum durum* Desf.). Gen. Appl. Plant Physiol. 35: 64-74.

Viana JMS (2000). Components of variation of polygenic systems with digenic epistasis. Genet. Mol. Biol. 23:883-892.

Waldron BL, Cai X, Friesen TL, Halley S, Stack RW, Xu SS (2008). Evaluation of fusarium head blight resistance in tetraploid wheat (L.). Crop Sci. 38:805-811.

Investigation of DNA changes in wheat (*Triticum aestivum* L.) induced by cadmium using random amplified polymorphic DNA (RAPD) analysis

Akbar Azimi[1], Farajollah Shahriari[1], Amir Fotovat[2], Reza Kazemi Qale[3] and Khodaverd Hajizade Agje[4]

[1]Department of Plant Breeding and Biotechnology, Ferdowsi University of Mashhad, Iran.
[2]Department of Soil Science, Ferdowsi University of Mashhad, Iran.
[3]Department of Horticultural Science, Islamic Azad University of Abhar, Iran.
[4]Department of Chemistry Engineering, Sahand University of Tabriz, Iran.

In recent years, many plants have been used as bioindicators or living indices and several experiments have been conducted to evaluate the genotoxic effects of environmental pollution on plant species. Plants such as Arabidopsis and barley have been used as biological indicators by several researchers and have been able to identify DNA changes using molecular markers. This study was conducted on wheat as an important cereal that provides human daily food and has an important role in human health. The purpose of study was to evaluate the effect of different concentrations of cadmium on amounts of the soluble proteins, investigation of growth and possible changes to the structure of DNA using random amplified polymorphic DNA (RAPD) marker. In the present study, wheat seedlings were used for detection of genotoxic effects of cadmium contamination in the range of 0 to 120 mgl^{-1}. The limiting effects of cadmium on length of root and plant height and total soluble proteins in root were quite evident by increase of cadmium concentration. Change in DNA was observed using RAPD marker as there was change in number of bands, present or absent of bands at the range of above 5 mgl^{-1} cadmium concentration. In the previous studies conducted by several researchers, variation in DNA bands was reported at levels above 30 mgl^{-1}, whereas in the present study, band variation was observed at the concentration of 5 mgl^{-1}. This study also showed that DNA stability is highly affected by cadmium pollution at >5 mgl^{-1} which was identified by RAPD markers.

Key words: Bioindicator, cadmium, genotoxic, random amplified polymorphic DNA (RAPD) marker, *Triticum aestivum* L.

INTRODUCTION

In the last several decades, cadmium (Cd) has been used extensively as materials and/or intermediates in the chemical industry and agriculture for electroplating, metallurgy, painting, plastic production, etc., and is being released into the biosphere, and largely found in the water and soil (Baryla et al., 2001; Liu et al., 2005;

Shahrtash et al., 2010). Through food chain, Cd accumulation could pose a direct threat to human health and cause diseases such as hypertension, angiopathy, kidney and bone function decay (Liu et al., 2005). The toxic and genotoxic effects of cadmium have been demonstrated in different plant and animal species and it has been shown that cadmium caused change in DNA and cellular structures (Conte et al., 1998; Liu et al., 2005; Shahrtash et al., 2010). However, the molecular mechanism responsible for the genotoxicity of cadmium remains unclear. It has been suggested that it may involve direct interaction of Cd^{2+} with DNA through the binding of Cd at G, A and T bases (Valverde et al., 2001; Hossain and Huq, 2002; Anastassopoulou, 2003). Furthermore, recent studies indicate that Cd acts as a mutagen primarily by direct inhibition of an essential DNA mismatch repair, resulting in a high level of genetic instability (Andrea, 1994; Jin et al., 2003). Cadmium cellular toxicity and genotoxicity may also be mediated indirectly; cells under oxidative stress display various dysfunctions due to lesions caused by reactive oxygen species (ROS, e.g. O_2^-, H_2O_2 and OH^-) to lipids, proteins and DNA (Waisberg et al., 2003; Joseph, 2009). Advantages of measuring effects of genotoxic chemicals directly on DNA are mainly related to the sensitivity and short response time. Recently, advances in molecular biology have led to the development of a number of selective and sensitive assays for DNA analysis in eco-genotoxicology. DNA based techniques [RFLP, QTL, random amplified polymorphic DNA (RAPD), AFLP, SSR and VNTR] are used to evaluate the variation at the DNA sequence level. RAPD of these techniques can be used to detect genotoxicity and differences in RAPD profiles can clearly be shown when comparing DNA fingerprints from untreated and treated individuals to genotoxic agents (Savva, 1996; Conte et al., 1998; Theodorakis and Bickham, 2001; Enan, 2006; Atienzar and Jha, 2006; Liu et al., 2007; Cenkci et al., 2009). The aim of this study was to detect DNA damage induced by Cd using the RAPD technique. Detection of the genotoxic effect involves comparison of RAPD profiles of the root tip DNA generated by control and treated wheat seedlings.

MATERIALS AND METHODS

Plant material, growth and total soluble protein level test of seedlings

Seeds of Cascogen wheat cultivar provided by Seed and Plant Improvement Institute of Mashhad, were sterilized by using incubation in 5% (w/v) sodium hypochlorite, soaked in distilled water, incubated for three days in a dark growth chamber at temperature of 18 ± 1°C and germinated to primary roots of 1 cm long. Uniformly 100 germinated seeds were selected and transferred to Petri pots (length 18 cm, width 10 cm and height 7 cm) containing 0.5 kg sandy soil of 1 to 3 mm (washed with 1 N HCl) with 200 ml of distilled water or test solution. The growth inhibition test was performed with the above wheat plantlets exposed to 0, 5, 10, 15, 20, 30, 60, 90, 120 mgl^{-1} Cd (in the form of $CdCl_2.H_2O$ with purity of 99.9% (Merk, Germany) for nine days, respectively. Petri pots were incubated in a growth chamber at temperature of 20 ± 1°C and a 16-h-day-8-h-night photoperiod. After nine days of incubation, the root and shoot length was measured using a ruler and total soluble protein level of root-tips in plantlets were measured with Bradford method (1976) by using nanodrop (Thermo scientific nanodrop 2000 Spectophotometry, U.S.A). The data were expressed as mean ± HSD. Inhibitory rate (IR, %) of the above indices was calculated using the following formula: IR = (1- x/y) × 100, where x and y are the average values detected in the control and each sample treated, respectively. Three replicate were set up for each treatment and comparison experiments were carried out as duplicate. The differences between mean values of treated plantlets and the controls were statistically investigated using HSD tests.

DNA extraction and RAPD procedures

After nine days of growth, approximately 1 cm root tips of seedlings were collected, ground in liquid nitrogen, and total genomic DNA was extracted using DNA purification kit (Germany, 5 prime). Purified DNA concentration and integrity of total genomic DNA in each sample were estimated by nanodrop (Thermo scientific nanodrop 2000 Spectophotometry., U.S.A). The PCR amplification was carried out with 15 10-base pair random primers (Bioneer-company) and genomic DNA as the template (Table 1). PCR were performed in a reaction mixture of 25 µl containing approximately 75 ng of the genomic DNA dissolving in sterile TE, 10X PCR buffer (2/5 µl), 2 mM $MgCl_2$, 0.2 mM of each dNTP, 2 µl of 10 µM primer and 1 U Taq DNA polymerase. Amplifications were performed in a DNA thermocycler (Biometra., U.S.A) programmed for 4 min at 95°C (initial denaturing step), 35 consecutive cycles each consisting of 4 min at 94°C (denaturing), 1 min at 34°C (annealing), 2 min at 72°C (extension), and followed by 1 cycle for 10 min at 72°C (final extension step). Negative controls with water, without any template DNA, were always included to monitor contamination. PCR reaction products were mixed with one-sixth volume of gel loading buffer (Fermentas, Germany), and then separated on 1.1% agarose gel using Tris-Borate- EDTA (TBE) buffer and GeneRuler DNA ladder (Fermentas, Germany). DNA bands were stained with ethidium bromide, visualized and photographed under UV light. The size and intensity of each amplification product was automatically estimated using the UV-Transilluminator image analyzer system.

RAPD profile and data analysis

All amplifications were repeated twice in order to confirm the reproducible amplification of scored fragments. Only reproducible and clear amplification bands were scored for the construction of the data matrix. Polymorphism observed in RAPD profiles included disappearance and/or appearance and variation in band intensity when comparison with untreated control treatments were evaluated (Liu et al., 2005; Cenkci et al., 2009). Change in soluble protein level and growth rate values were calculated as a percentage of their control (Liu et al., 2005).

Statistical analysis and computations

The growth index experiments and total soluble protein content was calculated according to randomized complete block designed with three replicates. Raw data were imported to Microsoft Excel 2010 program for calculation. The SPSS (Version 17.1) program was used for analysis of variance and comparison of the means was performed by HSD's method at P< 0.05.

Table 1. Sequences and temperature (°C) of the 15 primers used in this experiment.

Number of primer	Sequences of primer	Temperature (°C)
A.z 1	5´-CAATCGCCGT-3´	32
A.z 2	5´-TCGGCGATAG-3´	32
A.z 3	5´-GAAGCGCGAT-3´	32
A.z 4	5´-GCTGCGTGAC-3´	34
A.z 5	5´-CGACTCACAG-3´	32
A.z 6	5´-CCACCCGAGG-3´	36
A.z 7	5´-TACCGTCGGA-3´	32
A.z 8	5´-CGACAGTCCC-3´	34
A.z 9	5´-CTTGACGGGG-3´	34
A.z 10	5´-TCCCTCGTGC-3´	34
A.z 11	5´-CAGGCCCTTC-3´	34
A.z 12	5´-AATCGGGCTG-3´	32
A.z 13	5´-GTCCCGACGA-3´	34
A.z 14	5´-ACGGTACCAG-3´	32
A.z 15	5´-AATCGGGCTG-3´	32

Figure 1. Effect of Cd on wheat seedling length after nine days of treatment.

RESULTS

Effect of Cd on root and shoot growth in wheat seedlings

Cadmium treatment showed toxic effect on seedling growth and root growth of wheat seedlings in all treatment as compared to the control (Figure 1). Seedling and root length of wheat were significantly ($p<0.05$) decreased at 10, 15, 20, 30, 60, 90, 120 mgl^{-1} of cadmium treatment as compared to untreated seedlings. The inhibitory rate of root and shoot growth is given in Table 2. Toxicity effect of cadmium on plant has been previously described in few studies (Shafiq et al., 2008; Liu et al., 2005).

Effect of Cd on total soluble protein content of root tips of wheat seedlings

The data for total soluble protein level in root tips of seedlings are shown in Table 3. Total soluble protein levels in seedlings decreased at 5, 10, 15, 20, 30, 60, 90, and 120 mgl^{-1} Cd, but significant ($P<0.05$) reduction was obtained at concentrations above 10 mgl^{-1} Cd, and inhibited substantially respectively when compared with control plantlets. Results show that there was a negative correlation between Cd concentration and total soluble protein content in wheat seedlings, with a correlation coefficient of (r^2) 0.825.

DNA extraction

Since extracted genomic DNA was in varied concentration, it is needed to equal the concentration of DNA for our experiment. So genomic DNA concentration was equaled by using nanodrop and separated on agarose gel. Nanodroping and electrophoresis data showed that purity grade of DNA is suitable for RAPD. The integrity of the genomic DNA extracted is shown in Figure 2.

Table 2. Effect of Cd on root and shoot length of wheat seedlings after 9 days of treatment.

Cd concentration (mgl^{-1})	Root		Plant length	
	Root length (cm)	Inhibitory rate (%)	Plant length (cm)	Inhibitory rate (%)
0	17.5[a]	-	13.2a	-
5	17[a]	2.8	13a	1.5
10	15.9[b]	9.1	12.1b	8.3
15	14.75[c]	15.7	11.3b	14.3
20	13.3[d]	24	9.3c	29.5
30	12.25[e]	30	9.3d	29.5
60	10.3[f]	41.1	8.6e	34.8
90	9[g]	48.5	8.3f	52.2
120	7.5[h]	57.1	3.8g	71.2

*Average comparison done by using HSD test ($p<0.05$).

Table 3. Effect of Cd on total soluble protein content of root tips seedlings after 9 days of treatment.

Cd concentration (mgl^{-1})	Total soluble protein of root-tips	
	Soluble protein level (ng/µl)	Inhibitory rate (%)
0	444.37[a]*	-
5	439.97[a]	1.11
10	408.93[b]	8.33
15	378.4[c]	14.93
20	360.27[d]	19.10
30	328.77[e]	26.15
60	301.77[f]	32.47
90	290.97[g]	34.98
120	265.5[h]	40.24

*Average comparison done by using HSD test ($p<0.05$).

RAPD profile analysis

Aimed at verifying the genetic effect of Cd tested, the RAPD analysis was performed on DNA extracted from the roots. Fifteen (15) 10mer oligonucleotide primers of 70 to 75% GC content were utilized for screening the wheat genome for changes, whilst only 13 primers generated specific and stable results. The total number of bands was 80 for untreated control treatments. All 13 primers produced same RAPD profiles for the untreated roots seedlings. On the other hand, RAPD profiles showed substantial differences between untreated control and treated seedlings with apparent changes (disappearance and/or appearance) in the number and size of amplified DNA fragments for different primers. The changes in RAPD profiles were summarized for wheat roots of treated seedlings in comparison to their controls (Tables 4 and 5). The maximum change in RAPD profiles (disappearance b and/or appearance a) was obtained in 120, 90, 60, 20, 15 and 30 mgl^{-1} of Cd concentration, respectively, when compared with control plantlets. Also, maximum change in band intensity (decrease d or increase c) was obtained in 90, 120, 60, 15, 20 and 30 mgl^{-1} of Cd, as compared to untreated seedling. In this experiment, change in band intensity (decrease and increase) greatly occurred in high Cd concentrations. For Az.1 primer, disappearances of bands were more evident in 120, 30 and 20 mgl^{-1} concentration of Cd (Figure 3). With Az.2 and Az.4 primers, no change was observed in RAPD profiles (Figure 3). With Az.3 primer, new bands appeared in 20, 30, 60, 90 and 120 mgl^{-1} of Cd concentration (Figure 3). For Az.6, Az.5 and Az.13 primers, disappearance of bands occurred in 5 mgl^{-1} of Cd concentration (Figure 3). Disappearance of bands was also detected in 15, 20, 60,90 and 120 mgl^{-1} with Az.7 primer (Figure 3). Disappearance and decrease in band was observed in 15, 20 and 30 mgl^{-1}, and increase in band intensity was detected in 60, 90 and 120 mgl^{-1} (Figure 3). With Az.12 primer, band intensity was only an appearance of DNA change (Table 5) (Figure not shown). Appearance of new bands in low molecular weight was obtained in Az.13 profiles above 15 mgl^{-1} Cd (Table 5)

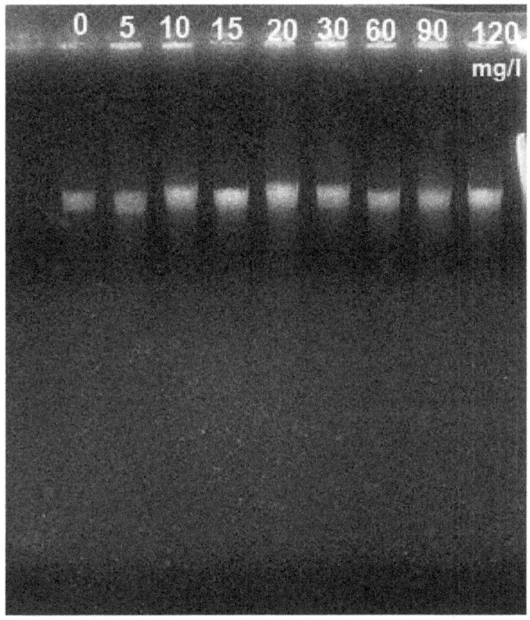

Figure 2. Genomic DNA of about 100 ng extracted by using DNA purification kit (Germany, 5 PRIME) in root-tips of wheat.

Table 4. Changes in number of appearing and disappearing bands in control, and of polymorphic bands and varied bands in Cd-contaminated wheat seedlings.

Number of primer	Cd concentration (mgl^{-1})																
	0	5		10		15		20		30		60		90		120	
		a	b	a	b	a	b	a	b	a	b	a	b	a	b	a	b
A.Z1	13	0	2	0	2	0	1	0	7	0	6	0	0	0	2	0	7
A.Z2	3	1	0	1	0	1	0	1	0	1	0	1	1	1	0	1	0
A.Z3	5	0	1	0	0	0	1	2	0	2	1	2	2	3	0	3	0
A.Z4	2	0	0	0	0	0	0	0	0	0	0	0	0	0	0	0	0
A.Z5	4	0	1	1	0	0	0	0	0	0	0	1	1	1	0	0	0
A.Z6	10	0	1	0	2	0	1	0	1	0	2	0	0	0	3	0	3
A.Z7	5	0	0	0	0	0	3	0	3	0	0	0	0	0	0	0	2
A.Z8	6	0	0	0	0	1	1	0	0	0	0	0	0	0	0	0	1
A.Z9	7	0	0	0	0	0	0	0	0	0	0	0	0	1	0	1	0
A.Z10	-	-	-	-	-	-	-	-	-	-	-	-	-	-	-	-	-
A.Z11	-	-	-	-	-	-	-	-	-	-	-	-	-	-	-	-	-
A.Z12	7	0	0	0	0	0	0	0	0	0	0	0	0	0	0	0	0
A.Z13	3	0	0	0	0	0	0	1	0	1	0	1	1	0	2	1	1
A.Z14	8	0	0	1	0	4	0	0	2	0	1	0	0	6	0	6	0
A.Z15	7	0	0	0	0	0	2	0	0	0	0	0	0	0	2	0	2
Total bands	80	1	5	3	1	7	9	4	13	4	10	5	14	12	9	12	15
a + b		6		4		16		17		14		19		21		27	

a, Indicates appearance of new bands; b, disappearance of normal bands; a + b, polymorphic bands.

(Figure not shown). The number of disappearance of RAPD bands was more at two Cd concentrations (90 and 120 mgl^{-1}) and appearance of new bands was greater at three Cd concentrations (15, 90 and 120 mgl^{-1}) for Az.14 primer. Absent of bands was particularly obvious for wheat exposed to 60, 90 and 120 mgl^{-1} Cd for primer Az.15 (Table 4) (Figure not shown). Decrease or increase in bands intensity generated by different primers is given

Table 5. Changes of total bands intensity in control, and polymorphic bands and varied bands in Cd-contaminated wheat seedlings.

Number of primer	Cd concentration (mgl^{-1})																
	0	5		10		15		20		30		60		90		120	
		c	d	c	d	c	d	c	d	c	d	c	d	c	d	c	d
A.Z1	13	1	0	2	2	2	3	1	0	5	0	4	1	3	1	2	0
A.Z2	3	0	1	1	0	1	0	0	1	1	0	0	1	0	1	2	0
A.Z3	5	2	0	0	0	2	0	0	0	0	0	1	0	1	0	1	0
A.Z4	2	0	0	0	0	0	0	0	0	0	0	0	0	0	0	0	0
A.Z5	4	0	0	0	1	1	0	2	0	0	0	1	0	0	2	0	1
A.Z6	10	2	0	0	0	4	0	5	0	4	0	4	0	5	0	5	0
A.Z7	5	0	0	0	0	0	0	0	0	0	0	0	0	0	0	1	0
A.Z8	6	1	0	1	0	2	0	4	0	2	0	2	0	2	0	5	0
A.Z9	7	2	0	0	0	2	0	0	4	0	3	4	0	0	4	4	0
A.Z10	-	-	-	-	-	-	-	-	-	-	-	-	-	-	-	-	-
A.Z11	-	-	-	-	-	-	-	-	-	-	-	-	-	-	-	-	-
A.Z12	7	0	0	5	1	5	0	5	1	5	1	5	1	5	1	1	0
A.Z13	3	1	0	0	0	1	1	0	1	0	1	0	1	1	0	0	0
A.Z14	8	1	0	1	0	2	2	6	0	6	0	4	0	0	7	0	7
A.Z15	7	0	0	0	0	5	0	0	1	0	1	4	0	4	0	4	0
Total bands	80	8	1	15	4	27	7	23	8	23	6	29	4	21	17	25	8
a + b		9		19		34		31		29		33		38		33	
a + b + c + d		15		23		50		48		43		52		59		60	

c, Decrease in band intensities; d, increase in band intensities; a + b + c + d, varied band.

in Tables 4 and 5. In Tables 4 and 5, data shows that there is a direct relationship between Cd concentration and band changes, thus suggesting nonrandom interaction between DNA and Cd.

Comparison of RAPD profiles, root growth and soluble protein content of roots in barley seedlings under cadmium stress

In Figure 4, the genomic template stability, a qualitative measure reflecting changes in RAPD patterns, was used to compare the modifications in RAPD profiles with reductions in root and shoot growth and soluble protein content of root tips in wheat seedlings. Following exposure to increasing Cd, root and shoot growth and soluble protein content of root tips in wheat seedlings decreased gradually. In contrast, the genomic template stability decreased after exposure to Cd too, but stabilized basically after 15 mgl^{-1} Cd of exposure. The plateau effect was ascribed to multiple changes in RAPD profiles (for example appearance, disappearance of new bands), which tend to counterbalance each other. In other words, the disappeared band was compensated by the low frequency of newly appearing bands at 15 mgl^{-1} Cd concentration for primers used in this experiment.

DISCUSSION

In the present study, the application of RAPD marker analysis is reported for an assessment of cadmium genotoxicity in wheat. This technique has been used by many researchers for determination of genotoxic effect of heavy metal (Conte et al., 1998; Atienzar et al., 2002; Liu et al., 2005; Gupta and Sarin, 2009; Al-Qurainy, 2009; Liu et al., 2009; Cenkci et al., 2009; Taspinar et al., 2009; Shahrtash et al., 2010; Sameer and Qari 2010; Al-Qurainy et al., 2010). The concentration of Cd used in the present study induced some changes in DNA template which was detected by RAPD marker. Cadmium is neither an essential microelement for the growth of plants and animals, nor does it participate in the processes of cell metabolism. However, it is readily taken up probably through zinc transporter proteins (Suzuki, 2005; Fusconi et al., 2007), and it is toxic even at very low concentrations. Cadmium interacts with the root cell wall and, at the cellular level, it induces numerous physiological and metabolic disturbances (Liu et al., 2004; Suzuki, 2005; Fusconi et al., 2007). Cadmium enhances the production of reactive oxygen species that may damage major classes of macromolecules in plant cells, causing DNA strand breaks and increases membrane lipid peroxidation (Andrea, 1994; Fusconi et al., 2007). Metal ions can directly bind to DNA or indirectly through hydrogen bonding of the coordinating water molecules surrounding the metal ions. Metal binding to the bases usually disrupts base pair hydrogen bonding and destabilizes the double helix. The binding of metals to DNA and RNA also influences indirectly the

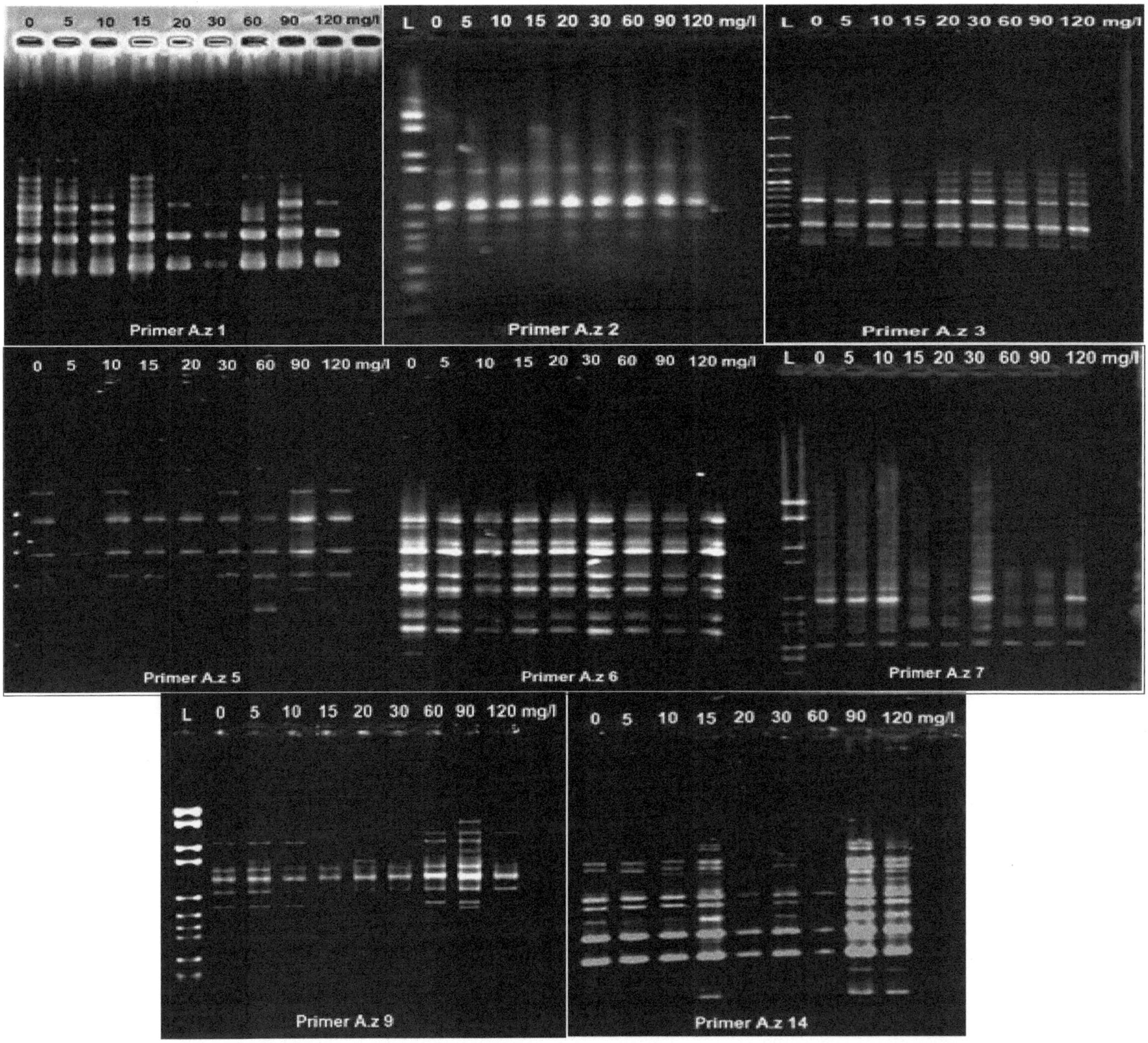

Figure 3. RAPD profiles of genomic DNA from root-tips of wheat seedlings exposed to varying Cd concentrations.

sugar conformation. As a result of this change in sugar puckering, the two helical conformations of A-DNA and B-DNA or RNA are characterized by the orientation of the bases with respect to the axis of the double helix (Valverde et al., 2001; Hossain and Huq, 2002; Anastassopoulou, 2003; Oliveira et al., 2008). This may be the reason why some metal ions influence DNA synthesis and the replication process. Like other metal ions, cadmium has direct interaction with DNA (Fuente et al., 2004; Oliveira et al., 2008). In this study, DNA damage was shown by RAPD profiles via disappearance or appearance and alteration in intensity (decrease or increase) of bands. Bands intensity in some primers under different concentration of cadmium showed some changes (increase or decrease). Changes observed in the DNA profiles such as modifications in band intensity and loss of bands may be due to the changes in oligonucleotide priming sites leading to genomic rearrangements, and less likely to point mutations or DNA damage in the primer binding sites which can block or reduce the efficiency of DNA polymerization in the PCR reaction (Atienzar et al., 2002; Liu et al., 2005; Gupta and Sarin, 2009; Gao et al., 2010). Appearance of new PCR products occurred because some oligonucleotide priming sites could become accessible to oligonucleotide primers after structural change or

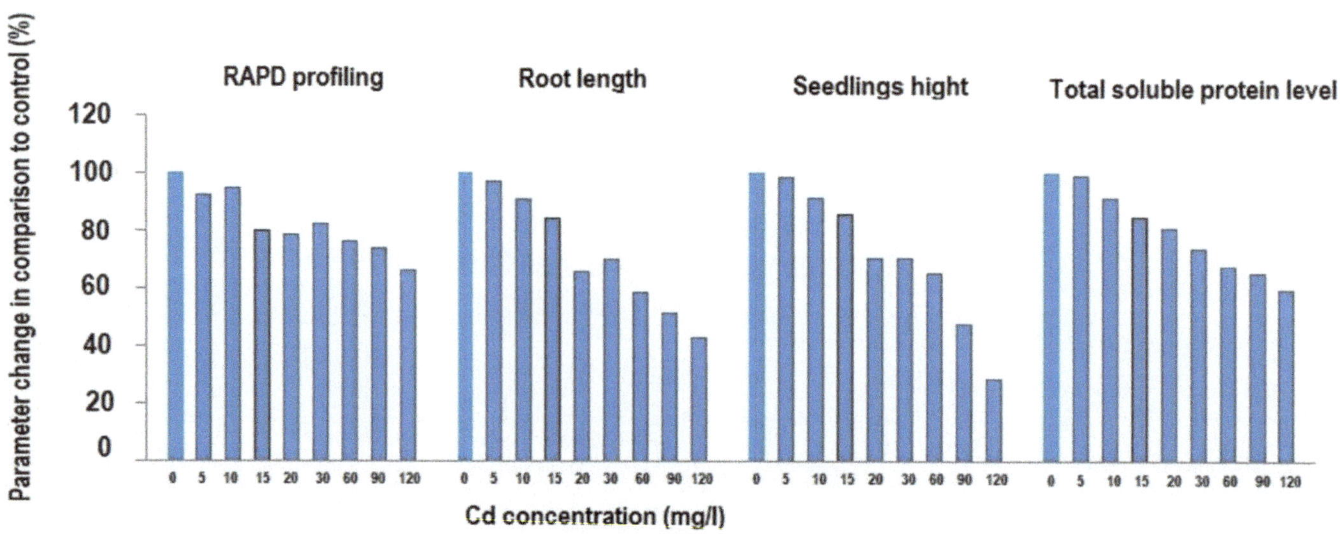

Figure 4. Comparison of root growth, plant height, total soluble protein level and genomic DNA template stability in root tips of wheat seedlings exposed to different Cd concentrations.

because some changes in DNA sequence have occurred due to mutations (resulting in new annealing events), and/or large deletions (bringing two pre-existing annealing sites closer), and/or homologous recombination (juxtaposing two sequences that match the sequence of the primer) (Atienzar et al., 1999; Liu et al., 2005). Trends in disappearing of RAPD bands in profiles in all primers were not similar. However, the number in high concentration was greater than in low concentration.

Disappearance of bands are likely to be due to changes in oligonucleotide priming sites, originating from rearrangements and less likely from point mutations and DNA damage in the primer binding sites (Liu et al., 2005; Enan, 2006; Liu et al., 2009). The disappearance of a normal RAPD product may be related to the events such as DNA damage (example, single and double strand breaks, modified bases, a basic sites, oxidized bases, bulky adducts, DNA–protein cross-links), point mutations and/or complex chromosomal rearrangements induced by genotoxins (Atienzar et al., 1999, 2000). The low concentration of Cd (5 and 10 mgl^{-1}) can also induce mutation and produce polymorphism in RAPD profiles. It was found that three informative 10-mer primers, Az.5, Az.12 and Az.13, may have great potential for detecting cadmium induced genomic alterations in low concentration. In Tables 4 and 5, data shows that there is a relative relationship between Cd concentration and RAPD profiles polymorphism, suggesting a nonrandom interaction between DNA and Cd.

Our finding on comparison between 'untreated' and 'treated' genomes show that RAPD analysis can be used for evaluation of toxicity on plants caused by environmental pollutants. On the basis of these considerations, we could suggest that RAPD technique is a powerful tool for measuring qualitative and quantitative genotoxic activities produced by cadmium. This method can be applied to a wide range of bioindicator organisms and may become a universal methodology to identify target genes for specific genotoxic agents .

In conclusion, the heavy metal such as Cd, can damage the seedlings of wheat and cause mutations even at low concentration. Since, wheat is immensely used as an important plant that provides human daily food and has an important role in human health; therefore, biomarkers are necessary for detection of heavy metals pollution in this species and its growing area to make human health free of hazardous materials. Thus, RAPD fingerprints appear to be stable in detection of genotoxic doses of those plants which are growing in heavy metal polluted soils.

ACKNOWLEDGEMENTS

We would like to thank Shahrbanoo Bustani (Biotechnolo-Technician of Ferdowsi University of Mashhad). Our special thanks to Ferdowsi University of Mashhad, Iran, for the financial supports of this project.

REFERENCES

Al-Qurainy F (2009). Toxicity of heavy metals and their molecular detection on *Phaseolus vulgaris* (L.). Aust. J. Basic Appl. Sci. 3(3):3025-3035.

Al-Qurainy F, Abdulhafed A, Alameri SK (2010). RAPD profile for the assessment of genotoxicity on a medicinal plant; *Eruca sativa*. J. Med. Plant. Res. 4(7):579-586.

Anastassopoulou J (2003). Metal–DNA interactions. J. Mol. Struct. 651-653:19-26.

Andrea H (1994). Role of DNA repair inhibition in lead- and cadmium-induced genotoxicity. Environ. Health Perspect. 102:45-50.

Atienzar FA, Conradi M, Evenden AJ, Jha AN, Depledge MH (1999). Qualitative assessment of genotoxicity using random amplified polymorphic DNA: comparison of genomic template stability with key fitness parameters in *Daphnia magna* exposed to benzo[a]pyrene. Environ. Toxicol. Chem. 18:2275-2282.

Atienzar FA, Jha AN (2006). The random amplified polymorphic DNA (RAPD) assay and related techniques applied to genotoxicity and carcinogenesis studies: A critical review. Mutat. Res. 613:76-102.

Atienzar FA, Paola V, Awadhesh NJ, Michael HD (2002). Evaluation of the random amplified polymorphic DNA (RAPD) assay for the detection of DNA damage and mutations. Mutat. Res. 521:151-163.

Baryla A, Carrier p, Frank F, Coulomb C, Sahut C, Havaux M (2001). Leaf chlorosis oilseed rape plants *(Brassica napus)* grown on cadmium-polluted soil: Causes and consequences for phothosynthesis and growth. Planta 212(5-6): 696-709.

Bradford MM (1976). A rapid and sensitive method for the quantitation of microgram quantities of protein utilizing the principle of protein-dye binding analytical. Biochemistry 72:248-254.

Cenkci S, Mustafa Y, Ibrahim H, Cig E, Muhsin K, Ahmet B (2009). Toxic chemicals-induced genotoxicity detected by random amplified polymorphic DNA (RAPD) in bean (*Phaseolus vulgaris* L.) seedlings. Chemosphere 76:900-906.

Conte C, Mutti I, Puglisi P, Regina G, Maestr E, Marmiroli N (1998). DNA fingerprinting analysis by a PCR based method for monitoring the genotoxic effects of heavy metals pollution. Chemosphere 37:2739-2749.

Enan MR (2006). Application of random amplified polymorphic DNA (RAPD) to detect the genotoxic effect of heavy metals. Biotechnol. Appl. Biochem. 43(3):147-154.

Fuente LDM, Hernanz A, Navarro R (2004). IR and Raman study on the interactions of the 5-GMP and 5-CMP phosphate groups with Mg (II), Ca (II), Sr (II), Ba (II), Cr (III), Co (II), Cu (II), Zn (II), Cd (II), Al (III) and Ga (III). J. Biol. Inorg. Chem. 9(8):973-986.

Fusconi A, Gallo C, Wanda C (2007). Effects of cadmium on root apical meristems of *Pisum sativum* L. Cell viability, cell proliferation and microtubule pattern as suitable markers for assessment of stress pollution. Mutat. Res. 632:9-19.

Gao Y, Zhou P, Liang M, Yue-E Z, Wan-J S (2010). Assessment of effects of heavy metals combined pollution on soil enzyme activities and microbial community structure: modified ecological dose–response model and PCR-RAPD. Environ. Sci. 60:603-612.

Gupta M, Sarin BN (2009). Heavy metal induced DNA changes in aquatic macrophytes: Random amplified polymorphic DNA analysis and identification of sequence characterized amplified region marker. J. Environ. Sci. 21:686-690.

Hossain Z, Huq F (2002). Studies on the interaction between Cd^{2+} ions and nucleobases and nucleotides. J. Inorg. Chem. 90:97-105.

Jin YH, Clark AB, Slebos, RJC, Al-Refai H, Taylor JA, Kunkel TA, Resnick MA, Gordenin DA (2003). Cadmium is a mutagen that acts by inhibiting mismatch repair. Nat. Genet. 34(3):326-329.

Joseph P (2009). Mechanisms of cadmium carcinogenesis. Toxicol. Appl. Pharmacol. 238:272-279.

Liu D, Jiang W, Gao X (2004). Effects of cadmium on root growth, cell division and nucleoli in root tips of garlic. Physiol. Plant. 47:79-83.

Liu W, Li PJ, Qi XM, Zhou QX, Zheng L, Sun TH, Yang YS (2005). DNA changes in barely (*Hordeum vulgare*) seedlings induced by cadmium pollution using RAPD. Chemosphere 61:158-167.

Liu W, Yang YS, Li PJ, Zhoua QX, Xiea LJ, Hana YP (2009). Risk assessment of cadmium-contaminated soil on plant DNA damage using RAPD and physiological indices. J. Hazard. Mater. 161:878-883.

Liu W, Yang YS, Zhou Q, Xie L, Li P, Sun T (2007). Impact assessment of cadmium contamination on rice (*Oryza sativa* L.) seedlings at molecular and population levels using multiple biomarkers. Chemosphere 67(6):1155-1163.

Oliveira SCB, Corduneanu O, Oliveira-B AM (2008). In situ evaluation of heavy metal DNA interactions using an electrochemical DNA biosensor. Bioelectrochemistry 72:53-58.

Sameer H, Qari M (2010). DNA-RAPD Fingerprinting and Cytogenetic Screening of Genotoxic and Antigenotoxic Effects of Aqueous Extracts of *Costus speciosus* (Koen.) JKAU Science 22(1):133-152.

Savva D (1996). DNA fingerprinting as a biomarker assay in ecotoxicology. Toxicol. Ecotoxicol. News 3:110-114.

Shafiq M, Iqbal MZ, Masol A (2008). Effect of lead and cadmium on germination and seedling growth of *Leucaena leucocephala*. J. Appl. Sci. Environ. Manag. 12(2):61-66.

Shahrtash M, Mohsenzadeh S, Mohabatkar H (2010). Cadmium-induced genotoxicity detected by the random amplification of polymorphism DNA in the maize seedling roots. Cell. Mol. Biol. 2(1):42-48.

Suzuki N (2005). Alleviation by calcium of cadmium-induced root growth inhibition in Arabidopsis seedlings. Plant Biotechnol. 22(1):19-25.

Taspinar MS, Guleray A, Nalan Y, Serap S, Ozkan A, Sedat B (2009). Evaluation of selenium effect on cadmium genotoxicity in *Vicia faba* using RAPD. J. Food. Agric. Environ. 7(3&4):857-860.

Theodorakis CW, Bickham JW, Lamb T, Medica PA, Lyne TB (2001). Integration of genotoxicity and population genetic analyses in kangaroo rats (*Dipodomys merriami*) exposed to radionuclide contamination at the Nevada Test Site, USA. Environ. Toxicol. Chem. 20(2):317-326.

Valverde M, Trejo C, Rojas E (2001). Is the capacity of lead acetate and cadmium chloride to induce genotoxic damage due to direct DNA–metal interaction. Mutagenesis 16(3):265-270.

Waisberg M, Joseph P, Hale B, Beyersmann D (2003). Molecular and cellular mechanisms of cadmium carcinogenesis. Toxicology 192:95-117.

An improved method of DNA extraction from plants for pathogen detection and genotyping by polymerase chain reaction

Kamal Sharma*, R. Bhattacharjee, Alieu Sartie and P. Lava Kumar

International Institute of Tropical Agriculture, Oyo Road, PMB 5320, Ibadan, Nigeria.

Polymerase chain reaction (PCR)-based applications in plant molecular biology and molecular diagnostics for plant pathogens require good quality DNA for reliable and reproducible results. Leaf tissue is often the choice for DNA extraction, but the use of other sources such as tubers, stems, or seeds, is not uncommon. The extraction of DNA from different tissue sources often requires different protocols. In this study, a simple protocol was established for the extraction of DNA from leaves, tubers, stems, seeds and even fungal mycelia. The protocol is simple and suitable for high-throughput DNA extraction using automated tissue grinders. It yielded large quantities of DNA (0.4 µg to 2 mg DNA from 100 mg tissue) of high quality from seeds of maize, soybean, and cowpea, tubers of yam, tuberous roots of cassava, and leaf tissues of banana and plantain, yam, cassava, maize, okra, mango, and other species. DNA was successfully used for the detection of fungal and viral pathogens and the genotyping of yam and cassava by PCR.

Key words: DNA isolation, plant tissues, PCR amplification, pathogen detection, high throughput DNA extraction.

INTRODUCTION

Extraction of a suitable quantity and quality of DNA is a critical step in molecular applications, such as polymerase chain reaction (PCR), sequencing, and cloning and restriction analysis. Several protocols have been established for DNA extraction, some of which are applicable on wide range of materials and some are specific to a particular tissue type (Weishing et al., 1995; Amani et al., 2011; Carrier et al., 2011; Pervaiz et al., 2011; Akkurt, 2012). Generally, freshly harvested leaves are preferred for extraction of DNA (Zhang and Stewart, 2000). In addition, extraction of DNA from preserved leaf tissues, seeds, stems, and tubers is not uncommon. However, the presence of polysaccharides, tannins and other metabolites with high affinity to DNA and protein in these tissues makes it difficult to extract good quality DNA for further utilization in molecular techniques. Polysaccharides inhibit the action of *Taq* polymerase, thus affecting the PCR reaction process (Rogers and Bendich, 1985; Wulff et al., 2002). Similarly, mucilage, a highly viscous secondary metabolite composed of a polar polymer of glycoprotein and present in tubers, seeds, and stems, also co-precipitates with DNA and inhibits the action of *Taq* polymerase (Jose and Usha, 2000; Ghosh et al., 2009). The presence of mucilage also hinders the accurate pipetting of DNA, leading to volumetric errors (Ghosh et al., 2009).

A number of studies have been carried out in the past to establish protocols for the extraction of good quality

*Corresponding author. E-mail: kamalsbhardwaj@gmail.com.

Abbreviations: LTP, Low throughput; **HTP,** high throughput; **DEB,** DNA extraction buffer; **PCR,** polymerase chain reaction; **ITS,** internal transcribed spacer; **BSV,** *Banana streak virus*; **ACMV,** *African cassava mosaic virus*; **EACMCV,** *East Africa cassava mosaic Cameroon virus*; **MSV,** *Maize streak virus*; **SSR,** simple sequence repeat; **PEG**, polyethylene glycol.

DNA from seeds (Kang et al., 1998), tubers (Wulff et al., 2002), mycelia (Chi et al., 2009), and other tissues. In this study, several available protocols were evaluated to extract good quality DNA from a wide array of tissues, with the aim of developing a technically simple, rapid, and cost-effective protocol for manual and high-throughput DNA extraction using automated tissue grinders.

MATERIALS AND METHODS

Tissue materials and PCR targets

Seeds of cowpea, soybean, and maize, tubers of yam (*Dioscorea* sp.), stems and tuberous roots of cassava, and leaves of mango, banana, maize, and okra were collected from the experimental fields in IITA, Ibadan, Nigeria. In addition, mycelia of pure cultures of *Colletotrichum gloeosporioides,* isolated from anthracnose-affected yam leaves and *Cercospora* sp. isolated from gray leaf spot-affected maize leaves were also included. For genotyping by PCR, DNA extracted from leaves (banana, cassava, okra, and mango), tubers (yam and cassava), stems (cassava), and seeds (soybean, cowpea, and maize) of all species were tested using oligonucleotide primers specific to nuclear ribosomal DNA or simple sequence repeat (SSR) markers (Table 1).

For pathogen diagnostics by PCR, DNA extracted from cassava leaves, stems, and tubers, both healthy and damaged by cassava mosaic disease (CMD), was tested for *African cassava mosaic virus* (ACMV) or *East African cassava mosaic Cameroon virus* (EACMCV). Leaves of maize and banana/plantain affected by streak disease were tested for *Maize streak virus* (MSV) or *Banana streak virus* (BSV). DNA extracted from pure mycelia of *C. gloeosporioides* and *Cercospora* sp., from maize was used for the amplification of nuclear ribosomal DNA for fungal genotyping. DNA was also extracted from FTA® Classic Cards (Whatman International Inc.) (Borman et al., 2006), a paper-based system designed to fix and store nucleic acids directly from fresh tissues pressed into the FTA® paper. Leaf sap of CMD-affected cassava leaves was applied on FTA® cards and allowed to dry at room temperature. The cards were stored in envelopes at room temperature until used for DNA extraction and subsequent analysis for ACMV/EACMCV by PCR.

DNA extraction protocol

Reference was made to several DNA extraction protocols to develop a high-throughput, rapid, and cost-effective method for large-scale DNA extraction from a wide range of tissue materials. The method described here was developed based on the protocols described in Kang et al. (1988) and Wuff et al. (2002) by introducing the following modifications:

1. Instead of samples being freeze-dried, seeds or tubers were heat treated at 65°C for 30 min;
2. The phenol: chloroform: isoamyl alcohol (25:24:1) step was removed;
3. CTAB was replaced by SDS as the detergent in the extraction buffer and
4. PEG (40,000 mol. wt.) was added in the extraction buffer.

Tissues were homogenized with sterile pestles and mortars or in 1.5 ml microfuge tubes using handheld micro tube pestles for manual extraction or by using GenoGrinder2000 for high-throughput (HTP) extraction (Bhattacharjee et al., 2009). For DNA extraction by HTP, leaves, stems, or mycelia were stored overnight at –80°C; seeds and tubers were treated at 65°C for 30 min before being ground. About 100 mg of tissue (leaves, tubers, stems, seeds, or mycelia) was ground following both the manual extraction and HTP methods, and samples were incubated at 65°C for 30 min in a water bath after the addition of 500 µL of DNA extraction buffer (DEB; 100 mM Tris, pH 8, 10 mM EDTA, 1% SDS, 2% PVP, 1 M NaCl, and 1% β-mercaptoethanol) supplemented with 0.05 mg/ml proteinase K and 4% (w/v) PEG (mol. wt. 40 000). The samples were then subjected to centrifugation at 13 000 rpm for 10 min in a tabletop centrifuge and the supernatant was collected into fresh microtubes. To each sample vial two-thirds volume of isopropanol (v/v) was added and incubated at –20°C for 1 h for DNA precipitation. The samples were centrifuged at 13 000 rpm for 10 min and the supernatant was discarded to obtain the DNA pellet. Each DNA pellet was further washed with 500 µL of 70% ethanol and centrifuged at 13 000 rpm for 5 min. The supernatant was discarded and traces of ethanol were removed by air drying the DNA pellet for about 15 min at room temperature. The DNA was re-suspended in 100 µL of TE buffer (10 mM Tris, pH 8, and 1 mM EDTA), treated with RNAase A (10 µg/100 µL of DNA sample), and stored at –20°C until further use. DNA was quantified in the NanoDrop ND1000 (NanoDrop Technologies) spectrophotometer by taking readings at 260 and 280 nm, following the manufacturer's instructions. DNA quality was assessed by electrophoresis of an aliquot of 1 µL DNA in an 1% agarose gel using TAE buffer, as described in Sambrook and Russell (2001).

DNA amplification by PCR

The DNA extracted by manual extraction and HTP methods from various sources were tested by PCR amplification using different primer sets listed in the Table 1. PCR for pathogen detection was performed in a total volume of 25 µL, consisting 5.0 µL of 10x PCR buffer (Promega, Madison, WI, USA), 0.5 µL of 10 mM dNTPs, 0.5 µL of 10 pmol of each primer, 1.5 µL of 25 mM $MgCl_2$, 0.5 units of *Taq* polymerase (Promega), and 10 ng of template DNA. The thermal cycle conditions for PCR using various primer sets are given in Table 1. The amplified products were electrophoresed on 1% TAE agarose gels containing 0.5 mg/ml ethidium bromide, and visualized on a UV transilluminator (Sambrook and Russell, 2001).

For genotyping by PCR, DNA extracted from leaves and tubers of cassava were subjected to PCR amplification using SSR markers (Table 1). The total volume of PCR reaction was 10 µL, containing 10 ng of genomic DNA, 1.0 µL of 10× reaction buffers, 0.3 µL of 50 mM $MgCl_2$, 0.8 µL of 2.5 mM dNTPs, 2 µL of each primer and 0.06 U of Platinum *Taq* polymerase (Invitrogen). The PCR products were electrophoresed on 1.2% agarose gel containing 0.5 mg/ml ethidium bromide, and visualized on a UV transilluminator. PCR was also performed by serially diluting 1 µg/ml DNA from $1{:}10^2$ to $1{:}10^{10}$ (v/v) in TE buffer, and 2 µL of it was used in PCR to assess the end-point detection of pathogens as a measure to assess the sensitivity of PCR.

RESULTS AND DISCUSSION

DNA quantity and quality

To date, many simplified DNA extraction methods have been developed for fungi and plant tissues, each having its own advantages and disadvantages (Cenis, 1992; Thomson and Henry, 1995; Liu et al., 2000; Cassago et al., 2002; Guo et al., 2005; Chi et al., 2009). In this study, a DNA extraction protocol was standardized, which is a modification of the methods reported by Kang et al. (1998) and Wulff et al. (2002). The protocol developed

Table 1. Details of PCR targets, oligonucleotide primers and PCR conditions used in the study.

Target	Species*	Amplicon size (bp)	Primer name and sequence (5' to 3')	Thermal cycling conditions	Reference
Nuclear ribosomal rDNA (universal)	All tested in present study	550	ITS1:TCCGTAGGTGAACCTGCGG ITS4: TCCTCCGCTTATTGATATGC	One cycle of 94°C/5 min ; 35 cycles of 94°C/30 s, 55°C/1 min., 72°C/1.5 min ; one cycle of 72°C/7 min; and hold at 4°C.	White et al 1990
Nuclear ribosomal rDNA (species-specific)	Yam	500	ITS1:TCCGTAGGTGAACCTGCGG YA(I)F:CCCTTTGTGAACATACC	One cycle of 94°C/5 min ; 35 cycles of 94°C/30 s, 55°C/1 min., 72 °C/1.5 min ; one cycle of 72°C/7 min; and hold at 4°C.	ITS1 (White et al 1990); YA(I)F (Sharma et al 2009)
Maize streak virus (MSV)	Maize	1400	MSV215-234: CCAAAKDTCAGCTCCTCCG MSV1770-1792: TTGGVCCGMVGATGTASAG	One cycle of 94°C/5 min ; 35 cycles of 94°C/30 s, 52°C/1 min., 72°C/1.5 min ; one cycle of 72 °C/7 min; and hold at 4°C.	Palmer and Rybicki 2001
Banana streak virus (BSV)	Banana	520	BadnaFP: ATGCCITTYGGIITIAARAAYGCICC Badna RP:CCAYTTRCAIACISCICCCCAICC	One cycle of 94°C/5 min ; 35 cycles of 94°C/30 s, 50°C/1 min., 72°C/1.5 min ; one cycle of 72°C/7 min; and hold at 4°C.	Yang et al 2003
African cassava mosaic virus (ACMV) / *East African cassava mosaic Cameroon virus* (EACMCV)	Cassava	ACMV-380 EACMCV- 650	OjaRep-F: CRTCAATGACGTTGTACCA EACMVRep-R: GGTTTGCAGAGAACTACATC ACMVRep-R: CAGCGGMAGTAAGTCMGA	One cycle of 94°C/5 min, 52°C/2min, 72°C/3min; 35 cycles of 94°C/30 s, 52°C/1 min., 72°C/1.5 min ; one cycle of 72°C/7 min; and hold at 4 °C.	Alabi et al 2008
EST SSR	Cassava	230, 280, 300	AGC-129 TCT140 AGC163	One cycle of 94°C/1 min, 52°C/2min, 72°C/3min; 35 cycles of 94°C/30 s, 52 °C/2 min., 72°C/1.33 min; one cycle of 72°C/7 min; and hold at 4°C.	de Bang et al 2010
EST SSR	Yam	210, 280, 400	Dab2C05 (F and R)- CCCATGCTTGTAGTTGT-F TGCTCACCTCTTTACTTG-R DalA01 (F&R)- AACTATAATCGGCCAGAGG-F TGTTGGAAGCATAGAGAATT-R	One cycle of 94°C/4 min; 34 cycles of 94°C/30 s, 55°C/30 s, 72C/1 min; one cycle of 72°C/7 min; and hold at 4°C.	Tostain et al 2006.

Table 2. Concentration of DNA extracted using three different protocols.

Sample	This paper			As per Kang et al. (1995)			As per Wulff et al. (2002)		
	*DNA (ng/µl)	$A_{260/280}$	$A_{260/230}$	DNA (ng/µl)	$A_{260/280}$	$A_{260/230}$	DNA (ng/µl)	$A_{260/280}$	$A_{260/230}$
Yam tuber	1147.9 ±117.0	1.8 ±0.0	0.5 ±0.4	63.4 ±9.1	1.1 ±0.1	0.3 ±0.0	222.3 ±0.9	1.3 ±0.0	0.4 ±0.0
Cassava stem	506.5 ±36.2	1.7 ± 0.2	0.4 ±0.3	38.7 ±11.3	1.3 ±0.2	-12.3 ±21.7	147.4 ±4.5	1.5 ±0.0	0.5 ±0.0
Cowpea seed	4538.4 ±232.7	1.7 ±0.2	0.7 ±0.6	1154.7 ±73.3	0.7 ±0.0	0.3 ±0.0	27.0 ±0.5	0.9 ±0.0	0.1 ±0.0
Soybean seed	3896.6 ±522.1	1.6 ±0.4	0.6 ±0.5	499.9 ±33.4	0.9 ±0.0	2.2 ±0.8	1289.9 ±170.2	1.1 ±0.0	0.4 ±0.0
Maize leaf	870.3 ±32.4	1.8 ±0.1	0.9 ±0.1	39.3 ±1.7	0.9 ±0.0	0.2 ±0.0	67.8 ±1.4	1.1 ±0.0	0.2 ±0.0
Banana leaf	4417.0 ±334.3	1.9 ±0.1	0.7 ±0.9	nt	nt	nt	nt	nt	nt
Mango leaf	917.4 ±30.0	1.9 ±0.0	0.3 ±0.3	nt	nt	nt	nt	nt	nt
Okara leaf	851.9 ±28.3	1.8 ±0.1	1.0 ±0.1	nt	nt	nt	nt	nt	nt
Cg* (*mycelia*)	1860.9 ±548.9	1.7 ±0.2	0.4 ±0.4	nt	nt	nt	nt	nt	nt

*Concentration estimated using NanoDrop Spectrophotometer ND-1000; nt, not tested; Cg, *Colletotrichum gloeosporioides*.

has several advantages:

a. The number of steps is minimal and thus large numbers of samples can be processed/day/person;
b. It can be efficiently used to obtain a sufficient amount of DNA from 100 mg of seed, tuber, or leaf tissues;
c. The protocol does not use liquid nitrogen and toxic chemicals such as phenol;
d. It is cost-effective protocol ($2 to $3/sample depending on manual extraction or HTP is used; and
e. Facilitates good quality DNA from different plant tissues. This protocol was successfully used to isolate DNA from several plant species, including banana, yam, okra, banana, mango and even fungal mycelia, from which genomic DNA isolation is difficult due to the presence of a high amount of mucilage, polyphenols and proteins (Cassago et al., 2002).

This protocol advocates the incubation of seeds/tubers in hot water for 30 min, which helps in softening the tissues for easy grinding (Kang et al., 1998); detergent, CTAB was been replaced with SDS to obtain good quality DNA (Edwards et al., 1991; Keb Llanes et al., 2002; Matasyoh et al., 2008). Generally, CTAB is used to avoid co-purifying polysaccharides from plant tissues; SDS breaks up the lipids in the membranes. Since seed and tuber samples carry high amounts of polysaccharides and mucilage, SDS was used in the extraction buffer. The extraction buffer contains proteinase K and PEG which helped in removing proteins and phenolic components, thus excluding from the protocol the organic solvents (such as phenol and chloroform).

The results from NanoDrop spectrophotometer showed that the protocol described in this study yielded a good quantity and quality of DNA (Table 2, Figure 1A). The concentration of DNA ranged between 506.5 and 4538.4 ng/100 mg of seeds and tubers; from 13.3 to 1176.24 ng of DNA/ 50 mg of mycelia; and from 766.98 to 1581.92 ng/ 100 mg of seeds (Table 2). The A260/280 ratio was 1.23 to 2.26 for DNA extracted from mycelia, and 1.61 to 1.92 for DNA extracted from seeds. It was observed that the addition of PEG in the extraction buffer increased the quality of DNA; this may be due to the effective removal of polysaccharides present in tissue samples. PEG is regularly used for tissue samples which are known to contain high amounts of polysaccharides or mucilage (Sharma et al., 2008).

PCR amplification

The DNA extracted from anthracnose-affected yam leaf samples, uninfected controls, and purified cultures of *C. gloeosporioides* using both manual extraction and HTP methods was amplified using universal rDNA primers and primers specific to *C. gloeosporioides*. Amplification of 550 bp DNA with universal rDNA primers and 500 bp DNA from the anthracnose-affected leaves and a pure culture of *gloeosporioides* confirmed that the quality of DNA extracted using the protocol was suitable for PCR amplification (Figures 1B and 2A).

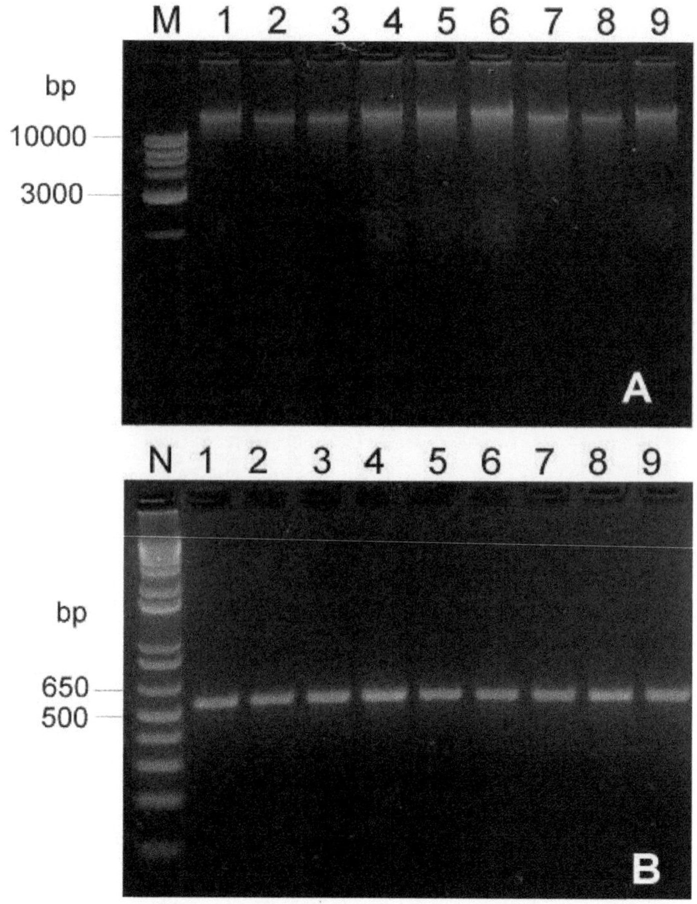

Figure 1. Resolution of DNA (A) and nuclear rDNA ITS segment amplified by ITS-1/ITS-4 primers in PCR (B). Lanes 1, Yam tuber; 2, Cassava tuber; 3, Cowpea seeds; 4, Soybean seeds; 5, Maize seeds; 6, Banana leaf; 7, Mango leaf; 8, Okra leaf; 9, *mycelia of C. gloeosporioides;* Lane M, 1 kb DNA marker (New England Biolab); and Lane N, 1kb plus DNA standard (Invitrogen). Position of molecular weight markers indicated in base pairs (bp).

DNA extracted from BSV-infected leaves of banana was amplified by ITS (Figure 2B; Lanes 1 to 2) used as controls and BSV-specific primers (Figure 2B; Lanes 3 to 5). The BSV-specific band was also observed with an asymptomatic banana leaf which could correspond to the BSV genome integrated in the host genome. The DNA samples extracted from healthy and infected leaves and tubers of cassava were amplified by ITS primers resulting in a 550 bp amplicon. ACMV/EACMCV-specific primers amplified a 390 bp amplicon specific to ACMV in CMD-affected cassava leaf, but not in healthy cassava. EACMCV specific primers did not result in amplification, indicating that the sample was free from EACMCV (Figure 2C). MSV specific amplicon resulted in PCR of MSV infected maize (Figure 2D).

The DNA extracted from FTA^R cards spotted with cassava leaf extract resulted in the amplification of a ~390 bp product of ACMV and a ~650bp product of EACMCV (Figure 2E), confirming the suitability of the protocol for extracting DNA even from FTA^{TM} cards, which are used for sample preservation during field surveys and subsequent laboratory diagnosis of pathogens (Borman et al., 2006). Pathogens detection in serially diluted DNA differed between the samples (Table 3) (Figure 3). For instance, ACMV was detected up to 10^{-6} dilution of 1 µg/µL DNA extracted from leaves, but up to 10^{-3} and 10^{-5} dilution in DNA extracted from stems and tubers, respectively (Figure 3). These results could be a reflection of variable concentration of ACMV in different tissues of cassava. Interestingly, PCR amplification of rDNA ITS regions using the DNA extracted from yam and banana leaves resulted from 10^{-4} DNA dilution (Table 3). High concentration of leaf polyphenols in yam and banana leaves might be interfering with PCR reactions at low dilutions ($<10^{-4}$). This result suggests that appropriate DNA dilution is critical to successful PCR, particularly

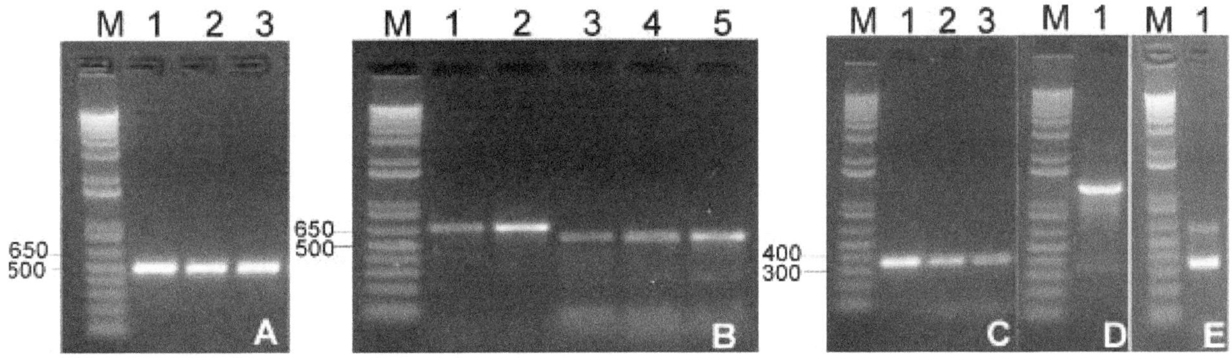

Figure 2. A. Detection of *C. gloeosporioides* in yam leaf DNA using ITS1 and YA(I)F (Lanes 1 to 3); B. Amplification of banana rDNA using ITS1/ITS4 primers (Lanes 1 and 2), and *Banana streak virus* (Lanes 3 to 5); C. Detection of *African cassava mosaic virus* (ACMV) in cassava mosaic disease-affected cassava leaf (Lane 1), stem (Lane 2) and tuber (Lane 3); D. Detection of *Maize streak virus* in the virus infected maize leaf using; E. Detection of ACMV and *East African cassava mosaic virus* in DNA extracted from FTA® Classic card spotted with the infected cassava leaf sap. Lane M, 1 kb DNA marker (NewEngland Biolab). Position of molecular weight markers indicated in base pairs (bp).

Table 3. Serial dilutions PCR using specific primer for pathogens.

Target*	DNA dilution**								
	10^{-2}	10^{-3}	10^{-4}	10^{-5}	10^{-6}	10^{-7}	10^{-8}	10^{-9}	10^{-10}
ITS amplification using ITS1/ITS4 primers in *C. gloeosporioides* mycelia	+	+	+	+	+	+	nt	nt	nt
ITS amplification in yam leaf using ITS1/ ITS4	-	-	+	+	+	+	nt	nt	nt
ITS amplification in maize leaf using ITS1/ ITS4	+	+	+	+	+	-	-	-	-
ITS amplification in maize seed using ITS1/ ITS4	+	+	+	+	+	+	+	-	-
ITS amplification in cassava leaf using ITS1/ ITS4	+	+	+	+	+	+	-	-	-
ITS amplification in cassava tuber using ITS1/ ITS4	+	+	+	+	+	-	-	-	-
Detection of BSV in banana leaf	-	-	+	+	+	nt	nt	nt	nt
Detection of MSV in maize leaf MSV infected leaf using MSV	+	+	+	+	+	nt	nt	nt	nt
Detection of ACMV/EACMCV in cassava leaves	+	+	+	+	+	nt	nt	nt	nt
Detection of ACMV/EACMCV in cassava stems	+	+	-	-	-	nt	nt	nt	nt
Detection of ACMV/EACMCV in cassava tubers	+	+	+	+	-	nt	nt	nt	nt

*Primer details and PCR conditions for each target is provided in Table 1. **Starting DNA concentration, 1 µg/µl; +, PCR amplification; -, no amplification; nt, not tested.

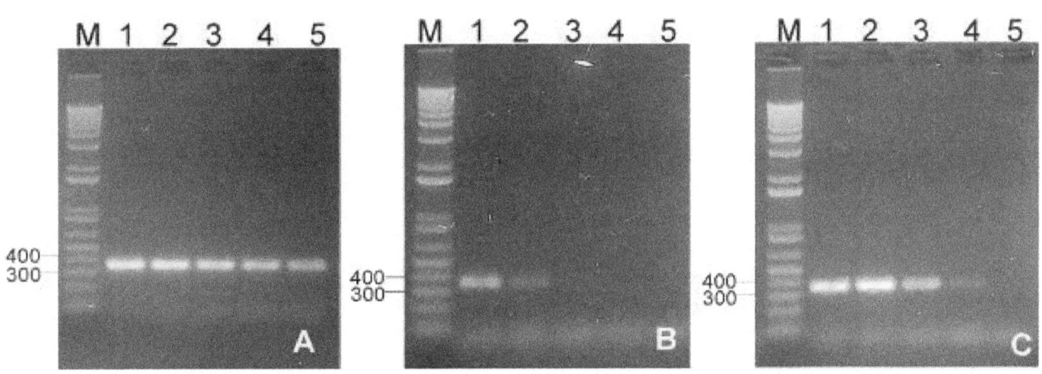

Figure 3. Detection of *African cassava mosaic virus* (ACMV) in DNA extracted from cassava leaves (A), stems (B) and roots (C). Lanes 1 to 5, DNA dilution from 10^{-2} to 10^{-6}. AMCV was detected up to 10^{-6} dilution in leaves (A), up to 10^{-3} dilution in stems (B) and up to 10^{-5} dilution in tubers (C). Lane M, 1 kb DNA marker (New England Biolab). Position of molecular weight markers indicated in base pairs (bp).

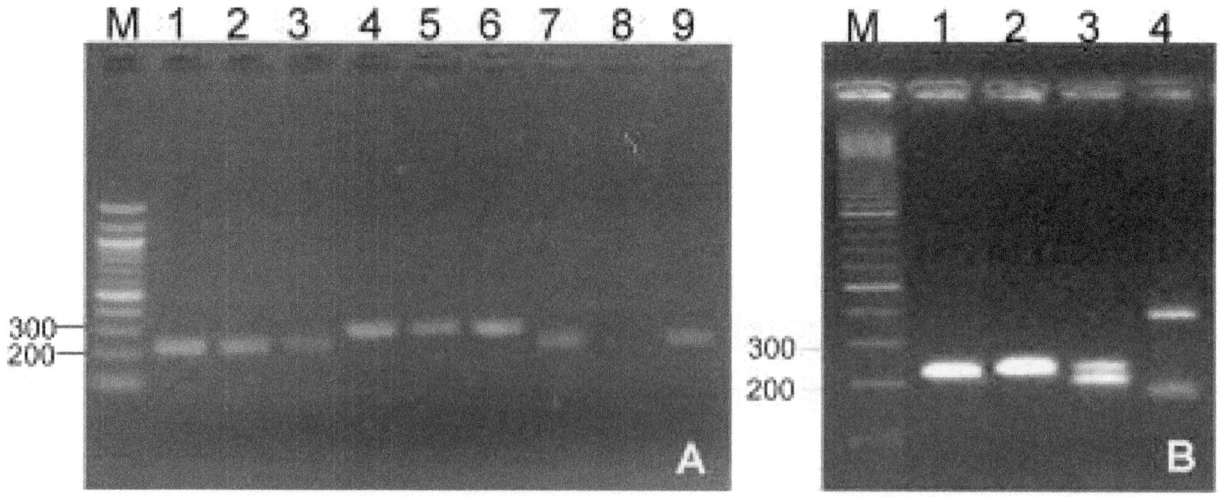

Figure 4. A. Amplification of cassava genomic DNA by PCR using SSR primers. Lanes 1, cassava leaf; 2. cassava stem; 3. cassava tuber; 4. cassava leaf; 5. cassava stem; 6. cassava tuber; 7. cassava leaf; 8. cassava stem; 9. cassava tuber. Lane 1-3 using SSR marker AGC129 forward and reverse; Lane 4-6 using SSR marker TCT140 forward and reverse; Lane 7-9 using SSR marker AGC163 forward and reverse. Lane M, 100bp DNA standard (NewEngland Biolab). B. Amplification of yam genomic DNA by PCR using SSR primers. Lanes: 1, yam leaf; 2, yam tuber; 3, yam leaf; 4, yam tuber. Lane 1 to 2 using SSR markerDa1A01 forward and reverse; Lane 3-4 using SSR marker Dab2C05 forward and reverse. Lane M, 100bp DNA standard (Promega). Position of molecular weight markers indicated in base pair (bp).

when plant samples rich in polyphenols are used as source materials. The SSR primers used for genotyping using DNA extracted from leaves and stems of cassava and leaves and tubers of yam resulted in amplification (Figure 4). By validating the improved DNA extraction protocol on diverse samples, this study demonstrated the suitability of the technique as a simple and cheaper alternative to extract DNA for PCR applications from many plant species and tissue sources.

REFERENCES

Akkurt M (2012). Comparison between modified DNA extraction protocols and commercial isolation kits in grapevine (*Vitis vinifera* L.). Genet. Mol. Res. 11(3):2343-2351.

Alabi OJ, Kumar PL, Naidu RA (2008). Multiplex PCR method for the detection of *African cassava mosaic virus* and *East African cassava mosaic Cameroon virus* in cassava. J. Virol. Met. 154:111-120.

Amani J, Kazemi R, Abbasi AR, Salmanian AH (2011). A simple and rapid leaf genomic DNA extraction method for polymerase chain reaction analysis. Iran. J. Biotechnol. 9(1):69-71.

Bhattacharjee R, Ferguson M, Gedil M, Dumet D, Ingelbrecht I (2009). Field collection, preservation and large scale DNA extraction procedures for cassava (*Manihot esculenta* Crantz.). Afr. J. Biotechnol. 8:3424-3430.

Borman AM, Linton CJ, Miles SJ, Campbell CK, Johnson EM (2006). Ultra-rapid preparation of total genomic DNA from isolates of yeast and mould using Whatman FTA filter paper technology - a reusable DNA archiving system. Med. Mycol. 44:389-398.

Carrier G, Santoni S, Rodier-Goud M, Canaguier A, de Kochko A, Dubreuil-Tranchant C, This P, Boursiquot JM, Cunff LL (2011). An efficient and rapid protocol for plant nuclear DNA preparation suitable for next generation sequencing methods. Am. J. Bot. 98(1):e13-e15.

Cassago A, Panepucci RA, Baião AMT, Henrique-Silva F (2002). Cellophane based mini-prep method for DNA extraction from the filamentous fungus *Trichoderma reesei*. BMC Microbiol. 2:14.

Cenis JL (1992). Rapid extraction of fungal DNA for PCR amplification. Nucleic Acids Res. 20:2380.

Chi MH, Park SY, Lee YH (2009). A Quick and Safe Method for Fungal DNA Extraction. Plant Pathol. J. 25:108-111.

de Bang TC, Raji AA, Ingelbrecht I (2010). A multiplex microsatellite marker kit for diversity assessment of large cassava (*Manihot esculenta* Crantz) germplasm collections. Plant Mol. Biol. Rep. 3:655-662.

Edwards K, Johnstone C, Thompson C (1991). A simple and rapid method for the preparation of plant genomic DNA for PCR analysis. Nucleic Acid Res. 19:1349.

Ghosh R, Paul S, Ghosh SK, Roy A (2009). An improved method of DNA isolation suitable for PCR-based detection of begomoviruses from jute and other mucilaginous plants. J. Virol. Met. 159:34-39.

Guo JR, Schnieder F, Abd-Elsalam KA, Verreet JA (2005). Rapid and efficient extraction of genomic DNA from different phytopathogenic fungi using DNAzol reagent. Biotechnol. Let. 27:3-6.

Jose J, Usha R (2000). Extraction of geminiviral DNA from a highly mucilaginous plant (*Abelmoschus esculentus*). Plant Mol. Biol. Rep. 18:349-355.

Kang HW, Choyg, Yoon UH, Eun MY (1998). A Rapid DNA Extraction Method for RFLP and PCR Analysis from a Single Dry Seed. Plant Mol. Biol. Rep. 16:1–9.

Keb Llanes M, Gonzalez G, Chi-Manzanero B, Infante D (2002). A rapid and simple method for small scale DNA extraction in Agavaceae and other tropical plants. Plant Mol. Biol. Rep. 20:299a-299e.

Liu D, Coloe S, Baird R, Pedersen J (2000). Rapid mini-preparation of fungal DNA for PCR. J. Clin. Microbiol. 38:471.

Matasyoh LG, Wachira FN, Kinyua MG, Muigai AWT, Mukiama TK (2008). Leaf storage conditions and genomic DNA isolation efficiency in *Ocimum gratissimum* L. from Kenya. Afr. J. Biotechnol. 7(5):557-564.

Palmer KE, Rybicki EP (2001). Investigation of the potential of *Maize streak virus* to act as an infectious gene vector in maize plants. Arch. Virol. 146:1089-1104.

Pervaiz ZH, Turi NA, Khaliq I, Rabbani MA, Malik SA (2011). A modified method for high-quality DNA extraction for molecular analysis in cereal plants. Genet. Mol. Res. 10(3):1669-1673.

Rogers SO, Bendich AJ (1985). Extraction of DNA from milligram amounts of fresh, herbarium, and mummified plant tissues. Plant Mol.

Biol. 5:69-76.
Sambrook J, Russell DW (2001). Molecular Cloning: a Laboratory Manual. Cold Spring Harbor Laboratory Press, New York.
Sharma K, Ayodele M, Bandyopadhyay R, Asiedu R, Kumar PL (2009). Molecular characterization of *Colletotrichum gloeosporioides* responsible for anthracnose of yam and cassava in Nigeria, and development of a diagnostic PCR assay. In: 15th Triennial Symposium of International Society for Tropical Root Crops, International Potato Centre, Lima, Peru. 104 pp.
Sharma K, Mishra AK, Misra RS (2008). A simple and efficient method for extraction of genomic DNA from tropical tuber crops. Afr. J. Biotechnol. 7:1018-1022.
Thomson D, Henry R (1995). Single-step protocol for preparation of plant-tissue for analysis by PCR. Biotechniques 19:394-400.
Tostain S, Scarcelli N, Brottier P, Marchand JL, Pham JL, Noyer JL (2006). Development of DNA microsatellite markers in tropical yam (*Dioscorea* sp.). Mol. Ecol. Notes 6:173-175.
Weishing K, Nybom H, Wolff K, Meyer W (1995). DNA isolation and purification. In: DNA fingerprinting in plants and fungi, CRC Press, Boca Raton, Florida. pp. 44-59.
White TJ, Bruns T, Lee S, Taylor J (1990). Amplification and direct sequencing of fungal ribosomal RNA genes for phylogenetics. In: Innis et al. (eds), PCR Protocols, Academic Press, London. Pp. 315-322.
Wulff EG, Torres S, Vigil EG (2002). Protocol for DNA extraction from potato tubers. Plant Mol. Biol. Rep. 20:187a–187e.
Yang IC, Hafner GJ, Dale JL, Harding RM (2003). Genomic characterization of taro bacilliform virus. Arch. Virol. 148:937-949.
Zhang J, Stewart JM (2000). Economical and rapid method for extracting cotton genomic DNA. J. Cotton Sci. 4:193-201.

Taxonomy of *Streptomyces* strains isolated from rhizospheres of various plant species grown in Taif region, KSA, having antagonistic activities against some microbial tissue culture contaminants

Mohamed S.H. [1,2], El-Helafiy Seham S.D.[1], Ismail Mona A.[1] and Sadik A.S.[1,3]

[1]Department of Biology, Faculty of Science, Taif University, P.O. Box 888, Taif, Kingdom of Saudi Arabia.
[2]Soil, Water and Environmental Research Institute, Agricultural Research Center, 9 Gamaa St., P.O. Box 12619, Giza, Egypt.
[3]Department of Agricultural Microbiology, Faculty of Agriculture, Ain Shams University, P.O. Box 68 Hadayek Shubra, 11241, Cairo, Egypt.

This work was designed to identify and taxonomically classify *Streptomyces* strains isolated from the rhizospheres of various plant species; banana, rose, pomegranate and grape plants, having antagonistic activity against some microbial (bacteria and fungi) tissue culture contaminants. *Streptomyces* strains with the most potent antagonistic traits were identified using sequencing of 16S rRNA gene amplicons generated by PCR of DNA and blasting the sequences into GenBank. The *Streptomyces* isolates had antagonism against five identified fungi (*Aspergillus fumigates, Aspergillus flavus, Aspergillus nidulans* (1), *Aspergillus nidulans* (2), *Cladosporium herbarum*) and against five out of the 11 Gram positive bacilli bacterial contaminants. The three selected *Streptomyces* isolates (14, 15 and 17) were identified as strains of *Streptomyces noboritoensis*, *Streptomyces albolongus* and *Streptomyces griseorubiginosus,* respectively. Isolate 14 for which 1443 nucleotides were obtained was compared with eight universal *Streptomyces* strains and was classified as a new species of gray *Streptomyces*, to be named *Streptomyces noboritoensis* SSMA2-KSA strain. This study recommends conducting further studies on the use of streptomycetes in the biological control in a large scale production.

Key words: Taxonomy, *Streptomyces*, microbial tissue culture contaminants, antagonistic activities, 16S rRNA.

INTRODUCTION

Plant tissue culture technology has allowed for the propagation of large number of plants from small pieces of stock plants in a relatively short period of time (Daniel, 1998). Microbial contamination is a constant problem, which often compromise development of all *in vitro* techniques. About 31 microorganisms from ten different plant cultivars growing in micro-propagation have been isolated, identified and characterized, with yeasts, *Corynebacterium* sp. and *Pseudomonas* sp. being predominant (Leggatt et al., 1994). Odutayo et al. (2004) reported associating the following bacteria *Pseudomonas syringae* pv. *phaseolicoli*, *Bacillus licheniformis*, *Bacillus subtilis*, *Corynebacterium* sp. and *Erwinia* sp. with the contamina-

*Corresponding author. E-mail: dratefsadik@gmail.com.

tion of *Hibiscus cannabinus* and *Telfaria occidentalis* in Nigeria. 19 microbial contaminants (including 11 bacteria and eight fungi) were found associated with the tissue culture plants and the laboratory environments (Odutayo et al., 2007).

Preventing or avoiding microbial contamination of plant tissue cultures is critical to successful micropropagation. Epiphytic alKl endophytic organisms can cause severe losses to micropropagated plants at each stage of growth (Cassells, 1991; Leifert et al., 1991). Bacterial contaminants are often difficult to detect because they remain mostly within the plant tissue (Debergh and Vanderschaeghe, 1988; De Fossard and De Fossard, 1988; Viss et al., 1991). Contaminated plants may have no visible symptoms, reduced multiplication and rooting rates or may die (Leifert et al., 1989, 1992). *Streptomyces* sp. is Gram-positive filamentous bacteria that produce and secrete a wide array of biologically active compounds including antibiotics, hydrolytic enzymes and enzyme inhibitors. These characteristics make streptomycetes attractive candidates for biological control agents against soil-borne plant pathogens (Samac and Kinkel, 2001; Hassan et al., 2011). Attempts have been made to develop *Streptomyces* species as fungal root disease control agents, since *Streptomyces* spp. are capable of producing a remarkably wide spectrum of antibiotics as secondary metabolites (Franklin et al., 1989). A rapid method for identifying filamentous soil-actinomycete genera was developed based on 16S rRNA gene (Moran et al., 1995; Cook and Meyers, 2003; Mohamed et al., 2012; Shori et al., 2012). Significantly, the genus *Streptomyces* could be differentiated from all other actinomycete genera.

This study aimed at isolating and identifying some streptomycetes from rhizosphere-soils of some plants used for propagation *via* tissue culture technique and evaluating their filtrates abilities to be used as biopesticides for *in vitro* control of the growth of plant-tissue culture contaminants (bacteria, fungi or yeasts).

MATERIALS AND METHODS

Soil samples collection

Rhizosphere soils of banana, rose, pomegranate and grape plants were collected from different private farms of Taif area. Samples were collected in clean plastic pages. At each rhizosphere, soil samples were randomly collected from five sites around the roots in clean plastic pags and thoroughly mixed together to form one representative sample.

Isolation, purification and maintenance of streptomycetes

Isolation (by plate technique) and purification (by streaking technique) of streptomycetes-like isolates (SLI) was carried out as described by Mohamed (1998) using starch nitrate agar medium (Waksman and Lechevalier, 1961).

Collection of some contaminated-tissue culture plant materials from some tissue culture laboratories

Blank (non-cultivated) as well as contaminated tissue-culture jars containing bacteria-like microorganisms or fungi or both were collected from Tissue Culture Laboratory, Department of Biology, Faculty of Science, Female Branch, Qarwah, Taif University, KSA (Figure 1). The jars were cultivated with different plant materials, that is, banana, rose, pomegranate and grape.

Isolation, purification and identification of the microbial contaminants

In the case of bacteria, isolation, purification and maintenance of bacterial tissue culture contaminants were carried out by plate technique (Mohamed, 1998) using nutrient agar medium (Jacobs and Gerstein, 1960). Inoculated plates were incubated at 37°C for 2 days. Fungal tissue culture-contaminants, were maintained on potato glucose agar medium (Waksman and Lechevalier, 1961), and sent to The Clinical Plant Center, Department of Plant Pathology, Faculty of Agriculture, Ain Shams University, Cairo, Egypt, for identification for genus and/or species.

Antagonistic activities of the purified streptomycetes-like isolates

Antimicrobial activities of purified streptomycetes-like isolates were tested against the 5 identified fungi and the 11 bacterial isolates isolated as tissue culture-contaminants as described by Mohamed et al. (2012). Bacterial strains were cultivated on nutrient agar medium (Jacobs and Gerstein, 1960) and fungal strains on potato agar medium (Waksman and Lechevalir, 1961). Sterile discs of filter-paper Whatman of about 10 mm diameter were made and 3 to 4 discs were saturated with 0.1 ml of a supernatant. Petri-dishes were kept in a refrigerator for one hour to permit diffusion of the supernatant. Inoculated plates were incubated at 30±2°C for 24 to 48 h. The inhibition zones (mm) were determined as described by British Pharmacopoeia (1968). Discs treated with 0.1 ml of un-inoculated starch nitrate broth was used as a control.

Identification of the *Streptomycete* isolates showing high levels of antagonistic activities against tissue culture contaminants

The proposed key of Bergey's Manual of Determinative Bacteriology (1974) as detailed by Mohamed (1998) was used for identification of all purified streptomycetes-like isolates up to genus.

PCR isolation and sequencing of the 16S rRNA gene of a *Streptomyces* isolate

The DNA of the selected *Streptomyces* isolate was extracted as described by El-Domyati and Mohamed (2004) and used as a template for PCR-isolation of 16S rRNA gene using two universal primers (518F: 5' CCA GCA GCC GCG GTA ATA CG3'; 800R: 5' TAC CAG GGT ATC TAA TCC3') (Jacob, 2012). The PCR product of 16S rRNA was firmly packed and was delivered to Lab-Technology® (Macrogen agent in Egypt) on crushed ice. Sequencing was performed using the ABI PRISM BigDyeTM Terminator Cycle Sequencing Kits, ABI PRISM 3730XL Analyzer

Figure 1. Collection of tissue culture contaminants from blank jars (A) and cultivated jars (B). The presence of different fungi approximately covered the medium.

(96 capillary type) sequencer (Applied Biosystems), MJ Research PTC-225 Peltier Thermal Cycler, DNA polymerase (FS enzyme) (Applied Biosystems). Following the protocol provided by the manufacturer, single-pass sequencing was performed on each template using the same primer pairs described in the PCR step. The fluorescent-labeled fragments were purified from the unincorporated terminators with an ethanol precipitation protocol. The samples were re-suspended in distilled water and subjected to electrophoresis in an ABI 3730xl sequencer. The DNA sequences of the PCR product of the 16S rRNA gene of *Streptomyces* sp. (isolate 14) were aligned with the related nucleotide sequences of the universal isolates collected from http://www.ncbi.nlm.nih.gov/, http://www.ncbi.nlm.nih.gov/, using the DNA Star Software Package – Lasergene (Expert Sequence Analysis Software, USA) and MegaAlign program.

RESULTS AND DISCUSSION

Isolation, purification and maintenance of streptomycetes

The actinomycetes especially the strains of *Streptomyces* are the most common antibiotic-producing microorganisms found in soil (Ara et al., 2012). Taddei et al. (2006) reported that the genus *Streptomyces* was represented in nature by the largest number of species and varieties among the family Streptomycetaceae. In this study, 23 SLI were isolated from the rhizosphere-soil samples and 18 SLI were purified. A number of 5, 5, 5, 4 and 4 were derived from banana, rose, pomegranate and grape, respectively. At the level of colour series, isolates 9, 7 and 2 SLI belonged to gray, white and red series. The widespread distribution of actinomycetes in the rhizosphere and non-rhizosphere soils was confirmed *via* several reports of Mohamed (1998), Duangmal et al. (2005), El-Sherbiny (2006), Ismail (2006), Al-Askar et al. (2011), Atta et al. (2011), Mohamed et al. (2012) and Shori et al. (2012).

Isolation, purification and identification of the microbial contaminants from some contaminated-tissue cultural plant materials

George and Sherrington (1984) reported that although aseptic conditions are typically employed for *in vitro* growth "sterile" of plant cells, tissue or organs as they are separated from the mother plant on artificial/synthetic medium, many plant cultures do not stay aseptic *in vitro* as they get contaminated. Contamination with bacteria is generally considered the most serious (Horsch and King, 1983). There are much fewer publications describing yeast and fungal contaminants and their effects on plantlets grown *in vitro* (Enjalric et al., 1988). In this study however, fungal contaminants were the most common on jars to be used for tissue culture. Bacterial contaminants associated with the tissue-cultured plant material were also observed. 11 bacterial contaminants were isolated, purified and identified as Gram positive bacteria. In the case of fungal contaminants, 5 fungal isolates were cultivated on potato dextrose agar (PDA) medium and after appearing, its growth were sent to a specific private laboratory for further purification and identification using specific media. Results show that the fungal isolates were identified as *A. fumigatus*, *A. flavus*, *C. herbarum* and two different strains of *A. nidulans* (1 & 2), (Figures 1, 2 and 3).

Antibiosis activities of the purified streptomycetes-like isolates against the 16 identified bacterial and fungal contaminants

Antibiosis activities of the 18 SLI were determined against the 11 bacterial contaminants. Data show that 2 SLI (10 and 11) were not active against the 11 tested bacterial isolates, as no zone of inhibition were found around the discs saturated with filtrates of these SLI.

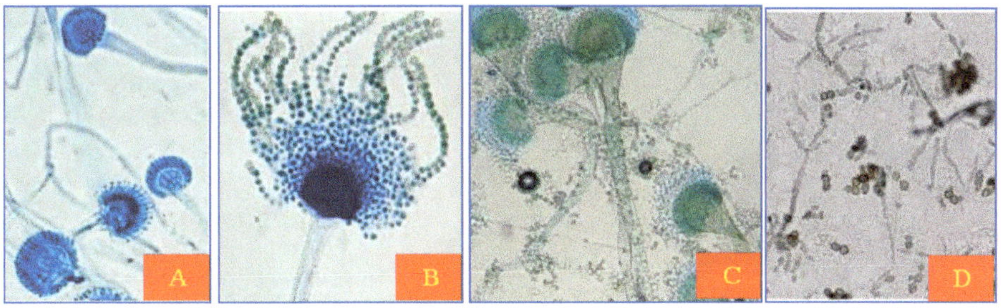

Figure 2. The four identified tissue culture contaminant fungi: *Aspergillus fumigates* (A), *Aspergillus flavus* (B), *Aspergillus nidulans* (1) and *Cladosporium herbarum* (D).

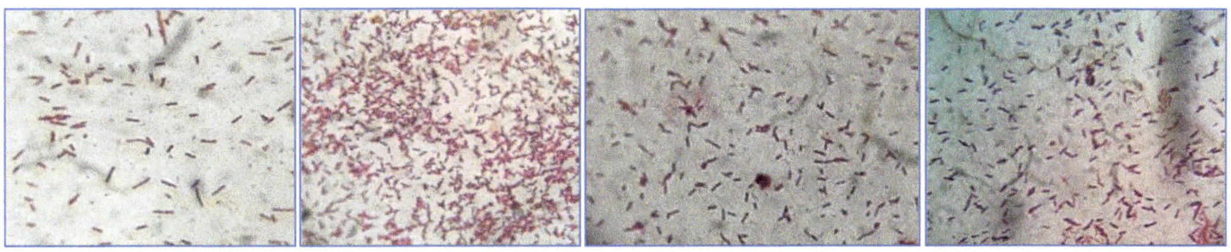

Figure 3. Four out of the eleven Gram positive tissue culture contaminant bacilli.

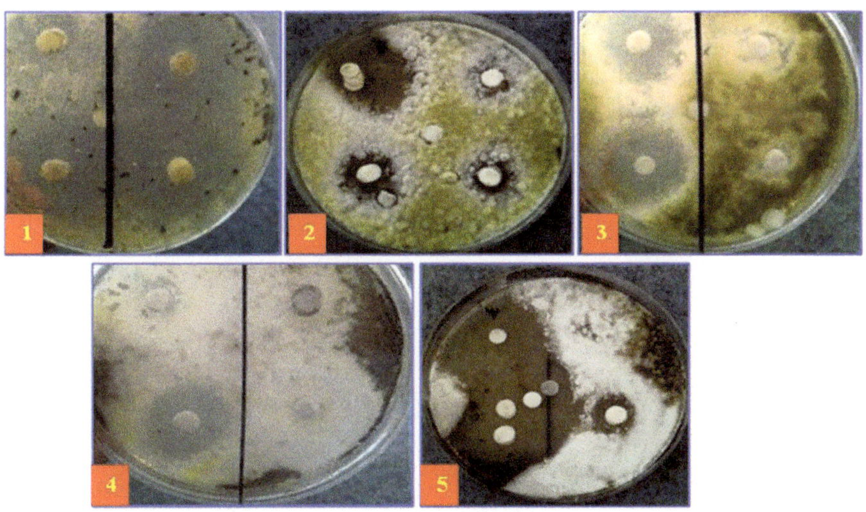

Figure 4. Antibiosis activity of some SLI against the five identified fungal isolates (1 to 5) identified fungal isolates (*A. fumigatus*, *A. flavus*, *A. nidulans* (1), *A. nidulans* (2), *C. herbarum*), 72 h post incubation.

While, 6 SLI were active against one bacterial isolate. A number of 8 SLI (2, 3, 4, 5, 7, 16, 19 and 21) showed antagonistic activities against two bacterial isolates. Two SLI (6 and 18) appeared with activities against 3 bacterial isolates. SLI 14, 15 and 17 were the highest isolates as they were active against 5 out of the bacterial isolates.

Regarding the antibiosis activities of the tested SLI against the five identified fungal isolates (*A. fumigatus*, *A. flavus*, *A. nidulans* (1), *A. nidulans* (2), *Cladosporium herbarum*), results show that thefive fungi were affected by 15, 4, 3, 7 and 10, respectively. Results also confirmed the high activities of the 3 selected SLI (14, 15 and 17), as they were able to inhibit the fungal growth of the 5 tested fungi. These results are in harmony with that of

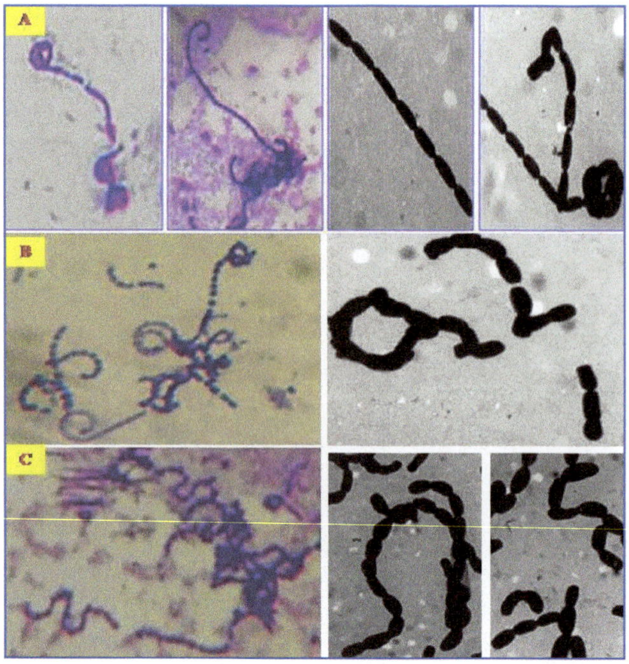

Figure 5. Microphotograph and electron micrograph of SLI No. 14 (A), 15 (B) and 17 (C) showing RF [RF: Rectus-Flexibilis (spores in straight (R) or flexuous (F) chains] and smooth spores surface.

Srividya et al. (2012), who identified a potent actinomycete isolate 9p with broad spectrum antifungal property OCA3; *Collectotrichum gleosporioides* OGC1, against four phytopathogens tested *Alternaria brassiceae Rhizoctonia solani* MTCC 4633 and *Phytophthora capsici* (Figure 4).

Identification of the *Streptomycete* isolates showing highly antagonistic activities

Streptomyces isolate 14

It was found that the *Streptomyces* isolate 14 belonged to the gray colour series, while the vegetative mycelium was pigmented with dark gray. This isolate had RF spore chain with smooth surface. Melanoid pigments were produced on tyrosine agar medium (Shinobu, 1958), peptone - yeast extract iron agar medium (Tresner and Danga, 1958) and tryptone-yeast extract broth medium (Pridham and Gottlieb, 1948). It also had very good growth on Cazpek's agar medium and actively utilized all added sugars as carbon sources for growth. This isolate showed antibacterial and antifungal activities. It was not sensitive to streptomycin antibiotic in the medium. This isolate was tolerant to NaCl up to concentration of 7%. According to the key proposed by Pridham and Tresner (1974), the experimental isolate 14 appeared to be related to *Streptomyces noboritoensis* although there was slight difference in use of L- Rhamnose and sucrose as sole sources of carbon. Therefore, isolate 14 could be considered a as strain of *S. noboritoensis* (Figure 5A).

Streptomyces isolate 15

Results clearly indicate that the *Streptomyces* isolate 15 belonged to the white colour series. Aerial spore chains belonged to section RF; the spores were characterized by smooth surface. Melanoid pigments were detected on the standard media used. This isolate was characterized by excellent growth on Cazpek's agar medium. The physiological characteristics showed that D-glucose, D-xylose, L-arabinose, L-rhamnose, D-mannitol, D-fructose, i-inositol and sucrose were used as carbon sources for growth.

In addition, this isolate showed antimicrobial activities and sensitivity to streptomycin (4 µg ml^{-1}) was observed. However, it was able to grow in the presence of 7% NaCl in the medium. Comparing the cultural, morphological and physiological characteristics of the *Streptomyces* spp. in Pridham and Tresner (1974) study with that of *Streptomyces* isolate 15, this isolate is very likely to be a strain of *Streptomyces albolongus* (Figure 5B).

Streptomyces isolate 17

Results of *Streptomyces* isolate 17 show that, this isolate has gray aerial mycelium (gray colour series), while the vegetative mycelium was pigmented with gray colour. It had straight and long spore chains (section RF) and the spores were characterized by smooth surface without any ornamentation. This isolate was also characterized by excellent growth on Cazpek's agar medium, actively utilized all added sugar, was tolerant to NaCl concentration up to 7%, sensitive to streptomycin (4 µg ml^{-1}) and antagonized some bacterial and fungal used. Considering the description keys proposed by Pridham and Tresner (1974), the tested isolate 17 was closely related *to Streptomyces griseorubiginosus* (Figure 5C).

PCR-isolation and sequencing the 16S rRNA gene of *S. noboritoensis* isolate 14

Identification of actinomycetes using microscopic techniques alone was not enough to ensure certainty. Biochemical methods would be the best method to identify actionmycetes to their species level. But this test consumes a lot of time and chemicals. With the advancement of technology in molecular study, primers had been developed by researchers to target specifically the 16S rRNA sequence of the actinomycetes. Identification of

```
GGACGAACGCTGGCGGCGTGCTTAACACATGCAAGTCGAACGGTGAAGCCCAGCTTGCTGGGTGGA
TCAGTGGCGAACGGGTGAGTAACACGTGAGCAACCTGCCCCCTACTTCGGGATAAGCCTTGGAAAC
GGGGTCTAATACCGGATATGAACGTCTGCCGCATGGTGGGTGTTGGAAAGTTTTTCGGTGGGGAT
GGGCTCGCGGCCTATCAGCTTGTTGGTGGGGTAATGGCCTACCAAGGCGTCGACGGGTAGCCGGCC
TGAGAGGGCGACCGGCCACACTGGGACTGAGACACGGCCCAGACTCCTACGGGAGGCAGCAGTGGG
GAATATTGCACAATGGGCGCAAGCCTGATGCAGCGACGCCGCGTGAGGGATGACGGCCTTCGGGTT
GTAAACCTCTTTCAGCAGGGAAGAAGCGTGAGTGACGGTACCTGCAGAAGAAGCGCCGGCTAACTA
CGTGCCAGCAGCCGCGGTAATACGTAGGGGCCGTTGTCCGGAATTATTGGGCGTAAAGAGCTCGTA
GGCGGTCTGTCGCGTCTGGTGTGAAATCCCATGGCTCAACTGTGGGCTTGCATCGGGTACGGGCAG
ACTGGAGTGCTGTAGGGGAGACTGGAATTCCTGGTGTAGCGGTGGAATGCGCAGATATCAGGAGGA
ACACCGATGGCGAAGGCAGGTCTCTGGGCAGTTACTGACGCTGAGGAGCGAAAAAGAGCGAACAGG
ATTAGATACCCTGGTAGTCCATGCCGTAAACGTTGGGCACTAGGTGTGGGGCCAGTTCCACTGGTT
CTGTGCCGTAGCTAACGCATTAAGTGCCCCGCCTGGGGAGTACGGCCGCAAGGCTAAAACTCAAAG
GAATTGACGGGGGCCCGCACAAGCGGCGGAGCATGCGGATTAATTCGATGCAACGCGAAGAACCTT
ACCAAGGCTTGACATGTACGAGAACGCCCCAGAGATGGGGTTCTCTTTGGACACTCGTACACAGGT
GGTGCATGGTTGTCGTCAGCTCGTGTCGTGAGATGTTGGGTTAAGTCCCGCAACGAGCGCAACCCT
TGTCTCATGTTGCCAGCGGGTTATGCCGGGGACTCATGGGAGACTGCCGGGGTCAACTCGGAGGAA
GGTGGGGATGACGTCAAATCATCATGCCCCTTATGTCTTGGGCTTCACGCATGCTACAATGGCCGG
TACAGAGGGCTGCGATACCGTGAGGTGGAGCGAATCCCAAAAAGCCGGTCTCAGTTCGGATCGGGG
TCTGCAACTCGACCCCGTGAAGTCGGAGTCGCTAGTAATCGCAGATCAGCAACGCTGCGGTGAATA
CGTTCCCGGGCCTTGTACACACCGCCCGTCAAGTCACGAAAGTTGGTAACACCCGAAGCTCACGGC
CCAACCCTTTGGGAGGGAGTGGTCGAAGGTGGGACTGGCGATTGGGACTAGTCGAAG
```

Figure 6. Final nucleotide sequence (1443 nts) of the PCR product of 16S rRNA gene amplified from the DNA of *S. noboritoensis*-isolate 14.

Table 1. Sequences producing significant alignments between *S. noboritoensis*-isolate 14 and the related overseas bacterial strains.

Accession	Description	Query coverage (%)	Max identity (%)
NR_041107.1	*Streptomyces noboritoensis*, strain NBRC 13065	99	89
AY999830.1	*S. noboritoensis*, strain KCTC 9060	97	88
AB184287.1	*S. noboritoensis*, strain NBRC 13065	99	89
NR_041144.1	*S. albolongus*, strain NBRC 13465	99	89
JN609385.1	*S. albolongus*, strain BC-32	97	89
AY999756.1	*S. albolongus*, strain JCM 4716	97	89
AB184425.1	*S. albolongus*, strain NBRC 13465	99	89
NR_042298.1	*S. griseorubiginosus*, strain :LMG 19941	99	88
AB706352.1	*S. griseorubiginosus*	99	89
AB184276.2	*S. griseorubiginosus*, strain: NBRC 13047	99	88

actinomycetes to genus level was made possible in a fast and accurate manner. This had been a great advancement in the area of identification as the ability to obtain the genus of the actinomycetes in just a few hours is now possible. In this study, actinomycetes from the farming soil of Agriculture Research Centre Semongok, Sarawak were isolated. The isolates were later tested for their bioactive compounds and selected isolates were later identified using primers targeting their 16S rRNA sequence (Jeffrey, 2008).

In this study, the nucleotide sequences of 16S rRNA gene were partially determined using the DNA template of the *Streptomyces* sp. (isolate 14). Results show the two sequences of both forward direction matched to each other. I A sequence of about 1443 nucleotides (Figure 6) was obtained and compared with four universal bacterial isolates as mentioned in Table 1 and results appendix (B). These bacteria are *S. noboritoensis*, strain NBRC 13065 (NR_041107.1); *S. noboritoensis*, strain KCTC 9060 (AY999830.1); *S. noboritoensis*, strain NBRC 13065 (AB184287.1); *S. albolongus*, strain NBRC 13465 (NR_041144.1); *S. albolongus*, strain BC-32 (JN609385.1); *S. albolongus*, strain JCM 4716 (AY999756.1); *S. albolongus*, strain NBRC 13465 (AB184425.1); *S. griseorubiginosus*, strain :LMG 19941 (NR_042298.1); *S. griseorubiginosus* (AB706352.1) and *S. griseorubiginosus*, strain: NBRC 13047 (AB184276.2).

Results in Table 1 show that the percent identities between the isolate of this study and that compared with the eight gray *Streptomyces* strains, collected from the GenBank, ranged between 88 and 89%. It is worth to mention that the *Streptomyces* strain of this study covered

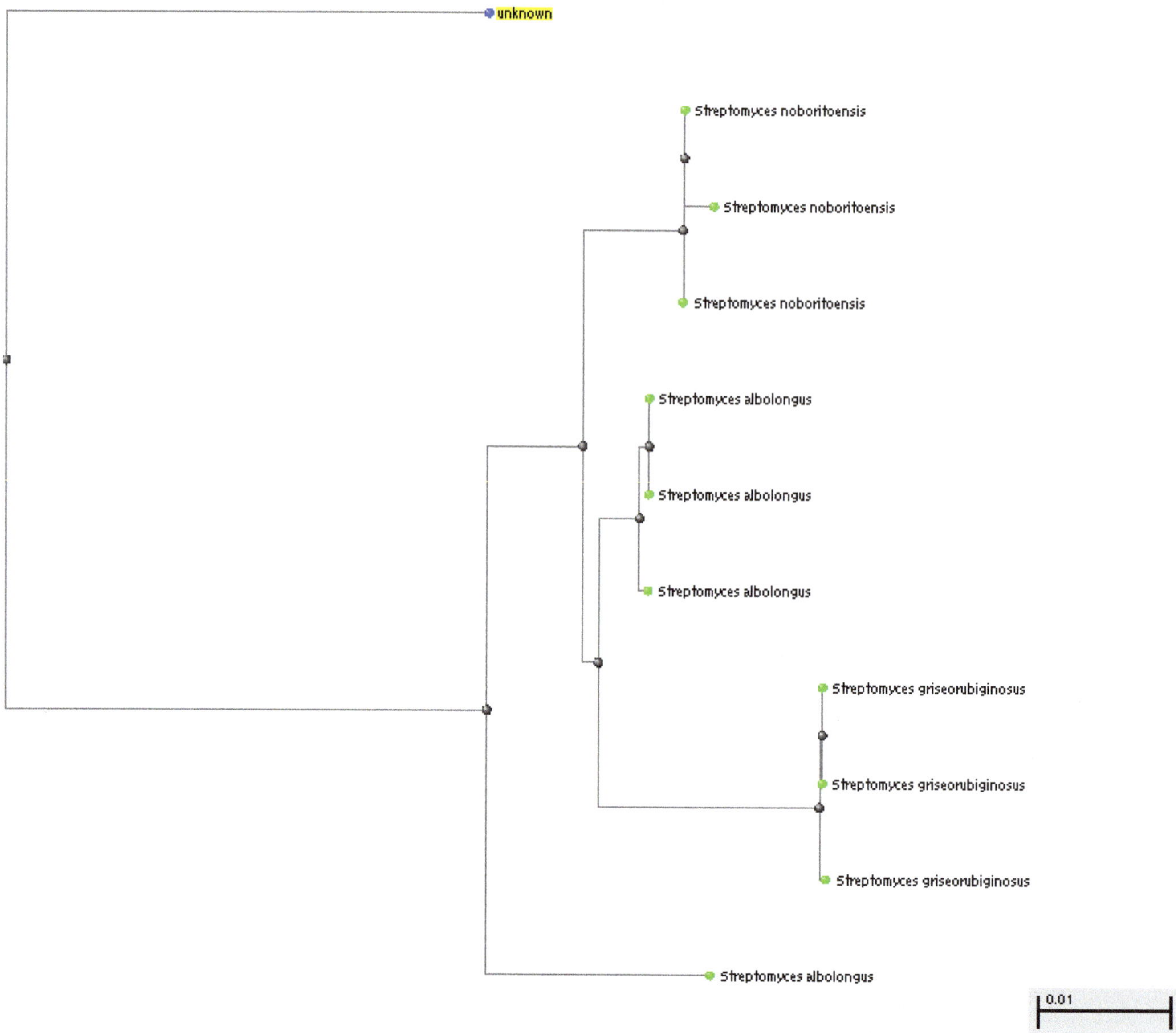

Figure 7. Phylogenetic tree of nucleotide sequence of the PCR product of 16S rRNA gene amplified from the DNA of *S. noboritoensis*-isolate 14 and the related universal bacteria strains.

from 97 to 99% of the sequences of the 8 overseas *Streptomyces* strains.

The phylogenetic tree (Figure 7) showed that the *S. noboritoensis* isolate 14 segregated into a separate cluster. Therefore, it could be classified as a new species of gray *Streptomyces*, and it was suggested to be named *S. noboritoensis* SSMA-KSA strain. These results are in agreement with that of other investigators, using the 16S rRNA gene to characterize Actinomycetes (Moran et al., 1995; Cook and Meyers, 2003; Song et al., 2004; Al-Askar et al., 2011; Jose et al., 2011; Mahasneh et al., 2011; Mohamed et al., 2012; Shori et al., 2012).

REFERENCES

Al-Askar AA, Abdul Khair WM, Rashad YM (2011). *In vitro* antifungal activity of *Streptomyces spororaveus* RDS28 against some phytopathogenic fungi. Afr. J. Agric. Res. 6(12): 2835-2842.

Ara I, Rizwana H, Al-Othman MR, Bakir MA (2012). Studies of actinomycetes for biological control of *Colletotrichum musae* pathogen during post harvest anthracnose of banana. Afr. J. Microbiol. Res. 6(17): 3879-3886.

Atta HM, Bahobail AS, El-Sehrawi MH (2011). Studies on isolation, classification and phylogenetic characterization of antifungal substance produced by *Streptomyces albidoflavus*-143. New York. Sci. J. 4(3):40-53.

Bergey's Manual of Determinative Bacteriology (1974). 8th ed. Buchanan RE. and Gibbons N.E. Williams and Wilkins Company,

Baltimore, USA.758, 764, 768 & 793.
Buchanan R.E. and Gibbons N.E. Williams and Wilkins Company, Baltimore, USA.758, 764, 768 & 793.
British Pharmacopoeia (1968). Biological assay of antibiotics. The Pharmaceutical Press, London, pp.1313-1317.
Cassells AC (1991). Problems in tissue culture: culture contamination. In: Micropropagation: Technology and Application. Kluwer Academic Publications, The Netherlands. Ed. By Deberh, R. Zimmerman. p. 31-44.
Cook AE, Meyers PR (2003). Rapid identification of filamentous actinomycetes to the genus level using genus-specific 16S rRNA gene restriction fragment patterns. International J. Syst. and Evol. Microbiol. 53: 1907-1915.
Daniel RL (1998). The many dimension of plant tissue culture research. Webmaster of Aggie Horticulture Publications, pp. 201-210.
De Fossard RA, De Fossard H (1988). Coping with microbial contaminants and other matters in a small commercial micropropagation laboratory. Acta Hort. 227: 346-351.
Debergh PC, Vanderschaeghe AM (1988). Some symptoms indicating the presence of bacterial contaminants in plant tissue culture. Acta Hort. 255: 77-81.
Duangmal K, Ward AC, Goodfellow M (2005). Selective isolation of members of the *Streptomyces violaceoruber* clade from soil. FEMS Microbiol. Lett. 245(2): 321-327.
El-Domyati FM, Mohamed SH (2004). Molecular genetic characterization of some Streptomyces isolates exhibiting different levels of resistance to the herbicide BASTA. Egypt. J. Gen. and Cytol. 33: 249-286.
El-Sherbiny GME (2006). Isolation, purification and characterization of an actinomycete isolate having the ability to produce an antimicrobial metabolite. Egypt. J. Microbiol. 41(1): 35-47.
Enjalric F, Carron MP, Lardet L (1988). Contamination of primary culture in tropical areas. The case of Hevea brasiensis. Acta Hort. 225: 25-65.
Franklin TJ, Snow GA, Barrett-Bee KJ, Nolan RD (1989). Antifungal, antiprotozoal and antiviral agents, p.137–161. In Biochemistry of rhizosphere with bacteria antimicrobial action, 4th ed. Chapman & Hall Ltd., New York.
George EF, Sherrington PD (1984). Plant Propagation by tissue culture, Eversley, Bassingstoke; Exegetices Ltd. pp. 88-93.
Hassan AA, El-Barawy AM and El Mokhtar NM (2011). Evaluation of biological compounds of *Streptomyces* species for control of some fungal diseases. J. Amer. Sci. 7(4): 752-760.
Horsch RB, King JM (1983). A convert contamination of cultured plant cells, elimination of a *Hyphomicrobium* spp. from cultures of *Datura innoxia* (Mill). Plant cell Tiss. Org. Cult. 2: 21-28.
Ismail A (2006). Numerical assessment of mycelium color in classification of some *Streptomyces* isolates. Int. J. Agric. and Biol. 8(6):872-875.
Jacob HJ (2012). Classification of halophilic heterotrophic bacteria thriving in the Jordanian dead sea littoral zone. J. Biol. Sci. 12: 246-252.
Jacobs MB and Gerstein MJ (1960). Hand book of Microbiology. D. Van Nostrand Co., Inc., New York, pp. 139-207.
Jeffrey LSH (2008). Isolation, characterization and identification of actinomycetes from agriculture soils at Semongok, Sarawak. Afr. J. Biotechnol. 7(20): 3697-3702.
Jose AP, Santhi SV, Jebakumar RD (2011). Phylogenetic-affiliation, antimicrobial potential and PKS gene sequence analysis of moderately halophilic *Streptomyces* sp. inhabiting an Indian saltpan. J. Bas. Microbiol. 51(4): 348-356.
Leggatt IV, Waites WM, Leifert C, Nicholas J (1994). Characterisation of microorganisms isolated from plants during micropropagation. In: Bacterial and Bacteria-like Contaminants of Plant Tissue Cultures. Ishs Acta Horticulturae, 225: *http://www.actahort.org/books/225/index.htm.*
Leifert C, Waites WM, Nicholas JR (1989). Bacterial contaminants of micropropagated plantcultures. J. Appl. Bacteriol. 67: 353 -361.
Leifert C, Ritchie JY, Waites W (1991). Contaminations of plant tissue and cell cultures. World J. Microbiol. Biotechol. 7: 452-469.
Leifert C, Camotta H, Waites WM (1992). Effect of combinations of antibiotics on micropropagated *Clematis, Delphinium, Hosta, Iris* and *Photinia*. Plant Cell Tiss. Org. Cult. 29: 153-160.
Mahasneh IA, Al-Thani RF, Abuannadi MA, Qandil MA (2011). A complete sequence of the 16S rRNA gene of a novel *Streptomyces coelicolor* (AB588124) (QU66c-2002) isolated from the soil of Qatar. Biotechnol. 10: 167-174.
Mohamed SH (1998). Role of actinomycetes in the biodegradation of some pesticides. Ph.D. Thesis, Agric. Microbiol., Dept. Agric. Microbiol., Faculty of Agric., Ain Shams University, Cairo, Egypt 151 pp.
Mohamed SH, Omran WM, Abdel-Salam SM, Al-Shehri AS, Sadik AS (2012). Isolation and identification of some halotolerant actinomycetes having antagonistic activities against some plant pathogens (*i.e.*, *Tobacco mosaic virus*, *Aspergillus* sp., *Fusarium* sp.) from soil of Taif Governorate, KSA. Pakist. J . Biotechnol. 9(1): 1-12.
Moran M, Rutherford L. Hodson R (1995). Evidence for indigenous *Streptomyces* populations in a marine environment determined with a 16S rRNA probe. Appl. Environ. Microbiol. 61(10): 3695-3700.
Odutayo OI, Oso RT, Akinyemi BO, Amusa NA (2004). Microbial contaminants of cultured *Hibiscus cannabinus* and *Telfaria occidentalis* tissues. Afr. J. Biotechnol. 3: 301-307.
Odutayo OI, Amusa NA, Okutade OO, Ogunsanwo YR (2007). Determination of the sources of microbial contaminants of cultured plant tissues. Plant Pathol. J. 6: 77-81.
Pridham TG, Gottlieb D (1948). The utilization of carbon compounds by some Actinomycetales as an aid for species determination. J. Bacteriol. 56: 107-114.
Pridham TG, Tresner HD (1974). Family Streptomycetaceae. In; Bergey's Manual of Determinative Bacteriology (1974), 758, 764, 768 & 793, (Buchanan R.E. and N.E. Gibbons, 8th Eds.), Williams and Wilkins Co., Baltmore, USA.
Samac DA, Kinkel LL (2001). Suppression of the root lesion nematodes (*Pratylenchus penetrans*) in alfalfa (*Medicago ativa*) by *Streptomyces* sp. Plant and Soil 235: 35-44.
Shinobu R (1958). Physiological and cultural study for the identification of soil actinomycetes species. Mem. Osaka Univ. B. Nat. Sci. 7: 1-76 (Modified).
Shori GBO, Mohamed SH, Abdel-Salam SM, Sadik AS (2012). Characterization of streptomycetes having antibiosis activities isolated from soil in western region of KSA. Pak. J. Biotechnol. 9(1): 21-38.
Song J, Chan LS, Kang JW, Beak HJ and Suh JW (2004). Phylogenetic analysis of *Streptomyces spp.* isolated from potato scab lesions in Korea on the basis of 16S rRNA gene and 16S-23S rDNA internally transcribed spacer sequence. IJSEM 54(1): 203-209.
Srividya S, Thapa A, Bhat DV, Golmei K and Dey N (2012). *Streptomyces* sp. 9p as effective biocontrol against chilli soilborne fungal phytopathogens. Europ. J. Exper. Biol. 2(1): 163-173.
Taddei A, Rodriguez MJ, Vilchez EM and Castelli C (2006). Isolation and identification of *Streptomyces* spp. from Venezuelan soils: Morphological and biochemical studies. 1. Microbiol. Res. 161(3): 222-231.
Tresner HD and Danga F (1958). Hydrogen sulfide production by *Streptomyces* as a criterion for species differentiation. J. Bacteriol. 76: 239-244.
Viss PR, Brooks EM, Driver JA (1991). A simplified method for the control of bacterial contamination in woody plant tissue culture. *In Vitro* Cell Dev. Biol. 27: 42.
Waksman SA and Lechevalier HA (1961). The actinomycetes Vol. II. Classification, identification and descriptions of genera and species. The Williams and Wilkins, Co. Baltimore, USA.

12

Microsatellite and mitochondrial DNA analysis of the genetic structure of Chinese horseshoe crab (*Tachypleus tridentatus*) in southeast China coast

Zhao Hong Weng[1,2,3], Yang Jie Xie[1], Zhi Qun Xiao[1], Zhi Yong Wang[1] and Jian Fang Gui[2]*

[1]Key Laboratory of Healthy Mariculture for the East China Sea, Ministry of Agriculture, Fisheries College, Jimei University, Xiamen 361021, China.
[2]State Key Laboratory of Freshwater Ecology and Biotechnology, Institute of Hydrobiology, Chinese Academy of Sciences, Wuhan 430072, China.
[3]Graduate School of Chinese Academy of Sciences, Beijing 100039, China.

Chinese horseshoe crab (*Tachypleus tridentatus*) is a Xiphosura animal of significant commercial importance and in danger of extinction in China. To better estimate how genetic structure can be used to obtain a conservation perspective of the species, genetic variation was examined in nine locations covering its distributing range in the coast of Chinese mainland using ten nuclear microsatellite DNA loci and mitochondrial DNA control region (CR) sequences. Moderate levels of genetic diversity were detected (expected heterozygosity from microsatellites was 0.635, haplotype diversity from mitochondrial DNA was 0.800) as a whole. Significant genetic differentiation was detected only by mitochondrial DNA (F_{ST} = 0.0693, P < 0.01), while microsatellite markers indicated nuclear genetic homogeneity of these locations. Probably, nuclear genetic homogeneity was caused by outbreeding among different groups due to artificial transporting. Very weak genetic differentiation indicates that reintroduction programs of the movement and mixing of horseshoe crab from different locations will result in minimal negative genetic effects. Upon four management units were inferred from the results of CR analysis, accordingly four or more nature reserves should be established to conserve this endangered animal along the Chinese coast. Haplotype network pattern indicated that *T. tridentatus* population in Chinese coast has undergone historic population expansion and very recent historic population recession. Mismatch distributions analysis also revealed existence of historic demographic expansion.

Key words: *Tachypleus tridentatus*, microsatellites, mitochondrial DNA, population structure, genetic diversity.

INTRODUCTION

The Chinese horseshoe crab (*Tachypleus tridentatus*) distributes mainly along the coast of the East and South China Sea, extending northward to the Pacific coast of Japan and southward to Southeast Asia (Sekiguchi, 1988; Yang et al., 2007). *T. tridentatus* was once abundant in the southeast coast of China, distributed from the estuary of Changjiang River to Beibu Bay (Sekiguchi, 1988; Liao and Li, 2001). At present, its resources have declined rapidly because of excess fishing (Hsieh and Chen, 2009) for seafood and wide use of amebocyte lysate in endotoxin tests (Swan, 2001) and habitat destruction for rapid economic development (Chiu and Morton, 1999; Liao and Li, 2001; Chen et al., 2004). It is a well accepted concept that the genetic diversity is a basic precondition for populations to react to environmental changes, and to keep their reproductive fitness, which will ensure the long term survival of species (Frankham et al., 2002; Allendorf

*Corresponding author. E-mail: jfgui@ihb.ac.cn.

and Luikart, 2007). Urgent attention is required to save this species from extinction in the near future, however, so far, little is known about its genetic diversity in Chinese coast in order to determine some effective measures to conserve this species.

Microsatellite has been considered as one of the efficient molecular markers that provide population genetic information, due to its abundance, randomly distributed in the genome, highly polymorphic, its codominant inheritance and ease of scoring (Guo and Gui, 2008; Kaya and Yildiz, 2008). Twenty-two microsatellite markers were developed from *Limulus polyphemus* (King et al., 2004), which revealed significant population difference among two Atlantic coast and one Gulf of Mexico site. Faurby et al. (2010) applied 12 microsatellite loci to analyze population dynamics in 1218 *L. polyphemus* sampled from 28 localities, covering its range along the west coast of the Atlantic. However, our result showed that no polymerase chain reaction (PCR) bands were achieved with these twenty-two pairs microsatellites primer and DNA extracted from *T. tridentatus* in PCR reaction in our previous study (not published), which indicated that those primer pairs from *L. polyphemus* are not suited for the application in *T. tridentatus*.

Currently, a few studies have been reported in the microsatellite determination in *T. tridentatus*, for instance, Nishida and Koike (2009) showed eight microsatellite markers with low diversity (number of allele from 2 to 4). Also, another eight polymorphic microsatellite markers were developed by Li et al. (2009), with rather low diversity. Although microsatellites are hypervariable nuclear markers, they are considered less appropriate for phylogeographical inference because of their tendency towards homoplasious mutations (Bryja et al., 2010; Hewitt, 2004).

In phylogeographical studies, analysis based on a single type of marker would lead to erroneous results and biased conclusions since different markers each have their own unique genealogy, different modes of inheritance, effective population sizes and sex-biased dispersal (Colbert et al., 2001; Hewitt, 2004). Fast-evolving markers, such as the mitochondrial control region has revealed much higher levels of genetic differentiation (Zink, 1997) and often help to discover clades with significant geographical information (Barrowclough et al., 2005). In conservation genetics, the use of nuclear gene information combining with the inferences from mtDNA sequence data is becoming necessary for a complete picture of the evolutionary forces shaping populations and species (Purvis et al., 2005; Shaw, 2002). More and more studies have applied the combination of microsatellites and mtDNA sequences, which is proved to be superior to that obtained through the use of separate sets of markers (Flanders et al., 2009; Bryja et al., 2010).

In this study, ten polymorphic microsatellite loci developed by our lab were used to analyze genetic diversity and structure of *T. tridentatus* from nine locations covering its distribution in Chinese coast. Furthermore, the control region sequence of mtDNA was used to analyze its genetic structure in seven locations. Combination analy-sis with these two sets of markers may give us more population genetic and phylogeographical information, upon which we can propose scientific conservation management strategy.

MATERIALS AND METHODS

Sampling and DNA extraction

The samples of Chinese horseshoe crabs (*T. tridentatus*) were collected from nine localities across its distribution range along the southeast coast of China during 2007 to 2008 (Figure 1, Table 1). The number of samples analyzed per locality varied due to sample availability and analysis success. The muscles from horseshoe crab were kept in 95% ethanol, then in 4°C refrigerator until DNA was extracted using standard proteinase K digestion/phenol-chloroform isolation procedure.

Genotyping in nine locations using ten microsatellite markers

The polymerase chain reaction (PCR) reactions were amplified in 10 μL reactions containing 1× PCR buffer, 2 mM $MgCl_2$, 0.2 mM forward primer, 0.2 mM reverse primer, 200 mM dNTPs, 0.1 U/μL Taq DNA polymerase (Invitrogen), and 50 ng of genomic DNA template. The used primers in this study are listed in Table 1. PCR was performed in a BioRad PTC0200 thermocycler (BioRad Lab, Inc., USA) with the following program: 94°C for 2 min, followed by 29 cycles of 94°C for 30 s, the primer specific annealing temperature (Table 1) for 20 s and 72°C for 30 s, with a final extension period of 7 min at 72°C, and maintained at 4°C until further use. The PCR products were separated on a 6% denaturing polyacrylamide gel and visualized by silver-staining. Product sizes were estimated by comparison with a 10 bp DNA ladder (Promega). A total of 371 individuals from nine locations (Figure 1, Table 1) were genotyped by ten polymorphic microsatlellite markers (CHR11, CHR45, CHR49, CHR51, CHR53, CHR57, CHR63, CHR77, CHR87, CHR91), which have been isolated using a fast isolation by Amplified fragment length polymorphisms (AFLP) of sequences containing repeats (FIASCO) protocol (Zane et al., 2002). They were also evaluated for polymorphism and heterozygosity in 39 individuals of Chinese horseshoe crab (Table 2).

Mitochondrial DNA amplifying and sequencing

Primers for PCR were designed using Primer 5.0 software according to mtDNA control region of *L. polphymus*. The primers used were, H-49: 5'-TTCTAGTATAATACATGGAA-3' and H-50: 5'-TTAGTGTAAAGGCACATT-3'. The final volume of 60 μL included 150 ng of template DNA, 200 μM dNTPs, 0.3 μM of each primer, 1.5 unit of Taq polymerase (Takara), 10× PCR buffer with 1.5 mM $MgCl_2$. PCR was performed in a BioRad PTC0200 thermocycler (MJ Research, Waltham, MA, USA) with the following program: 94°C for 5 min, followed by 35 cycles of 45 s at 94°C, 50 s at 40°C, and 1 min at 72°C; with a final extension period of 7 min at 72°C, and maintained at 4°C until further use. The results of PCR were checked in routine 2% agarose gel electrophoresis. The PCR products were purified using QIAGEN PCR Purification kit (50) (Biotools, Madrid, Spain). Samples were sequenced in both directions, using an ABI 310 automated sequencer (Applied Biosystems) of Invitrogen Biotechnology Company, Limited. Total of 116 individuals from seven locations (Table 1) were analyzed.

Data analysis

The frequency of Null alleles was calculated by MICRO-CHECKER

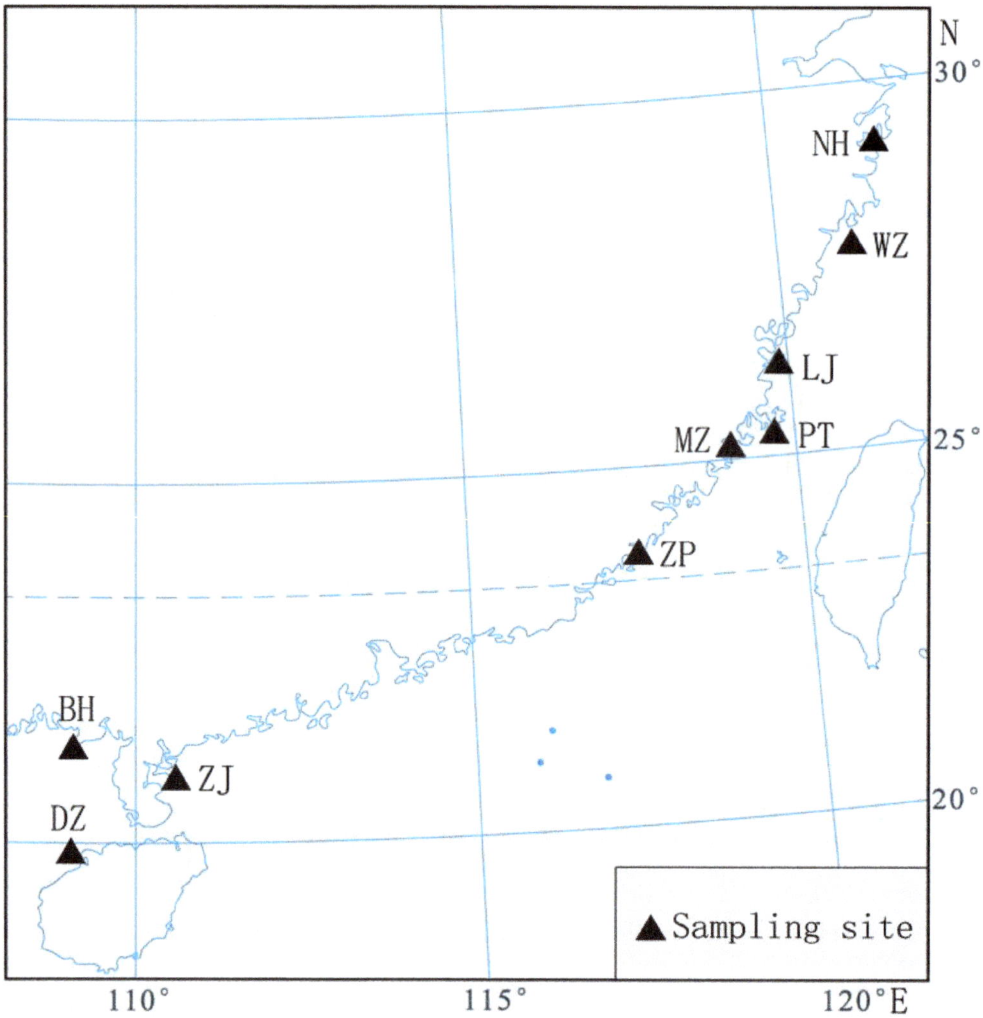

Figure 1. The map of sampling sites of *T. tridentatus* from Southeast coast of China. NH, Ninghai; WZ, Wenzhou; LJ, Lianjiang; PT, Pingtan; MZ, Meizhou Island; ZP, Zhangpu; ZJ, Zhanjiang; BH, Beihai; DZ, Danzhou.

2.2.3 (Oosterhout et al., 2004) using Brookfield algorithms (Brookfield, 1996). Microsatellite genetic diversity was quantified as number of alleles (N_A), average number of alleles per locus (A), effective number of alleles per locus (Ne) (estimates the reciprocal of homozygosity) (Hartl and Clark, 1989), F_{IS} (estimates for each allele) (Weir and Cockerham, 1984), observed heterozygosity (H_O), expected heterozygosity (H_E), which were estimated with GENEPOP version 4.0.1.0 (Rousset, 2008). The polymorphic information content (PIC) value was calculated using CERVUS 3.0 software package (Marshall et al., 1998; Kalinowski et al., 2007). PIC is a measure of informativeness related to expected heterozygosity and is likewise calculated from allele frequencies (Botstein et al., 1980; Hearne et al., 1992). Tests for Hardy-Weinberg equilibrium (HWE) and linkage disequilibrium (LD) were performed using GENEPOP. Genetic differences between locations were evaluated by calculating pairwise F_{ST} values and testing their significance by bootstrapping analysis (10,000 replicates) using Arlequin 3.0 (Excoffier et al., 2005). This program was also used to partition variation within and between locations using the AMOVA (Analysis of Molecular Variance) procedure.

The control region sequences were edited and aligned with MEGA 4.0 (Tamura et al., 2007). The nucleotide composition, numbers of variable sites and population diversity parameters were calculated using DNASP 4.10 (Rozas et al., 2003). A mantel test was performed with 10,000 permutations in Arlequin to assess the statistical significance in correlation between genetic and the geographical distances. Correlation between microsatellites and CR genetic diversity was calculated by SPSS 12.0. Phylogenetic relationships between the identified haplotypes were determined by calculating an unrooted haplotype median network using NETWORK 4.5.1.0 (Bandelt et al., 1999). Demographic history was examined using mismatch distribution (Slatkin and Hudson, 1991), and neutral test was conducted by Tajima`s *D*-test and Fu`s *Fs*-test based on mitochondrial DNA variation by Arliquin3.0.

RESULTS

Genetic diversity of *T. tridentatus* from microsatellite and mtDNA analysis

Most of these ten microsatellite loci were high polymorphic

Table 1. Sampling sites of *T. tridentatus* across its range of the east and south coast of China.

Locality	Province	Coordinates	Number of individual	
			Microsatellite analyses	Mitochondrial analyses
Ninghai (NH)	Zhejiang	29°9'29"N, 121°42'54"E	12	20
Wenzhou (WZ)	Zhejiang	27°57'04"N, 121°07'00"E	25	0
Lianjiang (LJ)	Fujian	26°13'56"N, 119°42'19"E	36	19
Pingtan (PT)	Fujian	25°25'33"N, 119°46'07"E	50	0
Meizhou Island (MZ)	Fujian	25°03'40"N, 119°04'39"E	50	16
Zhangpu (ZP)	Fujian	23°48'12"N, 117°33'53"E	50	15
Zhanjiang ((ZJ)	Guangdong	20°55'17"N, 110°33'32"E	50	13
Beihai (BH)	Guangxi	21°21'27"N, 109°05'44"E	50	16
Danzhou (DZ)	Hainan	19°51'33"N, 109°07'28"E	48	17

(PIC > 0.5) and only two loci (CHR53, CHR91) were deviated from Hardy-Weinberg equilibrium, no significant linkage disequilibrium was detected among them (Table 2), thus they were suitable for genetic analysis. Three hundred and seventy-one (371) horseshoe crab individuals from nine locations of *T. tridentatus* covering its range distributed in Chinese coast were genotyped by these polymorphic microsatellite loc. The results showed that nuclear genetic diversity is moderate, with a mean number of alleles over loci across nine locations ranging from 4.714 to 8.571 (7.018 averagely), and the mean effective numbers of alleles ranged from 2.915 to 3.837 (3.697 averagely) (Table 3). The observed heterozygosity (H_O) over all locations was 0.620 and ranged from 0.599 (ZP) to 0.701 (NH), while the H_E was 0.635 and varied from 0.598 (PT) to 0.655 (LJ).

In total, 18 polymorphic sites (accounting for 5.0% of the total number of sites), including nine parsimoniously informative sites, were detected based on 368 bp of full length mitochondrial control region (CR) from 116 individuals (Table 4), which were sampled from seven locations. The polymorphic sites were all transitions, without transversion or insertion/deletion. The average base composition of the control region sequence in *T. tridentatus* was in a strong adenine-thymine bias (43.3% thymine, 40.9% adenine, 6.7% cytosine and 9.1% guanine). The AT content (84.2%) was much higher than GC one (15.8%). Twenty-seven haplotypes were defined (H1 ~ H27) (Table 4) (GeneBank Accession Numbers JX437089-JX437115). The nucleotide diversity (*Pi*) over these locations was 0.00549 and ranged from 0.00207 to 0.00667, and haplotype diversity (*h*) was 0.80 and ranged from 0.524 to 0.889, which indicated moderate high levels of genetic diversity of this species. Nucleotide and haplotype diversities for different locations are shown in Table 5.

Four sample-specific haplotypes were in Ninghai locality, while three in Lianjiang, one in Meizhou, one in Zhanjiang, three in Beihai, but none in Zhangpu and Danzhou. H1 was the dominant haplotype in 49 individuals (43%), existing in all locations. The constructed phylogenetic relationships among these 27 haplotypes were constructed, yielding a complex network (Figure 2). Haplotypes such as H1, H20 and H2 were in the center of star-like networks, indicating that they were ancestral types. The star-like network indicated a likely existence of historic population expansion of horseshoe crab population. As shown in Figure 2, H20 was an ancestral type, but there were only two individuals in it, which indicated very recent historic population recession.

Comparison between microsatellites and CR genetic diversity

There was a significant correlation (r = 0.686, $P < 0.05$) between the values of *He* evaluated from microsatellite markers and *Pi* from CR, which stand for nuclear and mitochondrial genetic diversities, respectively (Figure 3). This result suggested that microsatellite data could be used to predict mitochondrial genetic diversity, and vice versa.

Analysis of molecular variance (AMOVA) analysis

The AMOVA analysis by microsatellite data partitioned 99.20% of the total genetic variation within location, and the value of the part among locations was only 0.80 with F_{ST} not significant ($P > 0$), which indicated there was not genetic differentiation among nine locations of *T. tridentatus*. However, the fixations index at the level of inter-locations was significant ($P < 0.01$) when analyzed with mtDNA sequences data. As shown in Table 6, significant genetic variance was discovered among locations, although most molecular variances were observed to occur within location (94.62%).

Pairwise F_{ST} analysis

Pairwise F_{ST} analysis was calculated for each pair of locations over ten microsatellite loci, with values ranging from -0.04084 to 0.02615, and no statistically significant difference existed among these locations ($P > 0$) (Table 7).

Table 2. Characterization of the ten polymorphic microsatellite loci used in this study.

Locus	Primer sequences(5'→3')	Repeat motif	Size range (bp)	T_a (°C)	N_A	PIC	H_O	H_E	Null allele frequence	F_{IS}	GenBank accession number
CHR11	F: CTCGTATCAAGACAAACAATCA R: AGAAAAGTCTACTGAAAGCACA	$(GT)_8$	146-168	46	9	0.732	0.744	0.768	0.008	0.026	JX443509
CHR45	F: CCACAAGCACCCATTGAAACAC R: ATCGGAACTGTGAACAACTACA	$(TG)_{15}$	195-205	50	7	0.652	0.641	0.702	0.030	0. 087	JX443503
CHR49	F: ATAACATGGCGGATTATTACAT R: GTTGCTGTTTAGTCCTGTTCAT	$(GT)_9(GA)_{10}$	145-197	48	22	0.932	0.974	0.948	0.	-0.030	JX443504
CHR51	F: CAAGCCGGGTCCCATCTTTGAT R:AGCGCTCAGTTAGACTTACCATTTAC	$(TG)_{15}$	160-184	51	11	0.840	0.872	0.868	0	-0.005	JX443506
CHR53	F: GTAGAATTATCTGTTATACACGCA R: TTACCATAACCTGGAGACAATA	$(CA)_{10}(AT)_9$	154-180	47	5	0.402	0.359	0.450	0.059#	0.204*	JX443505
CHR57	F: AAACATCAGCCTTATCTAACGG R: CGACAAGAAAGTAAAAGCAAAA	$(TG)_{14}$	74-90	48	8	0.725	0.769	0.770	0	0.001	JX443507
CHR63	F: TTGAAAGCCTATTCCTTTACGT R: AAGTGCAATGAAGAAGCAGTGT	$(TA)_5(AC)_{17}$	64-110	47	8	0.653	0.769	0.699	0	-0.101	JX443508
CHR77	F:CGAGATAACATAAAAGTATACAT R:TGGGATTCTGTTGTATTGTA	$(AT)_5(TG)_{13}$	118-130	45	7	0.703	0.821	0.749	0	-0.097	JX443510
CHR87	F: TCCAGGACAAACGTTAAAGAAT R: ACAGGTGATCTTTGAGTGTTGC185	$(TG)_{12}$	173-189	48	8	0.747	0.744	0.790	0.019	0.055	JX443511
CHR91	F:AGTCCAGAGGCCAGTTTCATAT R: TGGTTTCACTTTCCCAACAATC	$(GT)_{15}$	162-192	51	8	0.274	0.180	0.283	0.078#	0.370*	JX443512
Average					9.2	0.678	0.709	0.714			

T_a, Annealing temperature; N_A, number of alleles; PIC, polymorphic information content; H_O, observed heterozygosity; H_E, expected heterozygosity; F_{IS}, fix index; *indicates significantly deviating from Hardy–Weinberg equilibrium ($P < 0.05$); # indicates null allele existed (greated than 0.05).

As a result, F_{ST} indicated high level of gene flow between these locations, thus weak genetic structure of *T. tridentatus* along Chinese coast was revealed by the analysis with microsatellite markers. However, analysis according to control region of mitochondrion DNA showed significant difference

Table 3. Polymorphic characteristics of nine locations of *T. tridentatus* from ten microsatellite loci analysis.

Location	A	*Ne*	H_O	H_E
NH	4.714±2.215	2.915±1.151	0.701±0.234	0.623±0.212
WZ	6.286±2.828	3.401±1.626	0.660±0.264	0.640±0.222
LJ	6.714±2.812	3.517±1.722	0.647±0.266	0.655±0.201
PT	7.857±2.795	3.388±1.952	0.605±0.267	0.598±0.246
MZ	7. 714±3.039	3.399±1.963	0.609±0.247	0.619±0.214
ZP	7.286±2.690	3.565±2.126	0. 599±0.261	0.613±0.242
ZJ	7.286±2.628	3.511±1.938	0.609±0.251	0.619±0.253
DZ	7.429±2.225	3.633±1.841	0.637±0.236	0.647±0.223
BH	8.571±2.230	3.837±1.764	0.614±0.249	0.646±0.236
Species level	7.018±2.107	3.697±2.122	0.620±0.241	0.635±0.229

A, average number of alleles per locus; N_e, effective number of alleles per locus; H_O, observed heterozygosity; H_E, expected heterozygosity.

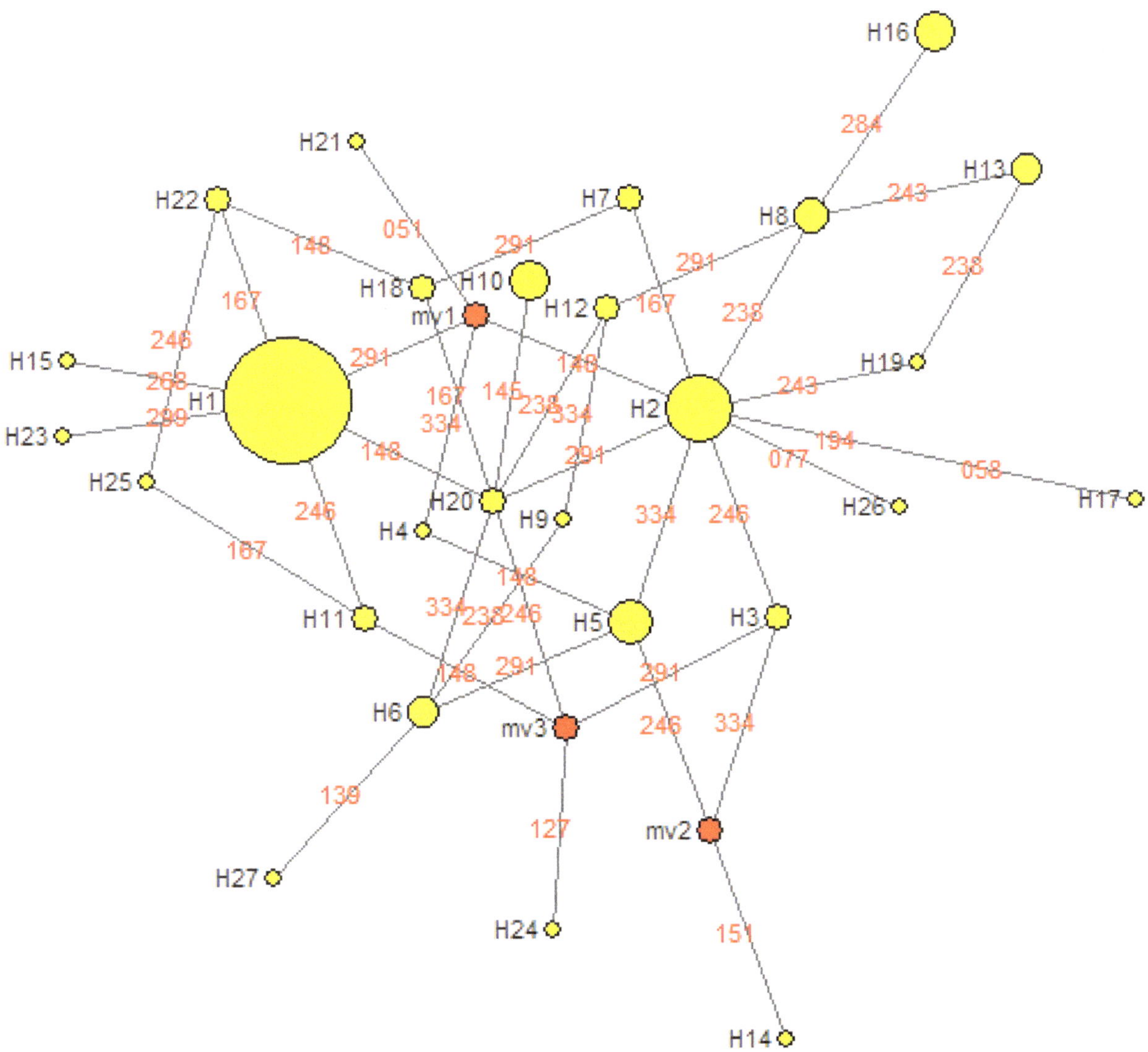

Figure 2. Median-Joining network showing phylogenetic relationships between 27 mtDNA control region haplotypes found in *T.tridentatus*. Each line represents a mutational step, numbers cross the lines represent the sites of nucleotide substitutions, circle areas depict proportions of haplotypes, yellow circles represent existing hyplotypes, red circles represent hypothesized sequences which were required to connect existing sequences within the network.

Table 4. Twenty-seven haplotypes, their mutation sites and their distribution in seven locations.

Haplotype	Variable nucleotide sites																		Distribution of haplotypes							
	51	58	77	127	139	145	148	151	167	194	238	243	246	268	284	291	299	334	NH	LJ	MZ	ZP	ZJ	BH	HN	total
H1	T	T	T	A	A	A	T	A	C	T	A	A	C	T	T	T	A	A	5	6	4	9	9	6	10	49
H2							C									C			2	3	7	1			2	15
H3							C						T			C			1		1					2
H4																C		G	1							1
H5							C									C		G	4	1	1					6
H6							C											G	3							3
H7							C		T							C								2		2
H8							C				G					C					2			1	1	4
H9							C				G							G						1		1
H10						G	C															3			2	5
H11													T							1		1				2
H12							C				G													2		2
H13							C				G	G				C				2					1	3
H14							C	G					T			C		G		1						1
H15														C						1						1
H16							C				G				C	C				1		1	1	1	1	5
H17		C					C			C						C								1		1
H18							C		T														1			1
H19							C					G				C								1		1
H20							C												1	1						2
H21	C															C				1						1
H22									T											1			1			2
H23																	G				1					1
H24				G			C						T						1							1
H25									T				T						1							1
H26			C				C									C			1							1
H27					G		C											G					1	1		2
Total																			20	19	16	15	13	16	17	116

among seven locations. NH population was significantly different from ZJ, two Beibu Bay locations (BH, DZ) and ZP (collected from semienclosed bay of Dongshan Bay) ($P < 0.01$). The significant difference also existed between BH and ZP, MZ and ZP, ZJ and BH (Table 7).

Mantel test

To test the degree of correlation between genetic variation and geographical distances, the Mantel test was used. The results indicate that genetic differentiation for *T. tridentatus* along Chinese mainland coast was not correlated with total geographical distance by the analysis with microsatellite markers ($r^2 = 0.026$, $P > 0.05$). However, converse result revealed by mtDNA sequences ($r^2 = 0.346$, $P < 0.05$), which indicated that the geographical distance is an important fact to restrict

Table 5. Genetic diversity parameters of mitochondrial control region (CR) among seven locations of *T. tridentatus*.

Location	Number of sample	Haplotypes observed	Haplotype diversity (*h*)	polymorphic site (*S*)	Mean number of pairwise difference (*k*)	Nucleotide diversity (*Pi*)
NH	20	10	0.895	7	2.074	0.00564
LJ	19	11	0.889	11	2.456	0.00667
MZ	16	6	0.767	6	1.525	0.00414
ZP	15	5	0.629	6	1.467	0.00399
ZJ	13	5	0.538	7	1.436	0.00390
BH	16	9	0.858	9	2.408	0.00654
HN	17	6	0.654	6	1.721	0.00468
Total	116	27	0.800	18	2.020	0.00549

Table 6. AMOVA analysis of genetic variation of *T. tridentatus*.

Parameter	Microsatellite analysis		Control region analysis	
Source of variation	Percentage of variation	F_{ST}	Percentage of variation	F_{ST}
Among locations	0.80	0.00803^{a}	5.38	0.0538^{b}
Within locations	99.20		94.62	

The superscript 'a' indicated $P > 0.01$ and 'b' for $P < 0.01$.

gene flow of *T. tridentatus*.

Demographic analysis

Fu`s *Fs* and Tajima`s *D* were used to test for selective neutrality and population equilibrium. Significant values of *Fs* and *D* can be due to factors other than selective effects, like population expansion or bottleneck (Fu, 1997; Tajima et al., 2007). The values of Tajima`s *D* were negative except for that of NH locality, but they were not significant ($P > 0.05$) for all locations. Fu (1997) has noticed that the *Fs* statistic was very sensitive to population demographic expansion which generally led to large negative *Fs* values. The values of *Fs* were also negative, furthermore they were significantly negative ($P < 0.02$) for three locations (NH, LJ and BH) and the whole data set, which indicated historic population expansion of horseshoe crab in the coast of China. Mismatch distributions analysis revealed that there was no significant difference between the observed and expected distribution under the expansion model, not only for each locality but also for the whole data set ($\tau = 2.625$; $\theta_0 = 0.005$; $\theta_1 = 9.023$; SSD = 0.025; $P = 0.074$) (Table 8, Figure 4). This indicated that the whole population of *T. tridentatus* in Chinese coast has undergone a historic demographic expansion, which was consistent with the result of star-like network of haplotypes and *Fs* test.

DISCUSSION

Genetic diversity of *T. tridentatus* distributed in Chinese coast

In general, genetic variation of nine locations in Chinese coast is still moderate according to allele numbers (7.018 averagely), H_E values (0.635 averagely) from microsatellite loci and *h* values (0.80 averagely) from mtDNA control regions in this study, although the horseshoe crabs are so rare in most of Chinese coast now. Compared with other population of *T. tridentatus*, it was higher than that of the population from Japan (H_E values 0.53 averagely) (Nishida and Koike, 2010), also higher than that of the population from Taiwan Strait (*h* values 0.626, averagely). The Chinese horseshoe crab (*T. tridentatus*) used to distribute mainly along the offshore of the East and South Sea of China, extending northward to estuary of Changjiang River and southward to Beibu Bay. The samples in this study cover the historical distribution region in China. The moderate genetic diversity of *T. tridentatus* may be due to historical abundance

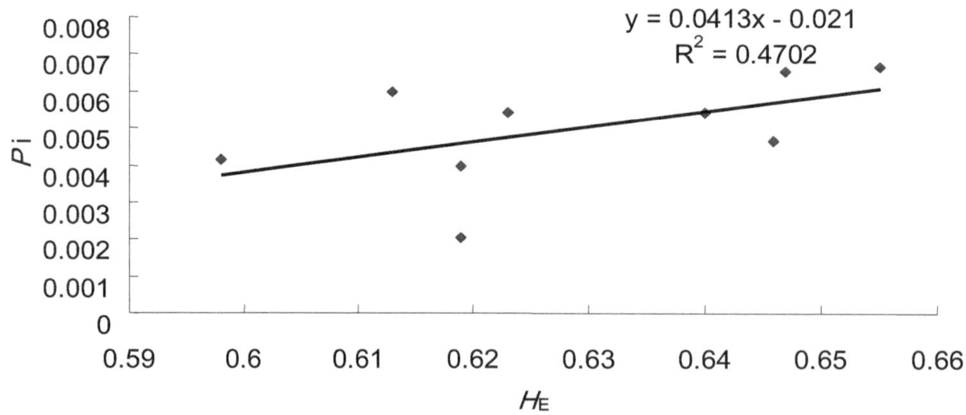

Figure 3. Genetic diversity comparison between microsatellites and mitochondrial loci (r = 0.686, *P* = 0.044). Each location is characterized by its expected heterozygosity (*He*, microsatellites) and nucleotide diversity (*Pi*, mtDNA).

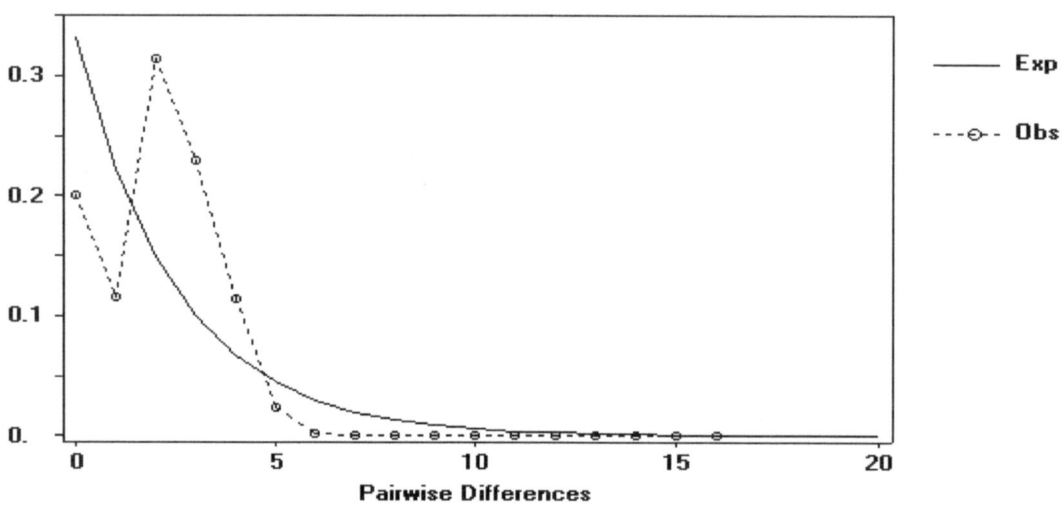

Figure 4. Mismatch distributions of CR haplotyes. Observed frequency distribution (dotted line) for the number of pairwise nucleotide differences among the total individuals of *T. tridentatus*. Solid line shows the expected distribution given a population expansion.

in resources and high in genetic diversity along China's southeast coast.

According to our survey, Chinese horseshoe crabs were still quite abundant in Chinese coast even 30 years ago. During summer, there were once a large number of horseshoe crabs spawning in many beaches several decades ago. But in recent years, the number of Chinese horseshoe crab in China drastically declined, almost disappeared in most sea water (weng et al., 2012). Accordingly, the reduction of the genetic diversity of Chinese horseshoe crab will be inevitable. It is surely necessary to take urgent and effective measures to protect the horseshoe crab from extinction in Chinese sea.

Genetic structure in *T. tridentatus*

The result of population differentiation tests at microsatellite loci indicated no genetic differentiation among different locations in Chinese coast, and none evolution significant unit was found. However, mtDNA markers revealed distinct subpopulation among these locations. Therefore, it is necessary to combine the analysis of microsatellites and mtDNA sequence in order to identify the population genetic and phylogeographical information. Actually, lack of congruence between nuclear and mitochondrial-based variation has also been previously reported, for example in terrestrial mammal the African elephant (Nyakaana and Arctander, 1999), cartilaginous fish, the blacktip shark (Keeney et al., 2005), teleost, the European bitterling (Bryja et al., 2010). This discrepancy between mtDNA and microsatellite data in these studies was thought to be due to male-biased gene flow which would lead to homogenization of nuclear alleles while the maternally inherited such as mitochondrial markers would remain restricted to native

Table 7. Genetic difference (F_{ST}) for mtDNA control region (above diagonal) among seven locations and for microsatellites (below diagonal) among nine locations.

Parameter	NH	WZ	LJ	PT	MZ	ZP	ZJ	DZ	BH
NH			0.02446		0.05343	**0.14898**	**0.1644**	**0.10272**	**0.06519**
WZ	0.0148								
LJ	-0.0034	0.0085			0.00391	0.04625	0.06327	-0.01534	-0.01436
PT	0.0262	-0.0154	0.0347						
MZ	-0.0037	0.0088	-0.0021	-0.0083		**0.20734**	**0.26066**	0.09328	0.0355
ZP	0.0049	-0.0031	-0.0106	-0.0408	0.0039		-0.00723	-0.02673	**0.08133**
ZJ	0.0152	0.0037	0.0139	-0.0312	0.0172	0.0039		0.02111	**0.08399**
DZ	0.0117	0.0065	0.0037	-0.0208	0.0103	0.0039	0.0076		0.00325
BH	0.0038	-0.0035	-0.0052	-0.0393	0.0009	0.0031	0.0063	-0.0089	

F_{ST} values are calculated with the method of Tajima and Nei (1984). Statistically significant values are shown in bold text ($P < 0.05$).

Table 8. Historical demography analysis based on control region variation in seven locations of *T. tridentatus*.

Location	NH	LJ	MZ	ZP	ZJ	DZ	BH	Total
Tajima`s *D*	0.168	-0.789	-0.536	-0.718	-1.378	-0.102	-0.412	-1.125
P_D	0.608	0.257	0.323	0.278	0.060	0.497	0.384	0.190
Fs	-4.421	-5.204	-1.432	-0.571	-0.863	-0.981	-3.394	-19.849
P_{Fs}	**0.001**	**0.002**	0.134	0.323	0.202	0.245	**0.017**	**0.003**

P_D, P_{Fs}, P_r, refer to *P*-value of Tajima`s *D*, *Fs*, Raggedness index, respectively. Statistically significant values are shown in bold text ($P < 0.05$).

localities (Nyakaana and Arctander, 1999). However, for horseshoe crab, the female is larger than male generally, and the male does not seem to be more movable or migrate farther distance than the female. Therefore, it needs more research to find out the reason for the inconformity in horseshoe crab.

According to the result revealed by CR, the NH location was significantly different from those sampling from South sea and Beibu Bay such as ZP, ZJ, BH and DZ. This can be interpreted by the fact that the geographical distance played an important role in genetic differentiation. Interestingly, ZP population was significantly different from MZ, although their distance is only about 200 km. The ZP population was sampled from semienclosed bay, Dongshan Bay (500 km in diameter), and the geographical barriers may play an important role in genetic differentiation. Likewise, Leizhou Peninsula might separate ZJ population from BH for genetic exchange. Similar pattern was also observed on the distribution in *L. polyphemus*. There is a genetic break between the Gulf of Mexico and Atlantic populations, and Florida Peninsula may be the barrier for gene flow (Pierce et al., 2000). Yang et al. (2007) also discovered that geographical barriers played an important role in genetic subdivision of *T. tridentatus*.

However, it was difficult to be interpreted by geographical distance for the LJ locality, since the population was not significantly different from any other locations even BH (1,360 km in distance), DZ (1,310 km in distance) and ZJ (1,130 km in distance) as well. Unlike most migration marine animals, the horseshoe crab has no significant potential for long distance dispersal and gene flow (Pierce et al., 2000; Yang et al., 2007), since adults stay on the offshore sea bottom and migrate to intertidal areas for reproduction, trilobite larvae settle right after hatching, and juveniles spend their life stages at or near the natal beach for feeding (Sekeguchi, 1988; Chen et al., 2004). A factor contributing to the poor structure of Chinese horseshoe crab at microsatellite markers and confused population structure pattern at CR sequences may be the influence of artificial movements of individuals due to commercial fishing and discharging among the south-eastern coast of China. Due to the fact that the horseshoe crab is popular as delicious seafood in China, and with the sharp drop in resources around Chinese coast, the horseshoe crab was extensively transported from south coast to the north, most from Beibu Gulf to other places for commercial purposes. However, this activity is illegal, a large number of horseshoe crab were confiscated and then discharged to local coast, which may mix with native species or even take the place of the native individual. The phenomenon of mixture of native species may be common in Chinese coast, which have been frequently found in news. However, due to the lack of available data, it is difficult to estimate their exact contribution to the genetic structure of Chinese horseshoe crab.

In this study, the divergence was much more pronounced revealed by mtDNA than microsatellites. This may reflect increased rate of differentiation by genetic drift at the mitochondrial compared with the nuclear genome, which was caused by the smaller effective population size of the former genome (Haavie et al., 2000). As a fast-evolving marker, the mitochondrial control region indicated much higher levels of genetic differentiation in this study, and similar results can be found in other reports (Scribner et al., 2001; Kerth et al., 2002; Clemencet et al., 2005; Brito, 2007).

Conservation implications for *T. tridentatus*

A drastic decline has occurred in the size of the horseshoe crab population in Chinese southeast coast, so it is urgent to carry out effective management strategies, such as reintroducing the species into those coasts where the horseshoe crabs have disappeared, determining management units, establishing nature reserves and so on. The microsatellite survey indicated weak population structure of horseshoe crabs along Chinese coast. Taken alone, these data would not distinguish separated evolution significant units and mandate the species in Chinese coast as a single management unit. The weak genetic differentiation suggests that exchangeability of individuals between locations will make minimal negative genetic impacts, and this may facilitate re-introduction efforts in a region that is suffering the greatest impacts in terms of habitat loss and resources destruction. The economic developed area such as Zhejiang Province, where horseshoe crab resources are extremely scarce is especially necessary for the introduction of Chinese horseshoe crab from other locations for population rejuvenation. Meanwhile, when we introduce horseshoe crab to other locations, we should avoid introgression by mixing differentiated populations which has been revealed by CR marker. Therefore conservation strategies such as *ex situ* conservation should be carried out.

However, mitochondrial DNA survey displayed significant difference among these locations, four management units were implied including locations from Beibu Bay, ZP, MZ and NH. There are only six municipal nature reserves of horseshoe crab established in China mainland nowadays, all merely in Guangdong Province. Upon four management units we proposed above, four or more nature reserves except six existing reserves should be established accordingly. Dongshan Bay and Meizhou Island were expected to be selected as nature reserves of Fujian Province, Shanmen Bay where NH location was sampled as nature reserve of Zhejiang Province, and Beihai coast selected as the representative reserve of Beibu Bay. For the species were just regarded as the province-level of "key protected aquatic wildlife" in zhejiang, Fujian, Guangdong Province and Guangxi Zhuang autonomous region, another important conservation strategy is the need to promote horseshoe crab to be included in wildlife conservation species list in China.

In the future, more conservation strategies such as extensive survey of the population size, artificial propagation should be carried out.

ACKNOWLEDGEMENTS

We thank everyone who assisted in the collection of horseshoe crab for this study, especially Kaichong Xu, Weiwen Li, Ziqiu Pan, Qiang Lin. This work was financially supported by the National Natural Science Foundation of China (41106144), Project of the Open Foundation from Ocean Fishery Science and Technology in the Most Important Subjects of Zhejiang (20100210).

REFERENCES

Allendorf FW, Luikart G (2007). Conservation and the genetics of populations. Blackwell Publishing, Oxford, UK.

Bandelt HJ, Forster P, Rohl A (1999). Median-joining networks for inferring intraspecific phylogenies. Mol. Biol. Evol. 16:37-48.

Barrowclough GF, Groth JG, Mertz LA, Gultiérrez RI (2005). Genetic structure, introgression, and a narrow hybrid zone between northern and California spotted owls (*Strix occidentalis*). Mol. Ecol. 14:1109-1120.

Botstein D, White RL, Skolnick M, Davis RW (1980). Construction of a genetic linkage map in man using restriction fragment length polymorphisms. Am. J. Hum. Genet. 32:314-331.

Brito PH (2007). Contrasting patterns of mitochondrial and microsatellite genetic structure among Western European populations of tawny owls (*Strix aluco*). Mol. Ecol. 16:3423-3437.

Brookfield JFY (1996). A simple new method for estimating null allele frequency from heterozygote deficiency, Mol. Ecol. 5:453-455.

Bryja J, Smith C, Konecny A, Reichard M (2010). Range-wide population genetic structure of the European bitterling (*Rhodeus amarus*) based on microsatellite and mitochondrial DNA analysis. Mol. Ecol. 19:4708-4722.

Chen CP, Yeh HY, Lin PF (2004). Conservation of horseshoe crabs in Kinmen, Taiwan: strategies and practices. Biodivers. Conserv. 13: 1889-1904.

Chiu HMC, Morton B (1999). The biology, distribution and status of horseshoe crab, *Tachypleus tridentatus* and *Carcinoscorpius rotundicauda* (Arthropoda: Chelicerata) in Hong Kong: recommenddations for conservation and management. The Swire Institute of Marine Science, University of Hong Kong, Hong Kong.

Clemencet J, Viginier B, Doums C (2005). Hierarchical analysis of population genetic structure in the monogynous ant *Cataglyphis cursor* using microsatellite and mitochondrial DNA markers. Mol. Ecol. 14:3735-3744.

Colbert J, Danchin E, Dhondt AA, Nichols JD (2001). Dispersal. Oxford University Press, New York, USA.

Excoffier LGL, Laval G Schneider S (2005). Arlequin ver. 3.0：An integrated software package for population genetics data analysis. Evol. Bioinform. (Online) 1:47-50.

Faurby S, King TL, Obst M, Hallerman EM, Pertoldi C, Funch P (2010). Population dynamics of American horseshoe crabs—historic climatic events and recent anthropogenic pressures. Mol. Ecol. 19:3088-3100.

Flanders J, Jones G, Benda P, Dietz C, Zhang S, Li G, Sharifi M, Rossiter SJ (2009). Phylogeography of the greater horseshoe bat, *Rhinolophus ferrumequinum*: contrasting results from mitochondrial and microsatellite data. Mol. Ecol. 18:306–318.

Frankham R, Ballou JD, Briscoe DA (2002). Introduction to conservation genetics. Cambridge University Press, UK.

Fu YX (1997). Statistical tests of neutrality of mutations against population growth, hitchhiking and background selection. Genetics 147:915-925.

Guo W, Gui JF (2008). Microsatellite marker isolation and cultured strain identification in *Carassius auratus gibelio*. Aquacult. Int. 16:497-510.

Haavie J, Saetre GP, Moum T (2000). Discrepancies in population differentiation at microsatellites, mitochondrial DNA and plumage colour in the pied flycatcher-inferring evolutionary processes. Mol. Ecol. 9:1137-1148.

Hartl DL, Clark AG (1989). Principles of population genetics. 2nd ed. Sinauer Associates, Sunderland, MA.

Hearne CM, Ghosh S, Todd JA (1992). Microsatellites for linkage analysis of genetic traits. Trends Ecol. Evol. 8:288-94.

Hewitt GM (2004). The structure of biodiversity—insights from molecular. phylogeography. BMC Front. Zool. 1:1-16.

Hsieh HL, Chen CP (2009). Conservation program for the Asian horseshoe crab *Tachypleus tridentatus* in Taiwan: characterizing the microhabitat of nursery grounds and restoring spawning grounds. In: Tanacredi J T Botton M L, Smith DR eds., Biology and conservation of horseshoe crab. Springer, New York, USA.

Kalinowski ST, Taper ML, Marshall TC (2007). Revising how the computer program CERVUS accommodates genotyping error increases success in paternity assignment. Mol. Ecol. 16:1099-1006.

Kaya M, Yildiz MA (2008). Genetic diversity among Turkish native chickens, Denizli and Gerze, estimated by microsatellite markers. Biochem. Genet. 46:480-491.

Keeney DB, Heupel MR, Hueter RE, Heist EJ (2005). Microsatellite and mitochondrial DNA analyses of the genetic structure of blacktip shark (*Carcharhinus limbatus*) nurseries in the northwestern Atlantic, Gulf of Mexico, and Caribbean Sea. Mol. Ecol. 14:1911-1923.

Kerth G, Mayer F, Petit E (2002). Extreme sex-biased dispersal in the communally breeding, nonmigratory Bechstein's bat (*Myotis bechsteinii*). Mol. Ecol. 11:1491-1498.

King TL, Eackles MS (2004). Microsatellite DNA markers for the study of horseshoe crab (*Limulus polyphemus*) population structure. Mol. Ecol. Notes. 4:394-396.

Li QZ, Li Q, Liu J, Ni LH, Kong LF (2009). Isolation and characterization of microsatellite loci in the horseshoe crab (*Tachypleus tridentatus*). Conserv. Genet, 10:1879-1881.

Liao YY, Li XM (2001). Present situation of horseshoe crab resources in the sea area of China and tactics of preservation. Resources Sci. Sinica 23:53-57.

Marshall TC, Slate J, Kruuk LEB, Pemberton JM (1998). Statistical confidence for likelihood-based paternity inference in natural populations. Mol. Ecol. 7:639-655.

Nishida S, Koike H (2010). Isolation and characterization of compound microsatellite loci in the tri-spine horseshoe crab (*Tachypleus tridentatus*). Conser. Genet. Resour. (2Supplement) 31-33.

Nyakaana S, Arctander P (1999). Population genetic structure of the African elephant in Uganda based on variation at mitochondrial and nuclear loci: evidence for male-biased gene flow. Mol. Ecol. 8:1105-1115.

Oosterhout CV, Hutchinson WF, Wills DM, Shipley P (2004). MICRO-CHECKER: software for identifying and correction genotyping errors in microsatellite data. Mol. Ecol. Notes 4:535-538.

Pierce JC, Tan G, Gaffney PM (2000). Delaware Bay and Chesapeake Bay populations of the horseshoe crab *Limulus polyphemus* are genetically distinct. Estuaries 23:690-698.

Purvis A, Gittleman JL, Brooks T (2005). Phylogeny and conservation. Cambridge University Press, New York, USA.

Rousset F (2008). Genepop'007: a complete reimplementation of the Genepop software for Windows and Linux. Mol. Ecol. Resour. 8:103-106.

Rozas J, Sánchez-Delbarrio JC, Messeguer X, Rozas R (2003). DanSP, DNA polymorphism analyses by the coalescent and other methods. Bioinformatics 19:2496-2497.

Sekiguchi K (1988). Biology of Horseshoe crab. Science House, Tokyo.

Scribner KT, Petersen MR, Fields RL, Talbot SL, Pearce JM, Chesser RK (2001). Sex-biased gene flow in spectacled eiders (Anatidae): inferences from molecular markers with contrasting modes of inheritance. Evolution 55: 2105–2115.

Shaw KL (2002). Conflict between nuclear and mitochondrial DNA phylogenies of a recent species radiation: What mtDNA reveals and conceals about modes of speciation in Hawaiian crickets. P. Natl. Acad. Sci. USA. 99:16122-16127.

Slatkin M, Hudson RR (1991). Pairwise comparisons of mitochondrial-DNA sequences in stable and exponentially growing populations. Genetics 129:555-562.

Swan BL (2001). A unique medial product (LAL) from the horseshoe crab and monitoring the Delaware Bay horseshoe crab population. In: Tanacredi JT eds., *Limulus* in the limelight. Kluwer Academic/Plenum Publishers, New York, USA.

Tamura K, Dudley J, Nei M, Kumar S (2007). MEGA4: Molecular Evolutionary Genetics Analysis (MEGA) software version 4.0. Mol. Biol. Evol. 24:1596-1599.

Weir BS, Cockerham CC (1984). Estimating F-statistics for the analysis of population structure. Evolution 38:1358-1370.

Weng ZH, Xie YJ, Xiao ZQ, Huang LM, Li J, Wang SH, Zhang YZ (2012). Survey on resource distribution of Chinese horseshoe crab (*Tachypleus tridentatus*) in Fujian and other coast water of China. Chinese J. Zool. 47(3):40-48.

Yang MC, Chen CA, Hsieh HL, Chen CP (2007). Population subdivision of the tri-spine horseshoe crab, *Tachypleus tridentatus*, in Taiwan Strait. Zool. Sci. 24:219-224.

Zane L, Bargelloni L, Patarnello T (2002). Strategies for microsatellite isolation: a review. Mol. Ecol. 11:1-16.

Zink RM (1997). Phylogeographic studies of North American birds. In: Mindell DP eds., Avian Molecular Evolution. Academic Press, San Diego, California.

13

Biological parameters of the non-target pest *Aphis gossypii* Glover (Hemiptera: Aphididae) on genetically modified (GM) *Bt* cotton

Thiago Alexandre Mota, Maria Freire de Souza, Marcos Gino Fernandes, Paulo Rogério Beltramin da Fonseca, Samir Oliveira Kassab and Jean Carlo de Quadros

Programa de Pós-graduação em Entomologia e Conservação da Biodiversidade, Faculdade Ciências Biológicas e Ambientais, Universidade Federal da Grande Dourados (UFGD), 79804-970, Dourados, MS, Brasil.

In the present work, we aimed to evaluate: 1) the influence of the Cry1Ac protein expressed by the genetically modified cotton variety (*Bt*) NuOpal, on the biological parameters of a non-target pest, *Aphis gossypii*, reared under laboratory conditions; 2) the influence of plant age on aphid development. Cotton cultivars were grown following technical advice. In the laboratory, the aphids were separated into plastic containers including a cotton sheet, which was changed daily with the help of a moist cotton cloth. Observations were carried out daily, and the duration of the immature and adult stages, and offspring number, were recorded. Our results show that regardless of the differences in the duration of each instar, cultivars *Bt* and non-*Bt* and the different ages of the plant to *Bt*, no differences were seen in the total duration of these phases. It was only during the reproductive period that differences were observed among genotypes for 120 days. Regarding the number of offspring and longevity total, there were differences between the different ages of the *Bt* treatment.

Key words: Biosafety, transgenic, *Gossypium hirsutum,* genetically modified (GM) plant.

INTRODUCTION

Cotton, *Gossypium hirsutum* L. is one of the most economically important crops worldwide, susceptible to a myriad of arthropod pests, which attack all stages of plant development. With a production estimated at 1.417 t/ha of cotton lint, and 2.334 t/ha of seed cotton (Fnp Consultoria and Comércio, 2009), Brazil is the fifth largest cotton producer in the world. The country's high productivity, however, is constantly threatened by pests. In the Midwest region, for instance, insecticides are applied on average 12 to 20 times on each crop field (Tomquelski et al., 2007; Fontes et al., 2006).

One economically importantly pest of cotton is the cotton aphid *Aphis gossypii* Glover, 1877 (Hemiptera: Aphididae) (Gallo et al., 2002). This piercing and sucking insect lives in colonies on the underside of leaves, sucking sap from the plants, which they pierce with their sharp style (Santos, 2001). Feeding by the cotton aphid causes wrinkled, deformed leaves and buds, slows plant development and hinders plant growth, thereby reducing productivity (Silva, 1992). Secondary pest populations have increased in *Bt* cotton as they are not targeted; for example *A. gossypii* and *Bemisia tabaci* (Gennadius) (Cui and Xia, 2000; Liu et al., 2002).

Genetically modified plants expressing toxins derived from the bacterium *Bacillus thuringiensis* Berliner, 1911(*Bt*) have been cultivated over 200 million hectares since 1996 (James, 2009). Transgenic cotton expressing the Cry1Ac toxin, which targets the main cotton pests the pink bollworm *Pectinophora gossypiella* Saunders, 1844, the apple caterpillar, *Heliothis virescens* Fabricius, 1781 and the leaf worm *Alabama argillacea* Hübner, 1818 (Perlak et al., 2001), has been cultivated since 1996 in Australia, Mexico, Brazil Argentina, China, Colombia, India, Indonesia, South Africa and in the United States (James, 2009).

In Brazil, the *Bt* cotton was released for commercial farming in March 2005. Some restrictions, such as incorporation of exclusion zones, refuge areas, conventional cultivars corresponding to 20% of the total crop field, in addition to the generally recommended crop management practices, were enforced. *Bt* cotton plants, except for their tolerance to target insects, are theoretically equivalent to their non-transformed parental strain (CTNBio, 2005) in all agronomic and phenotypic characteristics.

The fact with the advantages of *Bt* crops are the reduced need to spray chemical insecticides for pest control, and the fact of reduced risk of adverse impacts on non-target species, some of which are natural enemies of agricultural pests. Transgenic crops are generally considered beneficial to human health and to the environment (Shelton et al., 2002; Naranjo, 2005; Huang et al., 2005; Cattaneo et al., 2006; Herdt, 2006). During the first ten years of transgenic cotton cultivation (1996 to 2006), a reduction in insecticide application on this crop was recorded worldwide (Brookes and Barfoot, 2008).

Despite the advantages mentioned above, the proportion of plant sucking insects in *Bt* cotton fields has increased, and is higher when compared with conventional cotton fields (Wilson et al., 1992; Fitt et al., 1994; Cui and Xia, 1998; Greene et al., 1999). It is possible that the *Bt* technology has an indirect, positive effect on the growth of non-target herbivore populations, for instance *A. gossypii* (Suji et al., 2008), in part because *Bt* cotton fields are sprayed less with conventional insecticides (Liu et al., 2005).

Some studies have demonstrated that *Bt* crops expressing the Cry1A toxins to control lepidopteran pests can have a direct effect on aphids (Schuler et al., 2001; Dutton et al., 2002). Contrasting with those results, Sisterson et al. (2004) found no significant differences in mortality among aphids fed transgenic cotton, and Sujii et al. (2008) found that *Bt* cotton does not significantly affect aphid life cycle, survival, fecundity and colony formation. Research carried out by Deng et al. (2003), however, put forth convincing evidence that aphid populations on *Bt* cotton are consistently larger than those on conventional cotton.

Aphids have an important role in food chains within agroecosystems, serving as food for several parasitoid and predator species (Lawo et al., 2009). Knowing the response prey/hosts populations to *Bt* crops is very important in understanding predator/parasitoid dynamics, hence natural enemy community responses.

Although previous studies have arrived at conflicting results, none has compared the toxicity of leaves from *Bt* cotton in different phenological stages.

Based on the idea that *Bt* plants may indirectly affect the non-target pests, the objective of this paper was to compare the life cycles of *A. gossypii* individuals reared on different phonological stages of two crops, the *Bt* cotton (NuOpal) and non-*Bt* cotton (DeltaOpal), under laboratory conditions.

MATERIALS AND METHODS

Seeds of the *Bt* and non *Bt* cultivars were sown in the experimental area of the Universidade Federal da Grande Dourados and grown according to agronomic best practice (Beltrão et al., 2003).

Cotton leaves containing *A. gossypii* were collected from non-*Bt* cotton, to create insect stocks. Insects were kept in the insectary of the Entomology Laboratory of the Faculdade de Ciências Biológicas e Ambientais, in a climate-controlled chamber, BOD at 25°C, 70% RH, and photophase of 12 h. Non-*Bt* cotton leaves were used as a substrate.

In the laboratory, individuals from the stock were isolated in 250 mL plastic containers closed with lids. In order to evaluate how the plant's phenological phase (at days 30, 60, 90 and 120 after emergence) affects insect development, we used a total of 60 nymphs from the first generation obtained in the laboratory (30 nymphs for *Bt*-cotton and 30 for non-*Bt*) up to 24 h old.

We used the following method to monitor the aphid's life cycle, from birth to death. We quantified the offspring of the first generation and collected the individuals with the aid of a fine-tipped brush to use them in the experiments. In order to prevent samples from drying out, we placed a moist cotton swab inside the container along with the same age leaf for each phenological phase which were changed every 24 h. We used a completely randomized design with 10 repetitions, each consisting of three insects.

The following variables were evaluated: duration of the nymph stages, duration of each instar, average longevity, offspring number and reproductive period. All variables obtained at the different ages of the plant were compared. The results obtained with the *Bt* cotton were compared with those obtained with conventional, non-*Bt* cotton. In order to compare the mean variables obtained from the two cultivars, the data obtained were subjected to analysis of variance and, when a significant difference among the means was noticed, they were compared using the t test at 5% probability. Data obtained for each phonological phase of the plant were also subjected to analysis of variance. When significant, they were compared using the Tukey's test at 5%.

RESULTS

Nymphs of *A. gossypii* created in *Bt* cotton leaves had longer duration when compared to the non-*Bt* cotton in the third and fourth instars for 90 and 120 days for the fourth instar. Exception was observed in duration of third instar with plant leaves at 120 days after emergence (DAE), and the duration for individuals created were statistically higher in leaves of non-*Bt* cotton (Table 1).

Table 1. Duration of nymph stages (days) (± SE) of *A. gossypii* leaves from *Bt* (NuOpal) and non-*Bt* (DeltaOpal) cotton, at different phonological stages.

Stage	Cultivar	Days after emergency					CV (%)
		30	60	90	120	F	
1st instar	NuOpal	1.30 ± 0.08a	1.23 ± 0.07a	1.23 ± 0.07a	1.30 ± 0.08a	0.22	35.77
	DeltaOpal	1.36 ± 0.08a	1.4 ± 0.09a	1.36 ± 0.08a	1.41 ± 0.09a	0.032	35.26
	t test	0.5399ns	-1.1198ns	-1.3868ns	-0.9034ns		
2nd instar	NuOpal	1.20 ± 0.07a	1.46 ± 0.09a	1.25 ± 0.08a	1.35 ± 0.10a	1.94	36.87
	DeltaOpal	1.25 ± 0.08a	1.41 ± 0.09a	1.14 ± 0.08a	1.26 ± 0.08a	16.615	36.68
	t test	-0.4492ns	0.9015ns	0.4025ns	0.6325ns		
3rd instar	NuOpal	1.16 ± 0.06a	1.03 ± 0.03b	1.14 ± 0.08b	1.04 ± 0.03b	4.8	30.13
	DeltaOpal	1.11 ± 0.05a	1.03 ± 0.03a	1.07 ± 0.04a	1.09 ± 0.07a	0.5907	28.13
	t test	0.5942ns	0.6942*	0ns	-0.4924*		
4th instar	NuOpal	1.26 ± 0.08a	1.25 ± 0.09b	1.03 ± 0.03b	1.04 ± 0.03b	3.39	33.85
	DeltaOpal	1.04 ± 0.03a	1.23 ± 0.11a	1.00 ± 0.0a	1.00 ± 0.0a	10.776	33.79
	t test	2.3556*	1*	0.1755ns	1*		
Total nymph	NuOpal	4.96 ± 0.15a	4.70 ± 0.13a	4.57 ± 0.12a	4.66 ± 0.10a	1.37	15.51
	DeltaOpal	4.84 ± 0.16a	4.88 ± 0.15a	4.62 ± 0.10a	4.76 ± 0.14a	0.5113	16.39
	t test	0.4906ns	-0.3009 ns	0.8275ns	0.4619ns		

Means followed by the same letter in each line do not differ according to the Tukey's test (α= 0.05). ns, t-test non-significant (α= 0.05) for both cultivars; * means differ among one another according to the t-test (α= 0.05).

The analysis of the effects of two cultivars analyzed separately by age of the plant showed differences only for nymphs reared on leaves of *Bt* cotton for the third instar; individuals from leaves of 30 days lasted the longest stage in comparison to others. For the fourth instar *Bt* treatment, there was a lesser duration of this phase, when compared with other ages.

When the duration of the reproductive phase of aphids reared on different phenological stages of the cotton plant were compared among one another, no significant differences were found (Table 2), with one exception: the reproductive phase of aphids reared on *Bt* and non-*Bt* leaves 120 DAE. showed significant differences. Also, when the duration of the reproductive phase of aphids reared on *Bt* and non-*Bt* cotton cultivars were compared, no significant differences were observed between the two treatments at the different ages of the plant. (Table 2)

With respect to offspring number, no significant differences were found when cotton cultivars were compared, because the average number of offspring was similar in both, at all phenological stages of the plant (Table 2).

However, analysis of variance revealed significant differences in the average number of aphid descendants when the parental generation was reared on different phenological stages of *Bt* plants: offspring production decreased with plant age (Table 2). The average number of aphid offspring resulting from parents reared on conventional cotton leaves of different ages was not significant.

DISCUSSION

Aphids spent on average one day in each of the four instars. Instar duration did not differ significantly among treatments, except for differences found for fourth instar larvae reared on *Bt* plants 90 days old, and for third instar larvae reared on plants 120 DAE.

Similar data for the duration of each instar were generated by Person et al. (2004), who used four varieties of conventional and Parajulee (2007) cotton to rear separated individuals directly in the plant, under temperature similar to the one used by us. Analyzing three generations, Liu et al. (2005) found that the *Bt* cotton did not affect the duration of the pre-reproductive period of *A. gossypii*. Divergent results were observed by Zhang et al. (2008), who found a shorter pre-reproductive phase for aphids reared on conventional cotton 5.9 days old, and on 6.3 days old GM cotton sold in China. Using

Table 2. Longevity (days) offspring number (± SE) of *A. gossypii* leaves from *Bt* (NuOpal) and non-*Bt* (DeltaOpal) cotton, at different phonological stages.

Stage	Cultivar	Days after emergency				F	CV (%)
		30	60	90	120		
	NuOpal	13.08 ± 0.60^{a}	11.81 ± 0.93^{a}	10.42 ± 0.76^{a}	10.27 ± 0.69^{a}	0.08	38.17
Reproductive period	DeltaOpal	12.8 ± 0.69^{a}	10.69 ± 0.09^{a}	10.16 ± 0.62^{a}	9.85 ± 0.88^{a}	22.406	41.44
	t test	0.2611^{ns}	0.7793^{ns}	0.2444^{ns}	0.3133*		
	NuOpal	45.95 ± 2.50^{a}	42.77 ± 3.6^{a}	33.77 ± 3.16^{a}	29.39 ± 2.1^{b}	3.84	45.93
Offspring number	DeltaOpal	39.84 ± 2.61^{a}	35.73 ± 6.93^{a}	33.45 ± 2.79^{a}	32.28 ± 3.18^{a}	15.484	47.69
	t test	1.5165^{ns}	1.29^{ns}	0.0716^{ns}	-0.6545^{ns}		
	NuOpal	18.66 ± 1.46^{a}	15.16 ± 1.19^{a}	14.16 ± 0.98^{a}	13.03 ± 1.32^{b}	4.24	46.54
Total longevity	DeltaOpal	15.4 ± 1.31^{a}	13.83 ± 1.24^{a}	12.33 ± 1.09^{a}	16 ± 1.01^{a}	15.623	46.31
	t test	1.6608^{ns}	0.773^{ns}	1.2436^{ns}	-1.5883^{ns}		

Temperature was 25 ± 1 °C, RH was 70 ± 10% and photo phase was 12 h. Means followed by the same letter in each line do not differ according to the Tukey's test (α= 0.05). ^{ns} t-test non-significant (α= 0.05) for both cultivars. *Means differ among one another according to the t-test (α= 0.05).

three conventional cotton cultivars and their transgenic isolines, Lawo et al. (2009) found no differences in the pre-reproductive phase of experimental cotton aphids.

Host plant characteristics, for instance appropriate nutritional contents and plant hormonal stimuli which instigate the insect to start feeding can influence the success of the development and reproduction of phytophagous insects (Fernandes et al., 2001). Suji et al. (2008) observed that the *Bt* cotton did not influence the choice of cultivar for colony formation, and the number of winged individuals produced in *Bt* and non-*Bt* plants.

In our study, the *Bt* cotton did not affect the average duration of the reproductive period of *A. gossypii*. This result is very similar to that obtained by Suji et al. (2008) in which the reproductive period (of cotton aphids) lasted 16.35 and 16.18 days on *Bt* and conventional cotton, respectively. Michelotto and Busoli (2003) found that aphid reproductive period lasted 15.5 days for specimens fed conventional cotton. In Liu et al. (2005), the reproductive phase of aphids in the first generation varied among three types of cotton. According to the authors, however, when feeding on *Bt* cotton, aphids do not ingest the insecticide protein.

Our results show that, regardless of the type of cotton, the average offspring number was similar among the different phenological ages of the host plant. Our data is not consistent with the results of Suji et al. (2008). Even though they did not observe significant differences in the average offspring number among cultivars (*Bt* and non-*Bt*), a higher total mean offspring per female, for *Bt* (47.26) when compared with non *Bt* (46.98), did result from their experiment. According to Michelotto and Busoli (2003), the number of offspring produced per female was 84 nymphs in the DeltaOpal cultivar. Pessoa et al. (2004) found a total production of aphids between 54 and 69 individuals. Similar values were observed by Funichello et al. (2012), who reported an average offspring of 63 for NuOpal and 59 for Deltaopal. However, Lawo et al. (2009), in India, observed that the total fertility of the cotton aphid is closer to that of the present study, between 22 and 38 nymphs, not differing significantly between treatments. In China, Zhang et al. (2008) compared the biological parameters of *A. gossypii* reared on *Bt* and non-*Bt* cotton, and obtained an average of 30 and 29 nymphs produced per female in the respective treatments.

We do not know why the number of offspring is inversely proportional to the age of the plant; it might be the result of less favorable conditions for the establishment of *A. gossypii* populations as plants age. According to Degrande (1998), the attacks by the cotton aphids begin 15 days after plant emergence, and are especially intense between 40 and 70 days, when populations are densest. Gonzaga et al. (1991) concluded that the vertical distribution of pest populations on the cotton plant vary with plant age. Aphids are detected up to 90 days after plant emergence, but are not found afterwards.

The results of this study suggest that the longevity of the cotton aphid was not affected by the type of cultivar. However, the average longevity of aphids in this study was lower than that reported by Suji et al. (2008). In their data, aphids lasted on average 20.47 days on *Bt* cotton and 20.98 days on non-*Bt* crop, respectively. Fuchinello et al. (2009) also did not observe any kind of effect of transgenic cotton on the biological parameters of aphids, finding a higher average life span for those on *Bt* crops (21.83) when compared with individuals reared on non-*Bt* cotton (24.40 days). Michelotto and Busoli (2003) observed that *A. gossypii* reared on the cultivar DeltaOpal lived, on average, 24.33 days. Lawo et al. (2009) evaluated the performance of aphids on plants in the three varieties of transgenic *Bt* cotton (Mech 12, Mech 162, Mech 184) most commonly grown in India, and their respective isolines. In their results, aphid longevity was 24, 21 and 24 days for the three cultivars, and 22.8, 21.5 and 25.3 days for *Bt* crops, respectively. As longevity is one of the biological characteristics that can be more easily affected by toxins ingested by the insect, we hypothesize that the pest is not much affected by the Cry1Ac toxin.

Literature data suggests that the two most common methods of rearing aphids, that is directly on the plant, or detaching the plant leaves and keeping them with the insect inside a laboratory chamber do not interfere with the ecology of these insects. However, after analyzing the extensive literature available (Michelotto and Busoli, 2003; Suji et al., 2008; Fuchinello et al., 2009; Lawo et al., 2009), we suggest that the total longevity of the cotton aphid was affected by the methodology adopted in this work.

It is possible that the amino acid contents on different parts of a plant species vary according to the phenological stage of plant cycle, and plant variety (Fernandes et al., 2001). When considering the different phenological phases of the *Bt* cotton at 30 and 120 days, there were significant differences in the average longevity of *A. gossypii*.

Even though toxins produced by *Bt* plants affect insect targets and some non-target insects (Dogan et al., 1996; Hilbeck et al., 1998; Dutton et al., 2002), these substances are often not detectable. Some authors consider that the *Bt* endotoxins expressed in plants are not found in the phloem of the plant, or have been detected in the aphid (Raps et al., 2001; Dutton et al., 2002). By contrast, Zhang et al. (2008) reported having found the toxin in aphids reared on *Bt* cotton. The authors, however, did not investigate the possible effects of the toxin on this pest.

The results of this study show that *Bt* cotton did not exert any influence on the biological parameters of *A. gossypii*, because those parameters were equivalent with those obtained from organisms reared on conventional cotton. It is important to highlight, however, the need to study the possible effects of *Bt* cotton o tri-trophic interactions, evaluating the possible accumulation of this toxin in natural enemies.

ACKNOWLEDGEMENTS

We thank the Fundação de Apoio ao Desenvolvimento do Ensino, Ciência e Tecnologia do Estado de Mato Grosso do Sul (FUNDECT), for the masters grant to the first author.

REFERENCES

Beltrão NEM, Araújo AE, Benassi AC, Do Amaral JAB, Severino LS, Cardoso GD (2003). Zoneamento e época de plantio para o algodoeiro no norte do Estado do Espírito Santo. Rev. Bra. Eng. Agríc. Ambiental. 7:99-105.

Brookes G, Barfoot P (2008). Global impact of biotech crops: Socio-economic and environmental effects, 1996–2006. AgBioForum 11:21-38.

Cattaneo MG, Yafuso C, Schmidt C, Huang CY, Rahman M (2006). Farm-scale evaluation of the impacts of transgenic cotton on biodiversity, pesticide use, and yield. Proc. Natl. Acad. Sci. 103:7571-7576.

CTNBio (2005). Comissão Técnica Nacional de Biossegurança. Disponível em: <http: //www.ctnbio.gov.br>.

Cui JJ, Xia JY (1998). Effect of early seasonal strain of *Bt* transgenic cotton on population dynamics of main cotton pests and their natural enemies. Acta Gossypii Sin. 10:255-262.

Cui JJ, Xia JY (2000). Effects of *Bt* (*Bacillus thuringiensis*) transgenic cotton on the dynamics of pest population and their enemies. Acta. Phy. Sin. 27:141-145.

Degrande PE (1998). Guia prático de controle das pragas do algodoeiro. Dourados, UFMS, 60 pp.

Deng SD, Xu J, Zhang QW, Zhou SW, Xu GJ (2003). Effect of transgenic *Bacillus thuringiensis* cotton on population dynamics of non-target pests and natural enemies. Acta. Ento. Sin. 46:1-5.

Dogan EB, Berry RE, Reed GL, Rossignol PA (1996). Biological parameters of convergent lady beetle (Coleoptera: Coccinellidae) feeding on aphids (Homoptera: Aphididae) on transgenic potato. J. Econ. Entomol. 89:1105-1108.

Dutton A, Klein H, Romeis J, Bigler F (2002). Uptake of *Bt*-toxin by herbivores feeding on transgenic maize and consequences for the predator *Chrysoperla carnea*. Ecol. Entomol. 27:441-447.

Fernandes AMV, Farias AMI, Soares E, Simão MMM, Vasconcelos D (2001). Desenvolvimento do Pulgão *Aphis gossypii* Glover (Hemiptera: Aphididae) em Três Cultivares do Algodão Herbáceo *Gossypium hirsutum* L. r. *latifolium* Hutch. Neotrop. Entomol. 30:467-470.

Fitt GP, Mares CL, Llewellyn DJ (1994). Field evaluation and potential impact of transgenic cottons (*Gossypium hirsutum*) in Australia. Biocontrol Sci. Technol. 4:535-548.

Fnp Consultoria and Comércio (2009) Algodão. In: Agrianual 2009: Anuário da agricultura brasileira. São Paulo, pp. 403-478.

Fontes EMG, Souza Ramalho F, Underwood E, Barroso PAV, Simon MF, Sujii ER, Pires CSS, Beltrão N, Lucena WA, Freire EC (2006). In: The cotton agricultural context in Brazil, In: Hilbeck A, Andow DA, Fontes EMG (Eds.). Environmental risk assessment of genetically modified organisms: methodologies for assessing *Bt* cotton in Brazil. Wallingford: CABI publishing, pp. 21-66.

Funichello M, Costa LL, Busoli AC (2012). Aspectos biológicos e tabela de vida de fertilidade de *Aphis gossypii* Glover (Hemiptera: Aphididae) em cultivares de algodoeiro DeltaOPAL e NuOPAL. Arq. Inst. Biol. 79:84-90.

Gallo D, Nakano O, Carvalho RPL, De Baptista GC, Berti Filho E, Parra JRP, Zucchi RA, Alves SB, Vendramin JD, Marchini LC, Opes JRS, Omoto C (2002). Entomologia Agrícola. Piracicaba: Fealq. 920 pp.

Gonzaga JV, Ramalho SF, Dos Santos WJ (1991). Distribuição do *Aphis gossypii* no algodoeiro nos sistemas de plantio solteiro e consorciado. Pesq. Agropec. Bras. 26:1839-1844.

Greene JK, Turnipseed SG, Sulllivan MJ, Herzog GA (1999). Boll damage by southern green stink bug (Hemiptera: Pentatanidae) and tarnished plant bug (Hemiptera: Miridae) caged on transgenic *Bacillus thuringiensis* cotton. J. Econ. Entomol. 92:941-944.

Hilbeck A, Moar WJ, Pusztai-Carey M, Bigler F (1998). Toxicity of

Bacillus thuringiensis Cry1Ab toxin to the predator *Chrysoperla carnea* (Neuroptera: Chrysopidae). Environ. Entomol. 27:1255-1263.

Huang J, Ruifa H, Rozelle S, Pray C (2005). Insect-resistant GM rice in farmers' fields: assessing productivity and health effects in China. Science 308:688-690.

James C (2009). Global Status of Commercialized Biotech/ GM Crops: 2008. ISAAA Brief No. 39. International Service for the Acquisition of Ag-biotech Applications, Ithaca, NY.

Lawo NC, Wackers FL, Romeis J (2009). Indian *Bt* cotton varieties do not affect the performance of cotton aphids. Plos One 4(3):1-9.

Liu WX, Wan FW, Guo JY (2002). Structure and seasonal dynamics of arthropods in transgenic *Bt*-cotton Field. Acta. Ecol. Sin. 22:729-735.

Liu XD, Zhai BP, Zhang XX, Zong JM (2005). Impact of transgenic cotton plants on a non-target pest, *Aphis gossypii* Glover. Ecol. Entomol. 30:307-315.

Michelotto MD, Busoli AC (2003). Aspectos biológicos de *Aphis gossypii* Glover 1877 (Hemiptera: Aphididae) em três cultivares de algodoeiro e em três espécies de plantas daninhas. Cienc. Rural 33:99-1004.

Naranjo SE (2005). Long-term assessment of the effects of transgenic *Bt* cotton on the abundance of nontarget arthropod natural enemies. Environ. Entomol. 34:1193-1210.

Parajulee MN (2007). Influence of constant temperatures on life history parameters of the cotton aphid, *Aphis gossypii*, infesting cotton. Environ. Entomol. 36:666-672.

Perlak FJM, Oppenhuizen K, Gustafson R, Voth S, Sivasupramaniam D, Heering B, Carey RA, Ihrig JK, Roberts JK (2001). Development and commercial use of Bollgard cotton in the USA nearly promises versus today reality. Plant. J. 27:489-501.

Pessoa LGA, Souza B, Carvalho CF, Silva M (2004). Aspectos da biologia de *Aphis gossypii* Glover, 1877 (Hemiptera: Aphididae) em quatro cultivares de Algodoeiro, em Laboratório. Ciênc. Agrotecn. 28:1235-1239.

Raps A, Kehr J. Gugerli P, Moar WJ, Bigler F, Hilbeck A (2001). Immunological analysis of phloem sap of *Bacillus thuringiensis* corn and of the nontarget herbivore *Rhopalosiphum padi* (Homoptera: Aphididae) for the presence of Cry1Ab. Mol. Ecol. 10:525-533.

Santos WJ (2001). Identificação, biologia, amostragem e controle das pragas do algodoeiro, pp. 181−226. In: Algodão: tecnologia de produção. dourados, embrapa Agropecuária Oeste & Embrapa Algodão. 296 pp.

Schuler TH, Denholm I, Jouanin L, Clark SJ, Clark AJ, Poppy GM (2001). Population scale laboratory studies of the effect of transgenic plants on nontarget insects. Mol. Ecol. 10:1845-1853.

Shelton AM, Zhao JZ, Roush RT (2002). Economic, ecological, food safety, and social consequences of the deployment of *Bt* transgenic plants. Ann. Rev. Entomol. 47:845-881.

Silva AL (1992). Controle Integrado das Pragas do Algodoeiro em Goiás. In: Manejo integrado de pragas e nematóides. Jaboticabal, Funep. pp. 23-51.

Sisterson MS, Biggs RW, Olson C, Carriére Y, Dennehy TJ, Tabashnik BE (2004). Arthropod abundance and diversity in *Bt* and non-*Bt* cotton fields. Environ. Entomol. 33:921-929.

Sujii ER, Togni PHB, Nakasu YET, Pires CSS, Fontes EMG (2008). Impacto do algodão *Bt* na dinâmica populacional do pulgão-do-algodoeiro em casa de vegetação. Pesq. Agropec. Bras. 43:1251-1256.

Tomquelski GV, Martins GLM, Papa G (2007). Efeitos dos indutores de resistência acibenzolar-s-metil e silício na biologia de *Alabama argilacea* (Lepidoptera: Noctuidae) em algodoeiro. Rev. de Agricultura. 82:170-175.

Wilson FD, Flint HM, Deaton WR, Fischhoff DA, Perlak FJ, Armstrong TA, Fuchs RL, Berberich SA, Parks NJ, Stapp BR (1992). Resistance of cotton lines containing a *Bacillus thuringiensis* toxin to pink bollworm (Lepidoptera: Gelechiidae) and other insects. J. Econ. Entomol. 85:1516-1521.

Zhang GF, Wan FH, Murphy ST, Guo JY, Liu WX (2008). Reproductive Biology of Two Nontarget Insect Species, *Aphis gossypii* (Homoptera: Aphididae) and *Orius sauteri* (Hemiptera: Anthocoridae), on *Bt* and non-*Bt* cotton cultivars. Environ. Entomol. 37:1035-1042.

14

The effects of three different grinding methods in DNA extraction of cowpea (*Vigna unguiculata* L. Walp)

Huaqiang Tan, Manman Tie, Li Zhang, Yanxia Zhu and Huanxiu Li

College of Horticulture, Sichuan Agricultural University, Ya'an 625014, Sichuan, China.

Rapid DNA extraction is a prerequisite for molecular studies. Generally, plant tissue is ground in liquid nitrogen to isolate DNA; but, liquid nitrogen is dangerous and volatile. Besides, liquid nitrogen is not always available in many developing countries. To investigate if high quality DNA could be obtained for downstream PCR analysis without liquid nitrogen, the cowpea DNA was extracted by Hexadecyl trimethyl ammonium bromide cetyl trimethylammonium bromide (CTAB) method and sodium dodecyl sulphate (SDS) method, respectively, each with three different grinding methods, including ground in liquid nitrogen, in preheated mortar and in non-preheated mortar. The DNA was compared according to their yield, purity, integrity and functionality. The results showed that high quality DNA could be obtained by three grinding methods both in CTAB method and SDS method. Without liquid nitrogen, grinding plant tissue in preheated or non-preheated mortar with extraction buffer to extract DNA is feasible.

Key words: Cowpea (*Vigna unguiculata*), grinding method, liquid nitrogen, DNA extraction.

INTRODUCTION

Cowpea, *Vigna unguiculata* (L.) Walp, Leguminosae (2n = 2x = 22), is an essential food crop in developing countries of the tropics and subtropics, especially in sub-Saharan Africa, Asia, and Central and South America (Singh et al., 1997), with an annual production of more than five million metric tons world-wide (FAO, 2010). Cowpea is a most versatile African crop, it feeds people, their livestock and because of its ability in nitrogen fixation, it improves soil fertility, and consequently helps to increase the yields of cereal crops when grown in rotation and contributes to the sustainability of cropping systems (Agbicodo et al., 2009). As new tools that dramatically enhanced the efficiency of plant breeding, DNA based molecular markers have been widely used in cowpea genetic analysis (Zannou et al., 2008; Malviya et al., 2012; Badiane et al., 2012), genetic map construction (Muchero et al., 2009; Xu et al., 2011; Ouedraogo et al., 2002), QTL analysis (Muchero et al., 2010; Andargie et al., 2011; Kongjaimun et al., 2012) etc.

Rapid DNA extraction with expectable quality is the prerequisite for molecular studies. If intact and high molecular weight DNA was not acquired, the PCR downstream analysis would face some problems. Generally, the procedure is to grind plant tissue in liquid nitrogen and transfer it to a preheated extraction buffer (Dellaporta et al., 1983; Saghai-Maroof et al., 1984).

The liquid nitrogen is used to freeze the plant tissue and blend it to powder, and minimize DNA degradation when grinding samples. However, liquid nitrogen is not only dangerous that needs particular attention when using it but also volatile, which will add the cost of experiment. What's more, continuous liquid nitrogen supply is a problem in many developing countries. If plant tissue could be ground in room temperature instead of in liquid nitrogen, then the DNA could be extracted in a shorter time at a lower cost, and molecular studies could be performed as usual even when liquid nitrogen is unavailable. Therefore, many researchers have reported some DNA isolation methods without liquid nitrogen. Ouenzar et al. (1998) eliminated the use of liquid nitrogen by grinding the 0°C precooled plant tissue in -20°C precooled mortar and pestle with precooled extraction buffer. Biswas et al. (2011) ground precooled finely chopped plant tissue and dried tissue in precooled mortar and pestle. Ferdous et al. (2012) ground Rice leaf and seed tissues with 600 μL extraction buffer by mortar and pestle in room temperature. Sharma et al. (2003) developed a protocol in which leaves were fixed in alcohol before CTAB DNA extraction, making liquid nitrogen unnecessary. Rajendrakumar et al. (2011) soaked the dehusked rice seed or grain in 600 ul extraction buffer for 30 to 40 min at 37°C in a sterile 1.5 ml micrcentrifuge tube and ground the sample using a sterile micro pestle till the tissue disintegrates.

According to Doyle et al. (1987), plant DNA could be extracted by grinding 0.5 to 1.0 g fresh leaf tissue in 60°C CTAB isolation buffer in a preheated mortar. In this research, the effects of three different grinding methods (including ground in liquid nitrogen, in preheated mortar with extraction buffer and in non-preheated mortar with extraction buffer) applied in CTAB method described by Doyle et al. (1987) and SDS method described by Dellaporta et al. (1983), respectively, were compared. The aim was to see if high quality DNA could be acquired by directly grinding plant tissue in extraction buffer without liquid nitrogen.

MATERIALS AND METHODS

Plant material

The cowpea cultivar (Cheng-jiang 7), which belonged to *Vigna unguiculata ssp. sesquipedalis*, was provided by the Chengdu Academy of Agriculture and Forestry Sciences, China. Seeds were grown under greenhouse conditions and leaves were harvested from two-week-old seedlings for DNA isolation.

DNA extraction

DNA was extracted using CTAB method and SDS method, respectively. 100 mg of the leaf tissues were weighed in an electronic balance.

CTAB method

Firstly, the leaf tissue were ground into a fine powder in liquid nitrogen by a pestle and mortar, transferred into a 1.5 ml microfuge tube, 500 ul CTAB extraction buffer was added (100 mM Tris-HCl, pH 8.0; 1.4 mM NaCl; 20 mM EDTA; 2% CTAB; 0.2% 2-mercapto-ethanol), and incubated in 60°C for 30 min. Secondly, the leaf tissue was put in 60°C preheated mortar and pestle, 500 ul.

CTAB extraction buffer was added, and ground for 30 s, transferred into a 1.5 ml microfuge tube, and incubated in 60°C for 30 min. Thirdly, the leaf tissue was ground in non-preheated mortar and pestle (room temperature) with 500 ul CTAB extraction buffer for 30 s, transferred into a 1.5 ml microfuge tube, and incubated in 60°C for 30 min. The next steps were completely followed by Doyle et al. (1987). Each grinding method was replicated ten times.

SDS method

The grinding methods were the same as for CTAB method descrybed above, but with 500 ul SDS extraction buffer (100 mM Tris, pH 8; 50 mM EDTA, pH 8.0; 500 mM NaCl, 10 mM mercaptoethanol). After the grinding, the mixture was transferred into a 1.5 ml microfuge tube, 50 ul 20% SDS was added and incubated in 65°C for 10 min. The next steps were completely followed by Dellaporta et al. (1983). Each grinding method was replicated ten times.

RNAse treatment

After DNA was dissolved in 100 ul TE buffer (10 mM Tris-cl pH 7.4, 1 mM EDTA, pH 8.0), RNAse treatment was carried out according to the method of Doyle et al. (1987).

DNA concentration and purity

The concentration of the extracted DNA was determined spectrophotometrically with an Eppendorf Biophotometer. 5 ul of the DNA was diluted 1:10 in Tris-EDTA buffer (pH 8.0). The DNA concentration and OD260/280 ratio were given by the machine. OD260/280 readings ratio were taken as an indicator of DNA purity (Sambrook et al., 1989).

Restriction analysis

DNA was restricted by 15 units of Hind III using approximately 1 ug of DNA. The reaction mixture was incubated at 37°C for 2 h. After wards, undigested and digested DNA were separated on 0.8% agarose gel at 90V for 100 min, then stained with ethidium bromide for 20 min, and photographed in SYNGENE-GeneGenius.

PCR amplication

Functionality of DNA extractions were tested with two types of PCR. First, a 195 bp sequence containing SSR derived from EST sequence of *V. unguiculata* (downloaded from NCBI EST database, http://www.ncbi.nlm.nih.gov/nucest) was amplified. Primers were designed by Primer3 software with Forward sequence: 5'-CCTTGGAAGGCTTTGCAACC-3', reverse sequence: 5'-ATTCCTCTGCCTGCTCCATG-3'. The primers were synthesized by Life Technologies (AB & Invitrogen). PCR (50 ul volumes) contained approximately 100 ng of genomic DNA, 0.25 ul Taq (5U/ul, Fermentas), 5 ul 10× PCR reaction buffer, 3 ul $MgCl_2$ (25 mM), 4 ul dNTP mixture(2.5 mM), 1 ul each primer (20 uM), and dH_2O up to 50 ul. The amplification was performed in a Peltier Thermal Cycler (BIO-RAD DNAEngine). Cycling conditions consisted of 5 min initial denaturation at 95°C, followed by 1 min denaturing at 95°C, 1 min annealing at 55°C and 1 min extension at 72°C repeated for 40 cycles and 5 min extension at 72°C. PCR products were subsequently separated with 2% agarose gel electrophoresis at 80 V for 90 min, then stained and photograped as described above.

Secondly, a 773 bp fragment from the chloroplast gene for the photosystem II protein D1 (psbA) of *V. unguiculata* was also amplified. Primers were designed from the psbA photosystem II protein D1 sequence of *V. unguiculata* taken from Genbank database (GeneID:

Table 1. DNA yields and OD 260/280 ratios obtained from different grinding methods in CTAB and SDS methods.

Extraction method	Grinding method	Yield (ug)	OD260/280
	Liguid nitrogen	4.70±0.13	1.80±0.03
CTAB	Room temp mortar	11.53±1.28	1.83±0.04
	Preheat mortar	13.80±1.13	1.84±0.02
	Liguid nitrogen	4.20±0.20	1.77±0.02
SDS	Room temp mortar	14.17±3.03	1.76±0.06
	Preheat mortar	14.86±3.97	1.73±0.03

Values are mean ± SE.

13080510, http://www.ncbi.nlm.nih.gov/gene/13080510) using Primer-BLAST software (http://www.ncbi.nlm.nih.gov/tools/primer-blast/): forward primer PD1_F with sequence 5'-GGCCAAGCAGCTAGGAAGAA-3', reverse primer PD1_R with sequence 5'-ACCAGCACCGAAAATCGTCT-3'. The primers were synthesized by Life Technologies (AB & Invitrogen). PCR (50 ul volumes) contained the same mixture and was performed with the same PCR machine as used for SSR amplification using the same settings. PCR products were subjected to 1.5% agarose gel electrophoresis at 75 V for 80 min, then stained and photograped as described above.

Data analysis

The data analysis was followed by Chen et al. (2010). The general linear model (GLM) was applied to test the effects of grinding methods and extraction methods on the DNA yield and on the OD260/280 ratio. Tukey's pairwise comparisons with the confidence interval of 95% were used to compare the yield or the ratios between the grinding methods and extraction methods. The statistical analyses were accomplished by using the MINITAB® software Release 14.20 (www.minitab.com).

RESULTS

Yield of the DNA extracted by the CTAB method and SDS method with three different grinding methods are listed in Table 1. With the same grinding method, for example, ground in liquid nitrogen, the DNA yield had no significant difference between CTAB and SDS extraction methods ($F = 4.26$, $df = 1$, $P = 0.066$). The same results were obtained with preheated mortar ($F = 0.30$, $df = 1$, $P = 0.595$) and non-preheated mortar ($F = 1.40$, $df = 1$, $P = 0.264$).

In CTAB method, the DNA yield had no significant difference between grinding the leaf tissue in preheated mortar and non-preheated mortar with 500 ul extraction buffer; both of which obtained significantly higher DNA yield than grinding in liquid nitrogen (Turkey's, 95%). In SDS method, the results were the same.

The assessment of the purity of a nucleic acid sample is often performed by a procedure commonly referred to as the OD260/280 ratio. A pure sample of DNA has the ratio at 1.8 (Chen et al., 2010). The mean OD260/280 ratios of DNA isolated by the three different grinding methods with both CATB and SDS methods had no significant difference (Turkey's, 95%). They were all close to 1.8, which indicated that all the DNA extractions were pure. This was further proved by agarose gel electrophoresis (Figure 1).

The integrity, that is presence of high molecular genomic DNA, was determined by electrophoresis on a 0.8% agarose gel. High molecular DNA bands with no smear were obtained from both CTAB method and SDS method by ground in liquid nitrogen (Figure 1; lanes 1 and 7), indicating that the DNA were intact and pure while DNA isolated by grinding the plant tissue in preheated and non-preheated mortar with extraction buffer in both CTAB and SDS methods showed high molecular DNA bands with little smear (Figure 1; lanes 3, 5, 9 and 11), demonstrating that the DNA were intact but there existed some degraded DNA.

To test the DNA function, the DNA was digested with restriction enzyme Hind III, and amplified in two PCR reactions. As shown in Figure 1, the DNA was completely digested with Hind III restriction enzyme, as evidenced by the characteristic "smearing" and the absence of the high molecular weight bands seen in the adjacent lane of undigested DNA.

Besides, the SSR (Figure 2) and the target fragment from the psbA photosystem II protein D1 sequence of *V. unguiculata* were successfully amplified with respect to all the DNA extractions (Figure 3).

DISCUSSION

From the data in Table 1 and agarose gel in Figure 1, DNA isolated by grinding plant tissue in liquid nitrogen was intact and pure while the DNA extracted by grinding leaf tissue in preheated and non-preheated mortar with extraction buffer degraded a little. However, as the DNA yield of ground in preheated and non-preheated mortar were twice as many as ground in liquid nitrogen, there exists considerable intact and high quality DNA which could be used in downstream PCR analysis. When grinding plant tissue in preheated or non-preheated mortar and pestle, the CATB or SDS extraction buffer could provide some protection to the plant tissue, and this is theoretically feasible. Tris maintains the pH of the solution

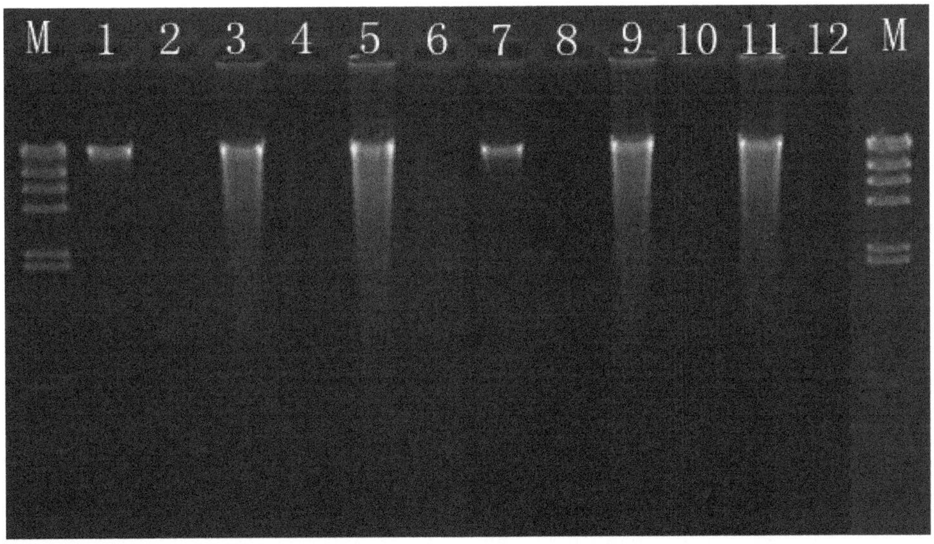

Figure 1. Agarose gel of undigested and digested cowpea DNA. The isolated DNA was digested by the restriction enzyme Hind III. Lanes designated (M) are lambda/Hind III molecular weight markers (Fermentas). Lanes 1-6, the DNA isolated by CTAB method in the order ground in liquid nitrogen, non-preheated mortar with extraction buffer, preheated mortar with extraction buffer; lanes 7-12, the DNA isolated by SDS method in the same grinding order as described above, alternating undigested and digested DNA.

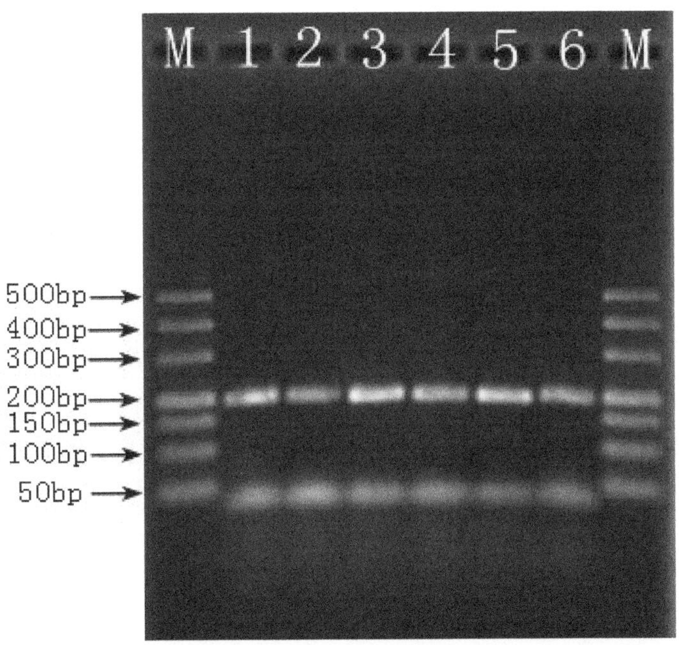

Figure 2. PCR amplification of the SSR from the cowpea DNA. Lanes marked (M) are 500 bp molecular weight markers (Fermentas). Lanes 1-6 are the SSR amplified from DNA isolated using CTAB + liquid nitrogen, CTAB + non-preheated mortar, CTAB + preheated mortar, SDS + liquid nitrogen, SDS + non-preheated mortar, SDS +preheated mortar, respectively.

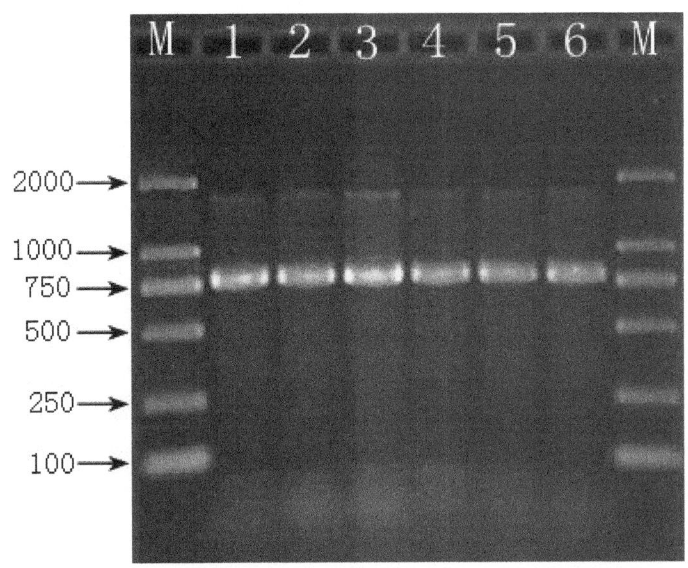

Figure 3. PCR amplification of the partial *psb*A gene from cowpea DNA. Lanes marked (M) are 2000 bp molecular weight markers (Fermentas). Lanes 1-6 are the *psb*A gene fragment amplified from DNA isolated using CTAB + liquid nitrogen, CTAB + non-preheated mortar, CTAB + preheated mortar, SDS + liquid nitrogen, SDS + non-preheated mortar, SDS + preheated mortar, respectively.

and interacts with the lipopolysaccharides present on the outer membrane which helps to permeabilize the membrane. NaCl provides Na^+ ions that will block negative charge from phosphates on DNA, thus contribute to the precipitation of DNA. EDTA binds with Mg-ion and nullifies the action of DNase. CTAB could facilitate in the lysis of cells so DNA can be released into the bulk of the solution. SDS is a strong anionic detergent that can solubilize the proteins and lipids that form the membranes. The extraction buffer was essential. The authors have tried grinding plant tissue in preheated or non-preheated mortar and pestle without any buffer or liquid nitrogen, as a result, negligible DNA was obtained. The reason maybe DNA was cut by the DNase or destroyed by the oxidized polyphenol, but when grinding plant tissue in mortar with extraction buffer, no matter what extraction protocol to choose, CTAB or SDS, there always exists some degraded DNA. This was consistent with the study of Rajendrakumar et al. (2011), who observed that grinding of seed/grain in extraction buffer results in distinct DNA degradation. Maybe during the grinding process, few plant tissue was exposed to the air and polyphenol was oxidized to quinone, or the extraction buffer was too less to cover the entire plant tissue, which means that the EDTA was not able to inhibit the activity of DNase. Although some modified methods obtained high quality genomic DNA without liquid nitrogen (Ouenzar et al., 1998; Biswas et al., 2011; Sharma et al., 2003), they added some other steps, spent more time or money. According to Doyle et al. (1987) and Zheng et al. (1995), fresh leaf tissue can be ground directly in extraction buffer. In this research, complete digestion with restriction endonuclease and successful amplification in PCR indicated that all the cowpea DNA extractions were all successful, which proved that grinding leaf tissue in mortar with extraction buffer to isolate DNA from cowpea is practicable. The mortar and pestle could be preheated or not. Add preheated extraction buffer to the mortar and grind plant tissue directly are able to obtain high quality DNA from cowpea which could be used in molecular analysis.

In this research, only cowpea was studied. Whether the grinding method can be successfully applied in other plant species or not needs to be confirmed by further experiment. It may not be able to obtain ideal results in some species. This may be due to the differences in levels of polysaccharides, fibers and associated anti-quality factors in these plant species (Sharma et al., 2010).

Conclusion

This research proved that grinding cowpea leaf tissue in preheated mortar with extraction buffer was able to obtain high quality DNA which was suitable for molecular study, thus eliminating the use of liquid nitrogen. In addition, after comparative analysis, grinding in non-preheated (room temperature) mortar was also practicable. This grinding method could be tried in other plant species, but whether it will work or not requires confirmation by further experiment.

ACKNOWLEDGEMENT

This work was supported by 'Shuang-Zhi Plan' of Sichuan Agricultural University.

REFERENCES

Agbicodo EM, Fatokun CA, Muranaka S, Visser RGF, Linden-van-der CG (2009). Breeding drought tolerant cowpea: constraints, accomplishments, and future prospects. Euphytica. 167:353-370.

Andargie M, Pasquet RS, Gowda BS, Muluvi GM, Timko MP (2011). Construction of a SSR-based genetic map and identification of QTL for domestication traits using recombinant inbred lines from a cross between wild and cultivated cowpea (*V. unguiculata* (L.) Walp.). Mol. Breeding. 28:413-420.

Badiane FA, Gowda BS, Cisse N, Diouf D, Sadio O, Timko MP (2012). Genetic relationship of cowpea (*Vigna unguiculata*) varieties from Senegal based on SSR markers. Genet. Mol. Res. 11:292-304.

Biswas K, Biswas R (2011). A modified method to isolate genomic DNA from plants without liquid nitrogen. Curr. Sci. 100:16-22.

Chen H, Rangasamy M, Tan SY, Wang H, Siegfried BD (2010). Evaluation of Five Methods for Total DNA Extraction from Western Corn Rootworm Beetles. Plos One. 5:1-6.

Dellaporta SL, Wood J, Hicks JB (1983). A Plant DNA Minipreparation: Version II. Plant Mol. Biol. Rep. 1:19-21.

Doyle JJ, Doyle JL (1987). A rapid DNA isolation procedure for small quantities of fresh leaf tissue. Phytochemical Bulletin. 19:11-15.

FAO (2010). FAO statistical databases. Online Available: http://faostat.fao.org.

Ferdous J, Hanafi MM, Rafii MY, Muhammad K (2012). A quick DNA extraction protocol: Without liquid nitrogen in ambient temperature. Afr. J. Biotechnol. 11:6956-6964.

Kongjaimun A, Kaga A, Tomooka N, Somta P, Shimizu T, Shu Y (2012). An SSR-based linkage map of yardlong bean (*Vigna unguiculata* (L.) Walp. *subsp.unguiculata Sesquipedalis* Group) and QTL analysis of pod length. Genome. 55: 81-92.

Malviya N, Sarangi BK, Yadav MK, Yadav D (2012). Analysis of genetic diversity in cowpea (*Vigna unguiculata* L.Walp.) cultivars with random amplified polymorphic DNA markers. Plant Syst. Evol. 298:523-526.

Muchero W, Diop NN, Bhat PR, Fentonb RD, Wanamakerb S, Pottorffb M (2009). A consensus genetic map of cowpea [*Vigna unguiculata* (L) Walp.] and synteny based on EST-derived SNPs. PNAS. 106:18159-18164.

Muchero W, Ehlers JD, Roberts PA (2010). QTL analysis for resistance to foliar damage caused by *Thrips tabaci* and *Frankliniella schultzei* (Thysanoptera: Thripidae) feeding in cowpea [*Vigna unguiculata* (L.) Walp.]. Mol. Breeding. 25: 47–56.

Ouenzar B, Hartmann C, Rode A, Benslimane A (1998). Date Palm DNA Mini-Preparation without liquid nitrogen. Plant Mol. Biol. Rep. 16:263-269.

Ouedraogo JT, Gowda BS, Jean M, Close TJ (2002). An improved genetic linkage map for cowpea (*Vigna unguiculata* L.) combining AFLP, RFLP,resistance traits. Genome. 45:175-188.

Rajendrakumar P, Sujatha K, Rao KS, Nataraj KP, Viraktamath BC, Balachandran SM, Biswal AK, Sundram RM (2011). A protocol for isolation of DNA suitable for rapid seed and grain purity assessments in rice. Rice Genetics Newsletter. 23: 92-95.

Saghai-Maroof MA, Soliman KM, Jorgensen RA, Allard RW (1984). Ribosomal DNA spacer-length polymorphisms in barley: Mendelian inheritance, chromosomal location, and population dynamics. PNAS. 81:8014-8018.

Sambrook J, Fritsch EF, Maniatis T (1989). Molecular cloning: A Laboratory Manual. Cold Spring Harbor Laboratory Press, Cold Spring Harbor, New York.

Sharma R, Mahla HR, Mohapatra T, Bhargava SC, Sharma MM (2003). Isolating Plant Genomic DNA without Liquid Nitrogen. Plant Mol. Biol. Rep. 21:43-50.

Sharma P, Joshi N, Sharma A (2010). Isolation of genomic DNA from medical plants without liquid nitrogen. Indian J. Exp. Biol. 48:610-614.

Singh BB, Chambliss OL, Sharma B (1997). Recent advances in cowpea breeding. In Singh BB, Mohan-Raj DR, Da-shiell KE, Jackai LEN, eds. Advances in cowpea research. Co-publication of International Institute of Tropical Agriculture (IITA), Ibadan, Nigeria and Japan International Research Centre for Agricultural Sciences (JIRCAS), Tsukuba, Ibaraki, Japan.

Xu P, Wu XH, Wang BG, Liu YH, Ehlers JD, Close TJ (2011). A SNP and SSR Based Genetic Map of Asparagus Bean (*Vigna. unguiculata ssp. sesquipedialis*) and Comparison with the Broader Species. Plos One. 6:1-8.

Zannou A, Kossou DK, Ahanchede A, Zoundjihekpon J, Agbicodo E, Struik PC (2008). Genetic variability of cultivated cowpea in Benin assessed by random amplified polymorphic DNA. Afr. J. Biotechnol. 7:4407-4414.

Zheng K (1995). PCR-based marker assisted selection in rice breeding. IRRI Discussion Paper Series No. 12. International Rice Research Institute, Manila, Philippines.

15

Prehydrolyzed dietary protein reduces gastrocnemial DNA without impairing physical capacity in the rat

Viviane Costa Silva Zaffani[1], Carolina Cauduro Bensabath Carneiro-Leão[1], Giovana Ermetice de Almeida Costa[1], Pablo Christiano Barboza Lollo[1], Emilianne Miguel Salomão[2], Maria Cristina Cintra Gomes-Marcondes[2] and Jaime Amaya-Farfán [1]

[1]Department of Food and Nutrition at the University of Campinas (UNICAMP), Campinas, Brazil.
[2]Department of Anatomy, Cell Biology and Physiology, at the University of Campinas (UNICAMP), Campinas, Brazil.

Previous studies have shown that hydrolyzed proteins exhibit antioxidant properties and may confer physical and physiological advantages when consumed by the exercising rat. The purpose of this study was to compare the effects of feeding either intact (I) and partly hydrolyzed (H) milk whey proteins on gastrocnemial DNA contents and protein metabolism in exercising Wistar rats. Protein synthesis and degradation, protein and DNA contents, and concentration of the serum insulin-like growth factor-1 (IGF1) were determined in six experimental groups according to the type of protein consumed [casein (C), whey protein isolate (I), hydrolyzed whey protein (H) and level of physical activity (sedentary (S) and trained (T)]. H produced significantly lower rates of protein synthesis and degradation and DNA contents in the gastrocnemius, while no differences were observed in the total muscle protein content and serum levels of IGF1. These results indicate that consumption of prehydrolyzed whey protein alters muscle metabolism resulting in less DNA, but maintains the muscle protein levels constant and sustain or improves physical performance, compared to the unhydrolyzed protein.

Key words: Dietary protein, hydrolyzed whey protein, peptides in muscle metabolism, phenylalanine, tyrosine, physical activity.

INTRODUCTION

Different food proteins may affect muscle metabolism in different fashions, as is known to occur when the milk whey proteins are compared with casein. One difference that has been reported between the whey protein concentrate compared with casein, is that the whey proteins stimulate the increase of fatty acid synthesis in the muscle, accompanied by a concomitant decrease in the liver, which was considered to be a positive effect on lipid metabolism (Morifuji et al., 2005). Hydrolyzing the protein alone could in itself alter the biological function of the protein, thus affecting metabolism (Meisel, 1998). It has been suggested, for instance, that the mere change of the physicochemical form in which the protein is presented to the animal suffices to influence the general metabolism, apparently as a result of the various peptides that are generated during partial enzymatic hydrolysis of the whey proteins (Morifuji et al., 2009; Nery-Diez et al., 2010; Faria et al., 2012). Alteration in glycogen metabolism has been perhaps the most often reported positive feature resulting from substituting a whey protein hydrolyzate for the native

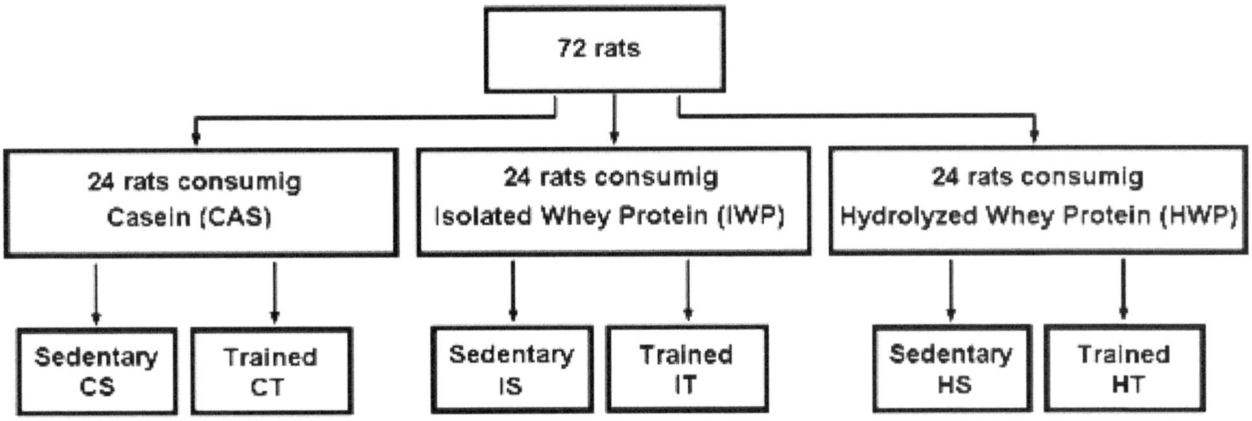

Figure 1. Grouping of animals by type of diet and physical activity. CS, casein diet and sedentary; CT, casein diet and trained; IS, whey protein-isolate diet and sedentary; IT, whey protein-isolate diet and trained; HS, whey protein-hydrolyzate diet and sedentary; HT, whey protein-hydrolyzate diet and trained.

protein (Morifuji et al., 2009; Faria et al., 2012). Morifuji et al. (2009) showed that the peptide Ile-Leu, from among seven possible dipeptides containing branched-chain amino acids from the milk-whey proteins, was capable of increasing the uptake of glucose by isolated rat muscle. The reports suggest that far from being unsafe, these alterations are consistent with the observed physiological advantages, such as greater endurance exercise performance, better post-exhaustion glycemic levels, and preserved blood total protein and albumin levels (Morifuji et al., 2009; Faria et al., 2012; Lollo et al., 2011).

The various advantages pointed out for the whey protein hydrolyzates are consistent also with a possible beneficial effect to the experimental animal, as has been the case of physical performance in rats undergoing endurance exercise (Morifuji et al., 2009; Pimenta et al., 2005) and promoting the synthesis of glycogen (Morifuji et al., 2009). Nery-Diez et al. (2010) additionally have suggested an anti-stress effect associated with the consumption of the whey protein hydrolyzate in rats undergoing endurance exercise.

Moreover, it has been observed that a whey protein hydrolyzate is capable of reducing the levels of creatine kinase in both exercising rats (Abecia-Soria, 2010) and soccer players (Lollo et al., 2011), which was originally understood to be the result of the antioxidant capacity of some bioactive peptides present in the hydrolyzate (Peng, Siong and Kong, 2009; Ren et al., 2008).

The present study was designed to investigate the possible effects produced by the consumption of a whey protein hydrolyzate, in comparison with those of a whey protein isolate. The comparisons included several parameters of muscle protein metabolism that included the rates of gastrocnemial protein synthesis and degradation, the concentration of serum insulin-like growth factor-1 (IGF1), and the gastrocnemial protein and DNA contents in Wistar rats. These parameters were assessed for the two types of protein under two states of physical activity, sedentary and exercised, using four experimental and two casein control groups.

MATERIALS AND METHODS

Biological assay

Male Wistar (144 twenty-one-days old, specific-pathogen free) rats, bred at the Multidisciplinary Center for Biological Research, University of Campinas, SP, Brazil, were housed (~22°C, 55% RH, inverted 12-h light cycle) in individual growth cages, with free access to commercial chow (Labina, Purina, Brazil) and water at all times, until they became six weeks old. At 42 days and still consuming the commercial chow, the rats were tested for their running aptitude in order to exclude those that did not pass the initial stage of the protocol. The selected animals were then segregated into six groups according to diet and physical activity (Figure 1). The research methodology was approved by the Ethics Committee on Animal Experimentation (CEEA-UNICAMP, protocol 1340-1/2007).

Experimental diets

The diets were isonitrogenous (approximately 14% protein, dry basis), isolipidic and isocaloric (approximately 360 kcal/100 g) and formulated following the recommendations of the American Institute of Nutrition, AIN93-M diet (Reeves and Nielsen, 1993). The three diets differed among themselves with respect to the nature of the protein source, considering that in one case the whey protein was prehydrolyzed (Hilmar product 8360) and in the other, the protein was not (Hilmar product 9410). The prehydrolyzed whey protein exhibited a degree of hydrolysis of approximately 12.5%. Both the whey proteins were gifts of Hilmar Ingredients (Dalhart, TX, USA). The standard reference protein was casein. Food consumption was determined every other day and body mass evolution monitored once per week.

Training protocol

The rats were selected in the treadmill beforehand, considering as not fit those that stood for 30 s without departing from the baseline. The animals began training on the day they started to consume the experimental diets and lasted for nine weeks following the progressive-resisted protocol of Hohl et al. (2009) Physical performance was determined by the last phase of this method, as the animals were brought to exhaustion. The sedentary rats were also subjected to a minimum of physical activity, running at the speed of 15 m/min for 10 min, three times a week.

Sample collection for analysis

In an attempt to minimize the consequences of both acute and chronic stress due to either training or diet, the animals were sacrificed by decapitation, following a 48-h recovery period. Immediately after sacrifice, the blood was collected in plain test tubes, centrifuged, and the serum stored at -20°C for the determination of serum IGF1. The gastrocnemius muscles, for the analysis of protein synthesis and degradation, were dissected, weighed and placed in a Krebs-Henseleit bicarbonate solution. The muscle tissues collected for protein determination were stored in liquid nitrogen until the time of analysis.

Determination of IGF1

The levels of serum IGF1 were determined by radioimmunoassay using the Lincoplex kit for rat/mouse IGF1 (cat # RMIGF1 87k), from Lincis ® Research Inc., according to the manufacturer's recommended procedure.

Muscle protein synthesis

The gastrocnemius muscles were placed in KHB solution (110 mM NaCl, 25 mM $NaHCO_3$, 3.4 mM KCl, 1 mM $CaCl_2$, 1 mM $MgSO_4$ and 1 mM KH_2PO_4, pH 7.4), supplemented with 5.5 mM glucose and 0.01% (w/v) of albumin. The muscles were pre-incubated for 30 min at 37°C with continuous oxygenation (95% O_2 and 5% CO_2), as described by Vary et al. (1998). Next, a new solution KHB supplemented with L-5µ Ci[^{3}H]-phenylalanine/mL was added for another 2 h. At the end of incubation, the tissues were homogenized in 10% of trichloroacetic acid (1:3 w/v), centrifuged at 10,000 x *g* for 15 min, at 4°C. The resulting granules were suspended in 1 M NaOH and incubated at 40°C for 30 min. The radioactivity from the β-emission was measured on a scintillation counter. The pattern of protein synthesis was calculated from the amount of radioactivity incorporated from phenylalanine during the 2 h of incubation, and was expressed in nanomoles of [^{3}H]-phenylalanine, per microgram of muscle protein/h.

Muscle protein degradation

Whole gastrocnemius muscles were placed in RPMI 1640 medium (Sigma-Aldrich, USA) and then pre-incubated for 30 min at 37°C, with continuous oxygenation (95% O_2; 5% CO_2), as described by Vary et al. (1998). This muscle was chosen rather than the soleus because of its significant involvement in the type of exercise imposed on the animals. The pattern of degradation was established by fluorimetric analysis, in terms of nanomoles of tyrosine released per microgram of muscle protein/h, as described by Waalkes and Udenfriend (1957).

Determining total muscle protein

Tissue samples weighing ~50 mg were submitted to protein extraction, in 250 µL of solution of 1M $HClO_4$ at 110°C for 30 min. Then the solution was centrifuged at 1006 *x g* for 30 min. The protein analysis was performed in accordance with the method of Lowry et al., (1951). We used 10 µL of the supernatant from extraction, adding 1 mL of distilled water and 0.9 mL of solution A (0.4 g of Na, K tartrate, 20 g Na_2CO_3, 1 g NaOH and 200 mL of distilled water), followed by incubation in a water bath (50°C) for 10 min. After this, 0.1 mL of solution B (0.4 g of Na,K tartrate, 200 mL of distilled water, 0.2 g of $CuSO_4.5H_2O$, 0.08 g NaOH) was added to the system and allowed to stand at room temperature for 10 min. Next, 3 mL of solution C (10 mL of reactive Folin Ciocalteau and 140 mL of distilled water) were added and the mixture placed in a 50°C water bath for 10 min. Then, the solution was allowed to cool to room temperature and held for reading in a spectrophotometer, set at 650 nm.

Determination of DNA in muscle

Muscle samples (~50 mg) were submitted to protein extraction, in 250 µL of 1M $HClO_4$ solution at 110°C for 30 min. Then the solution was centrifuged at 1006 x *g* for 30 min. The analysis of the content of DNA was carried out in accordance with the method of Giles and Myers (1965). We used 0.5 mL of extract, adding 500 µL of 1M $HClO_4$ and 1.0 mL of solution of diphenylamine. The solution was vortex stirred, and transferred to a water bath (30°C) for 12 h. After cooling to room temperature the reaction mixtures were read in the spectrophotometer set at 595 nm.

Statistical analysis

The results were subjected to statistical analysis using the software Statistical Package for the Social Sciences (SPSS), version 16.0. The data were tested for normality (Kolmogorov-Smirnov test) and homogeneity available therein. For parametric data, two-way analysis of variance was used and means were compared (Tukey test), considering the value of $P \leq 0.05$ as a criterion for statistical significance. In the case of data that did not show normal distribution, the non-parametric Kruskal-Wallis test was used, with statistical significance set at $P \leq 0.05$.

RESULTS AND DISCUSSION

The present study was designed to investigate the possible effects produced by the consumption of a whey protein hydrolyzate, in comparison with those of a whey protein isolate. This is the first time that consumption of a diet with a hydrolyzed whey protein is reported to result in lower amounts of muscle DNA without a detriment to physical performance. At the end of the experiment, the rats were 113 days old and had completed nine weeks of training and consumption of the experimental diets. Age is an important consideration for metabolic studies and in particular for the metabolism of proteins. In the normal Wistar male rat, both DNA and protein increase in skeletal muscle, between 4 and 44 days after birth due to the intense protein synthesis necessary for cell growth (Winick et al., 1972). From there on, the contents of DNA may remain constant until about day 95, while protein continues to rise till day 140. Thus, treatment with different dietary proteins in combination with training is likely to affect muscle composition by changing the contents of DNA and protein.

The results show no observable differences in protein content among treatments CS, IS, HS, CT, IT and HT in the gastrocnemius (Table 1). In spite of the progressive

Table 1. Protein and DNA of gastrocnemius muscle, for the six groups of rats.

Gastrocnemius	Groups of rats, according to both diet and level of activity					
	CS	IS	HS	CT	IT	HT
Muscle protein (mg/100 mg)	12 ± 2	11 ± 3	11 ± 2	12 ± 2	11 ± 1.9	11 ± 1
Total protein content (mg)	230 ± 44	207 ± 58	207 ± 43	234 ± 39	219 ± 38	207 ± 27
*Total protein content (mg/g BW)	0.55 ± 0.1	0.53 ± 0.14	0.51 ± 0.08	0.63 ± 0.12	0.58 ± 0.08	0.55 ± 0.05
DNA in muscle (µg/100 mg)	$23^a \pm 3$	$23^a \pm 5$	$17^b \pm 2.8$	$22^a \pm 1$	$21^{ab} \pm 3$	$17^b \pm 3.2$
Total content of DNA (µg)	$444^a \pm 78$	$419^a \pm 67$	$326^b \pm 55$	$427^a \pm 25$	$406^{ab} \pm 67$	$311^b \pm 80$
*Total DNA content (µg/g BW)	$1.06^a \pm 0.19$	$1.07^a \pm 0.16$	$1.00^a \pm 0.16$	$1.15^a \pm 0.06$	$1.08^a \pm 0.16$	$0.83^b \pm 0.19$
Mean weight of the animals ± SD (g)	455.42 ± 17.51	409.87 ± 8.63	441.03 ± 10.43	359.60 ± 5.02	384.73 ± 4.76	387.15 ± 8.63

*Total protein or DNA content normalized by body weight. The values represent means ± SD. CS, Control casein diet and sedentary; IS, whey protein isolate diet and sedentary; HS, hydrolyzed whey protein diet and sedentary; CT, control casein diet and trained; IT, protein isolate diet and trained; HT, hydrolyzed whey protein and trained. Means with different superscript letters along horizontal lines are different.

endurance training protocol to which the trained animals were subjected, the data suggest that neither protein concentration nor total muscle protein were affected by either physical activity or the state in which the dietary protein was administered to the animals. The weights of the gastrocnemius muscles also did not differ between groups. In the study of Pimenta et al. (2005), no significant changes in muscle protein content were also observed when comparing the whey protein isolate with the hydrolyzate (medium degree of hydrolysis ~8.5 - 9.0 %) in the diets of exercising young rats. Taken together, the results of this experiment suggest that the milk whey proteins, whether intact or prehydrolyzed, whether associated or not with physical endurance activity, did not result in an increase of the contents of muscle protein.

The DNA analyses, however, showed that by simply feeding the prehydrolyzed whey protein resulted in considerably lower concentrations and total contents of DNA in the gastrocnemius, as opposed to feeding the unhydrolyzed proteins, on a body weight basis (groups HS and HT were lowest; Table 1). Knowing that the levels of DNA resulting from consuming the hydrolyzate were significantly different from those of the isolate, the difference could then be due to the large number and nature of the peptides present in the prehydrolyzed protein, which were profusely released in the gut.

In previous experiments, we have shown that the performance of animals fed the hydrolyzate is equal or better than those fed either casein or the isolate protein (Nery-Diez et al., 2010; Pimenta et al., 2005). The fact that the content of DNA in the rats consuming the prehydrolyzed protein was consistently lower than in those consuming the whole proteins, did not mean that the hydrolyzate was disadvantageous to the muscle function because the muscle protein content did not differ between any two groups. In addition, a performance test done at the end of week eight gave the following exhaustion times (seconds): CT: $2{,}687 \pm 397^b$; IT: $2{,}511 \pm 70^b$ and HT: $3{,}266 \pm 779^a$, thus attesting to the good performance capacity of group HT.

Other investigations involving physical exercise and the consumption of hydrolyzed whey proteins have shown that protein hydrolyzates effectively increase liver and muscle glycogen stores (Morifuji et al., 2005; Pimenta et al., 2005), in contrast to the unhydrolyzed protein. By minimizing the demand for endogenous protein for energy use, it is possible that the demand for newly biosynthesized proteins be accordingly lower, thereby resulting in an alleviation of the need to increase the myonuclear DNA.

From the results of protein synthesis measured by the incorporation of [^{3}H]-Phe in the gastrocnemius (Figure 2), it was observed that the trained group consuming the hydrolyzate (HT) incorporated significantly less phenylalanine than any of the sedentary groups, or even the CT group. It should be noted that although the rate of synthesis in group HT was much lower than that in group HS at 48 h (Figure 2), the rate of synthesis of the HT was still commensurate with its rate of degradation (Figure 3), while the rate of synthesis in the sedentary was about 25 times greater. It is also pertinent to note that this state of equilibrium refers to a time period in which there is an inertial positive balance of muscle mass due to the acelerated metabolism of exercise, followed by the cessation of a high demand for protein synthesis (Rennie and Tipton, 2000).

Studies have shown that after consumption of the whey proteins, the appearance of amino acids in plasma is fast, abundant and transient; a profile associated with an increase in protein synthesis (Boirie et al., 1997). Furthermore, leucine, which is an abundant amino acid in whey proteins, has an anabolic effect by working along the intracellular pathways involved in stimulating the synthesis of proteins (Blomstrand et al., 2006). Both leucine and whey protein hydrolyzates have been recognized as being insulinotropic (Power et al., 2009) and, because of this property, in addition to its high protein nutritional value, the physiological response to produce more muscle protein was instead permuted for a higher protein sparing effect, in turn resulting in a lower rate of protein

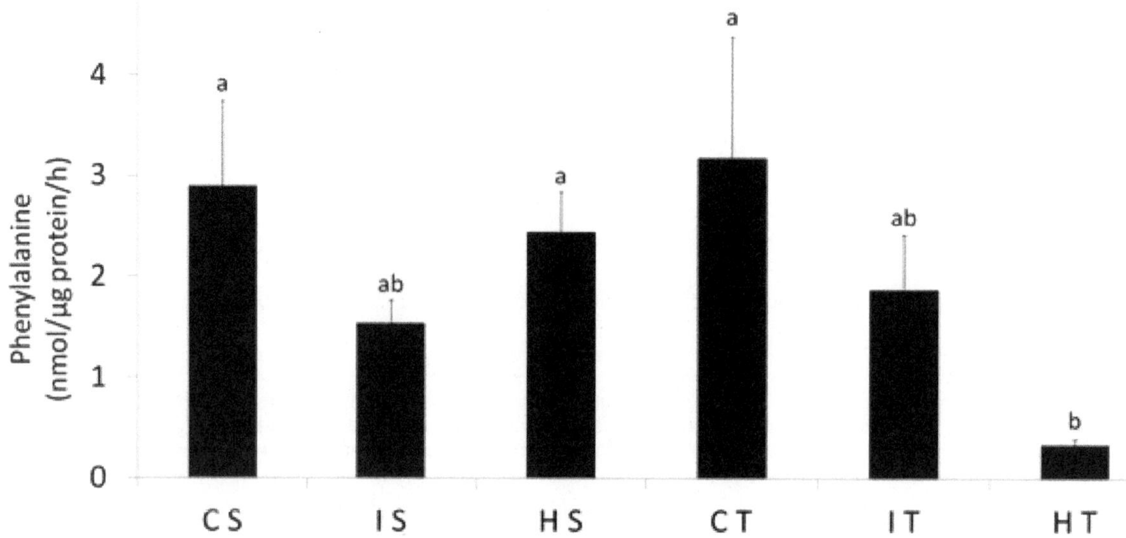

Figure 2. Phenylalanine incorporation in the gastrocnemius by groups. CS, Casein-Sedentary; IS, whey protein isolate-sedentary; HS, whey protein hydrolyzate-sedentary; CT, casein-trained; IT, whey protein isolate-trained; HT, whey protein hydrolyzate-trained). Values correspond to means ± SD. The non-parametric Kruskal-Wallis, with significance of $p < 0.05$ in A, and $p \leq 0.05$ in B was used.

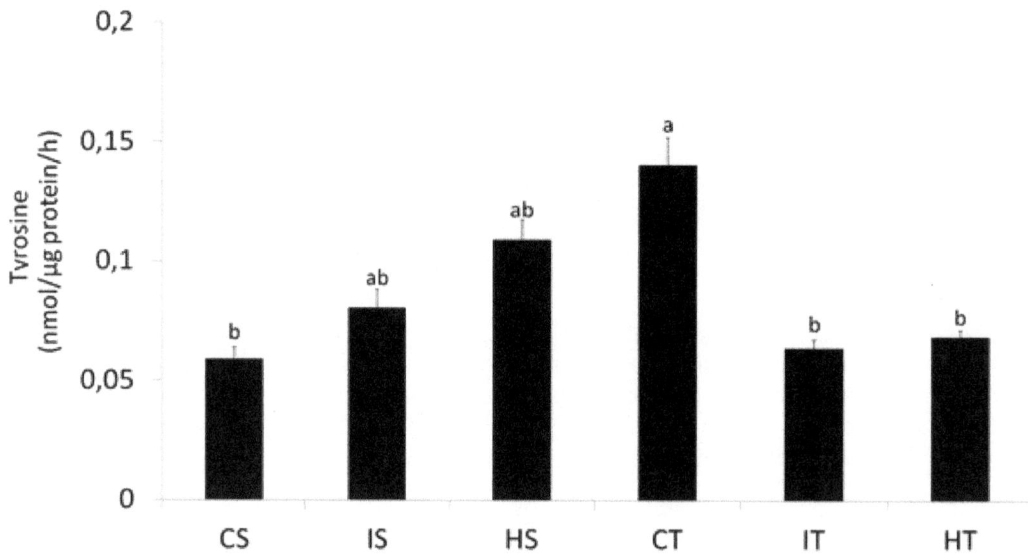

Figure 3. Tyrosine release from the gastrocnemius. (CS, Casein-Sedentary; IS, whey protein isolate-sedentary; HS, whey protein hydrolyzate-sedentary; CT, casein-trained; IT, whey protein isolate-trained; HT, whey protein hydrolyzate-trained). Values are means ± SD. The non-parametric Kruskal-Wallis analysis, at the 0.05 level of significance was used.

synthesis. One advantage of the milk whey proteins in physical activity is that their high leucine content usually produces a positive effect on muscle protein balance because, although endurance exercise may promote a negative balance (Johnson et al., 2004), leucine generally exerts the opposite effect, at both normal and supplemental levels (Garlick, 2005). Our results on the release of tyrosine from the gastrocnemius did not detect any considerable protein catabolism in the muscle (Figure 3), perhaps because leucine, besides stimulating protein synthesis, may also inhibit protein degradation (Garlick, 2005). Consistent with the above observations, the circulating serum IGF-1 levels showed no statistical difference between any two groups of the study (Figure 4), which means that levels were neither higher, nor lower than in the control. It has been suggested that

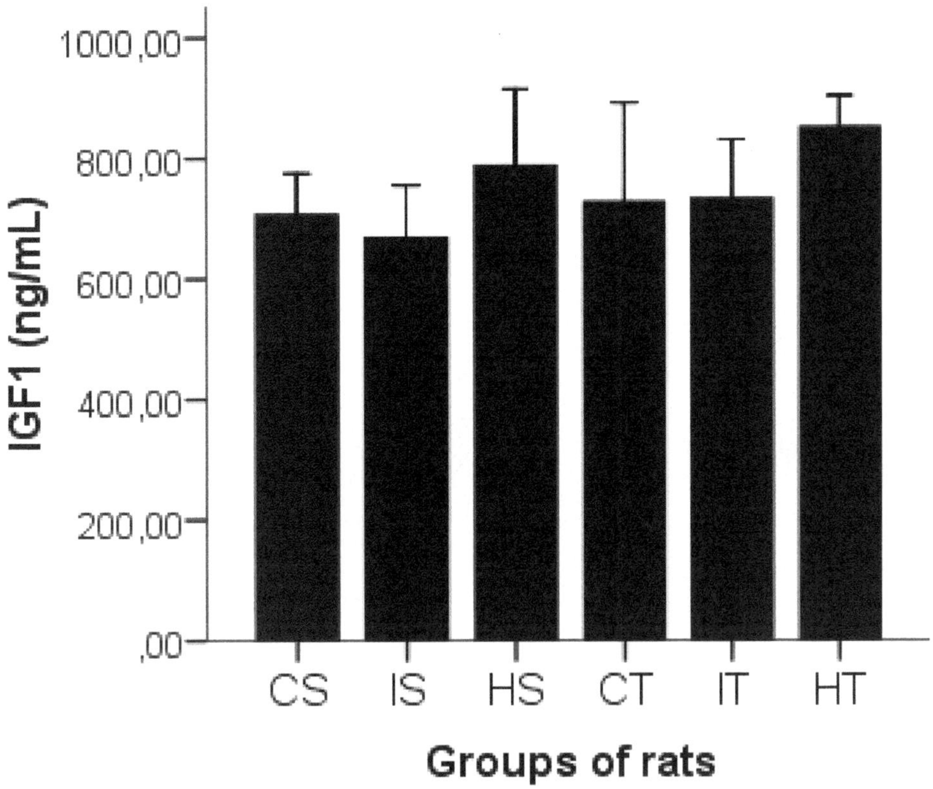

Figure 4. Free IGF-1 (insulin-like growth factor) determined in blood. CS, Casein-Sedentary; IS, whey protein isolate-sedentary; HS, whey protein hydrolyzate-sedentary; CT, casein-trained; IT, whey protein isolate-trained; HT, whey protein hydrolyzate-trained). The data for the six groups represent means ± SD. One-way ANOVA statistical analysis and Tukey, with significance <0.05.

reduction of IGF1 represents an adaptive response to DNA damage (Niedernhofer et al., 2006). On the other hand, heavy physical training is associated with elevated contents of DNA in the skeletal muscle, concomitant with elevated contents of IGF1 (Zentella and Massagué, 1992; Adams and Haddad, 1996). In this case, however, the good physical performance evidenced through the training protocol, in association with the unaltered levels of IGF1, may indicate that the animals were in normal health. Therefore, our results suggest that the observed low DNA levels and the invariant IGF1 levels did not reflect any pathological state or muscle damage, but rather an adaptive response to a muscle-protein sparing effect associated with the consumption of the hydrolyzed protein.

Conclusion

Data from this experiment suggest that consumption of a diet containing prehydrolyzed whey proteins as the sole source of protein can result in lower amounts of gastrocnemial DNA while maintaining normal levels of protein in the rat and the ability to perform physical work at above normal levels. Although this property appears to have a partial dependence on physical activity, the evidence presented here suggests that this is a distinctive characteristic of the hydrolyzate, not shared with the unhydrolyzed proteins, and that the health of the animal is not negatively affected.

Competing interests

The authors declare not to exist conflicts of interest of any type, whether financial or commercial, regarding the conception, objectives or materials used in this investigation.

ACKNOWLEDGEMENTS

The authors acknowledge the Brazilian Research Council for the scholarship granted to VCSZ, CCB, GEAC and PCBL, to FAPESP (Proc 09/00059) and to Hilmar Ingredients for the donation of the whey proteins.

REFERENCES

Adams GR, Haddad F (1996). The relationships among IGF-1, DNA content, and protein accumulation during skeletal muscle hypertrophy. J. App. Phys. 81:2509-2516.

Blomstrand E, Eliasson J, Karlsson HK (2006). Branched-chain amino acids activate key enzymes in protein synthesis after physical exercise. J. Nutr. 136:269S-273S.

Boirie Y, Dangin M, Gachon P (1997). Slow and fast dietary proteins differently modulate postprandial protein accretion. Proc of the Nat Acad of Sci of the USA 94:14930-14935.

Faria IC, Nery-Diez ACC, Lollo PCB, Amaya-Farfán J, Ferreira CS (2012). Myocardial glycogen is increased in sedentary rats fed milk whey protein digests when brought to exhaustion., Afr. J. Biotechnol. 11:2402-2405.

Garlick PJ (2005). The role of leucine in the regulation of protein metabolism. J. Nutr. 135:1553S-1556S.

Giles KW, Myers A (1965). An Improved Diphenylamine Method for Estimation of Deoxyribonucleic Acid. Nature, 206:93.

Hohl R, Ferraresso RL, De Oliveira RB (2009). Development and characterization of an overtraining animal model. Med. Sci. Sports Exerc. 41:1155-1163.

Johnson NA, Stannard SR, Thompson MW (2004). Muscle triglyceride and glycogen in endurance exercise: implications for performance. Sports Med. 34(3):151-164.

Lollo PCB, Amaya-Farfan J, Carvalho-Silva LBC. (2011). Physiological and Physical Effects of Different Milk Protein Supplements in Elite Soccer Players. J. Hum. Kinetics. 30:49-57

Lowry OH, Rosebrough NJ, Farr AL (1951). Protein measurement with the Folin phenol reagent. J. Biological. Chem. 193:265-275.

Meisel H (1998) Overview on Milk Protein-derived Peptides. Int. Dairy J. 8:363-373.

Morifuji M, Koga J, Kawanaka K (2009). Branched-chain amino acid-containing dipeptides, identified from whey protein hydrolysates, stimulate glucose uptake rate in L6 myotubes and isolated skeletal muscles. J. Nutr. Sci. Vitam 55:81-86.

Morifuji M, Sakai K, Sanbongi C (2005). Dietary whey protein increases liver and skeletal muscle glycogen levels in exercise-trained rats. Brit. J. Nutr. 93:439-445.

Nery-Diez AC, Carvalho IR, Amaya-Farfan J (2010). Prolonged ingestion of prehydrolyzed whey protein induces little or no change in digestive enzymes, but decreases glutaminase activity in exercising rats. J. Med. Food 13:992-998.

Niedernhofer LJ, Garinis GA, Raams A (2006). A new progeroid syndrome reveals that genotoxic stress suppresses the somatotroph axis. Nature 444:1038-1043.

Peng X, Xiong YX, Kong B (2009). Antioxidant activity of peptide fractions from whey protein hydrolyzates as measured by electron spin resonance. Food Chem. 113:196–201.

Pimenta FMV, Abécia-Soria MI, Auler F, Amaya-Farfán J (2005). Physical performance of exercising young rats fed hydrolyzed whey protein at a sub-optimal level. Int. Dairy J. 16:984-991.

Power O, Hallihan A, Jakeman P (2009). Human insulinotropic response to oral ingestion of native and hydrolysed whey protein. Amino acids 37:333-339.

Reeves PG, Nielsen FH, Fahey GC (1993). AIN-93 purified diets for laboratory rodents: final report of the American Institute of Nutrition ad hoc writing committee on the reformulation of the AIN-76A rodent diet. J. Nutr. 123:1939-1951.

Ren J, Zhao M, Shi J, Wang J, et al (2008). Purification and identification of antioxidant peptides from grass carp muscle hydrolysates by consecutive chromatography and electrospray ionization-mass spectrometry. Food Chem. 108:727–736.

Rennie MJ, Tipton KD (2000). Protein and amino acid metabolism during and after exercise and the effects of nutrition. Ann. Rev. Nutr. 20:457-483.

Vary TC, Dardevet D, Grizard J (1998). Differential regulation of skeletal muscle protein turnover by insulin and IGF-I after bacteremia. Am. J. Phys. 275:E584-593.

Waalkes TP, Udenfriend S (1957) A fluorometric method for the estimation of tyrosine in plasma and tissues. The J of Lab and Clinic Med 50:733-736.

Winick M, Brasel JA, Rosso P (1972). Nutrition and cell growth. Curr Concepts in Nutr 1:49-97.

Zentella A, Massague J. (1992) Transforming growth factor beta induces myoblast differentiation in the presence of mitogens. Proc. Natl. Acad. Sci. USA 89:5176-5180.

16

DNA fingerprinting of spore-forming bacterial isolates, using *Bacillus cereus* repetitive polymerase chain reaction analysis (Bc-Rep-PCR)

N. Du Rand and M. D. Laing

Discipline of Plant Pathology, School of Agriculture, Earth and Environmental Sciences, College of Agriculture and Environmental Sciences, University of KwaZulu-Natal, Private Bag X01, Scottsville 3201, Pietermaritzburg, South Africa.

Bc-repetitive extragenic palindromic polymerase chain reaction (Bc-Rep PCR) analysis was conducted on seven *Bacillus thuringiensis* isolates accessed from the Deutsche Sammlung von Mikroorganismen und Zellkulturen (DSMZ) culture collection and on five local isolates of entomopathogenic spore-forming bacteria. The five isolates were three strains of *B. thuringiensis,* one strain of *B. cereus* and one strain of *Brevibacillus laterosporus.* All five isolates were distinguished from each other using Bc-Rep PCR analysis. The three *B. thuringiensis* isolates were closely related. The isolate of *B. laterosporus* was not related to any of the *B. cereus* group isolates. Serotyping was also conducted on the five local isolates. However, only one of these isolates could be identified with serotyping and was identified as *B. thuringiensis* subsp. *kenyae.*

Key words: Bc-repetitive extragenic palindromic polymerase chain reaction, *Bacillus* sp., serotyping.

INTRODUCTION

The most commonly used technique for the sub-specification of varieties of *Bacillus thuringiensis* has been H-serotyping (Lecadet et al., 1999). This also has been the most common technique to date used for identifying and characterizing novel *Bacillus* strains (Lecadet et al., 1999). Since the classification by De Barjac and Frachon (1990), the numbers of serotypes has increased from 27 to 69 and is still increasing. Biochemical tests on their own have proven ineffective in classifying *B. thuringiensis* into serotypes (Keshvarshi, 2008). However, in situations of conflicting serotype results, they may be used for clarification (Lecadet et al., 1999).

H-serotyping has several major limitations, which have resulted in a search for better techniques for classifying *B. thuringiensis*. Firstly, H-serotyping is unable to differentiate between *B. cereus* and *B. thuringiensis*.

Secondly, phylogenic relationships between serotypes cannot be drawn (Lecadet et al., 1999). Thirdly, self-agglutinated and non-motile strains cannot be distinguished by H-serotyping (Lecadet et al., 1999). Lastly, only a few laboratories worldwide are able to perform a full H-serotyping assay (Lecadet et al., 1999).

Molecular techniques have been used as alternatives for typing *B. thuringiensis*. The most common techniques include ribosomal DNA restriction fragment polymorphism (RFLP) (Joung and Côte, 2001), Arbitrary Primer PCR technology (Brousseau et al., 1993) and amplified fragment length polymorphism (AFLP) (Pattanayak et al., 2000). Relatively few strains of *B. thuringiensis* have been analyzed using these techniques so far. Early results using these techniques suggest that the members of the *B. cereus* group and the *B. thuringiensis* group could be considered as a single species, and that *B. anthracis* is closely related to these two groups. Techniques such as the Reverse Transcriptase Sequencing of 16S rRNA have demonstrated that there are high levels of sequence similarity (>90%) amongst the members of the *B. cereus* group

*Corresponding author. E-mail: 962082021@stu.ukzn.ac.za.

(Ash et al., 1991). The only attribute that distinguishes these genera from each other is the functional genes that are carried on plasmids (Carlson et al., 1994; Helgason et al., 2000). These pathological and phenotypic attributes may easily be transferred between "species" through the horizontal transfer of plasmids (Helgason et al., 2000).

Methods such as the use of 16S to 23S ribosomal intergenic spacers Sequences (ISR) have failed to aid in the design of species-specific oligonucleotide probes to differentiate between members of this group (Bourque et al., 1995). Ash et al. (1991) stated that techniques using 16S rRNA to attempt to group members of the *B. cereus* group into one species should be supported with other techniques because similar results have been obtained in other genospecies. With more advanced techniques that use very specific primers, differences between these species have been highlighted. Strains of *B. thuringiensis* have a high degree of relatedness but are genetically distinct (Joung and Côte, 2001; Nakamura, 1998). The relatedness between these members is between 85 to 100%. In contrast to earlier research using 16S rRNA, the genetic relatedness between *B. thuringiensis* and *B. cereus* has been demonstrated to be between 60 to 70%, by using sensitive DNA techniques. In order for these two species to be seen as a single species, the relatedness needs to be greater than 70% (Joung and Côte, 2001; Nakamura, 1998). Research by Nakamura (1998) and Joung and Côte (2001) suggest that some serotypes of the *B. thuringiensis* groups should be reassigned. It has also been suggested that some members of *B. thuringiensis* are in fact members of *B. cereus*, but that, due to horizontal plasmid migration, these *B. thuringiensis* species have lost the ability to produce crystal proteins (Hill et al., 2004; Nakamura, 1998; Peruca et al., 2008). For this reason they may be identified as *B. cereus* due to the absence of the crystal protein.

A new technique for fingerprinting and potentially classifying novel strains of *B. thuringiensis* was developed by Reyes-Ramirez and Ibarra (2005). The technique is Repetitive Extragenic Palindrome Sequence-Based PCR analysis (rep-PCR). The technique originated from the designing of PCR primers from Rep sequences found in *Escherichia coli* Migula and *Salmonella typhimurium* genomes. This technique is based on the fact that Rep sequences have been shown to be common in prokaryote genomes and have been used in the design of primers specific to certain species or groups.

Reyes-Ramirez and Ibarra (2005) found a 26- base Rep that is common in the six members of the *B. cereus* group: CCCCACTGATTAAAGTTTCACTTTAT (accordingly named Bc-Rep). This was used to design two specific primers for the *B. cereus* group: 18-mer primer Bc-Rep-1 (5"-ATTAAAGTTTCACTTTAT-3") and 14-mer primer Bc-Rep-2 (5"-TTTAATCAGTGGGG-3"). These primers have been shown to be applicable to *B. thuringiensis*, *B. cereus, B. anthracis* and *B. mycoides.*

Analysis using this technique results in discrete and reproducible patterns. Almost all the serotypes of *B. thuringiensis* have been shown to have distinct patterns (Reyes-Ramirez and Ibarra, 2005). The aim of this study is thus, to determine how closely related the isolates were to each other, and how they compare from known serotypes.

MATERIALS AND METHODS

Bacterial samples and sample collection

Bacterial isolates were obtained from diseased insects such as *T. molitor*, Coleoptera adults and larvae of *Schizonycha* spp. collected in sugarcane producing areas in KwaZulu-Natal, (KZN), and insect rich environments such as compost, grain dust from grain storage bins and *T. molitor* cultures. Mushroom compost was sampled from a mushroom farm (Karkloof, KwaZulu-Natal), compost from an urban garden (Hillcrest, KwaZulu-Natal) and grain dust was sampled from a chicken grain storage facility (Pietermaritzburg, KwaZulu-Natal). Diseased *T. molitor* larvae were collected from the insect rearing facility (University of KwaZulu-Natal). Adult beetles and white grubs were collected from light traps and soil, respectively, from the sugarcane areas in the KwaZulu-Natal midlands, KZN. These were provided by the South African Sugar Research Institute (SASRI), Mount Edgecombe, KZN. All the samples were stored at 4°C in a refrigerator.

Isolation of endospore-forming bacteria

Isolation of *Bacillus* sp. was conducted using a similar pasteurization method as described by Ohba and Aizawa (1978). Approximately 0.5 g of each sample was suspended in 10 ml sterile distilled water. Suspensions were shaken vigorously for 30 s at full speed on a vortex shaker. Mixed suspensions were then left to stand for 1 h at room temperature. Suspensions were vortexed a second time at full speed for 30 s and then subjected to a pasteurization process in a preheated water bath for 10 min at 80°C. After pasteurization, 0.1 ml aliquots of the suspensions were plated out onto nutrient agar (Biolab, Biolab Diagnostics, 259 Davidson Road, Wadeville, Johannesburg) (Thiery and Frachon, 1997).

Plates were incubated for 24 h at 30°C in an incubator and examined for colonies with typical *B. cereus* morphology (fast growing, rhizoid, irregular, raised, smooth, and opaque white) or colonies with an 'ice crystal' appearance with a colony diameter larger than 2 mm (Damgaard et al., 1997; Prescott et al., 1999; Selvakumar et al., 2007). Various other white coloured bacterial colonies that predominated and were larger than 2 mm in diameter were also selected for microscopic evaluation. The number of colonies isolated from each isolate was estimated and counted using a colony counter. Selected colonies were then purified by sub-culture onto nutrient agar and incubated for 3 d at 30C in an incubator. Colonies were stored on nutrient agar slants.

Selected isolates were subjected to Gram staining, as well as, Coomassie Blue staining [Coomassie Blue stain 0.133% (w/v) and methanol 50% (v/v)] for 1 min. Slides were then rinsed gently for 30 s with distilled water and blotted dry with tissue paper (Kimberly-Clark, SA) (Ammons et al., 2002). Isolates were screened for the presence of crystal proteins using light microscopy with a 100 × oil immersion objective lens (Bernhard et al., 1997; Young et al., 1998). Crystal proteins stained dark blue (Ammons et al., 2002). Isolates were screened for size of cells, spore morphology and the presence of crystal proteins. Isolates without crystal proteins, but with cells broader than 0.9 μm, were also selected, in order to include *B. cereus* which does not produce crystal proteins. Rod-shaped bacterial cells containing oval spores were selected because this fits the criteria of *B. cereus, B. thuringiensis* and *B. laterosporus* cells (Thiery and Frachon, 1997). Selected cultures were purified by sub-culturing and were assigned numbers NDR1-

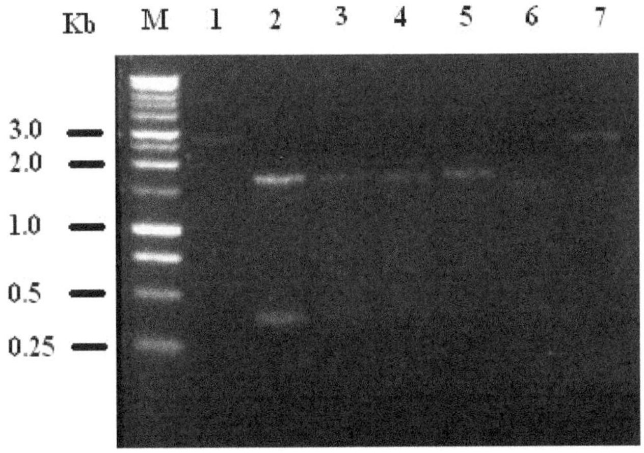

Figure 1. Bc-Rep-PCR fingerprint profiles of the DSMZ *Bacillus thuringiensis* strains. Lane 1, *B. thuringiensis* subsp. *thuringiensis*; Lane 2, subsp. *aizawai*; lane 3, subsp. *kurstaki*; lane 4, subsp. *kenyae*; lane 5, subsp. *nigeriae*; lane 6, subsp. *tochigiensis*; lane 7, subsp. *morrisoni*; lane M, 1 kb molecular weight markers (Promega).

NDR14.

Bacillus thuringiensis samples

Eight known subspecies of *B. thuringiensis* were obtained from the Deutsche Sammlung von Mikroorganismen und Zellkulturen (DSMZ) culture collection (www.dsmz.de). The subspecies were: *B. thuringiensis* subsp. *nigeriae, B. thuringiensis* subsp. *aizawai*, *B. thuringiensis* subsp. *morrisoni*, *B. thuringiensis* subsp. *israelensis, B. thuringiensis* subsp. *kurstaki, B. thuringiensis* subsp. *thuringiensis, B. thuringiensis* subsp. *kenyae* and *B. thuringiensis* subsp. *tochigiensis*.

Inoculum preparation

The isolates were used to inoculate 150 ml of sterile tryptone soy broth (TSB) (Biolab) (Merck, SA) and incubated in a shaker water bath (GFL 1083) at 250 rpm for 24 h at 30°C.

DNA extraction

The DNA was extracted from an overnight culture with Mobio UltraClean™ microbial DNA isolation kit (Mobio, USA).

DNA quantification

A test on the quality and quantity of DNA extracted was conducted electrophoretically as follows: 5 µL of DNA extract aliquots were loaded on a 1.2% horizontal agarose gel slab (Bio-Rad, SA). Samples were run in TAE at pH 8.0 (40 mM Tris-acetate and 1 mM EDTA). Ethidium bromide (Bio-Rad, SA) is a DNA stain that fluoresces under UV light. This was added to the tank of TAE buffer at pH 8.0 (40 mM Tris-acetate and 1 mM EDTA) in order to visualize the DNA bands. Gels were run at 90 V for 55 min.

Primers

The Bc-Rep sequence primers designed by Reyes-Ramirez and Ibarra (2005) were used. The primers were manufactured by Integrated DNA Technologies (IDT, SA) and were as follows: Direct 18 mer primer, 5"-ATT AAA GTT TCA CTT TAT-3" and a 14 mer reverse primer, 5"-TTT AAT CAG TGG GG-3".

Conditions for Bc-Rep-PCR amplification

The PCR mixtures were prepared as follows: 1 µg of template DNA, 15 pmol of each primer, 5 mM $MgCl_2$, 5 µL thermophilic DNA polymerase 10x reaction buffer, 1 µL 10 mM PCR nucleotide mix, 0.25 µl of 5 u/µL *Taq* DNA polymerase and nuclease free water to make the volume up to 50 µL (PCR Core Systems, Promega) (Whitehead Scientific, SA). The conditions of PCR amplification were as follows: 5 min initial denaturation at 94°C, followed by 34 cycles of 1 min denaturation at 94°C, 1 min annealing at 42°C and 1.5 min polymerization at 72°C. The amplification was completed with an extension step of 7 min at 72°C (Reyes-Ramirez and Ibarra, 2005). Amplifications were conducted in an Applied Biosystems Gene Amp PCR System 2400. Amplified samples were stored at -20°C (Conquest).

Electrophoresis

The Bc-Rep-PCR fragments were analyzed as follows: 5 µL aliquots of each of the amplified products were loaded on to a 1.2% agarose slab (110 x 140 mm). The gel was run in TAE buffer pH 8.0 (40 mM Tris-acetate and 1 mM EDTA) at 90V for 55 min. A 1 kb DNA ladder (Promega) (Whitehead Scientific, SA) was used as a molecular weight marker. A gel documentation system, Versa Doc 2000 (Bio-Rad), was used to photograph the gels. Quantity One Version 1.1 computer software was used to analyze the molecular weight patterns (Figures 1 and 2).

Bc-Rep-PCR analysis

The individual patterns of the polymorphic bands of each of the isolates were identified according to the migration rates. An estimation of the banding sizes was used to determine relatedness.

B. thuringiensis serotyping

B. thuringiensis isolates were sent to Dr M. Ohba at the Graduate School of Agriculture, Kyusha University Kyusha University, Hakozaki 6-10-1, Higashi-ku, Fukuoka, B12-8581 in Japan for serotyping. A standard method was used, similar to that developed by Thiery and Frachon (1997).

RESULTS

The Bc-Rep-PCR analysis of the seven DSMZ isolates did not produce many distinguishing bands. The bands from lanes 3 to 6 were very similar to each other (Figure 1) and were estimated at approximately 1.8 kb. Lanes 1 and 7 had similar bands to each other and were distinct from the bands in Lanes 3 to 6 (Figure 1). These bands were estimated at approximately 2.9 kb. Lane 2 had two bands, estimated to be approximately 1.8 and 0.35 kb (Figure 1). This demonstrated that *B. thuringiensis* subsp. *thuringiensis* and *B. thuringiensis* subsp. *morrisoni* are closely related to each. Similarly the *B. thuringiensis* subsp.

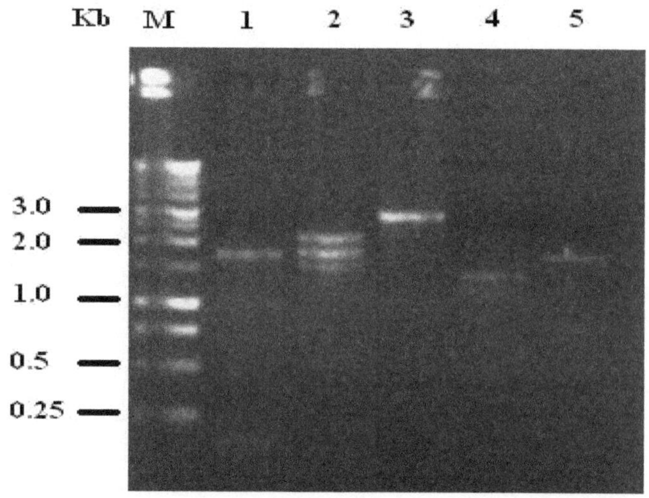

Figure 2. Bc-Rep-PCR fingerprint profiles of the five local, entomopathogenic bacterial isolates. Lane M, 1 kb molecular weight markers (Promega); lane 1, NDR 1 (Bt); lane 2, NDR 3 (Bt); lane 3, NDR 5 (*B. cereus*); lane 4, NDR 2 (*B. laterosporus)*; lane 5, NDR 11 (Bt).

from Lanes 2 to 6 could be regarded as being related to each other, with one common band, although the second band of *B thuringiensis* subsp. *aizawai* separated this subspecies from the other three (Figure 1).

The bands obtained from Bc-Rep-PCR of the five isolates (NDR1, NDR2, NDR3, NDR5, and NDR11) were found to be distinct from each other (Figure 2).This demonstrated that these isolates were not closely related to each other. NDR1 shared a common band with NDR3 and NDR11, at approximately 1.8 kb (Figure 2). Isolate NDR1 and NDR11 were also not distinguishable from each other. NDR3 had three bands at 2.0 kb, 1.8 kb and 1.6 kb respectively (Figure 2). NDR5 produced one clear band at approximately 2.9 kb (Figure 2). NDR4 also produced only one band at approximately 1.5 kb (Figure 2).

Bc-Rep PCR analysis was used to determine the relatedness of the five local entomopathogenic isolates, and to relate them to known serotype of *B. thuringiensis*. Overall, the DNA bands were not as clear and distinct as the bands obtained by Reyes-Ramirez and Ibarra (2005). The bands of the local isolates were markedly more distinct than the bands of the DMSZ isolates, which were not distinct. The bands of the DMSZ isolates were very similar to each other (Figures 1 and 2). The serotyping was done to identify the *B. thuringiensis* strains and demonstrated that this method is limited because only one strain could be identified (Table 1).

DISCUSSION

Serotyping has several limitations, as discussed earlier, which were apparent when only one of three local *B. thuringiensis* strains could be serotyped (Table 1). Such a limited technique cannot be used as the basis for classifying *B. thuringiensis* strains because not all strains can be identified using this technique. The technique of Bc-Rep-PCR analysis is a powerful alternative to seroty-ping that is applicable to all strains of *B. thuringiensis*. However, the technique is difficult to standardize due to the number of variables that require optimization for each isolate. Laboratory conditions may differ from each other and the sources of known isolates may differ. As a result, it will be difficult to produce the same banding patterns at different laboratories (Figures 1 and 2). Equipment, chemicals, conditions and isolates vary from laboratory to laboratory. Resources differ from country to country. It is essential that inter-laboratory calibrations using identical *B. thuringiensis* cultures should be undertaken to ensure that different laboratories generate similar results. Further research in terms of standardization of current laboratory protocols is required in order to determine whether certain variables may be eliminated to create a more streamline process. Problems arose when following the Bc-Rep-PCR protocol of Reyes-Ramirez and Ibarra (2005). They claimed that their technique is quick and easy to conduct. In practice, it has proven time consuming and laborious, and some runs were not successful. The magnesium chloride optimization for the DMSZ group was at 5 mM of $MgCl_2$, similar to that of Reyes-Ramirez and Ibarra (2005). However, the $MgCl_2$ optimization varied for each of the local isolates, ranging from 2 mM to 5 mM. Therefore, the quantity of 5 mM of $MgCl_2$ is not a universally applicable quantity. This is in spite of Reyes-Ramirez and Ibarra (2005) having stated that this fixed quantity is all that is required for optimization.

The Bc-Rep-PCR banding patterns of NDR1, *B. thuringiensis* subsp. *kenyae* (Figure 2), were compared to that of the DMSZ group *B. thuringiensis* subsp. *kenyae* (Figure 1). The banding patterns of these two subspecies of *B. thuringiensis* were found to be similar to each other (Figure 1 and 2). This demonstrates that these two isolates are the same organism but they produce different shaped crystal proteins (Table 1). The Bc-Rep-PCR banding pattern of the *B. thuringiensis* subsp. *morrisoni* isolate obtained from the DMSZ was not similar to that published by Reyes-Ramirez and Ibarra (2005). The banding patterns published by Reyes-Ramirez and Ibarra (2005) did not have a band at 3.0 Kb (Figure 1). This observation confirmed that the current protocol for Bc-Rep-PCR is not as reliable as claimed.

A further confounding factor is that various natural inhibitors may be present in the extracted DNA although, the manufacturer claims that their DNA isolation kit does eliminate inhibitors (Mobio) (Lamboy, 1994). The optimization with $MgCl_2$ does affect PCR reactions and this could be the main source of discrepancies (Lamboy, 1994). This could affect the binding of the primers to the template DNA hence resulting in relatively fewer bands being detected (Lamboy, 1994). Other unknown factors may affect the PCR reactions. These protocols are sensitive and the reproducibility of amplification patterns may be

Table 1. Serotypes and crystal protein shapes of the entomopathogenic bacterial isolates and the DSMZ *Bacillus thuringiensis* strains.

Bacterial sample	Crystal protein shape	Serotype
B. thuringiensis isolate NDR1	Bp$^{\alpha}$	4a, 4c
B. laterosporus isolate NDR2	Cs$^{\alpha}$	_
B. thuringiensis isolate NDR3	Tri$^{\alpha}$	Non-motile
B. cereus isolate NDR5	-$^{\beta}$	_
B. thuringiensis isolate NDR11	S$^{\alpha}$	Strong self agglutination
B. thuringiensis subsp. *thuringiensis*	S$^{\alpha}$	1
B. thuringiensis subsp. *azawai*	S$^{\alpha}$	7
B. thuringiensis subsp. *kurstaki*	Bp$^{\alpha}$	3a, 3b,3c
B. thuringiensis subsp. *kenyae*	S$^{\alpha}$	4a, 4c
B. thuringiensis subsp. *nigeriae*	U$^{\alpha}$	8a, 8d
B. thuringiensis subsp. *tochigiensis*	R$^{\alpha}$	19
B. thuringiensis subsp. *morrisoni*	S$^{\alpha}$	8a, 8b

$^{\alpha}$Bp, Bipyramidal; Cs, Canoe shaped; Tri, triangular; S, spherical, R, rhomboidal, U, undetermined; $^{\beta}$ Does not produce parasporal inclusion bodies.

affected by factors such as precise concentrations, cycling temperature and differences in parameters (Damiani et al., 1996).

DSMZ *B. thuringiensis* subspecies only distinguished between three of the seven subspecies (Figure 1). Four of the *B. thuringiensis* subspecies could not be distinguished from each other using this method (Figure 1). When comparing the results of these two groups of entomopathogenic spore-formers, *B. thuringiensis* subsp. *kenyae* produced a similar sized band as Isolates NDR1 and NDR11 (Figures 1 and 2). The common band shared between NDR1, NDR3 and NDR11 was not surprising because all three isolates are subspecies *B. thuringiensis* (Figure 2). NDR5 and NDR2 did not share any common bands with any of the isolates in Figure 2. This was expected because these two isolates are not *B. thuringiensis* subspecies. However NDR5 did share a common band with *B. thuringiensis* subsp. *thuringiensis* at 2.9kb (Figure 1 and 2). NDR5 is an isolate of *B. cereus* and it has been suggested that *B. cereus* and *B. thuringiensis* may be the same species of *Bacillus*, with the only difference being that *B. thuringiensis* produces crystal proteins (Bernhard et al., 1997; Priest et al., 2004).

The primers used for the Bc-Rep-PCR were meant to be *B. cereus* group specific. However, isolate NDR2 was an isolate of *B. laterosporus* that did not belong to this group. Hence it was surprising to find that the Bc-REP PCR generated clear bands for Isolate NDR2. This result may have been due to random REP-PCR amplification. This demonstrates another flaw in the technique as an alternative to serotyping for *B. thuringiensis* and closely related entomopathogenic bacteria. Only one of the isolates was identified with serotyping (Table 1). No prior research was found that reported toxicity by isolates of *B. thuringiensis* subsp. *kenyae* towards Coleoptera. Hence this isolate is the first of *B. thuringiensis* subsp. *kenyae* reported to show toxicity to this insect order. Serotypes such as *B. thuringiensis* subsp. *morrisoni* are toxic towards Lepidoptera, Diptera and Coleoptera.

Brevibacillus laterosporus was unfortunately not serotyped. Isolate NDR5 (*B. cereus*) could not be serotyped because the Graduate School of Agriculture, Kyusha University in Japan only serotypes *B. thuringiensis* isolates although *B. cereus* can be serotyped (Ohba, 2005). Serotyping is a technique for identifying strains of *B. thuringiensis*. However, its limitations have become apparent as more of *B. thuringiensis* and related bacteria are isolated (Reyes-Ramirez and Ibarra 2005). Some of the problems that have been highlighted include:

1. *B. cereus* and *B. thuringiensis* isolates sometimes have a strong cross reactivity (Ohba and Aizawa, 1986);
2. Serotyping is unable to discriminate between closely related intra-subspecies strains such as *B. thuringiensis* subsp. *entomocidus* and *B. thuringiensis* subsp. *subtoxicus* (Joung and Côté, 2001).
3. Different esterase patterns are present on different serotypes (Lysenko, 1983). Cross-reactions due to these patterns make it difficult to differentiate between the subspecies.
4. Serotyping is not a good indicator of pathogenic specificity (Kriege et al., 1987). *B. thuringiensis* subsp. *morrisoni* isolates have been found that are pathogenic to Lepidoptera, Diptera and Coleoptera (Thiery and Frachon, 1997).
5. Serotyping is a laborious and time-consuming technique that involves the production of flagellar antisera in rabbits (Thiery and Frachon, 1997).

Given these problems, many institutes have since abandoned serotyping. As such, serotyping is probably no longer valuable as a method of identifying *B. thuringiensis* strains. Reyes-Ramirez and Ibarra (2005) could not distinguish between various subspecies such as *B. thuringiensis* subsp. *amagiensis* and *B. thuringiensis* subsp. *seoulensis;* subsp. *sotto* and subsp. *Dakota.* These subspecies can however be identified from each other through serotyping (Reyes-Ramirez and Ibarra, 2005). Databases based on these techniques may thus be of limited value. However, the replacement technology of Bc-Rep-PCR is also not selective enough because different strains cannot be distinguished from each other by this method. This technique needs to be conducted in conjunction with serotyping to identify subspecies. However, this is not always applicable as not all subspecies are serotypable. A more sensitive and more stable technique is required for the classification of *B. thuringiensis* and related bacteria.

ACKNOWLEDGEMENTS

The authors are thankful to the following people: M. Ohba and K. Kagoshima for H-serotyping of some of the isolates and to the team from The South African Sugar Research Institute (SASRI) for the collection and delivery of the white grubs and adult beetles.

REFERENCES

Ammons D, Rampersad J, Khan A (2002). Usefulness of staining parasporal bodies when screening for *Bacillus thuringiensis*. J. Invertebr. Pathol. 79: 203-204.

Ash C, Farrow JAE, Dorsch M, Stackerbrandt E, Collins MD (1991). Comparative analysis of *Bacillus anthracis, Bacillus cereus,* and related species on the basis of reverse transcriptase sequencing of 16S rRNA. Int. J. Syst. Bacteriol. 41: 343-346.

Bernhard K, Jarrett P, Meadows M, Butt J, Ellis DJ, Roberts GM, Pauli S, Rodgers P, Burges HD (1997). Natural Isolation of *Bacillus thuringiensis:* worldwide distribution, characterization, and activity against insect pests. J. Invertebr. Pathol. 70: 59–68.

Bourque SN, Valero JR, Lavoie MC, Levesque RC (1995). Comparative analysis of the 16S to 23S ribosomal intergenic spacer of *Bacillus thuringiensis* strains and subspecies and of closely related species. Appl. Environ. Microbiol. 61 : 1623-1626.

Brousseau R, Saint-Onge A, Préfontaine G, Masson L, Cabana J (1993). Arbitrary primer polymerase chain reaction, a powerful method to identify *Bacillus thuringiensis* serovars and strains. Appl. Environ. Microbiol. 59: 114-119.

Carlson CR, Caugant DA, KolstØ AB (1994). Genotypic diversity among *Bacillus cereus* and *Bacillus thuringiensis* strains. Appl. Environ. Microbiol. 60: 1719-1725.

Damgaard PH, Hansen BM, Pedersen JC, Eilenberg J (1997). Natural occurrence of *Bacillus thuringiensis* on cabbage foliage and in insects associated with cabbage crops. J. Appl. Microbiol. 82: 253-258.

Damiani G, Amedeo P, Bandi C, Fani R, Bellizzi D, Sgaramella V (1996). Bacterial identification by PCR-based techniques. In Adolph KW (ed) Microbial Genome Methods, New York: CRC Press.

De Barjac H, Frachon E (1990). Classification of *Bacillus thuringiensis* strains. Entomophaga 35: 233-240.

Helgason E, Økstad OA, Caugant DA, Johansen HA, Fouet A, Mock M, Hegna I, KolstØ AB (2000). *Bacillus anthracis, Bacillus cereus,* and *Bacillus thuringiensis*- one species on basis of genetic evidence. Appl. Environ. Microbiol. 66: 2627-2630.

Hill KK, Ticknor LO, Okinaka RO, Assay M, Blair H, Bliss KA, Laker M, Pardington PE, Richardson AP, Tonks M, Beecher DJ, Kemp JD, KolstØ AB, Lee Wong AC, Keim P, Jackson PJ (2004). Fluorescent amplified fragment length polymorphism analysis of *Bacillus anthracis, Bacillus cereus,* and *Bacillus thuringiensis* isolates. Appl. Environ. Microbiol. 70: 1068-1080.

Joung KB, Côté JC (2001). A phylogenic analysis of *Bacillus thuringiensis* serovars by RFLP-based ribotyping. J. Appl. Microbiol.91: 279-289.

Keshvarshi M (2008). Isolation, Identification and Differentiation of Local *B. thuringiensis* Strains. J. Agric. Sci. Tech. 10: 493-499.

Kriege A, Schnetter W, Huger AM, Langenbruch GA (1987). *Bacillus thuringiensis* subsp. *tenebrionis,* Strain BI 256-82: a third pathotype within the H-serotype 8a8b. Syst. Appl. Microbiol. 9: 138-141.

Lamboy WF (1994). Computing genetic similarity coefficients from RAPD data: Correcting for the effects of PCR artifacts caused by variation in experimental conditions. Gen. Res. 4: 38-43.

Lecadet MM, Frachon E, Cosmao Dumanoir V, Ripouteau H, Hsmon S, Laurent P, Thiéry I (1999). Updating the H-antigen classification of *Bacillus thuringiensis.* J. Appl. Microbiol. 86: 660-672.

Lysenko O (1983). *Bacillus thuringiensis*: Evolutionary of a taxonomic concept. J. Invertebr. Pathol. 42: 295-298.

Nakamura LK (1998). *Bacillus pseudomycoides* sp. nov. Int. J. Syst. Bacteriol. 48: 1031-1034.

Ohba M, Aizawa K (1978). Serological identification of *Bacillus thuringiensis* and related bacteria isolated in Japan. J. Invertebr. Pathol. 32: 303-309.

Ohba M, Aizawa K (1986). Frequency of a crystalliferous spore-forming bacteria possessing flagellar antigens of *Bacillus thuringiensis.* J. Basic Microbiol. 26: 185-188.

Ohba M (2005). Personal communication.

Pattanayak D, Srinivasan K, Mandaokar AD, Shukla A, Bhalla R, Kumar PA (2000). AFLP fingerprinting and genotypic characterization of some serovars of *Bacillus thuringiensis.* World J. Microbiol. Biotechnol. 16: 667-672.

Peruca APS, Vilas-Boas GT, Arantes OMN (2008). Genetic relationships between sympatric populations of *Bacillus cereus* and *Bacillus thuringiensis*, as revealed by rep-PCR genomic fingerprinting. Memórias do Instituto Oswaldo Cruz **103**: 497-500.

Prescott LM, Harley JP, Klien DA (1999). Microbiology. Fourth Edition. New York: WCB McGraw-Hill.

Priest FG, Barker M, Baillie LWJ, Holmes EC, Maiden MCJ (2004). Population structure and evolution of *Bacillus cereus* group. J. Bacteriol. 186: 7959-7970.

Reyes-Ramirez A, Ibarra JE (2005). Fingerprinting of *Bacillus thuringiensis* type strains and isolates by using *Bacillus cereus* group-specific repetitive extragenic palindromic sequence-based PCR analysis. Appl. Environ. Microbiol. 71: 1346-1355.

Selvakumar G, Mohan M, Sushil SN, Kundu S, Bhatt JC, Gupta HS (2007). Characteristic and phylogenic analysis of an entomopathogenic *Bacillus cereus* strain WGPSB-2 (MTCC 7182) isolated from white grub, *Anomala dimidiate* (Coleoptera: Scarabaeidae). Biocontrol Sci. 17: 525-534.

Thiery I, Frachon E (1997). Identification, isolation, culturing and preservation of entomopathogenic bacteria. In Lacey LA (ed) Manual of Techniques in Insect Pathology, New York: Academic Press, pp 55-77.

Young JM, Chilcott CN, Broadwell A, Wigley PJ, Lecadet MM (1998). Identification of serovars of *Bacillus thuringiensis* Berliner 1915 in New Zealand. N. Z. J. Crop Hort. Sci. 26: 63-68.

Report of *Bufo tihamicus* karyotype from Saudi Arabia

A. H. Al-Shehri[1] and A. A. Al-Saleh[2]

[1]Ministry of Interior, Public Security, Criminal Evidence Administration, P. O. Box 271198, Riyadh 11352, Saudi Arabia.
[2]Department of Zoology, College of Science, King Saudi University, P. O. Box 2455, Riyadh 11451, Saudi Arabia.

This study gives a description of the Karyotype of the Tihama toad *Bufo tihamicus* from Saudi Arabia. Samples of males and females of *Bufo tihamicus* Parker were collected from Gazan Province of Kingdom of Saudi Arabia. The karyotype consists of a diploid number 2n = 22 and the fundamental number NF = 44 in both of the sexes. No heteromorphism in relation to sex chromosome pair was found.

Key words: Amphibian, karyotype, chromosome, *Bufo tihamicus*.

INTRODUCTION

There are few research on the distribution of frogs and toads in the kingdom of Saudi Arabia and neighboring country such as Yemen (Schmidt, 1953; Hass, 1957; 1961; Briggs, 1980,1981), but the work of Balletto et al. (1985) is considered a more precise survey dealing with the morphological and classification of Arabian anura. Nine nominal Anuran species were found, six of which are endemic to the peninsula (*Bufo arabicus, Bufo hadramautinus, Bufo scorteccii, Bufo dhufarensis Bufo tihamicus* and *Euphlyctis ebrenbergii*) and three of which occur in the Palaearctic region (*Bufo viridis, Hyla savignyi, Rana ridibunda*).

Recently, we started to investigate the amphibian chromosomes of the Arabian Peninsula and few paper have been published (Al-Shehri and Al-Saleh, 2005; 2008). In this paper we are going to describe the karyotype and for the first time, *B. tihamicus* from Kingdom of Saudi Arabia using bone marrow cells treated with colchicine *in vivo*.

MATERIALS AND METHODS

Samples of males and females of *B. tihamicus* Parker were collected from Gazan Province of Kingdom of Saudi Arabia. Each sample was injected intraperitoneal with 0.2 ml of colchicine solution (1 mg ML^{-1}) for 24 h before being killed. The bone marrow of the femur was flushed with 5 ml of 0.075 M KCl into centrifuge tube. The cell suspension was kept at room temperature for 15 min and a few drops of 1:3 glacial acetic acid and absolute methanol freshly prepared fixative were added. The cells were pelleted by centrifugation and the supernatant was discarded. A fresh fixative was added to the cells, suspended very well, left for 30 min and centrifuged again. The suspension, fixation and centrifugation of the cells were repeated three times. The slides were prepared by placing two drops of cell suspension on clean and very cold slide then, air-dried. The chromosomes were stained with 10% Giemsa stain in phosphate buffer pH 6.8. More than 20 metaphase chromosomes spreads from 5 males and 5 females were examined under bright field illumination, using 100X oil immersion objective and 10X eyepiece by ordinary Zeiss microscope. The nomenclature proposed by Levan et al. (1964) was followed for the classification of each homologous pair.

RESULTS

This study being the first of its kind, describe the karyotype of *B. tihamicus;* which has been collected from Gazan region, Kingdom of Saudi Arabia. The karyotype consists of a diploid number (2n) = 22 chromosomes. These chromosomes are nine pairs of meta-centric and two pairs of submeta-centric chromosomes (Figures 1 and 2).

The measurement of the length of the chromosomes was calculated for ten different karyotypes and the

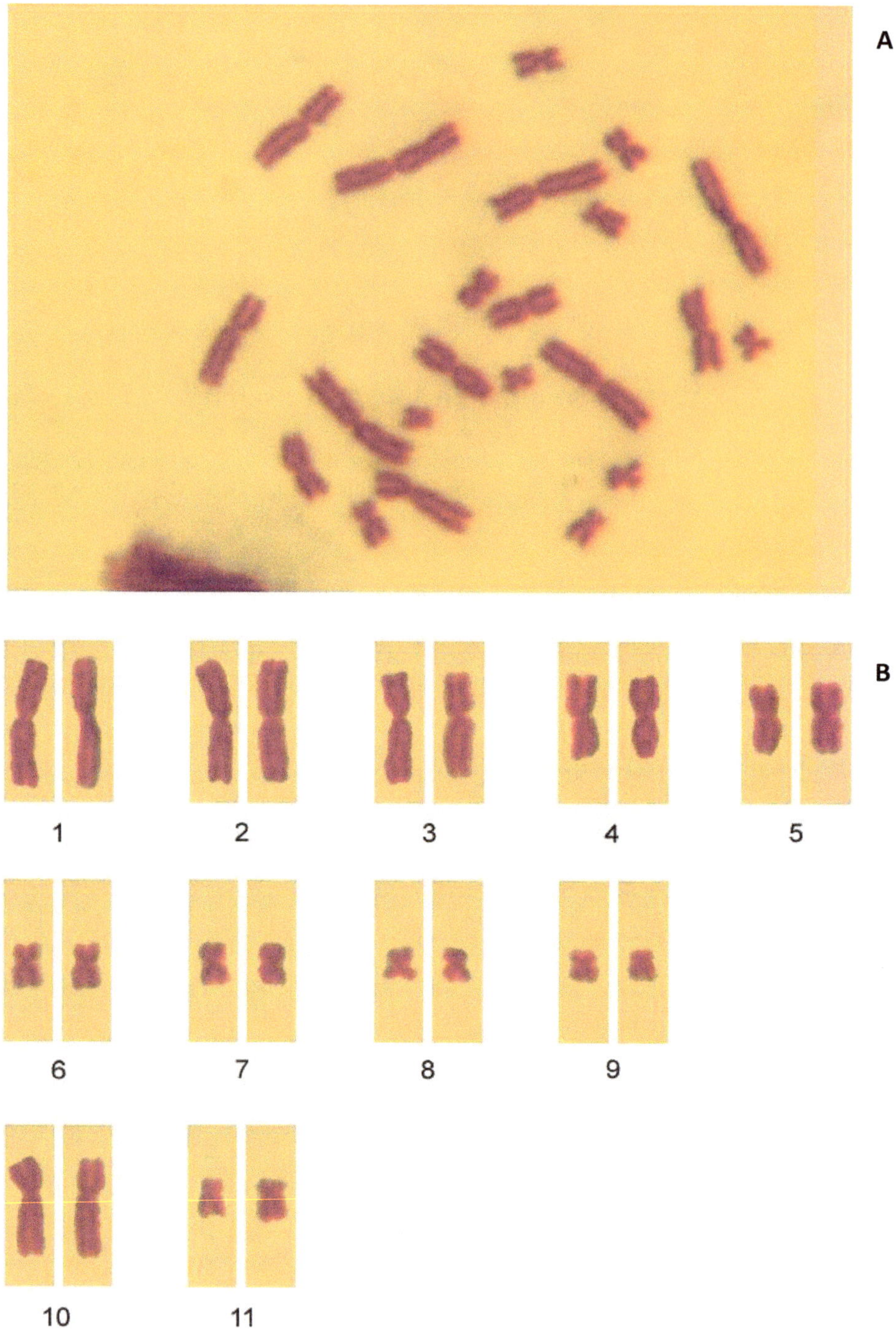

Figure 1. A, Photomicrograph of metaphase spreading from male *B. tihamicus*; B, Karyotype of male *Bufo tihamicus.*

relative length represented in Table 1. Meta-centric chromosomes are classified according to their length into three types: 1) Chromosomes 1-3 are large size chromosomes; 2) Chromosomes 4 and 5 are medium size chromosomes; 3) Chromosomes 6 to 9 are small size chromosomes.

The submeta-centric chromosomes are classified according to the length into 2 types: 1) Chromosome 10 is large size chromosomes; 2) Chromosomes 11 is small size chromosomes

The type of centromeres was determined according to the formula of Matthey and his findings is represented in Figure 3 and the fundamental number (FN) was calculated to be 44 (Table 2). We could not confirm the sex chro-

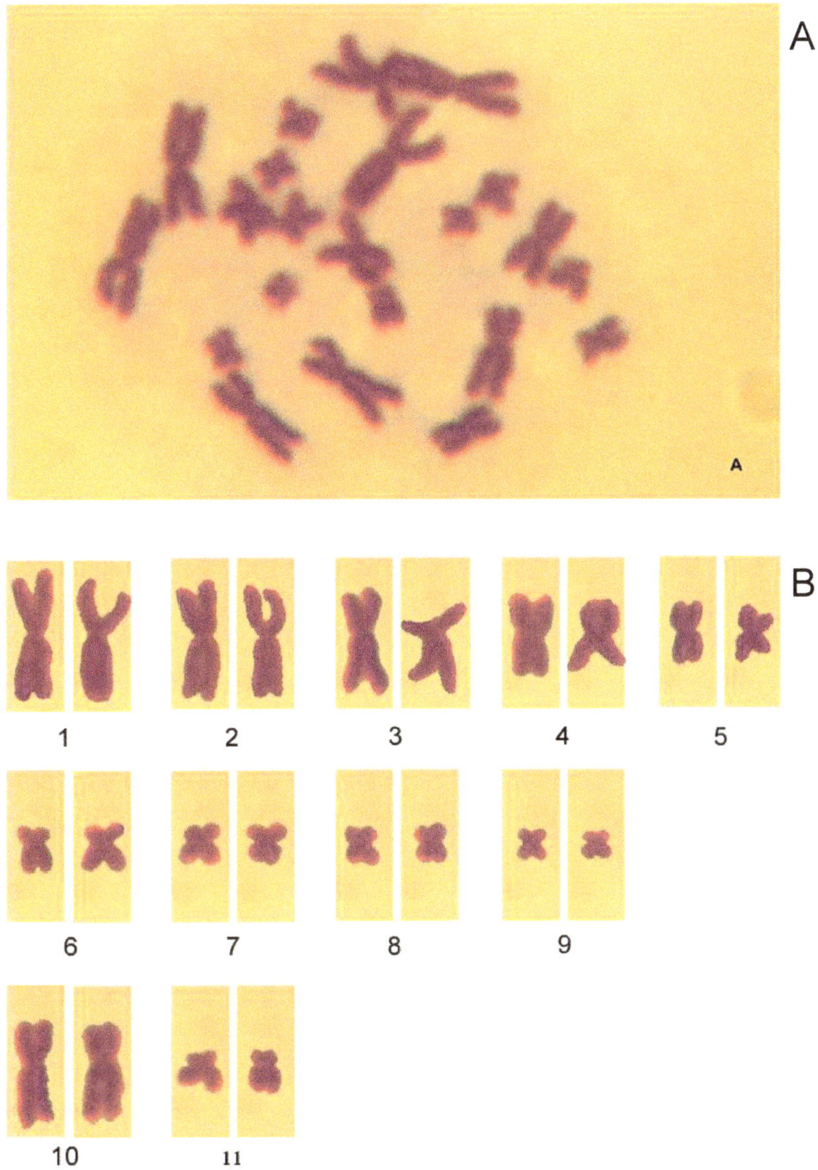

Figure 2. A, Photomicrograph of metaphase spreading from female *B tihamicus*; B, Karyotype of female *Bufo tihamicus.*

Table 1. Arm ratios and type of centromeres of *B. tihamicus.*

Chromosome number	Short arm p	Long arm q	Total length q + p	Arm ratio q/p	Type of centromere
1	4.61	4.99	9.60	1.2	M
2	4.22	4.61	8.83	1.1	M
3	3.46	4.61	8.07	1.3	M
4	3.07	3.07	6.14	1.0	M
5	2.30	3.07	5.37	1.3	M
6	1.92	2.30	4.22	1.2	M
7	1.54	1.92	3.46	1.3	M
8	1.15	1.54	2.69	1.3	M
9	1.15	1.15	2.30	1.0	SM
10	2.69	4.61	7.30	1.7	SM
11	1.15	2.30	3.45	2.0	SM

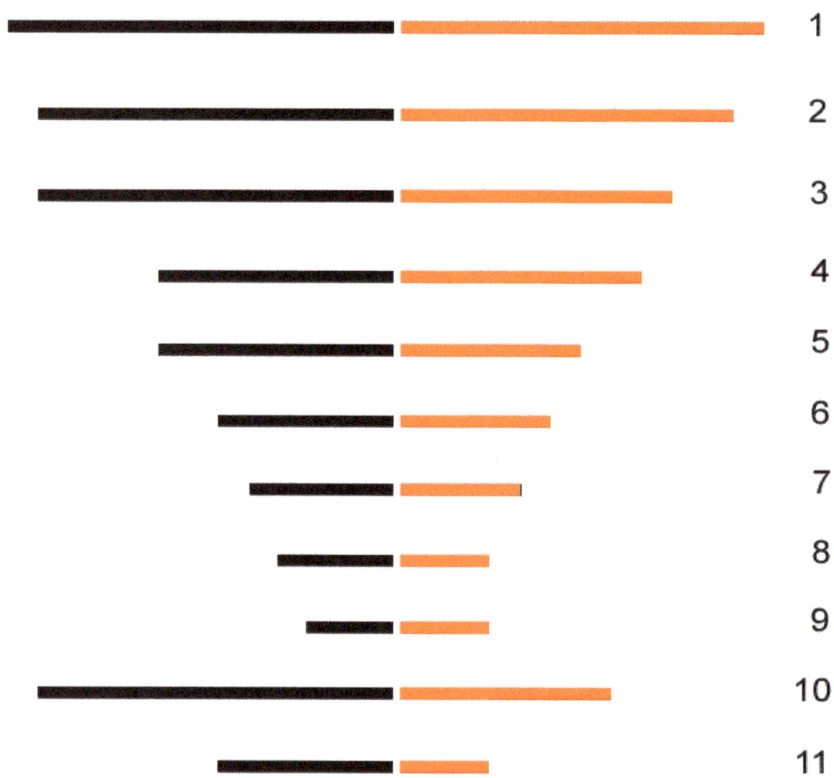

Figure 3. An idiogram of *B. tihamicus* constructed on the basis of chromosome numbers and the position of the centromeres.

Table 2. The fundamental number for *B. tihamicus.*

Type centromere	Haploid number N	Diploid number 2N	Fundamental number FN
M	9	18	32
SM	2	4	12
Total	11	22	44

M, Metacentric chromosome; SM, Submetacentric chromosome.

mosomes in this species.

DISCUSSION

The karyotype of *B. tihamicus* is asymmetric chromosomes consisting of diploid number of 22 chromosomes. This report is representing the karyotype of the toad *B. tihamicus.* The finding of 22 chromosome diploid number is in agreement with the bufonidae diploid chromosomes number which has been found to have karyotypic uniformity and consist of 22 chromosomes (Ullerich, 1966; Volpe and Gerhardt, 1968; Doyle and Beckert, 1970; Azevedo et al., 2003; Al-Shehri and Al-Saleh, 2008).

B. regularis from Saudi Arabia has 20 chromosomes and has a secondary constrictions on the short arms of chromosome pairs number 3 (Al-Shehri and Al-Saleh, 2008) while *B. tihamicus* has 22 chromosomes and do not have any secondary constrictions on the karyotype. It is well accepted that mammals are male heterogametic while birds are female heterogametic; but in amphibian the case is different (Nakamura, 2009). It has been stated that all amphibians have genotypic sex determination mechanism (Hayes, 1998; Wallace et al. 1999; Eggert, 2004), and this mechanism is common in vertebrates as well (Valenzuela, 2008). The sex determination can be classified as female homogametic/male heterogametic as in mammals or female heterogametic/male homogametic as in birds. However, sex chromosomes are often not heteromorphic all the time (Schmid and Steinlein, 2001) and this species have been found to be of the type of homogametic. The uniformity of sex chromosomes among toads seems to be accepted and there are several papers indicating such uniformity (Miura, 1995; Kasahara et al., 1996; Cavallo et al., 2002; Azevedo et al., 2003; Siripivasing et al., 200). Molecular and breeding experiments indicate that many anuran genera lack heteromorphic sex chromosomes (Hillis and green, 1990; Hayes, 1998).

An interesting finding by Hayes (1998) indicates that most amphibians lack morphologically distinguished sex chromosomes and suggests that the environment influences sex determination. Bergero and Charlesworth (2009) confirmed recently that heteromorphics sex chromosomes are a homologous pair in origin evolved from normal pair of homologous chromosomes.

Conclusion

We can stress the fact that this study not only confirm the karyotype for the first time and that the diploid number is 22 confirms that *B. tihamicus* is endemic to the Arabian Peninsula.

REFERENCES

Al-Shehri AH, Al-Saleh AA (2005). Karyotype of Amphibian in Saudi Arabia. 1. The karyotype of *Rana ridibunda*. J. Biol. Sci. 5:335-338.

Al-Shehri AH, Al-Saleh AA (2008). Karyotype of Amphibian in Saudi Arabia. 3. The karyotype of *Bufo regularis*. Asian J. Cell Biol. 3:67-71.

Azevedo MFC, Foresti F, Ramos PR, Jim J (2003). Comparative cytogenetic studies of *Bufo ictericus*, *BpParacnemis* (Amphibia, Anura) and an intermediate form in sympatry. Genet. Mol. Biol. 26:289-294.

Balletto E, Cherchi MA, Gasperetti J (1985). Amphibians of the Arabian Peninsula. Fauna of Saudi Arabia 7:318-392.

Bergero R, Charlesworth D (2009). The evolution of restricted recombination in sex chromosomes. Trends Ecol. Evol. 24:94-102.

Briggs J (1980). The green frog population of eastern Saudi Arabia. Abstract of Paper presented to annual joint meeting of the Society for the study of Amphibians and Reptiles – Herpetologists league, Milwaukee.

Briggs J (1981). Population structure of *Rana ridibunda* in the Al-Qatif Oasis. Proc. Saudi Biol. Soc. 5:333-345.

Cavallo D, Devita R, Eleuteri P, Borkin L, Emechenko V, Odiema G, Balletto, E, (2002). Karyological and flow cytometric evidence of triploid specimens in Bufo viridis (Amphibia Anura). Eur. J. Histochem. 16:159-164.

Doyle BW, Beckert WH (1970). Chromosome characteristics of bufonidae among species and within populations. Caryologia 23:145-154.

Eggert C (2004). Sex determination: the amphibian models. Reprod. Nutr. Dev. 44:539-549.

Hass G (1957). Some amphibians and reptiles from Arabia. Cal. Acad. Sci. 29:47-86.

Hass G (1961). On a collection of Arabian reptiles. Ann. Carnegie Mus. 36:19-28.

Hayes TB (1998). Sex determination and primary sex differentiation in amphibians: Genetic and developmental mechanisms. J. Exp. Zool. 281:373-399.

Hillis DM, Green DM (1990). Evolutionary changes of heterogametic sex in the phylogenetic history of amphibians. J. Evol. Biol. 3:49-64.

Kasahara S, Siva AP, Haddad CF, (1996). Chromosome banding in three species of Brazilian toads (Amphibia-Bufonidae). Rev. Brasil. Genet. 19:237-242.

Levan A, Fredga K, Sandbreg AA (1964). Nomenclature for centromeric position on chromosomes. Hereditas 52:201-220.

Miura I (1995). The late replication banding patterns of chromosomes are highly conserved in genera Rana, Hyla and Bufo (Amphibia, Anura). Chromosoma 103:567-574.

Nakamura M (2009). Sex determination in amphibians. Semen Cell Dev. Biol. 20:271-282.

Schmid M, Steinlein C (2001). Sex chromosome, sex-linked genes and sex determination in the vertebrate class amphibian. EXS 91:143-176.

Schmidt KP (1953). Amphibian and reptiles of Yemen. Fieldiana. Zoo, 34:253-261.

Siripivasing P, Chulalaksananukul W, Pariyanonth P, Kaewsri S, Sittigul S, Seatung N, Tanomtong A, (2008). The identification of the sex chromosome and karyo-type of four toad species (Genus Bufo) in Thailand bt T-lymphocyte cell culture. Cytologia 73:229-241.

Ullerich FH (1966). Karyotyp und DNS-Gehalt von *Bufo bufo*, *B. viridis*. *B. bufo* x *B. vindis* and *B. calamita* (Amphibia, Anura) Chromosoma 18:316-341.

Valenzuela N (2008). Sexual development and evolution of sex determination. Sex Dev. 2:64-72.

Volpe EP, Gerhardt BM (1968). Somatic chromosomes of the marine toad *Bufo marinus* (Linne). Copeia, 3:570-576.

Wallace H, Badawy GM, Wallacw BM (1999). Amphibian sex determination and sex reversal. Cell Mol. Sci. 55:901-909.

Phenotypic and molecular characteristics of rhizobia isolated from nodules of peanut (*Arachis hypogaea* L.) grown in Brazilian Spodosols

Maria do Carmo Catanho Pereira de Lyra[1], Ana Dolores Santiago de Freitas[2], Taciana Amorim Silva[3] and Carolina Etienne de Rosália e Silva Santos[2]

[1]Instituto Agronômico de Pernambuco (IPA), Laboratório de Genômica. Av. Gal. San Martin, 1371, Bongi. Recife, PE, CEP 50761-000, Brasil.
[2]Universidade Federal Rural de Pernambuco (UFRPE), Departamento de Agronomia, Laboratório de Microbiologia do Solo. Rua Dom Manoel de Medeiros, s/n, Dois Irmãos, Recife, PE, CEP 52171-900, Brasil.
[3]Universidade Federal do Amazonas (UFAM), Instituto de Ciências Biológicas. Rua Gen. Rodrigo Octávio Jordão Ramos, 3000, setor sul, Bloco D, Sala 08 Coroado I, Manaus, AM, CEP:69077-000, Brasil.

Peanut (*Arachis hypogaea* L.) is an important crop that can fix nitrogen through symbiosis with rhizobia. Rhizobia populations with diverse characteristics than those traditionally described as peanut microsymbionts have been found in tropical soils. With the objective of studying the diversity and phylogeny of these rhizobia, 22 bacterial strains were isolated from nodules of seven peanut varieties grown in Spodosols of Pernambuco State, Brazil. The isolates were examined in culture medium by means of some of their phenotypic characteristics and tested for intrinsic antibiotic resistance (IAR). DNA profiles were determined with the BOX-PCR and compared with 19 reference strains. All isolates showed rapid growth, and most of them acidified the culture medium. In general, high antibiotic resistance was observed to penicillin G, chloramphenicol and tetracycline and susceptibility was observed to neomycin, erythromycin and rifampicin. The analysis of their phenotypic characteristics, that is, colony morphology and IAR, provided little information about the phylogeny of the isolates. However, using compilation of phenotypic and molecular characteristics, we were able to observe a great diversity of these rhizobia and to reveal the presence of new species.

Key words: Antibiotic resistance, *Arachis hypogaea* L., BOX-PCR, colony morphology, diversity.

INTRODUCTION

Due to the importance of nitrogen for crop productivity and economic and environmental problems associated with the use of chemical fertilizers, nitrogen-fixing microorganisms have attracted significant attention, especially the ones capable of forming symbiosis with Leguminosae, generically known as rhizobia. To date, current taxonomy of these bacteria includes species belonging to several species of the genera *Rhizobium*, *Bradyrhizobium*, *Sinorhizobium*, *Mesorhizobium*, *Allorhizobium*, *Azorhizobium* (Liu et al., 2007), *Methylobacterium* (Kato et al., 2008), *Devosia* (Verma et al., 2009), *Burkholderia* (Bontemps et al., 2010), *Cupriavidus* (Liu et al., 2012), *Shinella* (Lin et al., 2008) and others. Using phenotypic and molecular

Table 1. Characteristics of the soils under peanut cultivation in Goiana municipalitiy, State of Pernambuco, Brazil.

Characteristic	Soil 1	Soil 2
Soil type	Spodosol	Spodosol
pH (H_2O)	4.5	5.0
P (mg dm^{-3})	5	3
Sand (g kg^{-1})	750	850
Silt (g kg^{-1})	100	150
Clay (g kg^{-1})	150	100
Ca^{2+}($cmol_c dm^{-3}$)	2.0	2.5
Al^{3+}($cmol_c dm^{-3}$)	1.5	1.0
H^{+}($cmol_c dm^{-3}$)	3.0	3.0
Mg^{2+}($cmol_c dm^{-3}$)	1.8	2.0
Na^{+}($cmol_c dm^{-3}$)	0	0
K^{+}($cmol_c dm^{-3}$)	86	86

methods, the presence of rhizobial populations with quite diverse characteristics has been verified in tropical soils (Martins et al.,1997; Lima et al., 2009), indicating the possibility of the existence of species that have not yet been described.

Traditionally, rhizobia isolated from peanut nodules (*Arachis hypogaea* L.) were categorized into the "cowpea miscellany" group (Allen and Allen, 1981), a very heterogeneous mixture comprising poorly characterized slow-growing strains that alkalinize the culture medium; these strains are known to belong to the genus *Bradyrhizobium* (Thies et al., 1991). However, Santos et al. (2007) reported the predominance of fast-growing bacteria that acidify the medium forming symbiosis with peanut grown in the soils of Northeastern Brazil. Previously, Taurian et al. (2006) also found that peanut forms symbiosis with fast-growing bacteria closely related to *Rhizobium giardini* and *Rhizobium tropici*.

Peanut is an important oilseed crop playing a significant role in the economy of various countries around the world. In the northeast region of Brazil, peanut is in great demand, and the cultivation of this legume is usually carried out on small farms with limited financial and technological resources, resulting in yields below the culture potential (EMBRAPA, 2006). Peanut nodulation by indigenous bacteria is considered sufficient, and inoculation is rarely performed. Studying the diversity of these bacteria is important in several aspects, both in terms of culture management, with the objective of facilitating the survival of more efficient and specific populations, and regarding the acquisition of adapted and efficient strains recommended for peanut inoculation.

Considering the economic potential of this agronomic crop and the lack of studies on native populations of peanut rhizobia in the soils of Northeastern Brazil, we isolated and characterized bacteria from peanut nodules grown in soils of the state of Pernambuco (07°33'38 S and 35°00'09 W, 13 m asl). The objective of this work was to study the diversity of these rhizobia and to assess the application of phenotypic and molecular methods in the characterization of these bacteria for future work on biological nitrogen fixation.

MATERIALS AND METHODS

Twenty-two (22) rhizobial isolates were obtained from root nodules of seven peanut varieties (L7, BR 1, Caipó, Jumbo, Tatu, IAC 8112 and Nativa) and cultured for 45 days in pots containing soil under peanut cultivation, collected from the layer of 0 to 20 cm in two sites of the municipality of Goiana (07°33'38 S and 35°00'09 W, 13 m asl), in the Zona da Mata physiographic zone of the state of Pernambuco, Brazil. The physical and chemical characteristics of the soils, determined according to the (EMBRAPA, 1997) methodology, were quite similar (Table 1). The isolation of rhizobia was performed according to the Vincent (1970) standard methodology on yeast mannitol agar medium (YMA) with 25 mgkg^{-1} (w/v) Congo red. The isolates were preserved at -80°C in peptone-glycerol medium. The isolates and reference strains are listed in Table 2.

For morphological characterization (Vincent, 1970), the isolates were transferred to YMA medium and grown at 28°C. The following culture parameters were considered: regarding growth period, pH alteration of growth medium, colony morphology (shape, size, border, transparency and surface) and amount of extracellular polysaccharides (EPS) (Xavier et al., 1998).

The isolates were tested for intrinsic antibiotic resistance (IAR) using six antibiotics at three different concentrations: chloramphenicol (2.5, 5 and 10 µg mL^{-1}), erythromycin (100, 200 and 400 µg mL^{-1}), penicillin G (100, 200 and 400 µg mL^{-1}), rifampicin (25, 50 and100 µg mL^{-1}), neomycin (100, 200 and 400 µg mL^{-1}) and tetracycline (1, 2 and 4 µg mL^{-1}). The test was performed in Petri dishes plated with solid tryptone yeast extract (TY) medium.

For molecular characterization, the isolates, along with the 19 reference strains listed in Table 2, were grown in YM medium (YMA without agar) for 48 h. Subsequently, DNA extraction was performed using the methodology described by Lyra et al. (2006). The amplification of genomic DNA was performed by PCR using the primer BOXA1R (Versalovic et al., 1994). The amplification was performed with an initial denaturation step at 95°C for 7 min followed by 30 amplification cycles, each cycle consisting of a 30 s phase at 90°C, a 1 min phase at 52°C and an 8 min phase at 65°C.

Table 2. Identification of rhizobia isolates from seven varieties of peanut (*Arachis hypogaea* L.) grown in two Spodosols in Goiana municipality, State of Pernambuco, Brazil, and type strains.

Isolate	Peanut variety	Type-strain	
		Strain	**Species**
LG32	L7	BR 29 (SEMIA 5019)	*Bradyrhizobium elkani*
LG29	BR 1	BR 11 (ATCC 10324)	*B. japonicum*
LG30	BR 1	BR 112 (USDA 205)	*Sinorhizobum fredii*
LG11	Caiapó	BR 7411(ATCC 9930)	*S. meliloti*
LG12	Caiapó	BR 525 (USDA 1037)	*S. medicae*
LG16	Caiapó	BR 526(USDA 4893)	*S. saheli*
LG40	Jumbo	BR 527 (USDA 4894)	*S. terangae*
LG7	TATU	BR 2406 (NGR 234)	*Rhizobium sp.*
LG8	TATU	BR 10016 (CFN 299)	*R. tropici* tipo IIA
LG41	IAC 8112	BR 322 (CIAT 899)	*R. tropici.*tipo IIB
LG42	Native	BR 10026 (CFN 42)	*R. etli*
LG39	L7	BR 10052 (ATCC 14482)	*R. leguminosarum* bv. *phaseoli*
LG34	BR 1	BR 7605 (LGM 6119)	*R. leguminosarum* bv. *trifolii*
LG35	BR 1	BR 10055 (HAMBI 540)	*R. galegae*
LG23	Caipó	BR 7801 (ATCC 33669)	*Mesorhizobium loti*
LG24	Caipó	BR 521 (USDA 3383)	*M. ciceri*
LG33	Caipó	BR 522 (USDA 3392)	*M. mediterraneum*
LG31	Jumbo	BR 524 (USDA 4779)	*M. huakuii*
LG37A	TATU	BR 5410 (ORS 571)	*Azorhizobium caulinodans*
LG37B	TATU		
LG43	IAC 112		
LG38	Native		

For the final step, an extra 16 min extension was performed at 65°C. The reagents and their final concentrations used in the reaction with a final volume of 10 µl were as follows: genomic DNA (10 to 40 ng), dNTP mix (0.3 mM), Taq DNA polymerase (1 U), $MgCl_2$ (2 mM), the BOX primer (2 mM), reaction buffer 10X (10%) and ultra pure water. The amplified products were stained with ethidium bromide, visualized using a UV transilluminator and photographed for further analysis for constructing dendrogram using bioinformatics tools.

From the phenotypic (culture and IAR) and molecular characteristics of the bacteria, matrices were constructed using the program NTSYS pc version 2.01 (Rohlf, 1973), and cluster analysis dendrograms were generated. Genetic distances were calculated by the Simple Matching coefficient (SM) using the unweighted pair-group method with arithmetic average (UPGMA) clustering method (Hart, 1983), and dendrograms were constructed using the SAHN method (sequential agglomerative hierarchical nested cluster analysis).

RESULTS

The 22 strains showed rapid growth (82% growing out in 24 h and 18% in 48 h) and most (91%) acidified the culture medium. Regarding colony characteristics, 18% of the isolates showed a yellowish color; 82% were beige; 9% were transparent; 36% were punctuate.

According to morphological characteristics, isolates were grouped into two main clusters, with the LG37B separated from the others due to its distinguished culture characteristics (Figure 1). In the first group, nine isolates (LG31, LG32, LG43, LG33, LG34, LG39, LG38, LG35 and LG37A) displayed 100% similarity, differing from other isolates of the same cluster by approximately 20%. The isolates LG40 and LG41 are with 100% DNA profile identity according to the morphological analysis. In cluster II, the isolates LG29 and LG30 are with 100% DNA profile identity and 60% similar to LG23. The dendrogram analysis did not show much variability among different isolates.

All 22 isolates were resistant to penicillin G and chloramphenicol at all tested concentrations (100, 200 and 400 µg mL^{-1} for penicillin; 2.5, 5 and 10 µg mL^{-1} for chloramphenicol). In the presence of tetracycline at a concentration of 1 µg mL^{-1}, 91% of the isolates grew on solid medium; this number fell to 73% at the concentrations of 2 and 4 µg mL^{-1}. In contrast, the isolates were susceptible to neomycin at all three tested concentrations (100, 200 and 400 µg mL^{-1}), with an overall index of 91% non-resistance. Only 14% of the isolates were resistant to the antibiotic erythromycin at the tested concentrations, with no change of the resistance index with increasing concentrations, and resistant to rifampicin at the concentration of 25 µg mL^{-1}. This resistance to rifampicin declined to 9% at the

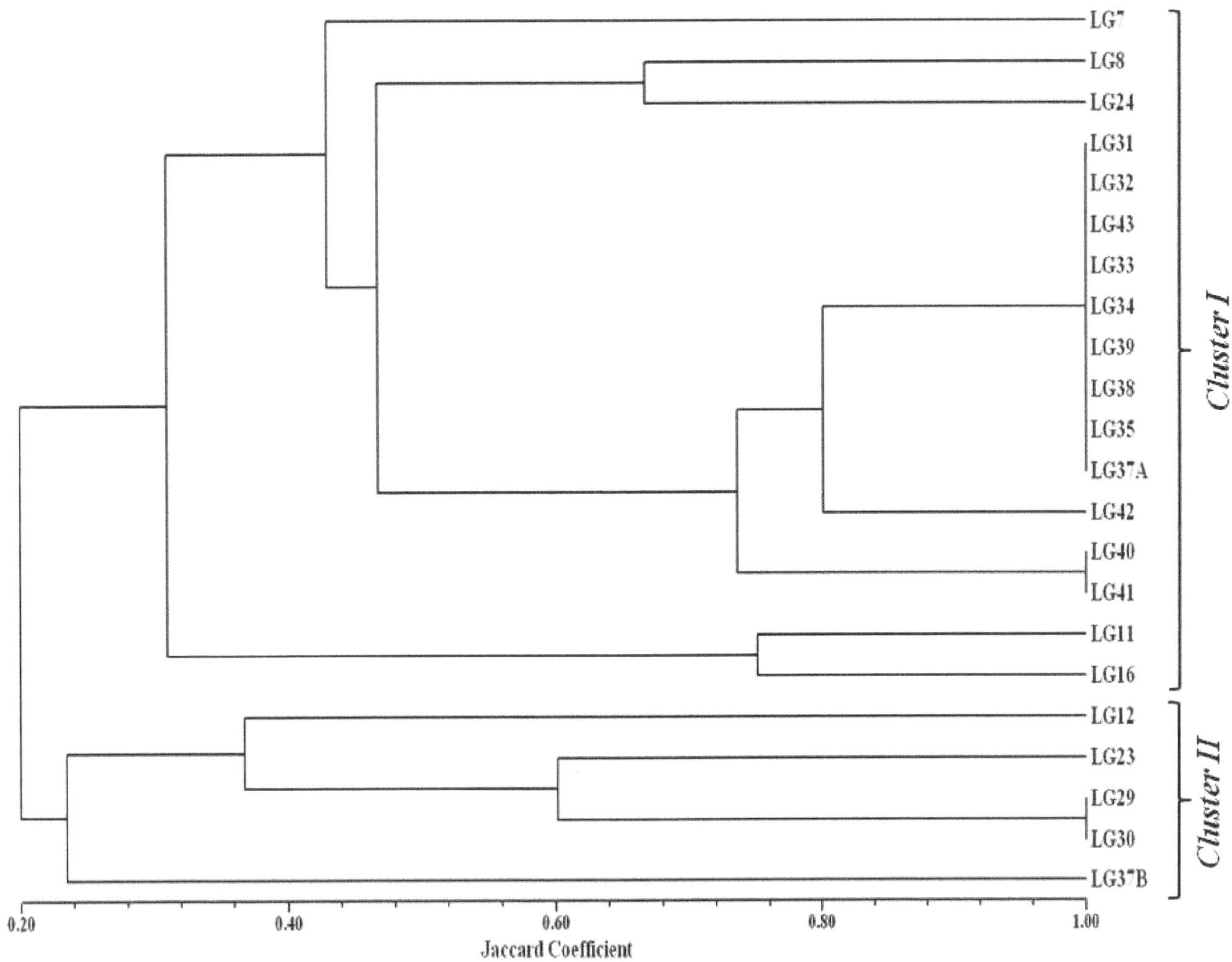

Figure 1. Similarity dendrogram constructed from the cultural characteristics of rhizobia isolated from peanut grown in two Spodosols in Goiana municipalitiy, State of Pernambuco, Brazil.

subsequent concentrations (50 and 100 µg mL^{-1}). The dendrogram constructed using the IAR data (Figure 2) was not effective in distinguishing classifications and grouping a large number of rhizobia within the same branch with maximum similarity indices.

Genetic characterization using the primer BOXA1R generated significant amounts of amplicons that are highly variable in size and number, both from the studied isolates (Figures 3 and 4) and from the reference strains (Figure 5). Most of the amplified fragments were shown as a band of 2000 to 200 bp. With this BOX marker, rhizobia (isolates and reference strains) were first grouped into four clusters (Figure 6). *Azorhizobium caulinodans* strain form external branch, with a genetic distance of 48% in relation to other strains and isolates.

Cluster I formed two major sub-clusters. In the first sub-cluster, all the strains of the genus *Sinorhizobium* were clustered, except *R. tropici* type IIA and type IIB, Rhizobium *etli* and the slow-growing bacteria *Bradyrhizobium elkanii* and *Bradyrhizobium japonicum* that remained in the same branch. In the second sub-cluster, the isolates LG23, LG30 and LG34, as well as the BR52 strain of *Mesorhizobium ciceri*, were not successfully amplified with the BOX element under the amplification conditions performed in this study. The other reference strains (*Rhizobium* sp. BR 2406, *Mesorhizobium ciceri* BR 521, *Rhizobium leguminosarum* bv. *Phaseoli* BR 10052, *Rhizobium galegae* BR 10055 and *Mesorhizobium mediterraneum* BR 524) and the isolates LG 40, LG 41, LG 37 B, LG 42, LG 43, LG 29, LG 35, LG 39, LG 37 and LG 38 showed high variability. Cluster II grouped the isolates LG 8 and LG 12, which appear with a genetic identity of 100%, and LG 7, together with the *Mesorhizobium huakuii* BR 524 strain. In cluster III, the bacteria *R. leguminosarum* bv. *trifolii* and *Mesorhizobium loti* appear with 100% DNA profile

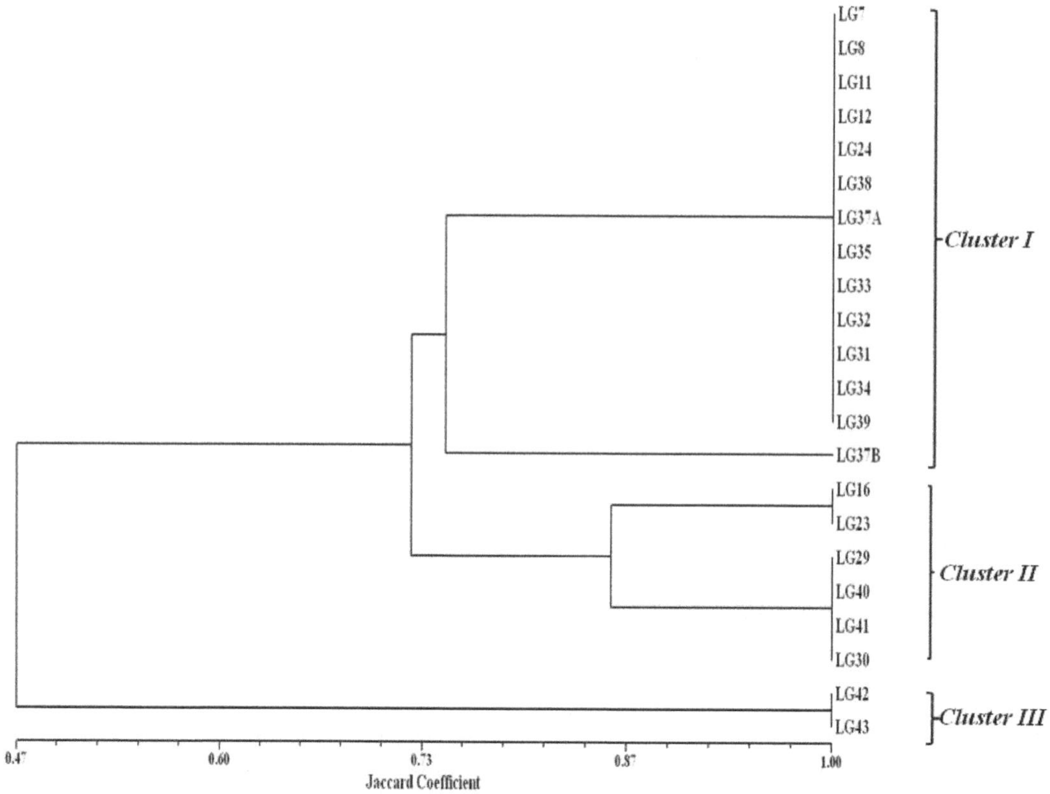

Figure 2. Similarity dendrogram constructed from the characteristics of antibiotic resistance of rhizobia isolated from peanut grown in two Spodosols in Goiana municipalitiy, State of Pernambuco, Brazil.

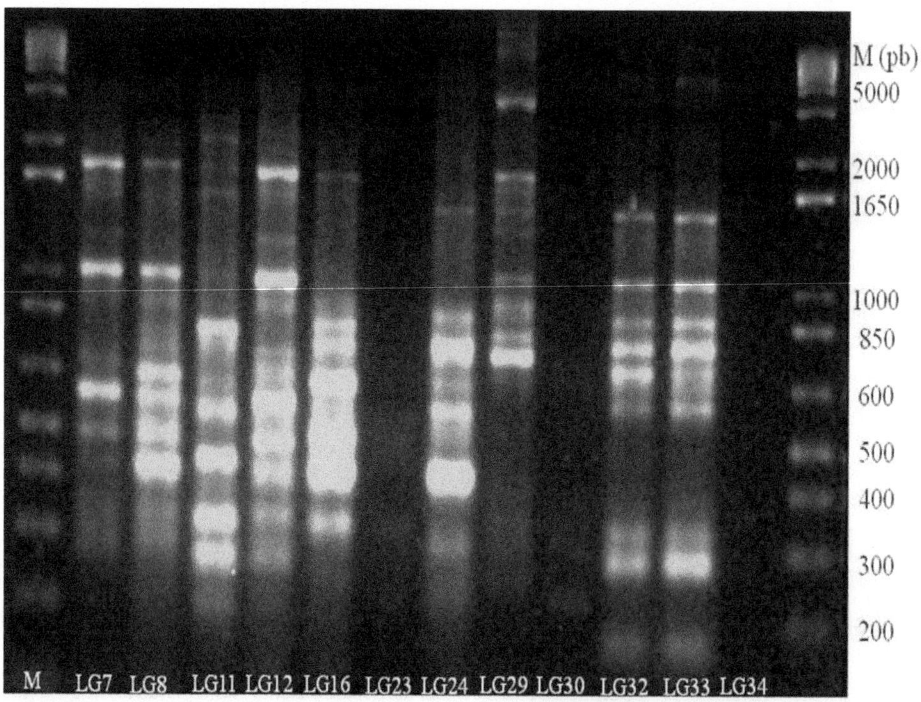

Figure 3. Profile of molecular DNA isolated LG 7, LG 8, LG 11, LG 12, LG 16, LG 23, LG 24, LG 29, LG 30, LG 32, LG 33 and LG 34, amplified with the primer BOXA1R in agarose gel 1%.

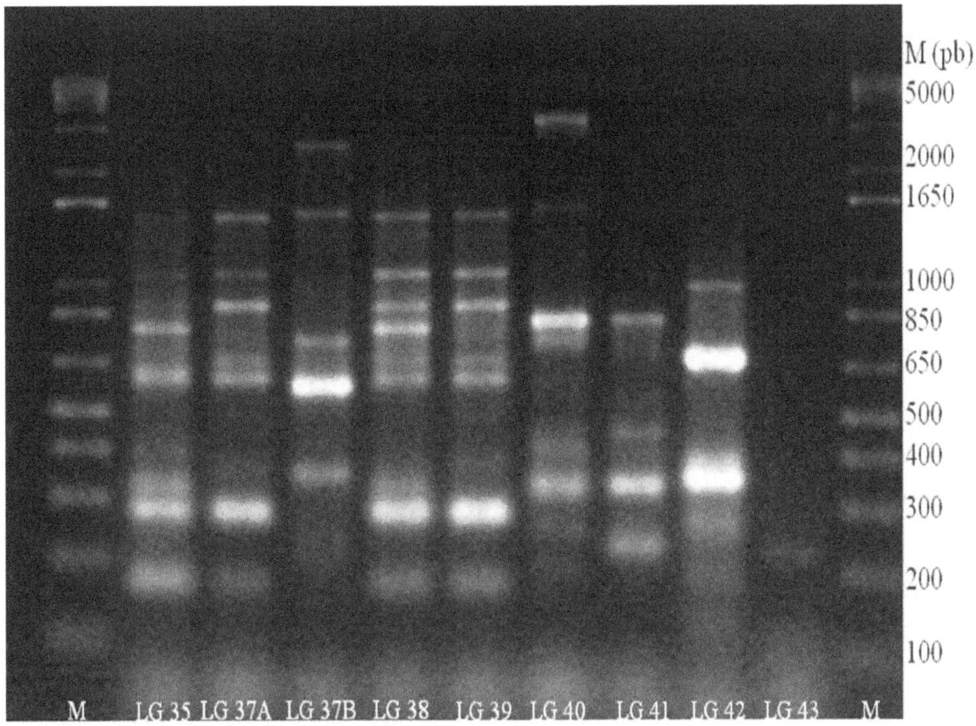

Figure 4. Profile of molecular DNA isolated from LG 35, LG 37 A, B LG 37, LG 38, LG 39, LG 40, LG 41, LG 42 and LG 43, amplified with the primer BOXA1R in agarose gel 1%.

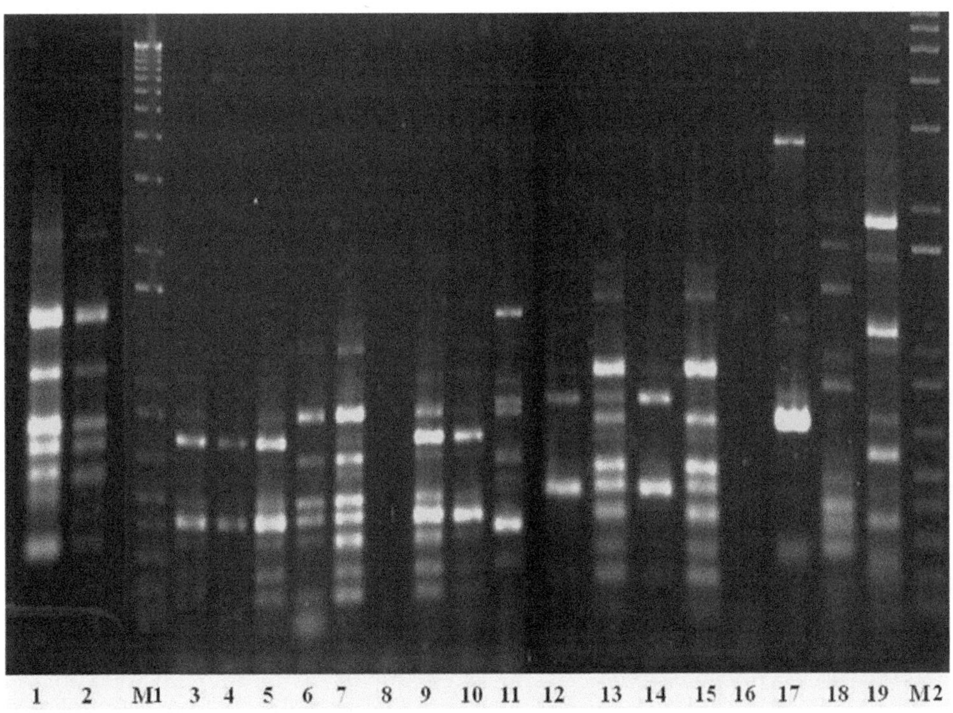

Figure 5. Profile of molecular DNA of reference strains amplified with the primer BOXA1R in agarose gel 1%. Legend: 1) BR29; 2) BR111; M1) Ladder 100 pb Invitrogen; 3) BR112; 4) BR7411; 5) BR525; 6) BR526; 7) BR527; 8) BR2406; 9) BR10016; 10) BR322; 11) BR10026; 12) BR10052; 13) BR7605; 14) BR10055; 15) BR7801; 16) BR521; 17) BR522; 18) BR524; 19) BR5410; M2) 1 Kb Plus Invitrogen.

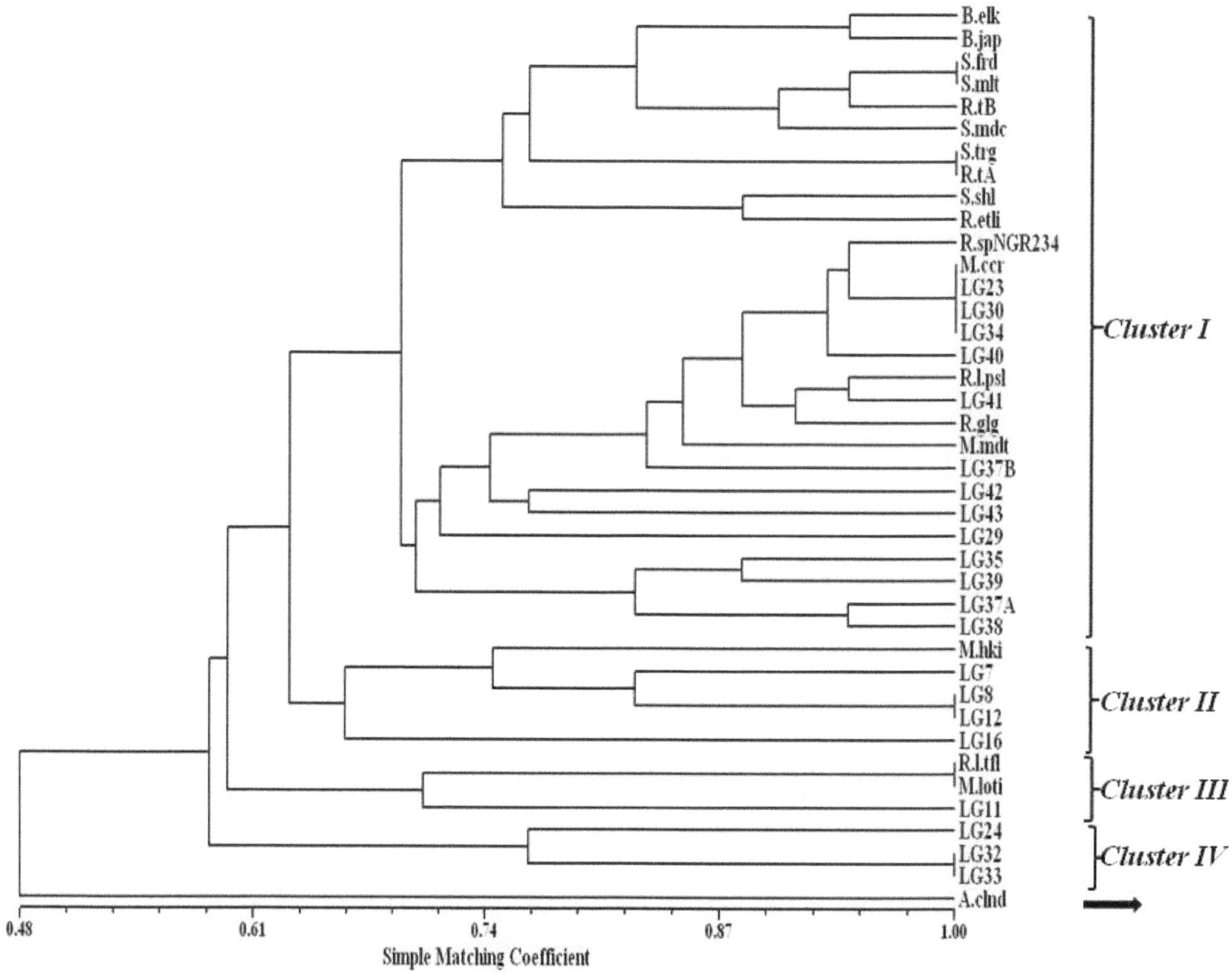

Figure 6. Dendrogram of similarity of molecular characteristics (profile of DNA amplified with primer BOXAR1 in agarose gel 1%) of reference strains and rhizobia isolated from nodules of peanut grown in two Spodosols in Goiana municipalitiy, State of Pernambuco, Brazil.

identity. In cluster IV, the isolates LG32 and LG33, appearing with 100% DNA profile identity, were grouped together with LG24, which showed a similarity of 77% to the previous two. Of the 22 isolates, 18 (85.71%) showed no similarity, a coefficient equal to 100%, to any of the reference strains used in this experiment.

The combination of the morphological, IAR and molecular data of these isolates (Figure 7) presented three clusters and two isolates with a monophyletic branch (LG16 and LG37B). Only the isolates LG32 and LG33 are with 100% DNA profile identity. In contrast, LG37B always appeared to be the most different.

DISCUSSION

According to morphological characteristics, isolates did not show much variability among different isolates. The 22 strains showed rapid growth according to the classification of Jordan (1984), with 82% growing out in 24 h and the remaining (18%) in 48 h. As expected for fast-growing rhizobia, most studied isolates (91%) acidified the culture medium. The pH changes generated by rhizobia in the culture medium are due to the preferential use of sugars by rapid-growth strains (Tan and Broughton, 1981), followed by excretion of organic acids and nitrogen compounds by slow-growing strains (Taurian et al., 2006). Other authors also isolated rhizobia with these growth characteristics in culture medium, in peanut growing in Brazilian (Santos et al., 2007) and Argentinean (Angelini et al., 2011; Nievas et al., 2012) soils. However, this finding is not in agreement with the current knowledge that the peanut crop is usually nodulated by slow-growing rhizobia (*Bradyrhizobium*). Pinto et al. (2004) confirmed the predominance of slow-growing bacteria in *Arachis pintoi* grown in the soils of the

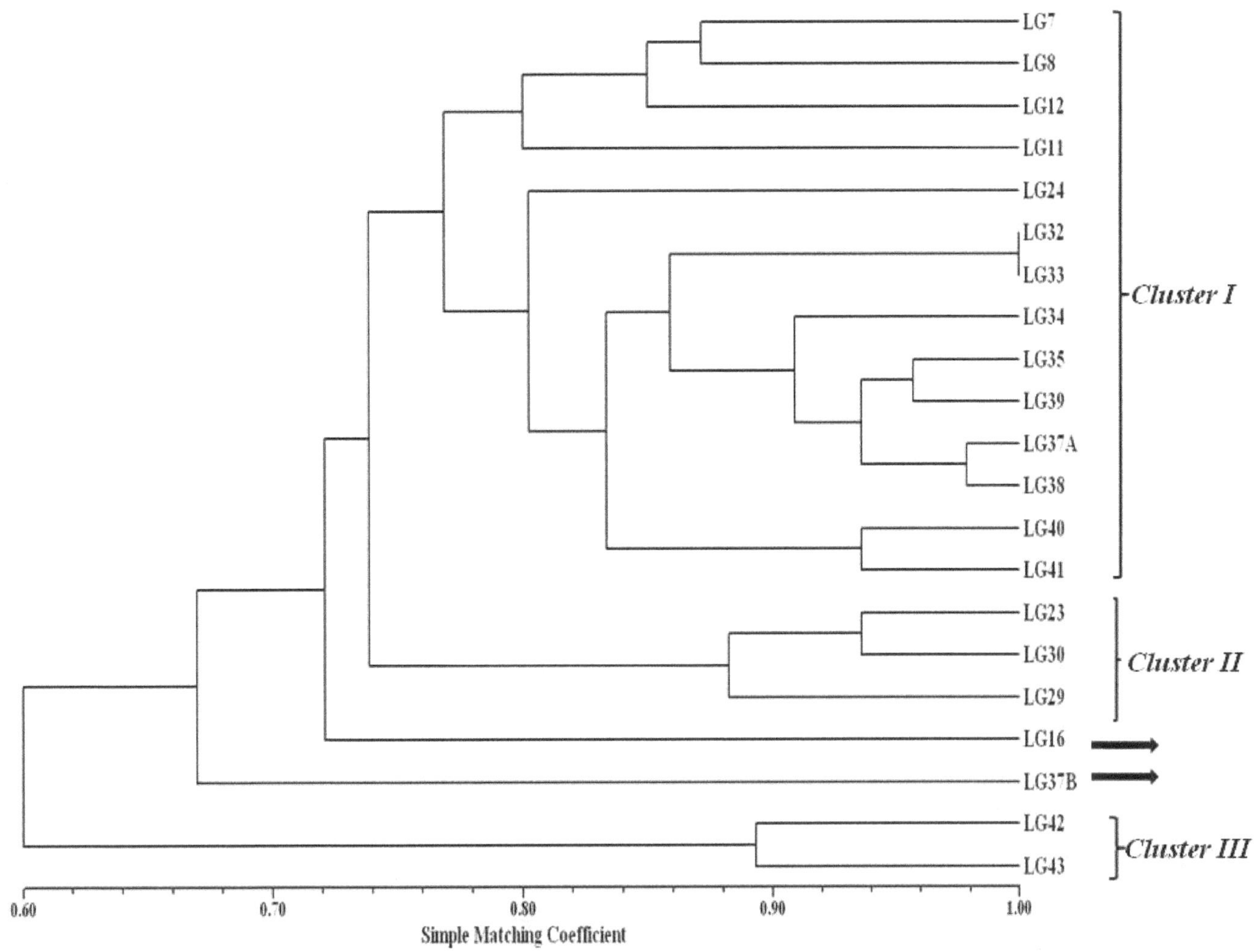

Figure 7. Similarity dendrogram resulting from the compilation of cultural characteristics, antibiotic resistance and molecular (DNA profile, amplified with primer BOXAR1 in agarose gel 1%) rhizobia from nodules of peanut grown in two Spodosols in Goiana municipalitiy, State of Pernambuco, Brazil.

Brazilian Cerrado (savannah). Taken together, these results indicate that peanut can form symbiosis with bacteria harboring diverse culture characteristics. As hypothesized by Angelini et al. (2011), in the field, peanut nodules are occupied by competitive strains. Thus, strains to be used for inoculation of peanuts in the Zona da Mata physiographic zone of the state of Pernambuco should be fast growing, to compete with the native strains.

The production of polysaccharide is a marked characteristic of rhizobia. In general, fast-growing strains produce more polysaccharides than the slow-growing ones (Teixeira et al., 2010). The strains that do not produce polysaccharides are called dry and are more common in slow-growing rhizobia (Eaglesham et al., 1990). In this study, nearly all isolates produced polysaccharide, except for the one that is classified as a dry colony. Polysaccharide can act as a protective barrier against biotic and abiotic factors, enabling the survival of the host organism in the soil (Teixeira et al., 2010; Xavier et al., 1998).

In general, the isolates demonstrated high resistance to the antibiotics penicillin G, chloramphenicol and tetracycline. Antibiotic resistance is a widely used parameter for rhizobium classification; it also serves as an indicator for the presence of actinomycetes in the soil from which rhizobia have been isolated (Martins et al., 1997). There is also a close correlation between antibiotic resistance and polysaccharide production (Fernandes Jr et al., 2012). Intrinsic antibiotic resistance is also one of the characteristics that can distinguish between strains of rhizobia and the pattern of IAR is useful in the strain identification (Rasul et al., 2012). However, the dendrogram constructed using the IAR data (Figure 2) was not effective in distinguishing classifications and grouping a large number of rhizobia within the same branch with maximum similarity indices. Similar observations have been made by Silva et al. (2002), stating that IAR is not a useful parameter for taxonomic purposes.

Genetic characterization using the primer BOXA1R generated significant amounts of amplicons that are highly variable in size and number, both from the studied isolates (Figures 3 and 4) and from the reference strains (Figure 5). It is possible that some isolates grouped in cluster I represent new rhizobia species capable of nodulating peanut. In cluster III, the bacteria *R. leguminosarum* bv. *trifolii* and *Mesorhizobium loti* had 100% DNA profile identity. However, these species are quite different in their genetic organization. In *R. leguminosarum* bv. *trifolii*, nodulation genes are located in the symbiotic plasmid pSym, while in *M. loti*, these genes are in the chromosome (Spaink et al., 1998). This result can be attributed to the fact that the BOX element is located in a very similar region in the two strains, despite the fact that they have distinctive genetic organization. Of the 22 isolates, 18 (85.71%) showed no similarity, a coefficient equal to 100%, to any of the reference strains used in this experiment, despite the fact that these rhizobia were isolated from plants of the same species and native from the soils of the same physiographic region. This observation further emphasizes the high degree of variability of rhizobia present in the tropics.

Comparing the dendrograms constructed according to the culture, IAR and molecular characteristics, we observed that the first two parameters were not effective in demonstrating the variability among the isolates. The clustering of a large number of isolates in the same branch obtained in the IAR dendrogram showed a high discrepancy with the results obtained from the morphological and molecular analyses. An example of this discrepancy is that the strain LG11, appearing alongside the isolates LG7, LG8, LG12, LG24, LG38, LG37A, LG35, LG33, LG31, LG32, LG34 and LG39 in the same branch, was shown far apart from them in the subsequent dendrograms; it was clustered with the isolate LG16 in the morphological dendrogram and was completely isolated in the molecular dendrogram.

The dendrogram resulting from the compilation of cultural characteristics, antibiotic resistance and molecular data of the isolates (Figure 7) presented three clusters and two isolates with an external branch (LG16 and LG37B). LG37B appeared to be the most different isolate and may be a novel species because it presents high diversity in morphology, IAR and molecular profiles as compared to other studied isolates. By combining the data together, the high variability among the isolates was even more striking.

According to the results shown in the constructed dendrograms, the number of taxonomic groups generated from the compilation of phenotypic and molecular characteristics (20 groups) was larger than the number generated from molecular data (18 groups), culture data (12 groups) and from the IAR data (5 groups). These results were expected because the compilation of several characteristics decreases the possibility of interpretation errors as compared to the analysis based on only one cellular character (Tan and Broughton, 1981).

Taking into account the definition of species (Stackebrandt and Goebel, 1994), one can consider that the isolates LG32 and LG33 belong to the same taxonomic group based on the constructed dendrograms because they showed maximum similarity in all dendrograms. Regarding environmental impact and morphological polymorphism, a relationship between the origin and the presented similarity of the isolates was observed. For example, in the first cluster of the morphological dendrogram, the isolates LG31, LG32, LG43, LG33, LG34, LG39, LG38, LG35 and LG37A appeared in the same branch, and all, except LG32, were native to the same soil in a peanut plantation. The same correlation was observed in the isolates LG40 and LG41, as well as in LG29 and LG30, which were native to the soil of the Umbú mill in Goiana, PE. The similarity between rhizobial isolates sharing the same microclimate and edaphoclimatic environment can be explained by the fact that bacteria develop similar adaptation strategies in a similar environment. It is unlikely to identify changes incorporated into the bacterial genome solely based on the morphological characterization. Morphological changes are more relevant to abiotic factors than to genetic factors (Straliotto and Rumjanek, 1999). Therefore, the rhizobia populations in our study may be recently established and may have not been completely adapted yet.

Bacteria from nodules of peanut plants cultivated in soils of the Zona da Mata physiographic zone of the state of Pernambuco have different characteristics from those described for the cowpea miscellaneous group. In these soils, the predominant bacteria are fast growing, produce acid in YMA medium and have attributes related to survival strategies (abundant exopolysaccharides and resistance to antibiotics). However, phenotypical characteristics provide little information on the diversity and phylogeny of these bacteria, while molecular analyses reveal large genetic diversity and differences among strain types, which indicates the possible occurrence of new species.

REFERENCES

Allen ON, Allen EK (1981). The Leguminosae: a source book of characteristics, uses, and nodulation. The University of Wiscosin Press, Madison, USA.

Angelini J, Ibáñez F, Taurian T, Tonelli ML, Valetti L, Fabra A (2011). A study on the prevalence of bacteria that occupy nodules within single peanut plants. Curr. Microbiol. 62:1752-1759.

Bontemps C, Elliott GN, Simon MF, Reis Jr FB, Gross E, Lawton RC, Elias Neto N, Loureiro MF, Faria SM, Sprent JI, James EK, Young JPW (2010). *Burkholderia* species are ancient symbionts of legumes. Mol. Ecol. 19:44-52.

Eaglesham ARJ, Ellis JM, Evans R, Fleischam DE, Hungria M, Hardy RWF (1990). The first photosynthetic nitrogen-fixing *Rhizobium*: characteristics. In: *Nitrogen fixation: achievements and objectives* (Gresshoff PM, Roth LE, Stacey G, Newton WE ed.). Chapman, New York, pp. 805-811.

EMBRAPA. Centro Nacional de Pesquisa de Solos (1997). Manual de métodos de análise de solos. EMBRAPA ,Rio de Janeiro, Brazil.

EMBRAPA. Centro Nacional de Pesquisa do Algodão (2006). Cultivo do Amendoim. http://sistemasdeproducao.cnptia.embrapa.br/FontesHTML/Amendoim/CultivodoAmendoim/index.html

Fernandes Júnior PI; Lima AA, Passos SR, Gava CAT, Oliveira PJ, Rumjanek NG, Xavier GR (2012). Phenotypic diversity and amylolytic activity of fast growing rhizobia from pigeonpea [*Cajanus cajan* (L.) MILLSP.]. Braz. J. Microbiol. 43:1604-1612.

Hart G (1983). The occurrence of multiple UPGMA phenograms. In: *Numerical Taxonomy* (Felsenstein J ed.). Springer-Verlag, New York, pp. 254-258.

Jordan DC (1984). Family III. Rhizobiaceae. In: *Bergey's Manual of Systematic Bacteriology* (Krieg NR, Holt JG ed.). Williams and Wilkins, Baltimore, pp. 234-254.

Kato Y, Asahara M, Goto K, Kasai H, Yokota A (2008). *Methylobacterium persicinum* sp. nov., *Methylobacterium komagatae* sp. nov., *Methylobacterium brachiatum* sp. nov., *Methylobacterium tardum* sp. nov. and *Methylobacterium gregans* sp. nov., isolated from freshwater. Int. J. Syst. Evol. Microbiol. 58:1134-1141.

Lima AS, Nóbrega RSA, Barberi A, Silva K, Ferreira DF, Moreira FMS (2009). Nitrogen-fixing bacteria communities occurring in soils under different uses in the Western Amazon Region as indicated by nodulation of siratro (*Macroptilium atropurpureum*). Plant Soil 20:1-19.

Lin DX, Wang ET, Tang H, Han TX, He YR, Guan SH, Chen WX (2008). *Shinella kummerowiae* sp. nov., a novel symbiotic bacterium isolated from root nodule of the herbal legume *Kummerowia stipulacea*. Int. J. Syst. Evol. Microbiol. 58:1409-1413.

Liu XY, Wei1 S, Wang F, James EK, Guo XY, Zagar C, Xia LG, Dong X, Wang YP (2012). *Burkholderia* and *Cupriavidus* spp. are the preferred symbionts of *Mimosa* spp. in Southern China. FEMS Microbiol. Ecol. 80:417-426.

Lyra MCC, López-Baena FJ, Madinabeitia N, Vinardell JM, Espuny MR, Cubo MT, Bellogín RA, Ruiz-Sainz JE, Ollero FJ (2006). Inactivation of the *Sinorhizobium fredii* HH103 *rhcJ* gene abolishes nodulation outer proteins (Nops) secretion and decreases the symbiotic capacity with soybean. Int. Microbiol. 9:125-133.

Martins LMV, Neves MCP, Rumjanek NG (1997). Growth characteristics and symbiotic efficiency of rhizobia isolated from cowpea nodules of the north-east region of Brazil. Soil Biol. Biochem. 29:1005-1010.

Nievas F, Bogino P, Nocelli N, Giordano W (2012). Genotypic analysis of isolated peanut-nodulating rhizobial strains reveals differences among populations obtained from soils with different cropping histories. Appl. Soil Ecol. 53:74-82.

Pinto PP, Paiva E, Purcino H, Passos RVM, Sá NMH (2004). Characterization of rhizobia that nodulate *Arachis pintoi* by RAPD analysis. Braz. J. Microbiol. 35:219-223.

Rasul A, Amalraj Eld, Kumar Gp, Grover M., Venkateswarlu B. (2012). Characterization of rhizobial isolates nodulating *Millettia pinnata* in India. FEMS Microbiol Lett. 336:148-158.

Rohlf FJ (1973). Algorithm 76. Hierarchical clustering using the minimum spanning tree. Comput. J. 16:93-95.

Santos CERS, Stamford NP, Borges WL, Neves MCP, Rumjanek NG, Nascimento LRS, Freitas ADS, Vieira IMMB, Bezerra RV (2007). Faixa hospedeira de rizóbios isolados das espécies *Arachis hypogaea*, *Stylosanthes guyanensis e Aeschynomene americana*. Rev. Bras. Cienc. Agrárias. 2:20-27.

Silva MLRB, Burtiy HA, Figueiredo MVB, Freitas NSA, Mergulhão ACES, Lyra MCCP (2002). Caracterización via RAPD (Random Amplified Polymorphic DNA) de aislamiento de rizobios de suelos ácidos y alcalinos en la región semi árida de Pernambuco. Rev. Argent. Microbiol. 34:186-192.

Spaink HP, Kondorosi A, Hooykaas PJJ (1998). The rhizobiaceae: molecular biology of model plant-associated bacteria. Kluwer, Dordrecht, Germany.

Stackebrandt E, Goebel BM (1994). Taxonomic note: a place for DNA-DNA reassociation and 16S rRNA sequence analysis in the present species definition in bacteriology. Int. J. Syst. Bacteriol. 44:846-849.

Straliotto R, Rumjanek NG (1999). Aplicação e evolução dos métodos moleculares para o estudo da biodiversidade do rizóbio. EMBRAPA-CNPAB, Seropédica, Brazil.

Tan IKP, Broughton WJ (1981). Rhizobia in tropical legumes. XIII Biochemical basis of acid and alkali reations. Rev. Argent. Microbiol. 34:186-192.

Taurian T, Ibañez F, Fabra A, Aguilar OM (2006). Genetic diversity of rhizobia nodulating *Arachis hypogaea* L. in Central Argentina soils. Plant Soil 282:41-52.

Teixeira FCP, Borges, WL Xavier GR, Rumjanek NG (2010). Characterization of indigenous rhizobia from caatinga. Braz. J. Microbiol. 41:201-208.

Thies JE, Bohlool BB, Singleton PW (1991). Subgroups of cowpea miscellany: symbiotic specificity within *Bradyrhizobium* spp. for *Vigna unguiculata*, *Phaseolus lunatus*, *Arachis hypogaea* and *Macroptilum atropurpureum*. Appl. Environ. Microbiol. 57:1540-1545.

Verma M, Kumar M, Dadhwal M, Kaur J, Lal R (2009) Devosia albogilva sp. nov. and Devosia crocina sp. nov., isolated from a hexachlorocyclohexane dump site. Int. J. Syst. Evol. Microbiol. 59:795-799.

Versalovic J, Schneider M, De Bruijn FJ, Lupski JR (1994). Genomic fingerprinting of bacteria using repetitive sequence-based polymerase chain reaction. Method. Mol. Cell Biol. 5:25-40.

Vincent JM (1970). A Manual for the Practical Study of Root Nodule Bacteria. IBP Handbook No 15. Blackwell, Oxford, England.

Xavier GR, Martins LMV, Neves MCP, Runjanek NG (1998). Edaphic factors as determinants for distribution of intrinsic antibiotic resistance in a cowpea rhizobia population. Biol. Fertil. Soils 27:386-392.

19

Effect of zeatin on the infection process and expression of MAPK-4 during pathogenesis of *Alternaria brassicae* in non-host and host *Brassica* plants

K. Kumar Marmath, P. Giri, G. Taj, D. Pandey and A. Kumar

Molecular Biology and Genetic Engineering, College of Basic Science and Humanities, G.B. Pant University of Agriculture and Technology, Pantnagar 263145, Dist. US Nagar, Uttarakhand, India.

Recent studies have revealed an important role of hormones in plant immunity. Cytokinins are phytohormones that are involved in various regulatory processes including plant defense. Zeatin a cytokinin antagonizes the effect of *Alternaria brassicae* pathotoxin in cell culture of *Brassica juncea*. Phytohormones are also the inducers of MAP Kinase signaling pathways which are the important signaling modules in eukaryotic cells. In this paper, an attempt was made to study the exogenous application of zeatin on the disease score, infection behavior of *A. brassicae*, expression pattern of MAPK-4 in the non host *Sinapsis alba*, host *B. juncea* susceptible c.v Varuna, *B. juncea* moderately tolerant c.v. Divya and transgenic *Brassica* to confer its role in plant immunity. We observed that high concentration of zeatin led to increased defense responses by delaying the infection process as well as significantly reducing the disease score. Semi-quantitative RT-PCR reveals that zeatin also increases the expression of MAPK-4 at early hours of infection. Our result supports that zeatin up regulates plant immunity via an elevation of MAPK-4 and clearly reflects that it antagonizes the effect of *A. brassicae*. The crosstalk between zeatin and MAPK signaling pathway may help plants fine-tune defense responses against *A. brassicae* in the disease Alternaria blight.

Key words: Alternaria blight, *Brassica*, disease score, MAPK, pathogenesis, infection process.

INTRODUCTION

Phytohormones have long been implicated in both biotic and abiotic interactions. Among the plant hormones, ethylene (ET), salicylic acid (SA) and jasmonate (JA) are known for differentially regulating defense responses against biotrophic and necrotrophic pathogens and are considered as the immunity hormones (Grobkinsky et al., 2011). The balance of hormonal crosstalk strongly influences the outcome of plant-pathogen interactions. In addition to these hormones, many researchers uncovered the role of several other hormones in plant defense like auxin, gibberellins, abscisic acid, brassinosteroid and cytokinins (CKs) (Spoel and Dong, 2008; Robert- Seilaniantz et al., 2007). CKs play an essential role in sustaining juvenility of plant tissues and have been investigated to understand the relationship between senescence and susceptibility towards several plant pathogens (Pogany et al., 2004). CKs also promote resistance against biotrophs by enhancing salicylic acid response (Choi et al., 2010). These are perceived by membrane-bound histidine kinase proteins similar to the two-component system of bacteria (Muller and Sheen, 2007; To and Kieber, 2008).

CKs are also produced by a range of various microbial pathogens, including *Pseudomonas syringae* (Akiyoshi et al., 1987; Morris et al., 1991), *Alternaria brassicae* (Tiwari 1993; Pandey et al., 2001) and leaf-mining insects

(Engelbrecht, 1968) causing the formation of green islands, galls, growth abnormalities and manipulation of primary carbon metabolism (Balibrea et al., 2004). For several pathogens, it was shown that this CK production is essential for infection (Crespi et al., 1994; Hwang et al., 2010). Based on the induction of sink metabolism by CKs (Ehness and Roitsch, 1997; Walters et al., 2008), it has been suggested that the host physiology is altered in response to CKs to allow the pathogen maximum access to nutrients in hemi biotrophic interactions (Walters et al., 2008). In contrast, the role of plant-derived CKs in defense responses against pathogens that are unable to produce CKs is largely unknown. Compared to auxin and abscisic acid signaling, our understanding of the roles of cytokinin in disease, and its interactions with other hormones, is relatively limited. Exogenous application of CKs has been reported to enhance plant resistance during some viral and fungal infections (Dekker, 1963; Clarke et al., 1998). However, the effect of zeatin on the tolerance of plants to disease caused by the necrotrophic fungus *A. brassicae* (Berk.) Sacc. has not been previously investigated. *A. brassicae* cause Alternaria blight (AB) disease in the economically important *Brassica* species, *Brassica juncea*. AB disease causes chlorotic and necrotic lesions mostly on aerial parts of the plant (Verma and Saharan, 1994) and is responsible for substantial reduction in seed yield and oil quality (Meena et al., 2010). Over the years, much effort has been made to understand the plant-pathogen/host specific toxin interaction during pathogenesis of AB of *Brassica*. Interestingly, *A. brassicae* toxin mediated signal transduction pathway was antagonized by a cytokinin (zeatin) in cell culture studies (Pandey et al., 2001). Phytohormones are also known to induce mitogen activated protein kinase (MAPK) cascade, a more conserved signalling pathway involved in plant response against fungal attack (Meskiene and Hirt, 2000). MAPK pathways are minimally comprised of three tiers MAP Kinase Kinase Kinase (MAPKKK), MAP Kinase Kinase (MAPKK) and MAP Kinase (MAPK). The *Arabidopsis thaliana* genome contains 20 MAPKs, 10 MAPKKs and 80 MAPKKKs which can engage in various MAPK modules to regulate responses of plant to different environmental and developmental signals. Recently, some members of the MAPK cascade have been reported to be involved in mediating defense response against various plant pathogens (Pitzschke et al., 2009). The best characterized MAPKs 3, 4 and 6 were activated by a diversity of stimuli including abiotic stresses and pathogens. Flagellin derived peptide flg22 triggers a rapid and strong activation of MAPK3, MAPK4 and MAPK6 (Droillard et al., 2004). MAPK4 and MAPK6 are also activated by harpin protein followed by the induction of pathogenesis- related (PR) genes (Desikan et al., 2001). A number of studies have been identified where MAPK4 is component of pathogen signalling cascade (Gao et al., 2008; Qiu et al., 2008; Pitzschke et al., 2009).

Arabidopsis MAPK4 has been involved in osmotic stress response pathway (Droillard et al., 2004) as well as implicated in plant defense regulation as *mpk4* knockout plants exhibited constitutive activation of salicylic acid (SA) dependent defenses, but failed to induce jasmonic acid (JA) defense marker genes in response to JA (Brodersen et al., 2006). Considering these observations, it was hypothesized that *A. brassicae* and its toxin may affect these MAPKs. Phytohormones are also the inducers of pathogenesis related proteins which also play important role in defense. Earlier, our laboratory has reported that overexpression of osmotin PR-5 in *Brassica* develop tolerance against *A. brassicae* (Taj et al., 2004). Since it is well known that CKs antagonize the effects of abscisic acid (ABA) and delay senescence and induce the expression of defense-related genes, we investigated whether CK (zeatin) can be used to develop the tolerance against *A. brassicae* in *B. juncea* c.v. Varuna, Divya, non host *Sinapsis alba* and transgenic *B. juncea* c.v. Divya harbouring osmotin gene. We also investigated the effect of exogenous zeatin on the infection process of *A. brassicae* and the expression pattern of MAPK4.

MATERIALS AND METHODS

Selection of non host/host genotypes

Seeds of a non-host (*S. alba*), susceptible host *B. juncea* cultivar (Varuna), moderately tolerant host *B. juncea* cultivar (Divya) and tolerant transgenic *B. juncea* c.v. Divya harbouring osmotin gene (Taj et al., 2004) were sown in plastic inserts (7.5 × 5 cm; 2 seeds per insert) containing mixture of soil, sand and vermicomposte in the ratio of 2:1:1. Plants were grown in the greenhouse (22°C day/18°C night; 16 h photoperiod), and watered at appropriate amount and time.

Alternaria brassicae

A. brassicae (Berk.) Sacc. was isolated from a diseased leaf of *B. juncea* cultivar 'Varuna' at Crop Research Centre (CRC), Pantnagar Uttarakhand, India. Pure single spore culture of *A. brassicae* was developed with the help of stereo microscope (Nikon make) and maintained on potato dextrose agar (PDA) slants at 4°C.

Inoculum preparation

A. brassicae was subcultured from the 7-day old culture on V-8 agar medium (10% V-8 juice, 0.02% $CaCO_3$ and 2% agar) and incubated at 22°C. A conidial suspension was prepared by scraping the mycelia and spores from surface of the actively growing fungal culture into autoclaved distilled water and filtered using four layered cheese cloth to remove most of the mycelia. The filtered spore suspension was centrifuged at 2000 x *g* for 5 min and resuspended in deionized water. This centrifugation was repeated one more time in order to ensure a clear spore suspension free of metabolites. After the final wash, supernatant was discarded and spores were resuspended in water containing 0.05% Tween-20 as an adhesive. The concentration of spore suspension was adjusted to 5×10^4

Table 1. The list of primers used for semiquantitative RT-PCR analysis.

Name of gene	Sense primer	Antisense primer	Expected product size (bp)
MAPK-4	GCTCTAACCAACCCTTAACTG	GTACCAGCGTGTAACAACGTA	228
Actin	ATTCTTACCCTCAAGTATCC	CATGATCTGAGTCATCTTCT	200

spores ml^{-1} using a haemocytometer (Sharma et al., 2007).

Sample preparation for light microscopy

10 µl suspension of *A. brassicae* spores were deposited separately on the leaf surface of all the plants (45 days old) with the drop plus agarose method (Giri, 2013, personal communication). For the control, 10 µl of sterile deionized water was applied. The inoculated plants were placed in humidity chamber (90 to 100% relative humidity). Inoculated plants were sampled at 3, 6, 9 and 12 h post-inoculation (hpi), and then daily for 4 day post-inoculation (dpi). Sampled leaves (area cut where drop was given) were decolourized in an acetic acid : ethanol : water (2:2:1) solution at 25°C for overnight. It was then washed with two changes of deionized water and stained with 1% cotton blue in lactophenol (Garg et al., 2010). Whole wet mounts of leaf piece on microscope glass slides were examined and photographed using a fluorescent microscope Nikon Eclipse 80i. The experiment was also done with exactly same procedure with *A. brassicae* spore suspension along with 1 ppm zeatin, to show the effect of zeatin on the infection process.

Sample preparation for molecular studies

The 45 days old plants were subjected to three kinds of treatments. In the first treatment, 10 µl volume of *A. brassicae* spore along with 10 µl of 1 ppm zeatin was applied onto the plants. In the second and third treatment, *A. brassicae* spore suspension (10 µl) and zeatin (10 µl) were applied individually, respectively. Inoculation was performed by drop plus agarose method (Giri, 2013, personal communication) of true leaves (2 leaves/plant) with a pipette tip. For the uninoculated controls, 10 µl of sterile deionized water was applied. The inoculated plants were incubated at 25°C with 90 to 100% relative humidity for 3 days in humid chamber and then transferred into growth room. Inoculated leaves were collected from each plant for each treatment at 3, 6 and 9 h post-inoculation for RNA isolation and stored at -80°C until use. Disease severity was assessed using a scale described by Conn et al. (1990) where a score of 0 indicates no symptoms; 1, small irregular spots covering <5% leaf area; 2, small irregular brown spots covering 5 to 10% leaf area; 3, symptom covering 10 to 20% leaf area; 4, symptom covering 20 to 30% leaf covering; and 5, symptom covering 30 to 50% leaf area covering. For RT-PCR analysis, leaves samples were subjected to RNA isolation using RNEasy plant minikit (Himedia Laboratories Private Limited India) as per the manufacturer instructions. RT-PCR analysis was performed with the help of One-Step RT-PCR (Qiagen, USA) using the gene specific primer of MAPK4 and internal control actin under the following PCR conditions: reverse transcription at 50°C for 30 min, initial PCR activation step at 95°C for 15 min followed by 35 cycles of amplification (94°C for 1 min, 59°C for 1 min and 72°C for 1 min) with final extension at 72°C for 10 min. After the completion of RT-PCR, the amplicons were analyzed by electrophoresing them in 1.8% agarose gel electrophoresis followed by quantification by using the spot densitometry tool of AlphaImager software. The sequences of all primers are described in Table 1.

RESULTS AND DISCUSSION

Conidial germination and fungal development on four different plants and effect of zeatin on the infection process of *A. brassicae*

The initial steps in the infection process of *A. brassicae* on four different plants: *S. alba*, *B. juncea* c.v. Varuna & Divya and transgenic *Brassica* were compared using light microscopy (Table 2). The effect of zeatin was also observed on the infection process of *A. brassicae*. It was observed that conidial germination was not evident during the interplay between *A. brassicae* and the leaf tissue of the non-host *S. alba* even upto the second day of infection (Figure 2: 1A to 1F). It suggests that non host might produce certain antifungal/antifungistatic compounds which are not suitable for the conidial germination. This is also supported by Kowalska and Niks (1999) for a resistant flax (*Linum usitatissimum*) genotype against *Melampsora lini* and by Blakeman and Sztejnberg (1973) in beetroot (*Beta vulgaris*) against *Botrytis cinerea*. At 3 days post inoculation (dpi), epidermal penetration was observed in *S. alba* (Figure 2: 1G), but it was not evident in zeatin plus pathogen treated leaves (Figure 2: 1N); the mycelium only grew at the surface of the plant. The penetration of fungal mycelium was much delayed; it might be due to the incompatible interaction, which showed resistance to *A. brassicae* attack (Sharma et al., 2002). Moreover, it also correlated with the disease scoring data of *S. alba* where no spot was observed at 1 day of post inoculation, but as the disease progresses, the penetration might have occurred as a result symptoms that appeared from 5 days onwards (Figure 1). In the case of Varuna cultivar, conidial germination began within 3 hpi (Figure 2: 2A) but conidial germination was not evident up to 6 hpi (Figure 2: 2H and 2I) in zeatin plus pathogen treated leaves. From this time point onwards, the conidial germination began and the infection proceeded. At initial hour of infection, multiple germ tubes appeared (Figure 2: 2B and 2C) but a bit late due to the exogenous application of zeatin in Varuna c.v. (Figure 2: 2J and 2K). Appresorium and infection thread are the two hyphal modifications observed in *A. brassicae* infection. Penetration of infection thread into the intercellular spaces was observed at 12 hpi (Figure 2: 2D) whereas, no such penetration was evident in the case of zeatin treated leaves in Varuna c.v. The process of infection behavior followed the same steps but at different time intervals. Exogenous application of zeatin delayed the infection process, as appresorium penetration through stomata

Table 2. Description of growth of *A. brassicae* and effect of zeatin on the infection process in the leaf surface of different plants.

Genotype		3 hpi	6 hpi	9 hpi	12 hpi	1 dpi	2 dpi	3 dpi
S. alba	*A. brassicae*	No spore germination	No SPore germination	No spore germination	No spore germination	No spore germination	No spore germination	Penetration of infection thread into intercellular spaces
	A. brassicae + *zeatin*	No Spore germination	No spore germination	No spore germination	No spore germination	No spore germination	No spore germination	No penetration
B. juncea c.v. Varuna	*A. brassicae*	Spore germinated	Germ tube elongation	Formation of several germ tubes	Penetration of infection thread into intercellular spaces	Penetration through stomata	Mycelial network	Proliferation of spores
	A. brassicae + *zeatin*	No spore germination	No Spore germination	Spore germinated	Formation of several germ tubes	No penetration	Penetration through stomata	No proliferation
B. juncea c.v. Divya	*A. brassicae*	Spore germinated	Germ tube elongation	Germ tube elongation	Penetration of infection thread into intercellular spaces	Sporulation on the surface	Penetration through stomata	Mycelial network
	A. brassicae + *zeatin*	No spore germination	Spore germinated	Germ tube elongation	Penetration of infection thread into intercellular spaces	Differentiation of hypha into infection thread	Appresoria approaches towards stomata	Penetration through stomata
Transgenic *Brassica*	*A. brassicae*	No spore germination	Spore germinated	Formation of several germ tubes	Formation of several germ tubes	Penetration of infection thread into intercellular spaces	Penetration through stomata	Mycelial network
	A. brassicae + *zeatin*	No spore germination	No Spore germination	Spore germinated	Germ tube elongation	Penetration of infection thread into intercellular spaces	Penetration through stomata	Penetration through stomata

was observed at 2 dpi (Figure 2: 2M) as compared to without zeatin where it was evident at 1 dpi (Figure 2: 2E) only. In both cases, penetration of appresorium was evident only through stomata. At late hours of infection process, proliferation of spores was observed (Figure 2: 2G) in contrast to the zeatin treated leaves where only mycelial network were observed (Figure 2: 2L and 2N). In the case of Divya cultivar, conidial germination also began within 3 hpi (Figure 2: 3A), but the length of the germ tube was significantly less as compared to the germ tube that emerged on Varuna cultivar (data not shown). Here also, zeatin delayed the conidial germination which was eventually initiated from 6 hpi onwards (Figure 2: 3I). Penetration of epidermal cells by fungal hyphae was observed at 12 hpi in both cases (Figure 2: 3D and 3K). At 1 dpi, the spores multiplied by budding on the host surface (Figure 2: 3E). This is a typical feature of fungi belonging to phylum ascomycota. This is an adaptation to augment the inoculums density, thereby enhancing the inoculums potential. But this type of augmentation of inoculums density was not evident on zeatin treated leaves suggested that zeatin somehow interferes in the infection process of *Alternaria brassicae.* At 2 dpi, appresorium penetration was observed (Figure 2: 3F) which is clearly evident at 3 dpi in zeatin treated leaves in Divya c.v. (Figure 2: 3N). In the case of transgenic *Brassica* harbouring PR-5 protein osmotin (Taj et al., 2004), there was a consistent delay in conidial germination which eventually started from 6 hpi onwards (Figure 2: 4B) and even later, that is, at 9 hpi in zeatin treated transgenic *Brassica* leaves (Figure 2: 4J). The infection process was slower as compared to Varuna cultivar. Here, appre-sorium penetration through stomata was evident at 2 dpi (Figure 2: 4F) and no proliferation of spores were observed. The findings of the present study suggested that zeatin obstruct the interplay of host and pathogen

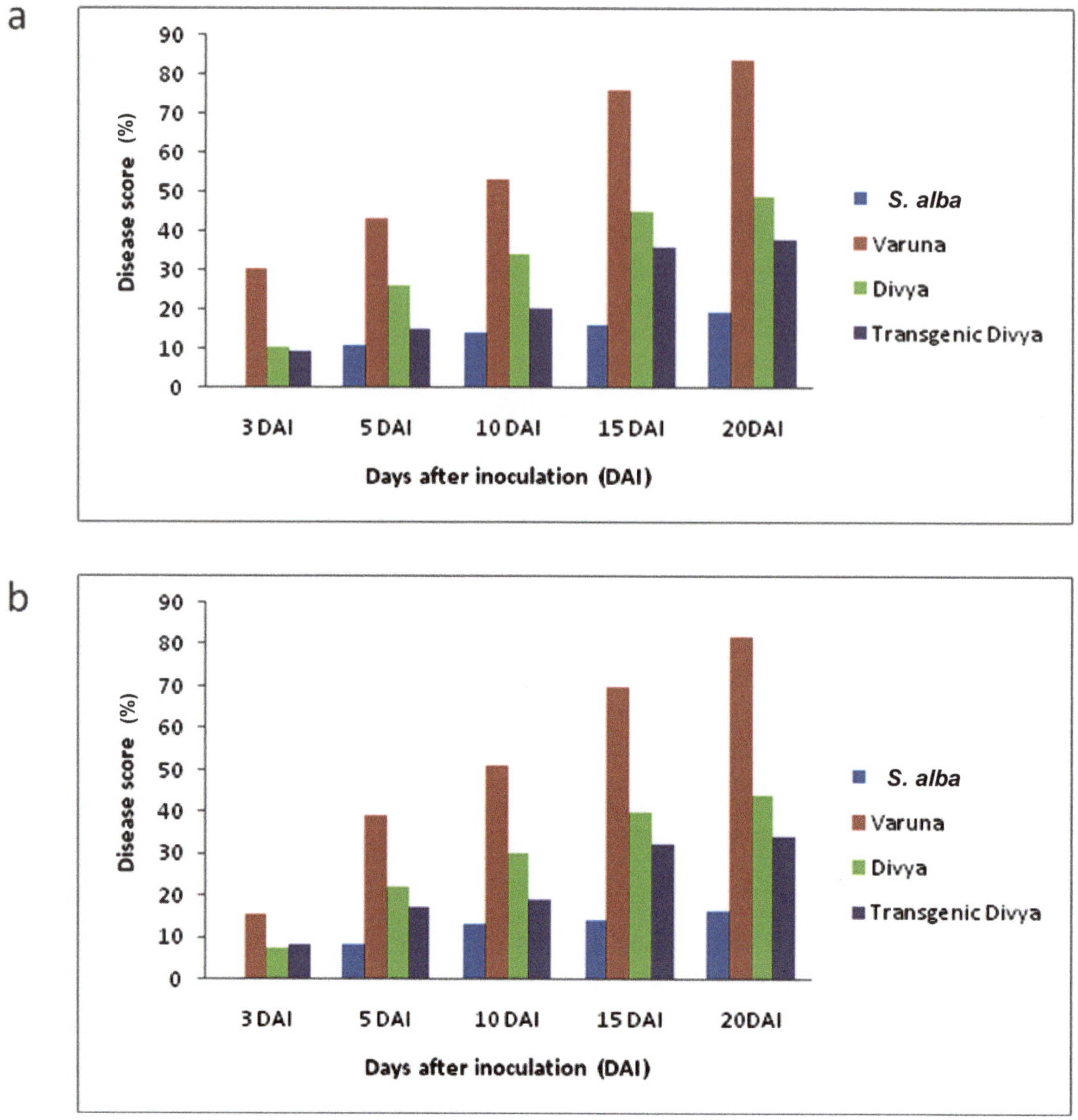

Figure 1. Disease severity score (%) induced by a) *A. brassicae*; b) Effect of zeatin on the disease severity score on leaves of non host *S. alba*, *B. juncea* c.v. Varuna & Divya and transgenic *Brassica* at 3, 5, 10, 15 and 20 days after inoculation (DAI).

and delays the infection process by inhibiting the conidial germination and fungal development. This is in accordance with the findings of Pandey et al. (2001) who reported that increased zeatin concentration antagonizes the effect of *A. brassicae* pathotoxin in cell culture of *B. juncea* c.v. Divya.

Response of host and non host plants against *A. brassicae* and *A. brassicae* along with zeatin treatment

The effect of zeatin was investigated on the disease score of all four different plants. A significant reduction in the disease score and on the appearance of the symptoms was observed when compared with the plants only sprayed with *A. brassicae* suspension and plants sprayed with *A. brassicae* suspension along with zeatin (Figures 1 and 3). *B. juncea* c.v. Varuna exhibited severe chlorosis and necrosis that spread from the spot that had been inoculated with *A. brassicae* spores, had irregular margins extending towards the periphery of the leaf suggesting that it was quite susceptible to *A. brassicae*. In contrast, the leaves of *S. alba*, transgenic *Brassica* and *B. juncea* c.v. Divya exhibited localized cholorosis and necrosis only around the inoculated region which clearly indicated that these hosts were more tolerant to the

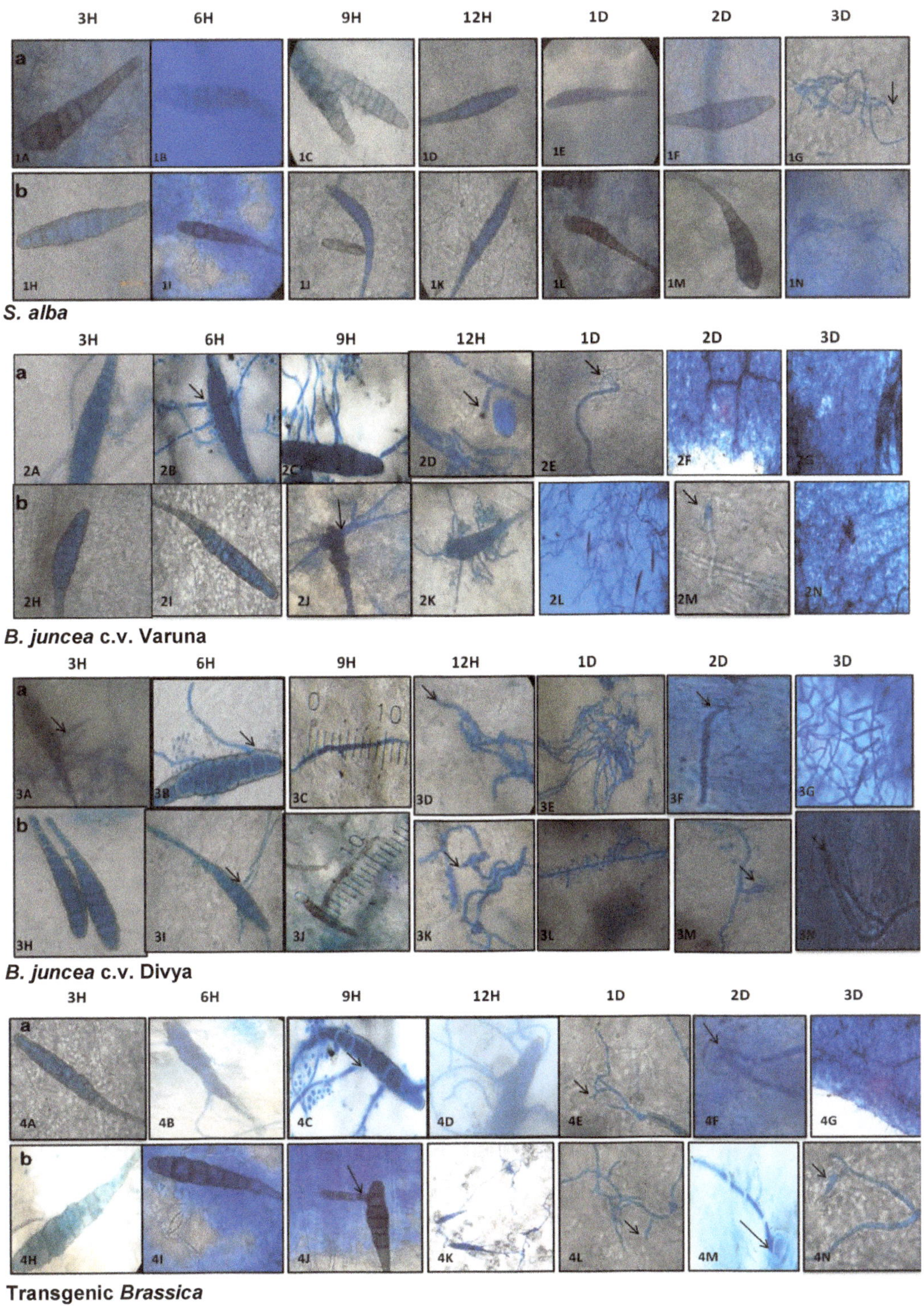

Figure 2. Light microscopic study of the leaves of non-host *S. alba*. 1A-1G, *A. Brassicae;* 1H-1N, *A. brassicae* together with zeatin; 2A-2G, *B. juncea* susceptible cultivar Varuna with *A. Brassicae;* 2H-2N, *A. brassicae* together with zeatin; 3A-3G, *B. juncea* moderately tolerant cultivar Divya with *A. Brassica;* 3H-3N, *A. brassicae* together with zeatin; 4A-4G, transgenic *Brassica* with *A. Brassicae;* 4H-4N, *A. brassicae* together with zeatin. (1A-1F), (1H-1M), (2H and 2I), 3H, (4A, 4H, 4I, conidium on the leaf surface; 2A, 2J, 3A, 3I, 4B, 4J, germ tube emerged from conidium; 2B, 3B, 3C, 3J, 4C, 4K- elongation of germ tube; 2C, 2K, 4D, formation of several germ tubes and their elongation;1N, fungal hyphae continued growth on the leaf surface; 1G, 2D, 3D, 3K, 4E, 4L, differentiation of hyphae into infection thread and its penetration through intercellular spaces; 2E, 2M, 3F, 3N, terminal hyphae leading to formation of appresoria and its penetration through stomata; 2F, 2L, 2N, 3G, 4G, mycelial network; 2G, excessive proliferation of the spores. Scale bar: 50 µm.

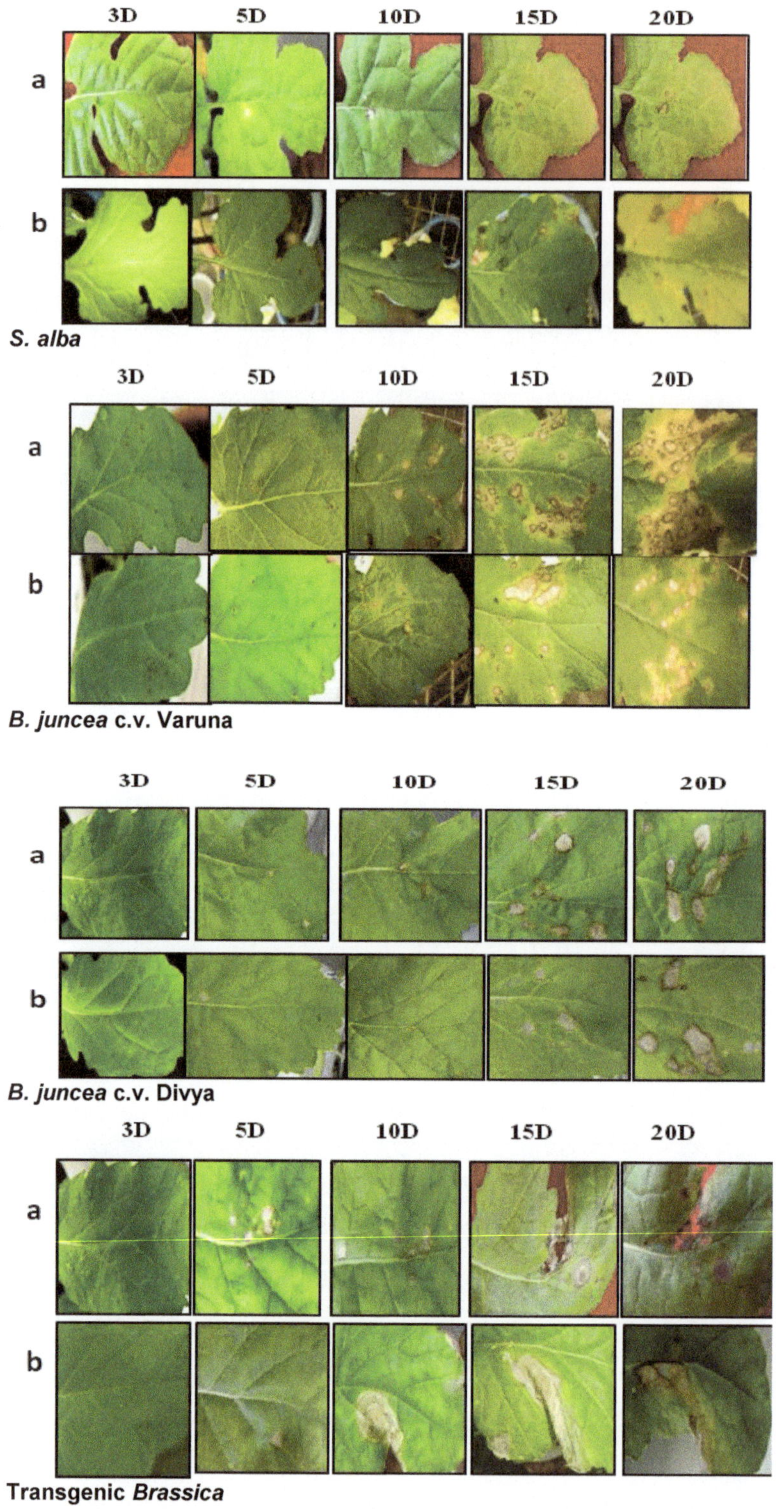

Figure 3. Response of non host *S. alba*, *B. juncea* susceptible c.v. Varuna, *B. juncea* moderately tolerant c.v. Divya and transgenic *Brassica* to (a) *A. brassicae* and (b) *A. brassicae* plus zeatin at 3, 5, 10, 15, 20 days after inoculation.

pathogen as compared to *B. juncea* c.v. Varuna. We can conclude that exogenous application of zeatin antagonizes the effect of *A. brassicae* and hence decreases the disease score imparting role in defence.

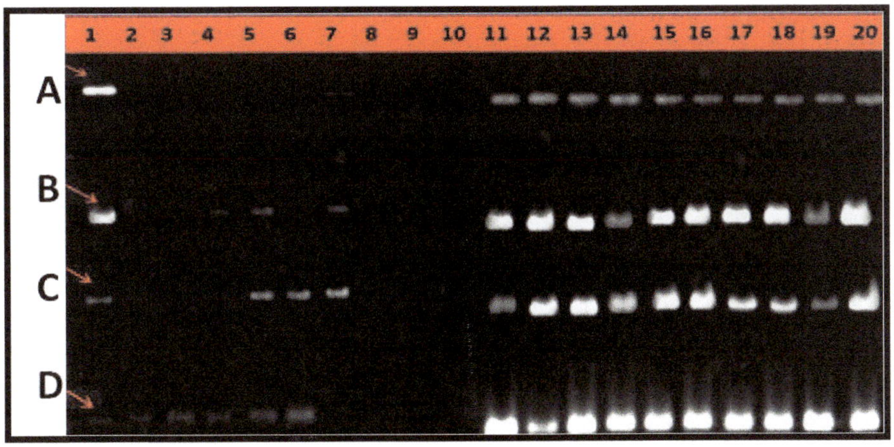

Figure 4. Determination of change in transcript profiling of MAPK4 gene by semi quantitative RT-PCR from *A. brassicae* infected leaves, zeatin sprayed leaves and *A. brassicae* along with zeatin sprayed leaves of A) *S. alba*, B) transgenic *Brassica*, C) *B. juncea* c.v. Divya, D) *B. juncea* c.v. Varuna. Lanes 1 to 4, Healthy, *A. brassicae* infected leaves, zeatin sprayed leaves and *A. brassicae* along with zeatin sprayed leaves at 3 h; lanes 5 to 7, *A. brassicae* infected leaves, zeatin sprayed leaves and *A. brassicae* along with zeatin sprayed leaves at 6 h; lane 8-10, *A. brassicae* infected leaves, zeatin sprayed leaves and *A. brassicae* along with zeatin sprayed leaves at 9 h; lanes 11 to 20, actin gene expression from *A. brassicae* infected leaves, zeatin sprayed leaves and *A. brassicae* togethjer with zeatin sprayed leaves at 3, 6, 9 h, respectively.

This strongly correlates with the infection behavior study where zeatin interferes with the conidial germination. Earlier, one group (Sharma et al., 2010) also reported that cytokinin provides resistance to canola plants against *Alternaria* pathogen.

Expression profiling of MAPK4 at different time intervals in different treatments by semi quantitative RT-PCR

Expression of mitogen activated protein kinase 4 (MAPK 4) was analyzed in three time points after infection with *A. brassicae* alone, *A. brassicae* along with zeatin and zeatin alone in four different plants namely *S. alba*, *B. juncea* cultivar Varuna & Divya and transgenic *Brassica* (Figure 4). In all the host plants, the expression level of MAPK 4 (228 bp) increased from 3 to 6 hpi and then decreased at 9 hpi. The transcript level of MAPK 4 was found to be highest at 6 hpi. In all the samples, actin was found to be constitutively expressed (Figure 5). This observation implies that at the early phase of infection, MAPK-4 gradually increases to strengthen plant defence, so as to restrict conidial germination. Upon looking up the expression pattern of MAPK 4 in all the plants, it was observed that the non-host *S. alba* and moderately tolerant Divya has the highest expression of MAPK 4 in the healthy samples in comparison with the treated samples indicating the involvement of MAPK 4 in other functions apart from the defense. Whereas, in transgenic *Brassica* and susceptible Varuna, the expression of MAPK 4 was found to be highest at 6 hpi in comparison with healthy samples. If we correlate the expression pattern of MAPK 4 with the conidial germination, we can say that in transgenic *Brassica* as the conidia started germinating at 6 hpi, the MAPK 4 expression was highest, although in susceptible Varuna, the conidial germination started at 3 hpi. In both host plants, Varuna had more expression of MAPK 4 at 3 hpi. This gives a clue that MAPK 4 expression triggers the conidial germination. This further supports that MAPK 4 playing role in defense as its expression started with the conidial germination. The exogenous application of zeatin in Varuna further enhanced the expression of MAPK 4 in comparison with healthy samples as well as pathogen treated samples at 3 and 6 hpi. This shows that zeatin enhanced the expression of MAPK 4 which in turn was not allowed to germinate the conidium as shown in the infection behavior of Varuna (Figure 2: 2H and I). Indeed, it is well known that CKs activates numerous plant defense response genes (Memelink et al., 1987; Smigocki et al., 1993; Schäfer et al., 2000). Surprisingly, no expression of MAPK 4 was found at 9 hpi in Varuna; it might be due to the fact that from this point of time, susceptible genes are expressed. In transgenic *Brassica* also, zeatin enhanced the expression of MAPK 4 at 6 hpi but at 9 hpi, the expression of MAPK 4 was low as compared to healthy sample; this further suggest that apart from MAPK 4, there is something which provides

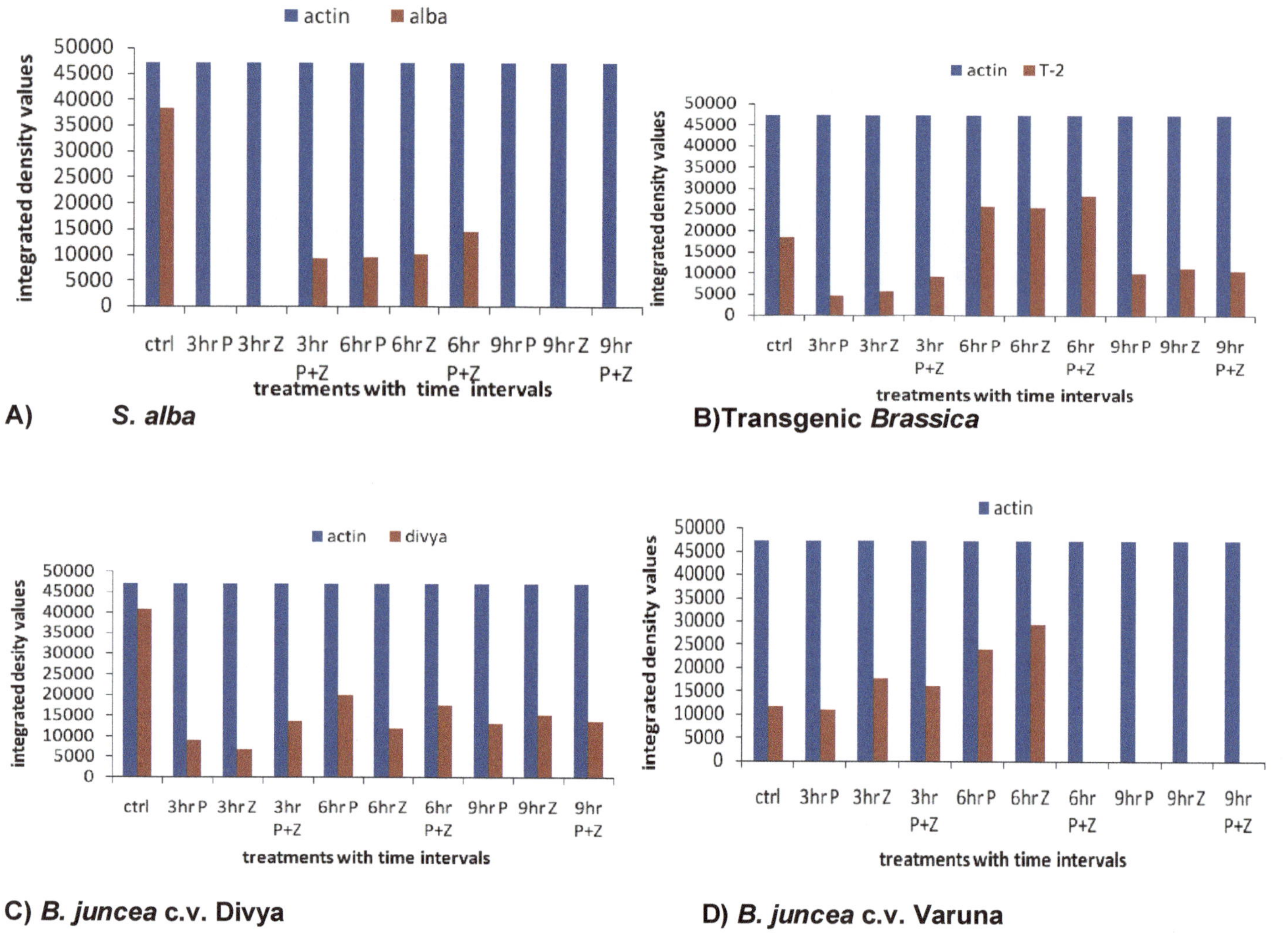

Figure 5. Densitometric analysis of the expression profile of MAPK4 gene from *A. brassicae* infected leaves, zeatin sprayed leaves and *A. brassicae* together with zeatin sprayed leaves of A) *S. alba*, B) transgenic *Brassica*, C) Divya, D) Varuna at different time intervals (3, 6 and 9 h).

tolerance in transgenic *Brassica*. This tolerance could be due to osmotin gene because at later stages of disease progression, transgenic *Brassica* exhibited localized bigger chlorotic and necrotic lesions around the inoculated region with no new spot (Figure 3) and it clearly indicates that the osmotin gene imparts some level of tolerance to *B. juncea* against *A. brassicae*. This was also reported earlier in our laboratory (Taj et al., 2004). In the treated samples, maximum expression of MAPK 4 was observed at 6 hpi in pathogen along with zeatin treatment and this implies that zeatin inhibits the conidial germination by enhancing the MAPK 4 expression, which further supports that both zeatin and MAPK 4 play role in defense. Zeatin also decreases the disease score in all the plants. The role of cytokinins has been investigated earlier and it was observed that cytokinins, serving as endogenous inducers for distinct classes of pathogenesis-related (PR) proteins, are necessary for the biosynthesis of SA and JA (Sano et al., 1994, 1995, 1996). Cytokinins are also known to delay senescence and can affect sensitivity of plants to pathogens. Moreover, *A. brassicae* and several other necrotrophic fungi are known to infect senescing plants due to increased susceptibility of the senescing tissue; exogenous application of CKs may aid in the delay and reduction of disease. Transgenic tobacco lines with higher CK levels were observed to be more tolerant to tobacco necrosis virus (TNV) (Pogany et al., 2004). CK also had a suppressive effect on the wildfire disease of tobacco caused by the bacterium *Pseudomonas tabaci* (Van Hall) (Lovrekovich and Farkas, 1963). In another study, CKs were observed to induce resistance of *Phaseolus vulgaris L.* to the white clover mosaic potexvirus (Clarke et al., 1998) and they also affected the growth of the fungus *Erysiphe cichoracearum* DC on leaf discs of tobacco (Cole and Fernandes, 1970). CKs have also been reported to enhance the resistance of barley to the fungal pathogen *Erysiphe graminis* f. sp. *hordei*. It is also supported by Sharma et al. (2010) who showed that zeatin antagonizes the effects of ABA produced by *A. brassicae* and delays the senescence and induces the expression of defense-related genes. The present study demonstrates that zeatin inhibits the *in vivo* growth of *A. brassicae* on the leaf surface and delays the infection process. To the best of our knowledge, even though CKs have been implicated in other host-pathogen interactions, this is the first direct demonstration of a protective role of zeatin against *A. brassicae- B. juncea* pathosystem. Apart from playing role in defense, MAPK 4 might also play role in other activities because in *S. alba* and Divya healthy samples, it showed maximum MAPK 4 expression. Our laboratory has identified that at the initial stage of *Alternaria* infection, all three well studied MAPKs viz. MAPK3 (Taj et al., 2011), MAPK6 (Tiwari, 2012, personal communication) and MAPK4 have been expressed, it is also well documented that these three kinases plays an important role in defense pathway. The expression of MAPK 3 and 6 is governed by the salicyclic acid pathway and MAPK4 is governed by the jasmonic acid pathway (Petersen et al., 2000). *A. brassicae* infection requires green tissue for sporulation which supports the view that this fungus is hemi biotrophic and behaves as both biotrophic as well as necrotrophic pathogen. At initial stage of infection, *A. brassicae* behaves like a biotroph but at later stages, it behaves as a necrotrophic pathogen. From this study, it is very difficult to say, at early stage which kinase among them is playing a prominent role in providing sustainable disease resistance; it might be a crosstalk among all these MAPKs and at later stages it could be a switch over from SA induced pathway (MAPK3/MAPK6) to JA induced (MAPK4) pathway. It could be hypothesized that resistance against *A. brassicae* may be carried out by the jasmonic acid pathway as it is a hemibiotrophic fungus. The earlier findings of our laboratory shows that LOX, AOC and OPR3 which are the upstream enzyme of the jasmonate biosynthesis pathway, were expressed after pathogen infection (Tej, 2012, personal communication) and also observed the co-expression of MAPK3 with LOX (Taj et al., 2011) which again support that SA and JA pathway have crosstalk at early infection of *A. brassicae.* This study gives a clue that cytokinin (zeatin) and MAPK4 both plays role in early defense. Our efforts are now directed towards understanding the role of zeatin induced defense genes in this pathosystem.

In conclusion, a more intensive study has to be carried out at the molecular level by taking later stages of disease progression.

ACKNOWLEDGEMENT

Authors are grateful to DBT for providing financial support in the form of R & D support under the Project Programme Mode Support in Agricultural Biotechnology.

REFERENCES

Akiyoshi DE, Regier DA, Gordon MP (1987). Cytokinin production by *Agrobacterium* and *Pseudomonas* spp. J. Bacteriol. 169:4242-4248.

Balibrea Lara ME, Gonzalez Garcia MC, Fatima T, Ehness R, Lee TK, Proels R, Tanner W, Roitsch T (2004). Extracellular invertase is an essential component of cytokinin-mediated delay of senescence. Plant Cell 16:1276-1287

Blakeman JP, Sztejnberg A (1973). Effect of surface wax on inhibition of germination of *Botrytis cinerea* spores on beetroot leaves. Physiol. Plant Pathol. 3:269-278.

Brodersen P, Petersen M, Nielsen HB, Zhu S, Newman MA, Shokat KM, Rietz S, Parker J, Mundy J (2006). *Arabidopsis* MAP kinase 4 regulates salicylic acid- and jasmonic acid/ethylene-dependent responses via EDS1 and PAD4. Plant J. 47:532-546.

Choi J, Huh SU, Kojima M, Sakakibara H, Paek KH, Hwang I (2010). The cytokinin-activated transcription factor ARR2 promotes plant immunity via TGA3/NPR1-dependent salicylic acid signaling in *Arabidopsis*. Dev. Cell 19:284-295.

Clarke SF, Burritt DJ, Jameson PE Guy PL (1998). Influence of plant hormones on virus replication and pathogenesis-related proteins in *Phaseolus vulgaris* L. infected with white clover mosaic potexvirus. Physiol. Mol. Plant Pathol. 53:195-207.

Cole JS, Fernandes DL (1970). Changes in the resistance of tobacco leaf to *Erysiphe cichoracearum* DC. Induced by topping, cytokinins and antibiotics. Ann. Appl. Biol. 66:239-243.

Conn KL, Pieterse JT, Van-loon LC (1990). A disease assessment key for Alternaria black spot in rapeseed and mustard. Can. Plant. Dis.Surv. 70:19-22.

Crespi M, Vereecke D, TemmermanW, VanMontagu M, Desomer J (1994). The fas operon of *Rhodococcus fascians* encodes new genes required for efficient fasciation of host plants. J. Bacteriol. 176:2492-2501.

Dekker JW (1963). Effect of kinetin on powdery mildew. Nature 197:1027-1028.

Desikan R, Hancock JT, Ichimura K, Shinozaki K, Neill SJ (2001). Harpin induces activation of the *Arabidopsis* mitogen-activated protein kinases AtMPK4 and AtMPK6. Plant Physiol. 126:1579-1587.

Droillard MJ, Boudsocq M, Barbier-Brygoo H, Lauriere C (2004). Involvement of MPK4 in osmotic stress response pathways in cell suspensions and plantlets of *Arabidopsis thaliana*: activation by hypoosmolarity and negative role in hyperosmolarity tolerance. FEBS Lett. 574:42-48.

Ehness R, Roitsch T (1997). Co-ordinated induction of mRNAs for extracellular invertase and a glucose transporter in *Chenopodium rubrum* by cytokinins. Plant J. 11:539-548.

Engelbrecht L (1968). Cytokinin in den gru¨ nen Inseln des Herbstlaubes. Flora 159:369-374.

Gao M, Liu J, Bi D, Zhang Z, Cheng F, Chen S, Zhang Y (2008). MEKK1, MKK1/MKK2 and MPK4 function together in a mitogen activated protein kinase cascade to regulate innate immunity in plants. Cell Res. 18:1190-1198.

Garg H, Li H, Sivasithamparam K, Kuo J, Barbetti MJ (2010). The infection processes of *Sclerotinia sclerotiorum* in cotyledon tissue of a resistant and a susceptible genotype of *Brassica napus.* Ann. Bot-London. 1-12 doi:10.1093/aob/mcq196.

Grobkinsky DK, Naseem M, Abdelmohsen UR, Plickert N, Engelke T, Griebel T, Zeier J, Nova´k O, Strnad M, Pfeifhofer H, Graaff EVD, Simon U, Roitsch T (2011). Cytokinins Mediate Resistance against *Pseudomonas syringae* in Tobacco through Increased Antimicrobial Phytoalexin Synthesis Independent of Salicylic Acid Signaling. Plant Physiol. 157:815-830.

Hwang HH,Wang MH, Lee YL, Tsai YL, Li YH, Yang FJ, Liao YC, Lin SK, Lai EM (2010). *Agrobacterium*-produced and exogenous cytokinin-modulated *Agrobacterium*-mediated plant transformation. Mol. Plant Pathol. 11:677-690.

Kowalska A, Niks RE (1999). Histology of Quantitative Resistance in Flax to the Flax Rust Fungus (*Melampsora lini*). Can. J. Plant Pathol. 21:354-360.

Lovrekovich L, FARKAS GL (1963). Kinetin as an antagonist to the toxic effect of *Pseudomonas tabaci*. Nature 198:710.

Meena PD, Awasthi RP, Chattopadhyay C, Kolte SJ, Kumar A (2010a).

Alternaria blight: a chronic disease in rapeseed-mustard. J. Oilseed *Brassica* 1:1-11.

Memelink J, Hoge JHC, Schilperoort RA (1987). Cytokinin stress changes the developmental regulation of several defence-related genes in tobacco. EMBO J. 6:3579-3583.

Meskiene I, Hirt H (2000). Mapkinase pathways molecular plug and play chips for the cell. Plant Mol. Biol. 42:791-806.

Morris RO, Jameson PE, Laloue M, Morris JW (1991). Rapid identification of cytokinins by an immunological method. Plant Physiol. 95:1156-1161.

Muller B, Sheen J (2007). *Arabidopsis* cytokinin signaling pathway. Sci STKE :cm5

Pandey D, Kumar A, Garg GK (2001). Antagonism between *Alternaria brassicae* toxin and zeatin in cell culture of *Brassica juncea* (L.) Czern and coss cv. Divya. Physiol. Mol. Biol. Plants 7(2):181-184.

Petersen M, Brodersen P, Naested H, Andreasson E, Lindhart U, Johansen B, Nielsen HB, Lacy M, Austin MJ, Parker JE (2000). Arabidopsis Map Kinase 4 Negatively Regulates Systemic Acquired Resistance. Cell 103:1111-1120.

Pitzschke, A., Schikora, A, Hirt H (2009). MAPK cascade signalling networks in plant defence. Curr. Opin. Plant Biol. 12:1-6.

Pogany M, Koehl J, Heiser I, Elstner E, Barna B (2004). Juvenility of tobacco induced by cytokinin gene introduction decreases susceptibility to Tobacco necrosis virus and confers tolerance to oxidative stress. Physiol. Mol. Plant Pathol. 65:39-47.

Qiu JL, Fiil BK, Petersen K, Nielsen HB, Botanga CJ, Thorgrimsen S, Palma K, Suarez-Rodriguez MC, Sandbech- Clausen S, Lichota J (2008). Arabidopsis MAP kinase 4 regulates gene expression through transcription factor release in the nucleus. EMBO J. 27:2214-2221.

Robert-Seilaniantz A, Navarro L, Bari R, Jones JD (2007). Pathological hormone imbalances. Curr. Opin. Plant Biol. 10:372-379.

Sano H, Ohashi Y (1995). Involvement of Small Gtp-Binding Proteins in Defense Signal-Transduction Pathways of Higher-Plants. Proc. Natl. Acad. Sci. USA. 92(10):4138-4144.

Sano H, Seo S, Koizumi N, Niki T, Iwamura H, Ohashi Y (1996). Regulation by cytokinins of endogenous levels of jasmonic and salicylic acids in mechanically wounded tobacco plants. Plant Cell Physiol. 37:762-769

Sano H, Seo S, Orudgev E, Youssefian S, Ishizuka K, Ohashi Y (1994). Expression of the Gene for a Small Gtp-Binding Protein In Transgenic Tobacco Elevates Endogenous Cytokinin Levels, Abnormally Induces Salicylic-Acid in Response to Wounding, and Increases Resistance to Tobacco Mosaic-Virus Infection. Proc. Natl. Acad. Sci. USA 91(22):10556-10560.

Schäfer S, Krolzik S, Romanov GA, Schmülling T (2000). Cytokinin-regulated transcripts in tobacco cell culture. Plant Growth Reg. 32:307-313.

Sharma G, Kumar VD, Haque A, Bhat SR, Prakash S, Chopra VL (2002). *Brassica* coenospecies: a rich reservoir for genetic resistance to leaf spot caused by *Alternaria brassicae*. Euphytica 125:411-417.

Sharma N, Rahman MH, Liang Y, Nat NVK (2010). Cytokinin inhibits the growth of *Leptosphaeria maculans* and *Alternaria brassicae*. Can. J. Plant Pathol. 32(3):306-314.

Sharma N, Rahman MH, Strelkov S, Thiagarajah M, Vipan K. Bansal VK, Kav NNV (2007). Proteome-level changes in two *Brassica napus* linesexhibiting differential responses to the fungal pathogen *Alternaria brassicae*. Plant Sci. 172:95-110.

Smigocki A, JR. Neal JW, Mccanna I, Douglass L (1993). Cytokinin-mediated insect resistance in *Nicotiana* plants transformed with the *ipt* gene. Plant Mol. Biol. 23:325-335.

Spoel SH, Dong XN (2008). Making sense of hormone crosstalk during plant immune responses. Cell Host Microbe 3:348-351.

Taj G, Agarwal P, Grant M, Kumar A (2011). Co-Ex-Pression and *In-Silico* Interaction Studies For Inter-Linking the Activation of MAPK3 And LOX Genes During Pathogenesis of Alternaria Blight (*Alternaria brassicae*) in *Brassica juncea*. J. Oilseed Brassica. 2(1):13-20.

Taj G, Kumar A, Bansal KC, Garg, GK (2004). Introgression of Osmotin Gene for Creation of Resistance Against Alternaria Blight By Perturbation Of Cell Cycle Machinery. Ind. J. Biotechnol. 3:291-298.

Tiwari JP (1993). Biochemical basis of resistance to A. brassicae in crucifers In: Lodha ML, Mehta SL, Ramagopal S, Srivastava GP. Eds. Advances in plant biotechnology and biochemistry. Indian Society of Agricultural Biochemists New Delhi, pp. 33-38.

To JP, Kieber JJ (2008). Cytokinin signaling: two-components and more. Trends Plant Sci. 13:85-92.

Verma PR, Saharan GS (1994). Monograph on *Alternaria* Diseases of Crucifers. Technical Bulletin 1994-6E. Saskatoon, Canada: Agriculture and Agri-Food Canada.

Walters DR, McRoberts N, Fitt BD (2008). Are green islands red herrings? Significance of green islands in plant interactions with pathogens and pests. Biol. Rev. Camb. Philos. Soc. 83:79-102.

Genes involved in sex determination and the influence of temperature during the sexual differentiation process in fish: A review

Cristina Helena Maria Moreira Vernetti[1], Marília Danyelle Nunes Rodrigues[2], Gutierrez, Harold Julian Perez[2], Calabuig, Cecilia Perez[3], Carla Giovane Ávila Moreira[4], Anthony A. Nlewadim[5] and Heden Luiz Marques Moreira[1]*

[1]Laboratório de Engenharia Genética Animal – LEGA/Departamento de Zoologia e Genética/Instituto de Biologia/Universidade Federal de Pelotas, Campus Universitário, s/nº. 96010-900 - Pelotas, RS, Brazil.
[2]Programa de Pós Graduação em Zootecnia, Faculdade de Agronomia Eliseu Maciel; Laboratório de Engenharia Genética Animal –LEGA/DZ/IB/UFPel, Brazil.
[3]Departamento de Ecologia – Universidade Federal do Rio Grande do Sul, Porto Alegre, RS, Brazil.
[4]Programa Pós-graduação em Genética e Biologia Molecular – Universidade Federal do Rio Grande do Sul, Porto Alegre, RS, Brazil.
[5]Fisheries and Aquatic Resources Management, College of Natural Resources and Environmental Management, Michael Okpara University of Agriculture, Umudike, Abia State, Nigeria.

This review attempts to group the recent hypotheses involved in the complex system of determination and sex differentiation in fish. Based on recent literature, we relate the key genes involved in the genomic cascade as the *Cyp19, Dmrt1, Sox9, Foxl2, Esr, Dax1, Sf1 and Amh1,* and still little known action of temperature on them. As the sex reversal is a highly desired process in fish farming aiming at obtaining male mono-sexual populations (due to weight gain of males), several techniques based on direct and indirect manipulation of phenotypic sex are being tested. Recent surveys show the use of temperature as alternative to the process of sex reversal. However, high or low temperatures have limited effect, in addition to there being a window of sex reversal in which temperature acts, varying from species to species. Thus, we draw a parallel with the role of temperature in the process of sex reversal and its effect on genes of the genomic cascade, which has been the subject of several studies that attempt to explain how temperature would be acting in this process. Intracellular receptors, such as those used for steroid hormones, act as transcription factors to regulate target genes moving between the nucleus and cytoplasm and, in the hormone absence, are linked to the complex of heat shock protein 90 kDa (Hsp90). Through this mechanism, it is possible to predict that fluctuations in temperature can influence the action of hormones, the increased transcription of genes involved in steroidogenesis and hence in sexual differentiation, becoming an alternative to explain where the temperature is acting. However, this literature review discusses the correlations between the genomic cascade, the action of intracellular receptors and the influence of temperature within this large system of determination and sex differentiation in fish.

Key words: Fish, gene, sex differentiation, temperature.

INTRODUCTION

Sexual differentiation in fish has been extensively studied, but little is known about the biochemical pathway responsible for sex reversal process and the true role of temperature in this process.

Many studies show that genes involved in sexual differentiation, many already sequenced, involve a genomic cascade: *Cyp19*; *Dmrt1; Sox9; Foxl2; Esr; Dax1; Sf1* and *Amh*, acting in a coordinated manner

(Baroiller et al., 2008, 2009; D'Cotta et al., 2007; Ijiri et al., 2008).

For some species, sex reversal is a highly desired process in fish farming aiming at obtaining monosex male populations, since males have higher growth rate and weight gain than females (Tachibana et al., 2004). In the case of the tilapia (*Oreochromis niloticus*), the fact that this species reach sexual maturity very early, usually between the 4th and 6th month of life, generates overcrowding in culture environments (Kubitza, 2000), and therefore very necessary to use techniques that control this population growth (Phelps and Okoko, 2011). In addition to sex reversal, other methods are used to control the size of the fish population. Jegede (2010) using *Hibiscus rosa sinesis* (4.0 g diet HLM/kg) showed disintegration and necrosis of spermatids in the testes as well as severe atretic follicles in the ovaries, revealing an effective inhibitor of reproduction in tilapia.

Overall, the most widely used and effective method for sex reversal from females to males in many fish species is the use of steroid hormones, which should start before gonadal tissue differentiation (Nakamura et al., 1998; Yamamoto, 1969). However, the use of hormones in animal production has been questioned at any stage of farming in such aspects as dosage and duration of use in animals intended for human consumption. Thus, alternatives influenced by consumer concerns about the residual effects of steroid hormones in the environment and human health are sought (Oliveira et al., 2008).

Some researchers have used variation of temperature as an alternative method (Baras et al., 2001; Tessemura et al., 2006). Many studies indicate the use of different water temperatures for sex reversal in fish; but where, how and when temperature is acting in the cascade of genes is a question not yet answered (Rougeot et al., 2008).

Thus, given the great number of investigations in this field through the study of genes involved in sex differentiation in fish and the use of temperature as a tool for sex reversal, there is the need to draw a picture of the genes involved in sex differentiation and relate the role of temperature in these processes in order to develop new alternatives for future studies. manipulation of the phenotypic sex of fish have been tested. They are aimed at the elimination of females in farm environments through sex reversal achieved using hormone treatments and genetic or environmental manipulation. According to

METHODS OF SEX REVERSAL

Several techniques based on direct and indirect Yamamoto (1969), treatment must begin before the onset of sexual differentiation.

Direct methods of sex reversal

Direct methods of sex reversal act on the physiological processes that determine masculinization, feminization or sterilization. These methods are the most widespread and used, such as the use of androgenic steroid hormones (Beadmores et al., 2001; Popma and Lovshin, 1996) which aims at eliminating or minimizing breeding in tanks, achieving higher production due to better growth of males, and obtaining more uniform lines.

The genotype determines the gene responsible for protein expression, which will act to determine the genotypic and phenotypic sex (Figure 1). There are thus two ways in the determination of females: the gene *Cyp19a1a* encodes a polypeptide and expresses a protein, the aromatase (Cyp19a1a), which catalyzes the denaturation (aromatization) of the A ring of androgens (C19) and converts them to estrogens (C18). In this process, methyl 19 is removed and this enzyme is linked to the membrane located in the endoplasmic reticulum of estrogen-producing cells of the ovaries, placenta, testis and adipose and brain tissues. Aromatase is encoded by the gene *Cyp19a1a* and acts in the formation of the complex NADPH-FERRIHEMOPROTEIN reductase in the cytochrome P-450 system. In males, when the gene *Cyp19a1a* for some reason is not expressed, the protein is not encoded and initiates the testis formation.

In European sea bass, Navarro-Martin et al. (2011) showed that methylation of the sea bass *cyp19a* promoter is the cause of the lower expression of *cyp19a* in temperature-masculinized fish, and may be an essential component of the long-sought-after mechanism connecting environmental temperature and sex ratios in vertebrate species with temperature-dependent sex determination.

The four direct methods of sex reversal can be used in the lots standardization (for both males and females). For reversal of genotypic females to phenotypic males, the commonly used method is the incorporation of steroid hormones in the feed (Phelps and Popma, 2000; Popma and Green, 1990). The hormone most often used in this process, due to its androgenic potential, easy availability and low cost, is the synthetic androgen 17-alpha-methyltestosterone (Phelps and Popma, 2000), which is a methylated derivative of testosterone (Piferrer and Donaldson, 1991).

According to Phelps and Okoko (2011), a dose greater than 240 $mgkg^{-1}$ of the hormone 17 alpha methyltestosterone could inhibit those endogenous hormones. Therefore, there was no efficient masculinization in genetic females. However, the reason for such reduction of efficacy has not been determined yet. The concentration of hormone 17 alpha methyltestosterone that showed the best performance

*Corresponding author. E-mail: heden@pq.cnpq.br.

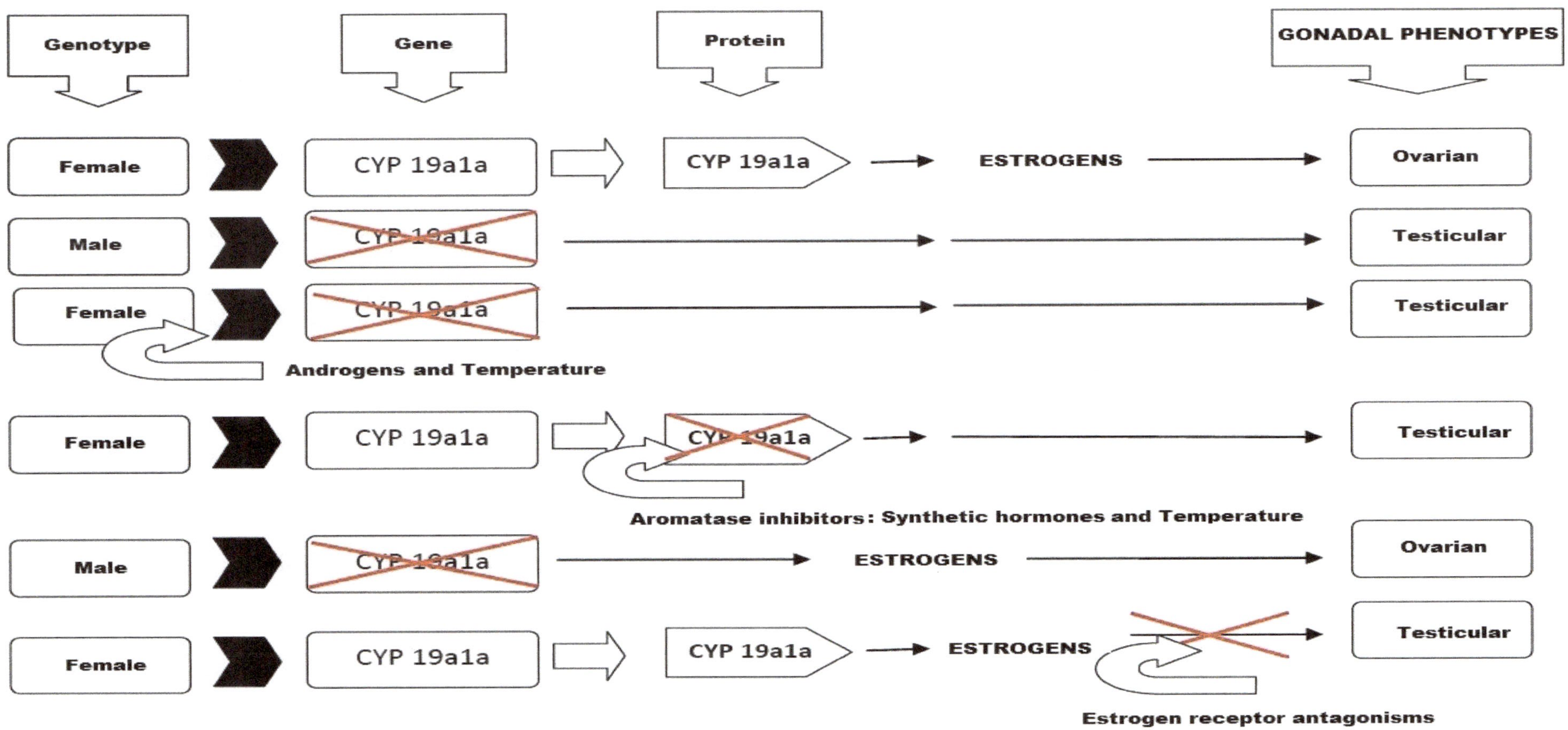

Figure 1. Sexual determination and direct treatment of sex reversal scheme adapted from Guiguen et al. (2010). The normal rout for ovarian development involves the *Cyp19a1* gene transcription and aromatase protein production that will act in synthesis of estrogens and ovarian development. In genotypic males, the blockage of *Cyp19a1* gene transcription leads to testicular development. In females, the use of androgens inhibits *Cyp19a1* gene synthesis and lead to testicular development, the same occurring when aromatase inhibitors are used. In males the blockade of *Cyp19a1* gene synthesis and use of estrogens leads to ovarian development. In females, even occurring the aromatase synthesis and estrogens production, the use of estrogens receptors inhibitors (antagonists) leads to testicular development.

in obtaining phenotypic males is between 15 and 60 mg kg^{-1} with percentage > 97% (Okoko and Ronald, 2011).

Enzyme inhibitors (Figure 1) are non-steroidal compounds that can interfere with the metabolismof endogenous steroid hormones before or during the period of sexual differentiation. They are similar to specific substrates of aromatase (testosterone and androstenedione) and generate competition for the enzyme active site, for example,

Fadrozole, Letrozole, Anastrozole and Vorozole (Bombardella et al., 2004). The antagonist estrogen-receptors (Figure 1) are compounds that prevent estrogens production. They are used for sex reversal, since they bind to specific intranuclear receptors of steroid hormones, preventing the production of estrogen and leading to the development of testes (Lehningher et al., 1995).

The normal route for ovarian development involves the *Cyp19a1* gene transcription and aromatase protein production that will act on the synthesis of estrogens and ovarian development. In genotypic teleost males, the block of *Cyp19a1* gene transcription leads to testicular development. In females, the use of androgens inhibits the synthesis of *Cyp19a1* gene and leads to testicular development, the same occurring when aromatase inhibitors are used. In males the blockade of *Cyp19a1* gene synthesis and use of estrogen leads to ovarian development. In females, even during the synthesis of aromatase and estrogen, the use of inhibitors of estrogen receptors (antagonists) leads to testicular development (Guiguen et al., 2009).

Indirect methods of sex reversal

The indirect treatments of sex reversal act directly on genetic processes or on environmental variables, which indirectly influence the physiological processes that determine the phenotypic sex (Donaldson, 1996). According to Baroiller (1995), sex determination is predominantly due to the existence of a large gene located on a pair of sex chromosomes.

Sexual differentiation requires a variety of biochemical pathways that involve many different proteins, for example, transcription factors, enzymes, receptors, second-messenger systems and others (Devlin and Nagahama, 2002). Whereas temperature can greatly influence both the structure and function of these molecules, sex reversal via temperature has been studied by several researchers as alternative to hormonal treatment (Azaza et al., 2008; Baras et al., 2000, 2001; Tsai et al., 2003; Wang and Tsai, 2000). These studies provide evidence that low water temperature (~25°C) favors female sex differentiation while high temperature (~35°C) favor male sex differentiation in most fish species.

GENES OF GENOMIC CASCADE INVOLVED IN SEXUAL DIFFERENTIATION

Cyp19 gene

The *Cyp19* gene encodes enzymatic proteins that have sequences with about 60% homology among each other. The key enzyme to balance the ratio of steroid hormones is the *Cyp19* aromatase (which is the terminal enzyme in the steroidogenesis pathway). The appropriate expression of this enzyme is essential for differentiation of sex and reproduction in vertebrates (Conley and Hinshelwood, 2001; Gelinas et al., 1998).

Using *in situ* hybridization, Harvey et al. (2003) demonstrated that the *Cyp19* gene is located on different chromosomes in the tilapia, *O. niloticus*. Chang et al. (2005) observed that aromatase isoforms have divergent distribution in tissues in response to exogenous estrogen and pattern of expression during gonad ontogeny.

The P450 cytochrome Aromatase (P450arom; *Cyp19*) is a product of the *Cyp19* gene, a terminal enzyme in the biosynthesis of estrogen that catalyzes the conversion of estrogen into androgen (Conley and Hinshelwood, 2001). Estrogen is essential for gonadal development and other diverse physiological processes ranging from normal growth to reproductive behavior. Aromatase, which is mainly expressed in the ovary, was designated *Cyp19a/P450aromA/Cyp19a1* (Harvey et al., 2003).

Two types of aromatase are known in tilapia: cerebral (*tCyp19b*) and ovarian (*tCyp19a*). The comparison of nucleotide sequences between genes *tcyp*19 and their corresponding cDNA clones indicate that *tcyp19a* and *tcyp19b* contain nine exons sequences and that DNA sequences at the boundaries between exon/intron follow the rule described by Breathnach and Chambon (1981). The coding region of exons of *tCyp19b* have 50, 71, 65, 64, 70, 65, 79, 67 and 60% homology with the *tCyp19a* corresponding exons, although there are great differences in nucleotide sequences among introns (Figure 2) (Chang et al., 2005). The similarity in genetic structure between *tcyp19a* and tcyp19b supports the suggestion that these two aromatase genes arose from a single ancestral gene in the early vertebrate evolution (Chiang et al., 2001b).

Binding sites of several transcription factors have been identified in P450arom mammalian and Atlantic Stingray genes (Ijiri et al., 2000), but not in the *Cyp19b* of zebrafish or goldfish. Using Sequence Motif Search (http:// motif.genome.ac.jp), TFSEARCH (http://www.cbrc.jp/researchdb/ TFSEARCH.html) and Dragon ERE Wnder (http://sdmc.Lit.org.sg/ERE-V2), which are promoters analyses tools available on the internet. Ijiri et al. (2000) analyzed the *tCyp19* promoter sequences for several transcription factors: steroidogenic transcription factor -1/Ad4 BP (SF-1/Ad4); GATA 4 transcription factor; Wilm (WT1-KTS), SRY, CRE, ERE (estrogen) and RAR-2-related to connection regions. These differences in the promoter/regulatory structure suggest that the expression of two *tCyp19* genes is regulated by different transcription factors (Ijiri et al., 2000).

Analysis of the promoter structure of aromatase isoforms points to the decisive role of *Cyp19a1a* gene for sexual differentiation of teleost (Trant, 1994; Tanaka et al., 1995; Fukada et al., 1996; Callard and Tchoudakova, 1997). Studies indicate that generally *Sry, Dmrt1, Sox9, Fgf9, Gata-4, Wt1-kts* and *Sf1* are genes involved in mammalian sex determination and key to the beginning of the sex differentiation cascade among the first days of embryogenesis (Parker et al., 1999). Peichel and Graves (2010) conclude that the homologies in sex determination in fish were due to independent re-use of particularly

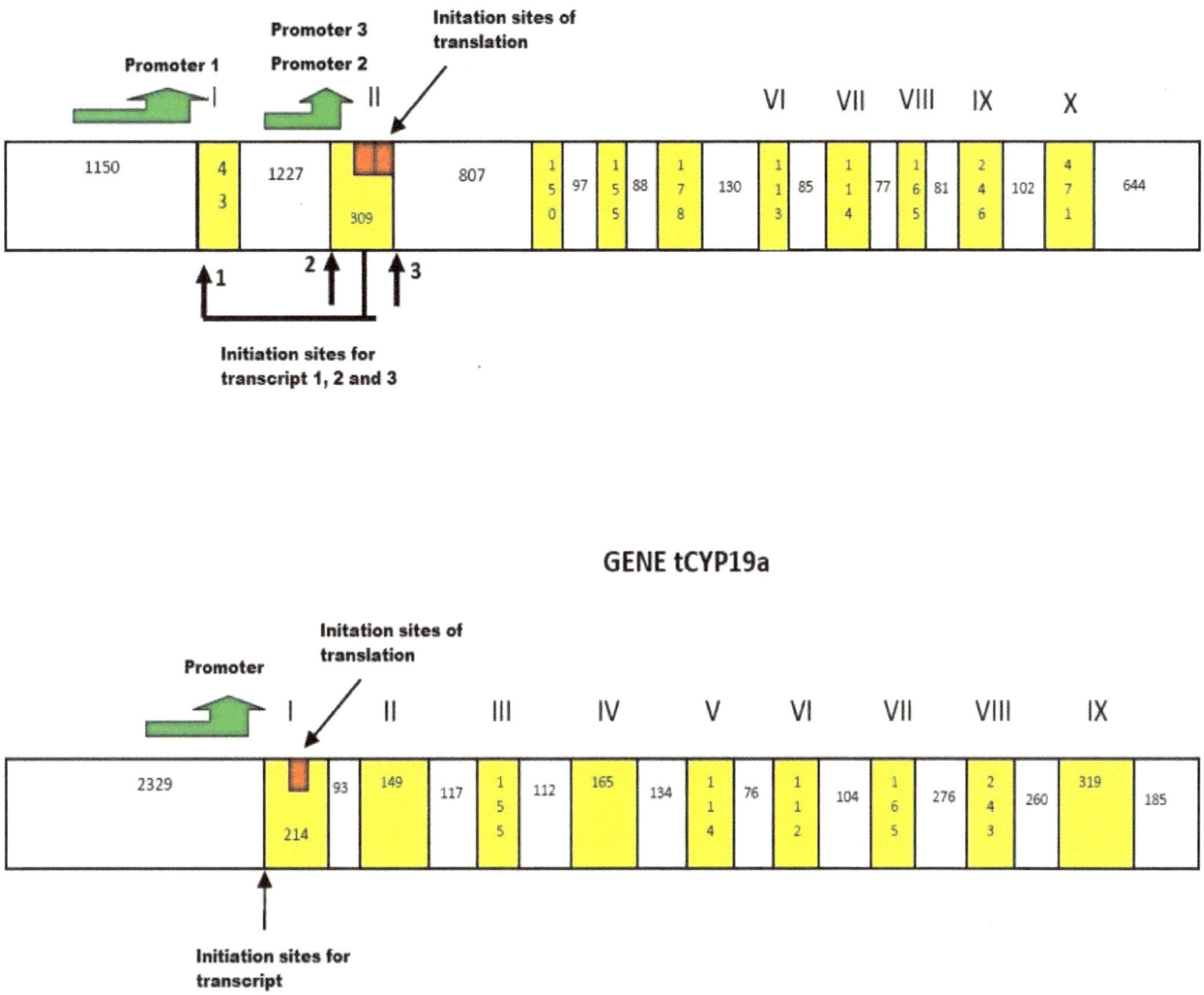

Figure 2. *tCYP19b* and *tCYP19a* gene schematic structure. Exon and intron regions are indicated by open and solid boxes, respectively. Boxes numbers represent fragments sizes. Transcription and translation initiation sites were sign by darts. Adapted from Chang et al. (2005).

handy genes (copies of *DMRT1* and *SOX3*), and of fish chromosomes that have become sex chromosomes multiple times.

Overall, gonadal aromatase *Cyp19a1a* is more abundant than that of cerebral (*Cyp19a1b*). The expression of two *Cyp19* genes have been investigated in the period between 0 and 41 days after fertilization (dpf), which is the expected time of sex determination and differentiation in zebrafish. The expression was higher after the outbreak in the period from 4 to 8 dpf. The *Cyp19a1b* expression pattern was different in two populations, suggesting association with sexual differentiation (Thorgaard, 1977).

Multiple forms of aromatase were found in teleosts including goldfish, zebrafish and sea bass (Blazquez and Piferrer, 2004; Chiang et al., 2001a, b; Gelinas et al., 1998; Kishida and Callard, 2001; Tchoudakova and Callard, 1998; Callard and Tchoudakova, 1997; Tong et al., 2001). According to Chang (2004), the *tCyp19a* was only detected in the ovary of tilapia fingerlings aged at 15, 20, 25 and 35 days.

Several studies confirm that estrogen is necessary for ovarian differentiation (Kobayashi et al., 2003; Kwon et al., 2000; Nakamura et al., 1998). It is possible that the level of *Cyp19a* gene expression control sex determination by regulating estrogen synthesis in teleosts (Fukada et al., 1996; Tanaka et al., 1995; Tchoudakova and Callard, 1998; Trant, 1994). The high level of *Cyp19a* expression in the ovary is consistent with the high *tCyp19a* levels of P450 aromatase activity and the production of 17 beta estradiol, which have been observed in teleost follicles during vitellogenesis (Kagawa et al., 1984; Tanaka et al., 1995; Trant et al., 2001). In zebrafish, the gene *Cyp19a* (2.1 kb) is expressed mainly in follicular cells during vitellogenesis (Chiang et al., 2001a; Kishida and Callard, 2001a, b; Trant et al., 2001). Similarly, *Cyp19a* was only expressed at high levels in the ovary during vitellogenesis in goldfish and bass (Blazquez and Piferrer, 2004; Gelinas et al., 1998; Tchoudakova and Callard, 1998). Based on the TATA box location, blocking the signals and transcription initiation sites, it is possible to deduce a possible promoter for the *tCyp19a* gene (Chang et al., 2005).

With northern blot, Yoshiura et al. (2003) demonstrated the parallel expression of *tCyp19a* and the *SF-1/Ad4BP* transcription factor during ovarian growth and showed that the HCG levels are simultaneously stimulated by the mRNA levels of *tCyp19a* and *SF-1/Ad4BP* in the follicle growth, suggesting that *Cyp19a* regulates the expression of gonadotropin by modulating the SF -1/Ad4. Kwon et al. (2001) observed a high expression of *tCyp19a* in female tilapia during early sexual differentiation, and reported that expression is greatly reduced in males between 15 and 27 days after hatching. In addition, Guiguen et al. (1999) reported a large increase in aromatase expression in the ovaries of rainbow trout prior to sexual differentiation.

Cyp19a inhibition in male teleost individuals can lead to bi-potential gonad to turn into a testis. In contrast, the down-regulation of genes responsible for sex determination in females deregulates the *Cyp19a* expression, which leads to high estrogen production and thus triggers the development of ovaries (Callard et al., 2001; Chang et al., 2005; Sawyer et al., 2006).

In addition, some studies have reported that estrogen may conversely inhibit the expression of sex-determining genes (Crews et al., 2001). With American turtle and trout, Crews et al. (2001) observed that treatment with estrogen, either directly or indirectly, represses the SF-1 leading to the development of female animals, and that inhibition of estrogen production caused by up-regulation of SF-1 leads to the development of males.

In Chang et al. (1997), cerebral aromatase designated *CypP450aromB/Cyp19a2* was described through a cDNA obtained from ovarian tissue of tilapia. The enzyme is expressed at high levels in brain tissue, strongly induced by estrogen and highly homologous to *Cyp19b* of goldfish and zebrafish, being designated in tilapia as *tCyp19b*. According to Chang (2004) part of the *tCyp19b* sequence shows 63.8% global homology with *tCyp19a* (Figure 2) and on average 60% homology with the aromatase of catfish, *Cyp19b* of goldfish, *Cyp19a* of goldfish, aromatase of medaka, *Cyp19b* and *Cyp19a* of zebrafish, *Cyp19b* of bass (Blazquez and Piferrer, 2004; Chang et al., 1997, 2005 Gelinas et al., 1998; Kishida and Callard, 2001; Tanaka et al., 1995; Tchoudakova and Callard, 1998; Callard and Tchoudakova, 1998; Trant, 1994).

The sequence of *tCyp19b* also shares homology with aromatase of frogs, chicken, rat, human placenta, swine placenta and ovary (Corbin et al., 1988; Harada, 1988; McPhaul et al., 1988; Simpson et al., 1987; Terashima et al., 1991).

Through real-time polymerase chain reaction (RT-PCR) with tilapia tissue, Chang (2005) detected the expression of two isoforms of P450arom in *tCyp19b* during sexual differentiation in all brain tissues collected from 1, 10, 15, 20, 25 and 35 days old and in gonads aged at 15, 20, 25 and 35 days. This study found no significant differences between the sexes and, therefore, the results suggest that *tCyp19b* is expressed at high levels in the brain and at low levels in other tissues. In studies with goldfish and sea bass, brain aromatase expression was detected at high levels in brain tissue (Blazquez and Piferrer, 2004; Gelinas et al., 1998; Tchoudakova and Callard, 1998), it was observed in zebrafish with predominant expression of brain aromatase occurring in the brain (Chiang et al., 2001a; Kishida and Callard, 2001; Trant et al., 2001).

According to Trant et al. (2001), the individual expression of *Cyp19a1b* during determination and sex differentiation distributed between the two groups (male and female) indicates that this gene may be associated with sexual differentiation in zebrafish. A high mRNA level detected of Cyp19b in brain tissue of the teleost suggests that the gene is involved in the production of neural-estrogen in the brain (Callard et al., 2001; Chang et al., 2005; Sawyer et al., 2006).

According to studies of Chang et al. (2005), analyzing the TATA box location, blocking the signals and transcription initiation sites, it is possible to deduce three possible promoters for gene *tCyp19b*. Prosecutors 1 and 2 of *tCyp19b* are located at the 5' end of exons I and II, respectively, suggesting that they are involved in regulating expression of transcription 1 and 2 respectively. The promoter 3 is located in the middle of exon II, suggesting that it is involved in the initiation of transcription 3. It is also possible that promoter 2 regulates the transcription of both 2 and 3, with alternative splicing, producing different transcriptions (Figure 2). Gelinas et al. (1998) also concluded that *in vivo* administration of estrogen and androgen in goldfish there were high mRNA levels of *tCyp19b* in the brain.

Evidence indicates that the differing responses of the two aromatase genes for estrogen are due to the presence or absence of the transcription factor ERE (estrogen responsive element) in their promoters.

According to Gelinas et al. (1998) working on tilapia, it is possible to suggest that the *tCyp19b* transcription is regulated by estrogen and preferentially expressed during neural estrogen biosynthesis, as consequence of estrogen binding to the ERE transcription factor in the promoter 1.

However, when investigating the aromatase expression in embryos and juveniles of zebrafish, Kishida and Callard (2001) and Trant et al. (2001) emphasized the important role of *Cyp19b* for sexual differentiation of gonads. Two groups have reported that estradiol, ethinyl estradiol and 17-methyltestosterone could increase the production of mRNA of *Cyp19b* in zebrafish embryos.

Blazquez and Piferrer (2004) also reported that *Cyp19b* expression was increased in brains of male sea bass after sexual differentiation (200 to 250 days after fertilization). These results suggest that *Cyp19b* is deeply involved in sexual differentiation of species.

On the other hand, the presence of ERE in the promoter region of *Cyp19b* gene suggests that the exogenous estrogen effect on the gonad development, at least partially, occurs through up-regulation of *Cyp19a* expression. Thus, Kishida and Callard (2001) and Trant et al. (2001) observed the increased expression of *Cyp19b* in zebrafish embryos after treatment with estrogen. The elevated expression of *Cyp19b* by exogenous estrogen has the same effect as the high natural expression of *Cyp19a* under physiological conditions on the ovary ontogenesis. Besides the high expression of *Cyp19b* in the period of sexual differentiation and later period of sex differentiation, which was observed in zebrafish and sea bass, it indicates that the *Cyp19b* gene activation is a consequence of sex determination and its expression is involved in gonadal development after differentiation (Blazquez and Piferrer, 2004; Kishida and Callard, 2001; Trant et al., 2001).

Kishida and Callard (2001), Trant et al. (2001) and Blazquez and Piferrer (2004) also reported that the mRNA levels of *Cyp19a* are much higher than that of *Cyp19b* in the early sexual differentiation of zebrafish and sea bass. In contrast, Gato-Kazeto et al. (2004) investigated the high and low expression of *Cyp19b* and found no correlation with sex in adulthood in zebrafish. In addition, a recent study on the expression of *Cyp19b* in the brain of adult zebrafish showed that levels of expression were the same in males and females (Resko et al., 2000).

Dmrt1 gene

Dmrt1 (Doublesex-and mab-3-related transcription factor-1) is a transcription factor expressed in XY gonads and Sertoli cell lineage before any sexual differentiation and it is essential for its occurrence (Raymond et al., 1998). Dmrt1 is a member of a specific family of transcriptional regulators - those that contain the DM domain DNA binding motif (Matson and Zarkower, 2012).

The gene *Dmrt1* has been cloned in a variety of invertebrates and vertebrates, including fish. In each one, the embryonic expression of Dmrt1 occurred in a sexually different way, where high levels were observed in the developing male gonad before sexual differentiation (Lei and Heckert, 2004).

In several species, such as frog, alligator, trout, birds and rats, *Dmrt1* gene is expressed at high levels in males compared to females, suggesting that high expression is required for testicular differentiation while low expression is compatible with the ovary differentiation (Raymond et al., 2000; Nanda et al., 2000).

The gene *Dmrt1* probably have an indispensable role in the event of sex determination and may be involved in testis development of teleost fish, which is analogous to its supposed role in some mammalian species (Loffler and Koopman, 2003). Works on Medaka indicate that *Dmrt1* regulates the spermatogonia differentiation (Kobayashi et al., 2004).

The *Dmrt1* expression at the beginning of the sexual differentiation period may indicate that the gene is involved in the signaling cascade in the early determination and sex differentiation in zebrafish (Hofsten and Olsson, 2005). In a genome-wide linkage analysis in zebrafish has implicated a locus containing *Dmrt1* as a major determinant of sex (Bradley et al., 2011).

In a recent model of gonadal development in medaka, *Dmrt1* expression was detected from 10 dph (post-hatching days), thus being the first gene to be differentially expressed in males and females (Kurokawa et al., 2007).

According to studies by Guan et al. (2000), *Dmrt1* was expressed in the testis of normal XY males and XX males in tilapia, indicating that this is a gene not connected to Y whose expression is correlated with testicular formation. These authors hypothesized that a Y chromosome signal (regulatory genes) acts in *Dmrt1* in males to promote the testis. In XX males, hormonal or environmental influence seem to compensated for the absence of a signal from the Y chromosome, which results in sex reversal of XX and up-regulation of *Dmrt1* expression.

Despite the studies performed to date, it is still unknown the major target genes of this important transcriptional factor. According to Wang et al. (2010), an over transgenic expression of *Dmrt1* in XX fish resulted in decreased expression of the aromatase gene; reduced serum levels of 17 beta estradiol; delayed development of ovarian cavity, varying the degrees of follicular degeneration, and even caused a partial to complete sex reversal. These results indicate that the aromatase *Cyp19* is one of the *Dmrt1* targets. It suppresses the feminization pathway by transcription repression of aromatase gene and estrogen production in the tilapia gonads and possibly other vertebrates.

Sox9 gene

The Sox gene family (SRY-related genes containing HMG) encodes an important group of regulators involved in development and sex determination. The HMGbox (high mobility group) that characterizes the Sox proteins is a binding domain with the DNA, and proteins encoded by Sox genes act as transcription factors. Using a comparative genomics approach, Bagheri-Fam et al.

(2010) showed that testis-specific enhancer within *Sox9* gene (TESCO) contains an evolutionarily conserved region (ECR) of 180 bp which is present in all vertebrates except fish.

In zebrafish, two genes were identified: *Sox9a* and *Sox9b*. Both contain the HMG and are able to bind to the recognition site AACAAAG in a similar way to murine *Sox9* (Chiang et al., 2001a).

The expression patterns of *Sox9a and b* are different in adult zebrafish. *Sox9a* shows a broad expression pattern and it is found in the brain, kidney, muscle, testis and pectoral fin, while *Sox9b* is only found in the ovary (Chiang et al., 2001a; Hofsten and Olsson, 2005).

Both, *Sox9a and b* are expressed during embryogenesis in cells involved in craniofacial and brain development (Chiang et al., 2001a). Furthermore, *Sox9a* appears as essential for chondrogenic development (Yan et al., 2002) and *Sox9b* is involved in the neural crest development (Li et al., 2002). The *SRY* gene is the member of the Sox-gene family. It is the determinant of Y male chromosome in most mammals (Kobayashi et al., 2004; Lei et al., 2007).

According to Berta et al. (1990) and Hacker (1996), *SRY* is a little conserved gene that seems to be unique in mammals. In contrast, *Sox9* is a conserved gene, present in all types of vertebrates. Like *SRY*, *Sox9* gene is required for testicular development in mammals, and its deficiency can result in sex reversal in human males. The *Sox9* expression is seen immediately after the SRY gene and may be a downstream effector. In mice, *Sox9* mediates the onset of *Amh* expression (anti-Mullerian hormone) in Sertoli cells (Yao and Capel, 2005). According to Chiang (2001a), *Sox9* expression during gonadal differentiation is up-regulated in the testes and down-regulated in the ovaries of mammals, birds and turtles.

The organization and function of Sox-family genes is poorly understood in other kinds of vertebrates and, despite the wide distribution of the *Sox9* genes in fish, the gene has been investigated only in a few ones (Chiang, 2001a; Zhou et al., 2002).

The expression levels of *Sox9* were similar from 9 to 29 dpf in XX and XY gonads of tilapia, getting stronger shortly after in males XY (Ijiri et al., 2008). In medaka, Nakamoto et al. (2005) found high levels of *Sox9a and b* in somatic cells surrounding the germinative cells during early sexual differentiation of gonads. Subsequently, during the early stages of development of testicular tubules (seminiferous tubules), the expression of *Sox9a* and *Sox9b* is maintained only in XY gonads, and it is markedly reduced in XX gonads. Some studies suggest that *Sox9* is not involved in the initial determination and sexual differentiation, but it is necessary for testicular tubules development (Nakamoto et al., 2005). Using both transgenic and chimeric *sox9b* medaka mutants, Nakamura et al. (2012) showed that medaka *Sox9b* is required for germ cell proliferation and also to survival, but not for testis determination or subsequent process of early testis differentiation.

Foxl2 gene

Foxl2 is a transcription factor involved in the gonad differentiation and ovarian function in various vertebrates (Wang et al., 2004). The *Foxl2* gene has a specific ovarian expression in mammals, chicken and rainbow trout (Loffler et al., 2003; Baron et al., 2004).

In tilapia, the *Foxl2* gene is expressed on the ninth day after fertilization in XX gonads at levels only slightly higher than in XY; subsequently, levels increase linearly in the XX ovaries (Ijiri et al., 2008). In the Medaka, *Foxl2* was expressed primarily in somatic cells that surround the germ cells in XX gonads immediately after the initiation of ovarian differentiation and is maintained in granules cells throughout the ovary development (Nakamoto et al., 2006).

Wang et al. (2007) through *in vitro* studies identified that *Foxl2* binds to the *Cyp19a* promoter and activates its transcription (Figure 3). Many studies have reported the involvement of *Foxl2* gene in the differentiation of females holding cell lineages, granule cells and the formation and/or maintenance of ovarian follicles in various vertebrates (Govoroun et al., 2004; Loffler et al., 2003; Nakamoto et al., 2006; Wang et al., 2004), thus confirming the *Foxl2* correlation with *Cyp19a* expression (Ijiri et al., 2008; Baroiller et al., 2008).

Estrogen receptors (*Esr*) gene

In vertebrates, estrogens (especially 17-beta-estradiol) play a critical regulatory role in many physiological processes, including growth, differentiation and homeostasis of the reproductive organs of males and females, in addition to influencing the functioning of skeletal, cardiovascular, immune and central nervous systems (Korach, 1994).

The first Er (*Er alpha*) was isolated from humans (Green et al., 1986). The presence of a second subtype, *Er beta*, has been found in mammals, birds, fish and amphibians. More recently, a third subtype, *Er beta2*, was discovered in teleost fish; and it is closely related to *Er beta* suggesting that it reflects in total or partial duplication of this gene in fish (Filby and Tyler, 2005).

Currently, the zebrafish nomenclature (Zebrafish Nomenclature Official Guideline; http://zfin.org) was followed to standardization; *Er alpha* has been termed *Esr1*, the subtype *Esr beta1* became *Esr2b* and the subtype *Er beta2* became *Esr2a.*

The biological significance of multiple *Esr* is far from being completely understood. *Esr1* and *Esr2* are structurally similar (Beato, 1989; Pakdel et al., 1989). They show different binding capabilities (Pakdel et al.,

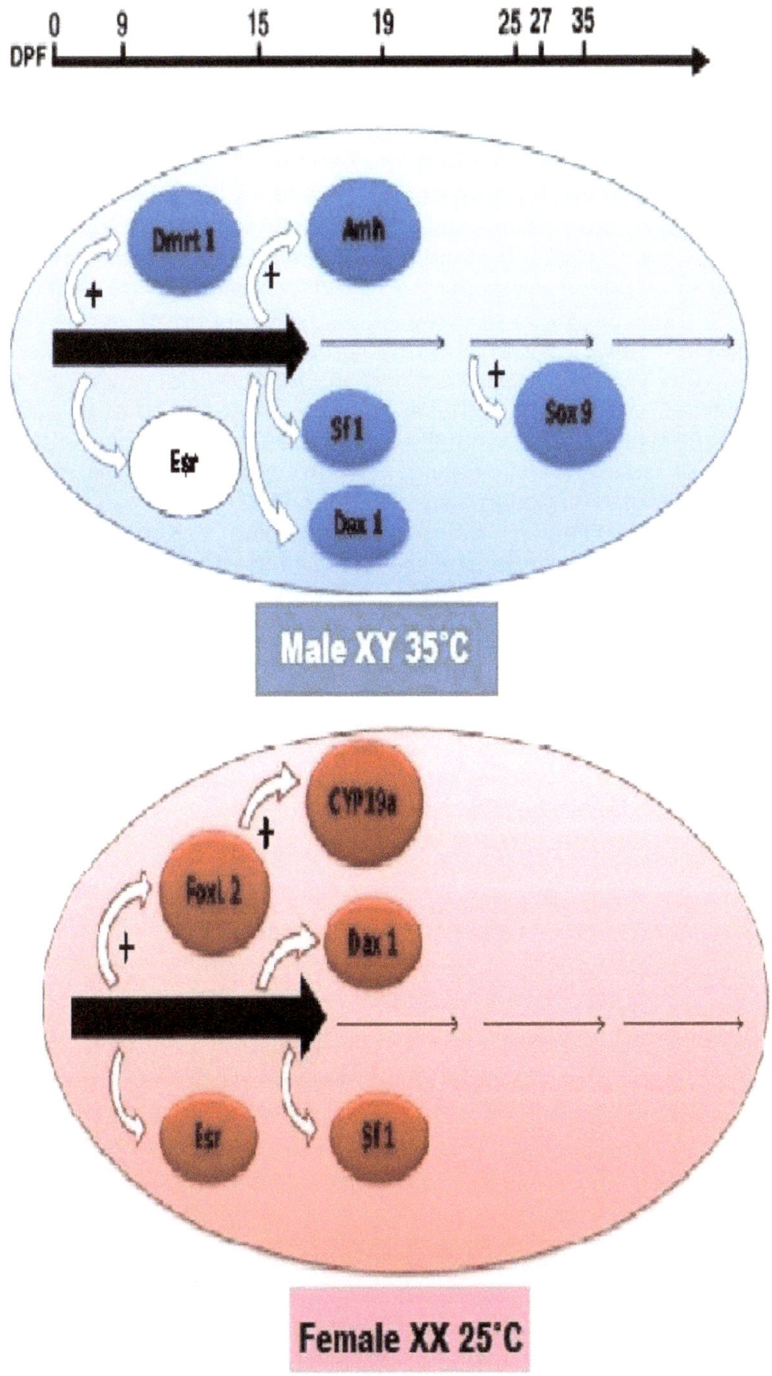

Figure 3. Scheme of genes involved in sexual differentiation and sex determination. Adapted from D'Cotta et al. (2007); Ijiri et al. (2008); Baroiller et al. (2008, 2009).

1989) and transcriptional properties (Filby and Tyler, 2005).

Esr levels are influenced by estrogen (Mommsem and Lazier, 1986), mediated by its effects on gene expression of estrogen receptor (Mackay et al., 1996; Pakdel et al., 1989). This positive control of *Esr* genes by estrogen ensures enough available receiver, stimulating cell types in response to the full presentation of estrogen (Devlin and Nagahama, 2002).

Distinct expression patterns of *Esr1* and *Esr2* have been documented in fish; seeming to vary among species. They exhibit a broad tissue expression, but greatly concentrated in the liver, gonads and with considerably lower levels of expression in the brain, pituitary, intestine and muscle (Filby and Tyler, 2005).

In tilapia, *Esr1* and *Esr2* genes are expressed in early male and female gonads between 10 and 15 days after hatching (Watanabe, 1999) showing different expression patterns that suggest an important role in the regulation of early sexual differentiation (Devlin and Nagahama, 2002). The results of Tsai et al. (2003) indicated that the brain aromatase, *Esr1* and *Esr2* are differentially regulated according to the environmental water temperature during the developmental period in tilapia.

Huang et al. (2008), studying the expression of three types of *Esr* in male tilapia, have shown that environmental estrogens affect the *Esr* expression and induce reproductive abnormalities in males. These authors also reported that the three types were maximally expressed during the early stage of recurrence, indicating that they play a role in testicular function and spermatogenesis mediated by *Esr*. Clearly, further studies are required to better understand the role of *Esr* in fish and its relationship with temperature.

Anti-Mullerian hormone (*Amh*) gene

In mammals, the anti-Mullerian hormone (*Amh*) is involved in the regulation of gonadal steroidogenesis by inhibiting the aromatase expression in developing gonads (Clemente et al., 1992) also presenting a complex regulation, which involves several factors, including the FTZ-F1 related to genes *Sf1, gata4, Sox9* and *WT1* (Santa Barbara et al., 1998; Rey et al., 2003). In addition, it negatively modulates the differentiation and function of Leydig cells (Racine et al., 1998) through the regulation of several enzymes involved in the steroidogenic pathway.

The anti-Mullerian hormone's (*Amh*) gene, responsible for Mullerian ducts regression in males, was found in fish that do not have Mullerian ducts, and its role is still unknown (Hofsten and Olsson, 2005). According to D'Cotta et al. (2007), the increased expression of the *Amh* gene in male gonads of tilapia is evident between 10 and 15 dpf and after 19 dpf (Ijiri et al., 2008). Hofsten et al. (2005) cloned the *Amh*'s cDNA of zebrafish and noted that this gene was expressed exclusively in the gonads.

By *in situ* hybridization, the *Amh* expression was found predominantly in Sertoli cells, testis and ovarian follicular layer. The regulation of *Amh* transcription in zebrafish has not been elucidated so far.

Sf1 gene

In mammals, two homologous genes to FTZ-F1 were found: *NR5A1* and *NR5A2*. Ad4BP/SF-1 proteins, members of the family of orphan nuclear receptors, have been identified as a specific steroidogenic transcription factor found in tissues (Morohashi et al., 1992, Morohashi and Omura 1990) such as pancreas, liver, intestine; in addition to being involved in cholesterol metabolism, which involves the synthesis of steroids and consequently in sex determination. The *ad4bp/Sf1* gene deletion resulted in lack of reproductive organs in mammals, suggesting that Ad4BP/SF-1 functions as a transcription factor involved in the regulation of enzymes involved in steroidogenesis as well as in tissue differentiation by regulating the *Dax1* gene (Kawabe et al., 1999; Luo et al., 1994). Such genes are regulators that have a key role in steroidogenesis and are involved in testicle determination during the course of sex determination (Ito et al., 1998; Ikeda et al., 1993; Lala et al., 1992; Sadovsky et al., 1995).

In zebrafish, four genes for FTZ-F1 have been identified (*ff1a*, *b*, *c* and *d*) (Hofsten and Olsson, 2005). According to Kuo et al. (2005), the genes *ff1b* and *ff1d* are of the same origin, emerged from the ancestral genes duplication. This hypothesis was supported by the expression patterns, overlap found in embryos of kidney and pituitary cells (Hofsten and Olsson, 2005).

Studies on fish that explain the relationship between P450arom and Ad4BP/SF-1 expression in response to stimulation of gonadotropins on the ovary growth and their relationships with temperature sensitivity are still scarce.

Dax1 gene

Dax1, member of the nuclear receptors family, is encoded and expressed in embryonic gonads of both sexes, but levels of expression have not been determined (Guiguen et al., 2010). According to Zanaria et al. (1994), the *Dax1* gene encodes an atypical protein that is part of the nuclear receptors family.

In rats, the *Dax1* expression was detected during the early stages of gonadal and adrenal development; its expression persisted in ovarian development but decreased in the testis coinciding with the *SRY* gene activation and testis differentiation (Swain et al., 1996). *Dax1* gene homologues were found in chickens (Smith et al., 1999; Smith and Sinclair, 2001) and in the American alligator, species with sex determination by temperature (TSD) (Western et al., 2000). Even though the *Dax1* gene expression has been verified in embryonic gonads of these species, the levels have not been determined, nor their relationship with temperature sensitivity.

THE TEMPERATURE IN THE PROCESS OF REVERSAL AND SEXUAL DIFFERENTIATION

So far, few studies have been performed to evaluate the effect of water temperature on sex ratio in fish.

In rainbow trout, experiments performed by Van den

Hurk and Lambert (1982) and Baroiller et al. (1999) found no effect of temperature on sex ratio. In salmon, Azuma et al. (2004) performed sex reversal of genetic females to functional males under 18°C (between 42 and 151 dpf).

Craig et al. (1996) however, found a strong feminization (62 to 84% female) in different salmon populations when eggs were exposed to higher temperatures (10.4 to 12.0°C and control: 8.3 to 9.7°C) during embryonic development (from 360 dpf). These results showed that, depending on the genetic background of the sockeye salmon population, temperature manipulation during incubation can change sex ratio. This is similar to the results of Magerhans et al. (2009), which also showed the partial ratios in breeding progenies when exposed to higher post-hatching temperature, producing females. However, the temperature regime varied between studies as in the distortion range in females' percentage: 20% (in the study of Magerhans et al., 2009) versus 28% of females (compared to control, in Craig et al., 1996).

The temperature is generally able to increase or decrease the proportion of females. In both studies (Craig et al., 1996; Magerhans et al., 2009) have proven that the sex ratios of offspring in treatments with temperature could not be attributed to the differential of high mortality of females or males.

High response repeatability to sex reversal with temperature treatments was proven in tilapia by Tessema et al. (2006). In addition, the study of Magerhans et al. (2009) provides evidence that both males and females contribute to the different sex ratios after treatment with temperature.

These studies show a clear paternal and maternal influence on the sex differentiation after treatment with temperature. Taken together, these results suggest that these species are heat sensitive and temperature can affect gonadal differentiation of the sex of individuals during the early stages of development.

The sensitivity of Nile tilapia (*O. niloticus*) to temperature during sexual differentiation is not seen in all progenies. Some breeders, males and females, produce offspring that exhibit high sensitivity to temperature thereby producing high proportion of males in their sexual rate, while others have produced non-sensitive and balanced sex rate.

The mode of temperature action on sexual differentiation during embryogenesis is a process that is still not understood. One hypothesis is that high temperature (35°C) can act directly on the brain aromatase gene (*Cyp19b*) to modify the path of sexual differentiation during embryogenesis (Morrison et al., 2001).

This hypothesis of "brain sexualization" is consistent with the presence of a rudiment of the brain 31 h after fertilization, as seen in Nile tilapia (Morrison et al., 2001), consistent with the idea that sexual differentiation occurs in the brain before occurring in the gonads (Arnold, 2004; Francis, 1992).

In Atlantic halibut (*Hippoglossus hippoglossus*), Matsuoka et al. (2006) also observed higher levels of aromatase gene expression in the brain at an earlier stage of development of the gonads, suggesting that sexual differentiation may begin in the brain before gonadal differentiation.

Another hypothesis is that the high temperature, applied during embryonic development, acts directly on somatic cells of future gonads and/or on the germ cells, which were already present (46 hpf post-fertilization hours) in embryos raised at 27 and 29°C (Morrison et al., 2001).

Sex differentiation by temperature

Numerous cases showed large deviations in experimental sex ratio of Nile tilapia, which cannot be predicted by a simple single-factor model. The segregation of DNA markers linked to sex (Cnaani et al., 2008; Lee et al., 2003, Shirak et al., 2006) confirmed what has been previously postulated on the existence of a single or main multi-allelic sexual determinant as well as an additional epistatic locus (or perhaps multiple loci) probably autosomal (Baroiller et al., 1995, 1996; Baroiller, 1997; D'Cotta, 2001; Hammerman and Avtalion, 1979; Mair et al., 1991, 1997).

Some studies show the sex determination system behavior as reported in zebrafish (Uchida et al., 2004), sole marble (Goto et al., 2000), gold (Kazeto-Goto et al., 2006) Japanese flounder (Yamamoto, 1999) and tilapia (Baroiller et al., 1995; Ovidio et al., 2002). These species also have a XX-XY sex determination system and the high temperature induces the XX fish masculinization. Despite the strong genetic basis for sex determination in tilapia, it is clear that other factors are responsible (Wang and Tsai, 2000) and the temperature seems to be a important factor in sex differentiation of species. On the other hand, in gold fish, zebrafish (Uchida et al., 2004), sole marble (Goto et al., 2000), Japanese flounder (Yamamoto, 1999), Loach (Nomura et al., 1998) and tilapia (Baroiller et al., 1995), TSD can occurs when extreme temperature override genetic sex determination (GSD). In gold fish, genetically female gonads developed into ovaries when fish were reared under 20°C and at 30°C almost all the gonads differentiated into testes. However, the sex ratio varied among the different groups even when reared at the same temperature (90.5 and 72.7% at 23°C). The production of a population in the same sex is useful for research on GSD/TSD as it is easy to judge the genotypic and phenotypic sex of the surviving individuals. However, high or low temperatures have a limited effect. Ospina-Бlavarez and Piferrer (2008), classify these results as being most likely due to thermal effects on GSD (GSD + TE) than for a real TSD.

An effect of temperature on sexual differentiation has also been observed in some flatfishes (Luckenbach et al.,

2003; Goto et al., 1999; Yamamoto, 1999). In eight families evaluated for TSD (Haffray et al., 2009), both for higher and lower temperatures (~ 35 and ~ 24°C, respectively), there was increased proportion of females. This interaction of temperature with the sex ratio has been reported in sea bass by Saillant et al. (2002).

An inter-family variation was found by Bezault et al. (2007), pointing to an important parental effect and genotype-environment interactions, as suggested by other studies (Abucay et al., 1999; Baroiller et al., 1995b; Tessema et al., 2006). Such parental effects were also found in other heat-sensitive species such as *Menidia menidia* (Conover and Heins, 1987) and *Odonthestes bonariensis* (Strussmann et al., 1996). Moreover, other experiments showed that high temperatures can have both feminizing and masculinizing effects, depending on the sex genotype (Abucay et al., 1999; Baroiller and D'Cotta, 2001; Kwon et al., 2002).

Wessels and Horstgen-Schwark (2007) provided evidence that an excess of males in groups treated by temperature can be selected for a quantitative trait. They were the first to determine the heritability for sensitivity to temperature in Nile tilapia. The heritability was 0.69, obtained from a selection experiment of two generations of high sensitivity to temperature.

A reproductive scheme of diallel crossing (5 Ч 5) showed that the males' percentage in tilapia was very different at inter-individual level, indicating that there was an important parental effect (Baroiller and D'Cotta, 2001). This was confirmed by an exhaustive investigation in two populations of tilapia by Tessema et al. (2006). Both, male and female, contribute to these parental genetic effects.

Within the Bouaké strain of Nile tilapia, the progeny test in females originating from classical treatment by temperature revealed no XY female by sex reversal (Baroiller, unpublished data), while other studies have demonstrated the presence of XX males in offspring treated with temperature (Baroiller et al., 1995). This suggests that the supposed feminization effect with high temperatures is probably restricted and peculiar to the genotype (eg, YY instead of XY) and/or population/strain/species.

In the work on goldfish of Goto-Kazeto et al. (2006) on the temperature-dependent sex differentiation, sex ratios of females were 94.6 to 100% at 15 to 20°C and 7% of females at 30°C, while in the latter, percentage of males increased. The masculinized fish were functional males. When using genotypic XX females, male fish emerged at high temperatures, being evidence of XX fish masculinization and not the result of a skewed sex ratio of fish survival.

In 1995, Baroiller et al. demonstrated that tilapia were sensitive to temperature during the critical period of sexual differentiation. It was possible to masculinize the XX offspring (100% females) with increasing temperature above 32°C, producing functional male phenotype. The use of mono-sex female populations, as well as the progeny testing for males treated by temperature, showed definitely the existence of distortions in the sex ratio of males, corresponding to a sexual inversion of genetic females (XX) for phenotypically functional males. These new males by temperature produce female or almost exclusively female progenies depending on the cross (Baroiller et al., 1995).

The limited extent of the temperature effect can be associated with the observation made by Yamamoto (1999) in Hiram, which was not possible to reverse genetic males into phenotypic females using temperature treatment. The small effect of temperature observed by Haffray et al. (2009) also suggests that an early evaluation of the effect of these treatments are made, since other external factors other than temperature could be involved in sex differentiation, but there is also an unknown factor of autosomal influence.

Following these results, the influence of environmental factors on sexual differentiation has been reported in over 60 different species of teleosts (Baroiller et al., 1999; Baroiller and D'Cotta, 2001; Conover, 2004; Godwin et al., 2003). Most of these studies concluded that the temperature sensitivity is impaired because the mechanisms of sex determination of these species could not be well characterized (e.g., Zebrafish, *Danio rerio, Tetraodon*) and physiological, genetic and ecological studies could not be performed to better understand the components of environmental sensitivity.

Periods of sensitivity to temperature

It has long been known that gonadal estrogens act as natural inducers of ovarian development in lower vertebrates (Nakamura et al., 1998; Yamamoto, 1969).

The enzyme aromatase, as mentioned earlier, catalyzes the conversion of androgens into 17-beta-estradiol (Baroiller et al., 1999) and if inhibited, it blocks the estrogen production so that there are sex reversals of females into males (Guiguen et al., 1999, 2010; Kwon et al., 2000).

In XX fish developed ovaries, the *Cyp19a* gene (ovarian aromatase) is up-regulated (D'Cotta et al., 2001; Ijiri et al., 2008; Kwon et al., 2001) such that Ijiri et al. (2008) found elevated *Cyp19a* levels in future ovaries, as early as 9 dpf and increased rapidly until 19 dpf.

The temperature when applied during the period of sexual differentiation induced down-regulation of *Cyp19a* in XX tilapia progenies (D'Cotta et al., 2001) that was observed at 17 dpf (Baroiller et al., 2008). Moreover, the expression levels of *Cyp19a* were correlated with the masculinization proportion by temperature (TM) in XX individuals (D'Cotta et al., 2008). On the other hand, studies on goldfish reported that reversal window is between 7 and 12 dpf (Kazeto-Goto et al., 2006). Thus, in order to achieve an effective sex control by temperature manipulation, treatments should begin within the sensitive period.

High temperatures could effectively masculinize some progenies if started about 10 days after fertilization (dpf) and maintained for 10 days (Baroiller et al., 1995; Wessels and Horstgen-Schwark, 2007; Tessema et al., 2006). However, if the treatment was applied for a period of 10 days, but started at 7 dpf, it had no effect on the sex ratio (Baroiller et al., 1995). This window to temperature sensitivity coincides with the gonadal sensitivity in relation to other external factors, especially hormones.

Data from Rouget et al. (2008) suggest a small window of sensitivity term in tilapia after fertilization (between 12 and 52 h post fertilization = 4 dpf), before the development of presumptive gonads. Interestingly, two similar windows were also identified for hormone treatments as temperature, hormone treatments or the use of aromatase inhibitors during sexual differentiation can overlap the sex genetic determination, producing sex reversal and functional phenotypes (Baroiller et al., 1999; Guiguen et al., 1999; Nakamura, 1975).

In tilapia (*Oreochromis mossambicus*), a treatment at 20°C for five days starting before 14 dpf induced feminization, while treatments of high temperature for five days at 32°C, initiated after 14 dpf, induced the masculinization (Wang and Tsai, 2000). However, it has not observed feminization with low temperatures ranging from 18 to 23°C in tilapia (*O. niloticus and Oreochromis aureus*) (Abucay et al., 1999; Baroiller et al., 1995; Desprez and Melard, 1998; Tessema et al., 2006).

Recently, Rougeot et al. (2008), studying the process of sexual differentiation in tilapia, demonstrated that even early treatment with elevated temperature applied 12 h after fertilization (hpf) and maintained for 52 ± 2 h (until hatching), can also induce significantly distorted sex ratios among males (8 to 27%, n = 4) in a true offspring of females (100% females in the control group). According to Rougeot et al. (2008), even the high mortality associated with early treatment for high temperature (35 to 36°C) must be considered, since these data suggest that there is a very short thermo-sensible window after fertilization (hpf between 24:52 (dpf = 4), long before gonads development. At this stage (12 hpf), tilapia larvae continue to have a rudimentary brain (31 hpf), as well as primordial germ cells (46 hpf) (Morrison et al., 2001).

Although the induction of deviations in sex ratios by environmental factors has been recognized in many species, the environmental sex determination (ESD) mechanisms are still unknown.

INTRACELLULAR RECEPTORS

According to Squires (2003), glucocorticoid (GR), mineralocorticoid (MR) and androgen (AR) receptors are found predominantly in the cytoplasm of cells, while thyroid hormone receptors (TR), estrogen (ER), progesterone (PR), retinoic acid (RAR) 1.25 and D3-hydroxy vitamin (VDR) are predominantly found in the nucleus.

Steroids and thyroid hormones operate via intracellular receptors. Toft and Gorski (1966) identified estrogen receptors, using hormone radiotracer and found that the receptor was only present in target cells. Receptors for steroid hormones act as transcription factors to regulate transcription of target genes (e.g. aromatase gene). These receptors move between the nucleus and cytoplasm and in the absence of hormone are linked to the 90kDa (Hsp90) heat shock protein complex (Figure 4). Hormone binding to the receptor results in Hsp90 complex release and translocation of the hormone-receptor complex to the nucleus. A dimer of the hormone-receptor complex then interacts with hormone response elements in specific genes that affect DNA transcription. This exposes sites on the DNA template, either directly or through influence of pre-existing repressor molecules to increase the initiation sites for RNA polymerase and increase the transcription (Figure 4). This action of steroid hormones occurs during a much longer time (hours) compared with peptide hormones (Squires, 2003).

Through this mechanism, it is possible to predict that fluctuations in temperature can influence the action of hormones, increasing the transcription of genes involved in steroidogenesis and hence in sexual differentiation, both in fish as in other groups.

CONCLUSIONS

Sex can be determined by multiple mechanisms in vertebrates. Despite the fact that the fate of the gonad is determined genetically, temperature can override it and switch the mechanism when the gonad is undifferentiated (Baroiller et al., 2009). Many genes involved in the sex-determining cascade have been identified and the existence of a complex SDS, combining GSD and thermo-sensitivity.

Research shows that the thermo-sensitive period is variable depending on the species, occurring before the onset of the first sex-specific differences, an active mitosis in the ovary (D'Cotta et al., 2001a; Ijiri et al., 2008; Nakamura et al., 1998). Thus, it can be assumed that high temperatures can act early on the brain under development and/or primordial segregation of germ cells (Rougeot et al., 2008).

As stated by Bull (2008), if the coexistence of TSD and GSD is an "accident" (infrequent relic) or adaptation, it remains being an important issue (Warner and Shine, 2008). To explain the coexistence of sex chromosomes and TSD, benefits of both systems must be combined (Bull, 2008). Therefore, it is possible to suggest that intracellular receptors are presented as key mechanism in the sexual differentiation of animals with TSD system;

however, no work demonstrating the action of this mechanism was performed, although many report that temperature is directly linked to sex determination.

TSD and GSD combined show that sex in fish is governed by the interactions of some components such as: a complex system of genetic determination, with some determinant loci and some minor genetic factors, as well as the influence of temperature. We need to understand how temperature and genetics match for sex determination in fish. For this, three important issues arise in relation to temperature effects: (1) does the temperature act in sex determination, sex differentiation or both? (2) are there different heat-sensitive stages in the cascade of sex determination depending on the species? and (3) what is (are) the true target-organ (s) (gonads, central nervous system, both organs) to the temperature effects?

Among so many studies, the role of temperature on sex determination in fish has not been fully elucidated. To the extent that future research will answer these questions, the use of molecular markers associated with temperature, combined with advances in reproductive techniques brings promising prospects for genetic improvement in aquaculture.

REFERENCES

Abucay JS, Mair GC, Skibinski DOF, Beardmore JA (1999). Environmental sex determination: the effect of temperature and salinity on sex ratio in *Oreochromis niloticus* L. Aquaculture 173:219-234.

Arnold AP (2004). Sex chromosomes and brain gender. Nat. Rev. Neurosc. 5:701-708.

Azaza MS, Dhranef MN, Kraпem MM (2008). Effects of water temperature on growth and sex ratio of juvenile Nile tilapia *Oreochromis niloticus* (Linnaeus) reared in geothermal waters in southern Tunisia. J. Therm. Biol. 33:98-105.

Azuma T, Takeda K, Doi T, Muto K, Akutsu M, Sawada M, Adachi S (2004). The influence of temperature on sex determination in sockeye salmon *Oncorhynchus nerka*. Aquaculture 234:461-473.

Bagheri-Fam S, Sinclair AH, Koopman P, Harley VR (2010). Conserved regulatory modules in the Sox9 testis-specific enhancer predict roles for SOX, TCF/LEF, Forkhead, DMRT, and GATA proteins in vertebrate sex determination. Int J Biochem Cell Biol. Mar. 42(3):472-477.

Baras E, Jacobs B, Mélard C (2001). Effect of water temperature on survival, growth and phenotypic sex of mixed (XX – XY) progenies of Nile tilapia *Oreochromis niloticus*. Aquaculture 92:187-199.

Baras E, Prignon C, Gohoungo G, Mйlard C (2000). Phenotypic sex differentiation of blue tilapia under constant and fluctuating thermal regimes and its adaptative and evolutionary implications. J. Fish Biol. 56:88-96.

Baroiller JF, Clota F, Geraz E (1995). Temperature sex determination in two tilapia, *Oreochromis niloticus* and the Red tilapia (Red Florida strain): effect of high or low temperature. In: Goetz, F. W.; Thomas, P.(Eds), Proceedings of the 5th International Symposium on the Reproductive Physiology of Fish, July 2-8, Austin, Texas, USA. pp. 158-160.

Baroiller JF, D'Cotta H (2001). Environment and sex determination in farmed fish. Comp. Biochem. Phys. Part C. Pharm. Toxic. 130:399-409.

Baroiller JF, D'Cotta H, Saillant E (2009). Environmental effects on fish sex determination and differentiation. Sex Dev. 3:118-135.

Baroiller JF, D'Cotta H, Wessels S, Bezault E, Hoerstgen-Schwark G (2008). Temperature effects on tilapia sex differentiation. Proceeding of the 6th International Symposium on Fish Endocrinology, Calgary, Canada, June 22–27.

Baroiller JF, Guiguen Y, Fostier A (1999). Endocrine and environmental aspects of sex differentiation in fish. Cell Mol. Life Sci. 55:910-931.

Baroiller JF, Nakayama I, Foresti F, Chourrout D (1996). Sex determination studies in two species of teleost fish, *Oreochromis niloticus* and *Leporinus elogans*. Zool. Stud. 35:279-285.

Baron D, Cocquet J, Xia X, Fellous M, Guiguen Y, Veitia RA (2004). An evolutionary and functional analysis of Foxl2 in rainbow trout gonad differentiation. J. Mol. Endocrinol. 33:705-715.

Beadmores JA, Mair JA, Lewis RI (2001). Monosex male production in finfish as exemplified by tilapia: applications, problems and prospects. Aquaculture 197:283-301.

Beato M (1989). Gene regulation by steroid hormones. Cell 56:335-344.

Berta P, Hawkins JR, Sinclair AH, Taylor A, Griffiths BL, Goodfellow PN, Fellous M (1990). Genetic-Evidence Equating Sry and the Testis-Determining Factor. Nature 348:448-450.

Bezault E, Clota F, Derivaz M, Chevassus B, Baroiller JF (2007). Sex determination and temperature-induced sex differentiation in three natural populations of Nile tilapia *(Oreochromis niloticus)* adapted to extreme temperature conditions. Aquaculture 272:S3-S16.

Blazquez M, Piferrer F (2004). Cloning, sequence analysis, tissue distribution, and sex-specific expression of the neural form of P450 aromatase in juvenile sea bass (*Dicentrarchus labrax*). Mol. Cell Endocrinol. 219:83-94.

Bombardella RA, Hayashi C, Meurer F (2004). Aplicaзго de mйtodos diretos e indiretos para a produзго de populaзxes monossexuais na tilapicultura. Arq. Cікп. Vet. Zool. 7:57-68.

Bradley KM, Breyer JP, Melville DB, Broman KW, Knapik EW, Smith JR (2011). An SNP-based linkage map for zebrafish reveals sex determination loci. G3 - Genes Genom. Genet. 1:3-9.

Breathnach R, Chambon P (1981). Organization and expression of eucaryotic split genes coding for proteins. Annual Rev. Biochem. 50:351-359.

Bull JJ (2008). Sex determination: are two mechanisms better than one? J. Biosci. 33:5-8.

Callard GV, Tchoudakova A (1997). Evolutionary and functional significance of two *Cyp19* genes differentially expressed in brain and ovary of goldfish. J. Steroid Biochem. Mol. Biol. 61:387-392.

Callard GV, Tchoudakova AV, Kishida M, Wood, E (2001). Differential tissue distribution, developmental programming, estrogen regulation and promoter characteristics of *cyp19* genes in teleost fish. J. Steroid Biochem. Mol. Biol. 79:305-314.

Chang X, Kobayashi T, Kajura H, Nakamura M, Nagahama Y (1997). Isolation and characterization of the cDNA encoding the tilapia (*Oreochromis niloticus*) cytochrome P450 aromatase (P450arom): changes in P450arom mRNA, protein and enzyme activity in ovarian follicles during oogenesis. J. Mol. Endocrinol. 18:57-66.

Chang X, Kobayashi T, Senthilkumaran B, Kobayashi-Kajura H, Sudhakumari CC, Nagahama Y (2005). Two types of aromatase with diverent encoding genes, tissue distribution and developmental expression in Nile tilapia (*Oreochromis niloticus*). Gen. Comp. Endocrinol. 141:101-115.

Chiang EF, Pai CI, Wyatt M, Yan YL, Postlethwait J, Chung B (2001a). Two sox9 genes on duplicated zebrafish chromosomes: expression of similar transcription activators in distinct sites. Dev. Biol. 231:149-163.

Chiang EF, Yan YL, Tong SK, Hsiao PH, Guiguen Y, Postlethwait J, Chung BC (2001b). Caracterization of duplicated zebrafish *cyp19* genes. J. Exp. Zool. 290:709-714.

Clemente N, Ghaffari S, Pepinsky RB, Pieau C, Josso N, Cate RL, Vigier B (1992). A quantitative and interspecific test for biological activity of anti-mullerian hormone: the fetal ovary aromatase assay. Development 114:721-727.

Conley A, Hinshelwood M (2001). Mammalian aromatase. Reproduction 121:685-695.

Conover DO (2004). Temperature-dependent sex determination in fishes. In: Valenzuela, N., Lance, V. (Eds.). Temperature-Dependent Sex Determination in Vertebrates Washington:Smithsonian Books. pp. 11-20.

Conover DO, Heins SW (1987). The environmental and genetic

components of sex ratio in *Menidia menidia* (pisces, Atherinidae). Copeia 3:732-743.

Corbin CJ, Graham-Lorence S, McPhaul M, Mason M, Mendelson CR, Simpson ER (1988). Isolation of a full-length cDNA insert encoding human aromatase system cytochrome P-450 and its expression in nonsteroidogenic cells. Proc. Natl. Acad. Sci. USA 85:8948-8952.

Craig JK, Foote CJ, Wood CC (1996). Evidence for temperature-dependent sex determination in sockeye salmon (*Oncorhynchus nerka*). Canadian J. Fish Aquatic Sci. 53:141-147.

Crews D, Fleming A, Willingham E, Baldwin R, Skipper JK (2001). Role of steroidogenic factor 1 and aromatase in temperature-dependent sex determination in the red-eared slider turtle. J. Exp. Zool. 290:597-606.

D'Cotta H, Fostier A, Guiguen Y, Govoroun M, Baroiller JF (2001). Search for the molecular mechanisms implicated in the thermosensitivity of sex gonadal differentiation of tilapia fish: differential approach and expression of candidate genes. J Exp. Zool., 290:574-585.

D'Cotta H, Pepey E, Pfennig F, Bienvenu D, Gutzeit HO, Volff JN, Wennning M, Baroiller JF (2007). *Sox9a, Sox9b* and *Amh* are up-regulated in the gonads during natural and temperature-induced tilapia male differentiation. Proceedings of the 8th International Symposium on Reproductive Physiology of Fish, June 3rd–8th, Saint-Malo, France.

D'Cotta H, Pepey E, Wessels S, Poonlaphdecha S, Reinelt B, Hoerstgen-Schwark G, Baroiller JF (2008). Temperature-induced male differentiation in the Nile tilapia: gonad gene expression using female monosex populations and divergent thermossensitive lines. Proceedings of the 20th International Congress of Zoology, S10 Sex determination & differentiation, Paris, August pp. 26-29.

Desprez D, Melard C (1998). Effect of ambient water temperature on sex determinism in the blue tilapia *Oreochromis aureus*. Aquaculture 162:1-2.

Devlin RH, Naghama Y (2002). Sex determination and sex differentiation in fish: an overview of genetic, physiological, and environmental influences. Aquaculture 208:191-364.

Donaldson EM (1996). Manipulation of reproduction in farmed fish. Anim. Reprod. Sci. 42:381-392.

Filby AL, Tyler CR (2005). Molecular characterization of estrogen receptors 1, 2a, and 2b and their tissue and ontogenic expression profiles in fathead minnow (*Pimephales promelas*). Biol. Reprod. 73:648-662.

Francis RC (1992). Sexual lability in teleosts: developmental factors. Q. Rev. Biol. 7:1-17.

Fukada S, Tanaka M, Matsuyama M, Kobayashi D, Nagahama Y (1996). Isolation, characterization, and expression of cDNAs encoding the medaka (*Oryzias latipes*) ovarian follicle cytochrome P-450 aromatase. Mol. Reprod. Dev. 45:285-290.

Gelinas D, Pitoc GA Callard, GV (1998). Isolation of a goldfish brain cytochrome P450 aromatase cDNA: mRNA expression during the seasonal cycle and after steroid treatment. Mol. Cell Endocrinol. 16:81-93.

Godwin J, Luckenbach JA, Borski RJ (2003). Ecology meets endocrinology: environmental sex determination in fishes. Evol. Dev. 5:40-49.

Goto R, Kayaba T, Adachi S, Yamauchi K (2000). Effects of temperature on sex determination in marbled sole Limanda yokohamae. Fish Sci. 66:400-402.

Goto R, Mori T, Kawamata K, Matsubara T, Mizuno S, Adachi S, Yamauchi K (1999). Effects of temperature on gonadal sex determination in barfin flounder Verasper moseri. Fish Sci. 65:884-887.

Goto-Kazeto R, Abe Y, Masai K, Yamaha E, Adachi S, Yamauchi K (2006). Temperature-dependent sex differentiation in goldfish: Establishing the temperature-sensitive period and effect of constant and fluctuating water temperatures. Aquaculture 254:617-624.

Govoroun MS, Pannetier M, Pailhoux E, Cocquet J, Brillard JP, Couty I, Batellier F, Cotinot C (2004). Isolation of chicken homolog of the *Foxl2* gene and comparison of its expression patterns with those of aromatase during ovarian development. Dev. Dyn. 231:859-870.

Green S, Walter P, Kumar V (1986). Human estrogen receptor cDNA: sequence, expression and homology to v-erb-a. Nature 320:134-139.

Guan G, Kobayashi T, Nagahama Y (2000). Sexually dimorphic expression of two types of DM (Doublesex/Mab-3)-domain genes in a teleost fish, the Tilapia (*Oreochromis niloticus*). Bioch. Biophys. Res. Commun. 272:662-666.

Guiguen Y, Baroiller JF, Ricordel MJ, Iseki K, Mcmeel OM, Martin SA, Fostier A (1999). Involvement of estrogens in the process of sex differentiation in two fish species: the rainbow trout (*Oncorhynchus mykiss*) and a tilapia (*Oreochromis niloticus*). Mol. Reprod. Dev. 54:154-162.

Guiguen Y, Fostier A, Piferrer F, Chang CF (2010). Ovarian aromatase andestrogens: A pivotal role for gonadal sex differentiation and sex change in fish. Gen. Comp. Endocrinol. 165:352-366.

Hacker A, Capel B, Goodfellow P, Lovellbadge R (1996). Expression of *Sry*, the moure sex-determining gene. Development 121:1604-1614.

Haffray P, Petit V, Guiguen Y, Quillet E, Rault P, Fostier A (2009). Successful production of monosex female brook trout *Salvelinus fontinalis* using gynogenetic sex reversed males by a combination of methyltestosterone immersion and oral treatments. Aquaculture 290:47-52.

Hammerman IS, Avtalion RR (1979). Sex determination in Sarotherodon (Tilapia): Part 2. The sex ratio as a tool for the determination of genotype a model of autosomal and gonosomal influence. Theor. Appl. Genet. 55:177-187.

Harada N (1988). Cloning of a complete cDNA encoding human aromatase: immunochemical identification and sequence analysis. Biochem. Biophys. Res Commun. 156:725-732.

Harvey SC, Kwon JY, Penman DJ (2003). Physical mapping of the brain and ovarian aromatase genes in the Nile Tilapia, *Oreochromis niloticus*, by Xuorescence in situ hybridization. Anim. Genet. 34:62-64.

Hofsten J, Larsson A, Olsson PE (2005). Novel steroidogenic factor-1 homolog (ff1d) is coexpressed with anti-Mullerian hormone (*Amh*) in zebrafish. Dev. Dyn. 233:595-604.

Hofsten J, Olsson PE (2005). Zebrafish sex determination and differentiation: Involvement of FTZ-F1 genes. Reprod. Biol. Endocrinol. 3:63-68.

Ijiri S, Berard C, Trant JM (2000). Characterization of gonadal and extra-gonadal forms of the cDNA encoding the Atlantic stingray (*Dasyatis sabina*) cytochrome P450 aromatase. Mol. Cell Endocrinol. 164:169-181.

Ijiri S, Kaneko H, Kobayashi T, Wang DS, Sakai F, Paul-Prasanth B, Nakamura M, Naghama Y (2008). Sexual dimorphic expression of genes in gonads during early differentiation of a teleost fish, the Nile tilapia *Oreochromis niloticus*. Biol. Reprod. 78:333-341.

Ikeda Y, Lala DS, Luo X, Kim E, Moisan MP, Parker KL (1993). Characterization of the mouse FTZ-F1 gene, which encodes a key regulator of steroid hydroxylase gene expression. Mol. Endocrinol. 7:852-860.

Ito M, Matsuda A, Yumoto K, Otomo A, Takahashi Y, Takamatsu N, Kanda H, Yamashita S, Shiba T (1998). cDNA cloning of a new member of FTZ-F1 subfamily from a rainbow trout. Biochem. Biophys. Acta 1395:271-274.

Jegede T (2010). Control of Reproduction in Oreochromis niloticus (Linnaeus 1758) using *Hibiscus Rosa-sinensis* (Linn.) Leaf Meal as Reproduction Inhibitor. J. Agric. Sci. 2:149-154.

Kawabe K, Shikayama T, Tsuboi H, Oka S, Oba K, Yanase T, Nawata H, Morohashi K (1999). Dax-1 as one of the target genes for Ad4BP/SF-1. Mol. Endocrinol. 13:1267-1284.

Kishida M, Callard GV (2001). Distinct cytochrome P450 aromatase isoforms in zebrafish (*Danio rerio*) brain and ovary are diverentially programmed and estrogen regulated during early development. Endocrinology 142:740-750.

Kobayashi T, Kajiura-Kobayashi H, Nagahama Y (2003). Induction of XY sex reversal by estrogen involves altered gene expression in a teleost, tilapia. Cytogen. Gen. Res. 101:289-294.

Kobayashi T, Matsuda M, Kajiura-Kobayashi H, Suzuki A, Saito N, Nakamoto M, Shibata N, Nagahama Y (2004). Two DM domain genes, DMY and Dmrt1, involved in testicular differentiation and development in the medaka, *Oryzias latipes*. Dev. Dynam. 231:518-526.

Korach KS (1994). Insights from the study of animals lacking functional estrogen receptor. Science 266:1524-1527.

Kuo MW, Postlethwait J, Lee WC, Lou SW, Chan WK, Chung BC (2005). Gene duplication, gene loss and evolution of expression domains in the vertebrate nuclear receptor NR5A (Ftz-F1) family. Biochem. J. 389:19-26.

Kurokawa H, Saito D, Nakamura S, Katoh-Fukui Y, Ohta K, Baba T, Morohashi K, Tanaka M (2007). Germ cells are essential for sexual dimorphism in the medaka gonad. Proc. Nat. Acad. Sci. USA. 104:16958-16963.

Kwon JY, Haghapanah V, Kogson-Hurtado LM, MCandrew BJ, Penman DJ (2000). Masculinization of genetic female nile tilapia (*Oreochromis niloticus*) by dietary administration of an aromatase inhibitor during sexual diferentiation. J. Exp. Zool. 287:46-53.

Kwon JY, MCandrew B, Penman DJ (2001). Cloning of brain aromatase gene and expression of brain and ovarian aromatase genes during sexual differentiation in genetic male and female Nile tilapia *Oreochromis niloticus*. Mol. Reprod. 59:359-370.

Kwon JY, McAndrew BJ, Penman DJ (2002). Treatment with an aromatase inhibitor suppresses high-temperature feminization of genetic male (YY) Nile tilapia. J. Fish Biol. 60:625-636.

Lala DS, Rice DA, Parker KL (1992). Steroidogenic factor I, a key regulator of steroidogenic enzyme expression, is the mouse homolog of fushi tarazu-factor I. Mol. Endocrinol. 6:1249-1258.

Lee BY, Penman DJ, Kocher TD (2003). Identification of a sex determining region in Nile tilapia (*Oreochromis niloticus*) using bulked segregant analysis. Anim. Genet. 34:379-383.

Lehningher AL, Nelson DL, Cox MM (1995). Princнpios da Bioquнmica. (3rd ed.). Sro Paulo: Sarver Editora de Livros Mйdicos Ltda.

Lei N, Heckert LL (2004). Gata 4 Regulates Testis Expression of Dmrt1. Mol.Cell Biol. 24:377-388.

Lei N, Hornbaker KI, Rice DA, Karpova T, Agbor VA, Heckert LL (2007). Sex-specific differences in mouse DMRT1 expression are both cell type- and stage-dependent during gonad development. Biol. Reprod. 77:466-475.

Li M, Zhao C, Wang Y, Zhao Z, Meng A (2002). Zebrafish sox9b is an early neural crest marker. Dev. Genes Evol. 212:203-220.

Loffler KA, Koopman P (2003). Sex Determination: the Fishy Tale of Dmrt1. Curr. Biol. 13:177-179.

Loffler KA, Zarkower D, Koopman P (2003). Etiology of ovarian failure in blepharophi-mosis ptosis epicanthus inversus syndrome: *FOXL2* is a conserved, early-acting gene invertebrate ovarian development. Endocrinology 144:3237-3243.

Luckenbach JA, Godwin J, Daniels HV, Borski RJ (2003). Gonadal differenciation and effects of temperature on sex differentiation in southern flounder (*Paralichthys lethostigma*). Aquaculture 216:315-327.

Luo X, Ikeda Y, Parker KL (1994). A cell-specific nuclear receptor is essential for adrenal and gonadal development and sexual differentiation. Cell 77:481-490.

MacKay ME, Raelson J, Lazier CB (1996). Up-regulation of estrogen receptor mRNA and estrogen receptor activity by estradiol in liver of rainbow trout and other teleostean fish. Comp. Biochem. Phys. 115C:201-209.

Magerhans A, Müller-Belecke A, Hцrstgen-Schwark G (2009). Effect of rearing temperatures post hatching on sex ratios of rainbow trout (*Oncorhynchus mykiss*) populations. Aquaculture 294:25-29.

Mair GC, Abucay JS, Skibinski DOF, Abella TA, Beardmore JA (1997). Genetic manipulation of sex ratio for the large-scale production of all-male tilapia, *Oreochromis niloticus*. Can. J. Fish Aquatic Sci. 54:396-404.

Mair GC, Scott AG, Penman DJ, Beardmore JA, Skibinski DOF (1991). Sex determination in the genus *Oreochromis*: I. Sex reversal, gynogenesis and triploidy in *O. niloticus* (L.). Theor. Appl. Genet. 82:144-152.

Matson CK, Zarkower D (2012); Sex and the singular DM domain: insights into sexual regulation, evolution and plasticity. Nat. Rev. Genet. 13:163-174.

Matsuoka MP, Van Nes S, Andersen O, Benfey TJ, Reith M (2006). Real-time PCR analysis of ovary- and brain-type aromatase gene expression during Atlantic halibut (*Hippoglossus hippoglossus*) development. Comp. Biochem. Phys. 144:128-135.

McPhaul MJ, Noble JF, Simpson ER, Mendelson CR, Wilson JD (1988). The expression of a functional cDNA encoding the chicken cytochrome P-450arom (aromatase) that catalyzes the formation of estrogen from androgen. J. Biol. Chem. 263:16358-16363.

Mommsen TP, Lazier CB (1986). Stimulation of estrogen receptor accumulation by estradiol in primary cultures of salmon hepatocytes. FEBS Lett. 195:269-270.

Morohashi K, Honda S, Inomata Y, Handa H, Omura T (1992). A common trans-acting factor, Ad4-binding protein, to the promoters of steroidogenic P-450s. J. Biol. Chem. 267:17913-17919.

Morohashi K, Omura T (1990). Tissue specific transcription of P450 (11beta) gene *in vitro*. J. Biol. Chem. 108:1050-1056.

Morrison CM, Miyake T, Wright Jr JR (2001). Histological study of the development of the embryo and early larva of *Oreochromis niloticus* (Pisces: Cichlidae). J. Morphol. 247:172-195.

Nakamoto M, Matsuda M, Wang DS, Nagahama Y, Shibata N (2006). Molecular cloning and analysis of gonadal expression of Foxl2 in the medaka, *Oryzias latipes*. Biochem. Biophys. Res. Commun. 344:353–361.

Nakamoto M, Suzuki A, Matsuda M, Nagahama Y, Shibata N (2005). Testicular Type Sox9 Is Not Involved in Sex Determination but Might Be in the Development of Testicular Structures in the Medaka, Oryzias Latipes. Biochem. Biophys. Res. Comm. 333:729-736.

Nakamura M, Chang XT, Kobayashi T, Kajiura H, Nagahama Y (1998). Gonadal sex diferentiation in fish. J. Exp. Zool. 281:362-372.

Nakamura S, Watakabe I, Nishimura T, Toyoda A, Taniguchi Y, Tanaka M (2012). Analysis of medaka sox9 orthologue reveals a conserved role in germ cell maintenance. PLoS One 7(1):229982.

Nanda I, Zend-Ajusch E, Shan Z, Grutzner F, Schartl M, Burt DW, Koehler M, Fowler VM, Goodwin G, Schneider WJ, Mizuno S, Dechant G, Haaf T, Schmid M (2000). Conserved Synteny Between the Chicken Z Sex Chromosome and Human Chromosome 9 Includes the Male Regulatory Gene Dmrt1: a Comparative (Re) View on Avian Sex Determination. Cytog. Cell Genet. 89:67-78.

Navarro-Martín L, Viñas J, Ribas L, Díaz N, Gutiérrez A, Di Croce L, Piferrer F (2011) DNA Methylation of the Gonadal Aromatase (cyp19a) Promoter Is Involved in Temperature-Dependent Sex Ratio Shifts in the European Sea Bass. PLoS Genet 7(12):e1002447. doi:10.1371/journal.pgen.1002447

Nomura T, Arai K, Hayashi T, Suzuki R (1998). Effect of temperature on sex ratios of normal and gynogenetic loach. Fish Sci. 64:753-758.

Oliveira PA, Hein G, Almeida DB, Costa MAP, Moreira CGA, Silva JC, Tavares RA, Teixeira F, Manzke VHB, Moreira HLM (2008). Avaliaзro da Reversro Sexual de Alevinos de Tilбpias (Oreochromis Niloticus) com o uso de Vorozol em estufa. In: XVI Congresso de Iniciaзro Cientнfica, 2008, Pelotas, Anais do XVI CIC da Universidade Federal de Pelotas. pp. 1-5.

Ospina-Álvarez N, Piferrer F (2008). Temperature-dependent sex determination in fish revisited: prevalence, a single sex ratio response pattern, and possible effects of climate change. PloS ONE 3:1-11.

Ovidio M, Desprez D, Mélard C, Poncin P (2002). Influence of sexual genotype on the behaviour of females (genotype WZ) and pseudofemales (genotype ZZ) in the tilapia Oreochromis aureus. Aquat. Liv. Res. 15:163-167.

Pakdel F, Le Guellec C, Vaillant C, Le Roux MG, Valotaire Y (1989). Identification and estrogen induction of two estrogen receptors (ER) messenger ribonucleic acids in the rainbow trout liver: sequence homology with other ERs. Mol. Endocrinol. 3:44-51.

Parker KL, Schedl A, Schimmer BP (1999). Gene interactions in gonadal development. Annu. Rev. Physiol. 61:417-433.

Peichel CL, Graves JAM (2010). Are homologies in vertebrate sex determination due to shared ancestry or to limited options? Genome Biology 2010, 11:205 http://genomebiology.com/2010/11/4/205.

Phelps RP, Okoko M (2011). A non-paradoxical dose response to 17a-methyltestosterone by Nile tilapia Oreochromis niloticus (L.): effects on the sex ratio, growth and gonadal development. Aquacult. Res. 42:549-558.

Phelps RP, Popma TJ (2000). Sex Reversal of Tilapia. In: COSTA-PIERCE, B.A.; RAKOCY, J. E. (Ed.). Tilapia aquaculture in the Americas. Louisiana: The World Aquacult. Soc. 2:34-59.

Piferrer F, Donaldson EM (1991). Dosage-dependent differences in the effect of aromatizable and nonaromatizable androgens on the resulting phenotype of coho salmon (*Oncorhynchus kisutch*). Fish

Physiol. Bioch. 9:145-150.
Popma TJ, Green BW (1990). Sex reversal of tilapia in earthen ponds: aquaculture production manual. Alabama: Auburn University, 15 pp.
Popma TJ, Lovshin L (1996). Wordwide Prospects for Commercial Production of Tilбpia, Internacional Center for Aquaculture and Aquatic Environments. Res. Dev. 41:23.
Racine C, Rey R, Forest MG, Louis F, Ferre A, Huhtaniemi I, Josso N, Di Clemente N (1998). Receptors for anti-mullerian hormone on Leydig cells are responsible for its effects on steroidogenesis and cell differentiation. Proc. Natl. Acad. Sci. 95:594-599.
Raymond CS, Murphy MW, O'Sullivan MG, Bardwell VJ, Zarkower D (2000). Dmrt1, a gene related to worm and fly sexual regulators, is required for mammalian testis differentiation. Genes. Develop. 14:2587-2595.
Raymond CS, Shamu CE, Shen MM, Seifert KJ, Hirsch B, Hodgkin J, Zarkower D (1998). Evidence for evolutionary conservation of sex - determining genes. Nature 391:691-695.
Resko JA, Pereyra-Martinez AC, Stadelman HL, Rosseli CE (2000). Region-specific regulation of cytochrome P450 aromatase messenger ribonucleic acid by androgen in brains of male rhesus monkeys. Biol. Reprod. 62:1818-1822.
Rey R, Lukas-Croiser C, Lasala C, Bedecarras P (2003). AMH/MIS: what we know already about the gene, the protein and its regulation. Mol. Cell Endocrinol. 211:21-31.
Rougeot C, Prignon C, Kengne CVN, Mйlard C (2008). Effect of high temperature during embryogenesis on the sex differentiation process in the Nile tilapia, *Oreochromis niloticus*. Aquaculture 276:205-208.
Sadovsky Y, Crawford PA, Woodson KG, Polish JA, Clements MA, Tourtellotte LM, Simburger K, Milbrandt J (1995). Mice deficient in the orphan receptor steroidgenic factor 1 lack adrenal glands and gonads but express P450 side-chain-cleavage enzyme in the placenta and have normal embryonic serum levels of corticosteroids. Proc. Natl. Acad. Sci. 92:10939-10943.
Saillant E, Fostier A, Haffray P, Menu B, Thimonier J, Chatain B (2002). Temperature effects and genotype–temperature interactions on sex determination in the European sea bass (*Dicentrarchus labrax* L.). J. Exp. Zool. 292:494-505.
Santa Barbara P, Bonneaud N, Boizet B, Desclozeaux M, Moniot B, Sudbeck P, Scherer G, Poulat F, Berta P (1998). Direct interaction of SRY-related protein SOX9 and steroidogenic factor 1 regulates transcription of the human anti-Mullerian hormonegene. Mol. Cell Biol. 18:6653-6665.
Sawyer SJ, Gerstner KA, Callard GV (2006). Real-time PCR analysis of cytochrome P450 aromatase expression in zebrafish: gene specific tissue distribution, sex differences, developmental programming, and estrogen regulation. Gen. Comp. Endocrinol. 147:108-117.
Simpson ER, Evans CT, Corbin CJ, Powell FE, Ledesma DB, Mendelson CR (1987). Sequencing of cDNA inserts encoding aromatase cytochrome P-450 (P-450AROM). Mol. Cell Endocrinol. 52:267-272.
Smith C, Sinclair A (2001). Sex determination in the chicken embryo. J. Exp. Zool. 290:691-699.
Smith C, Smith M, Sinclair A (1999). Gene expression during gonadogenesis in the chicken embryo. Gene 234:395-402.
Squires EJ (2003). Applied animal endocrinology. Wallingford: USA 252 pp.
Strussmann CA, Moriyama S, Hanke EF, Cota JCC, Takashima F (1996). Evidence of thermolabile sex determination in pejerrey. J. Fish Biol. 48:643-651.
Swain A, Zanaria E, Hacker A, Lovell-Badge R, Camerino G (1996). Mouse Dax1 expression is consistent with a role in sex determination as well as in adrenal and hypothalamus function. Nat. Genet. 12:404-409.
Tachibana L, Castagnolli N, Pezzato LE, Barros MM, Valle JB, Siqueira M (2004). Desempenho de diferentes linhagens de tilбpia do Nilo (Oreochromis niloticus) na fase de reversro sexual. Acta Scient. 26:305-311.
Tanaka M, Fukada S, Matsuyama M, Nagahama Y (1995). Structure and promoter analysis of the cytochrome P-450 aromatase gene of the teleost fish, medaka (*Oryzias latipes*). Biochem. J. 117:719-725.
Tchoudakova A, Callard GV (1998). Identification of multiple CYP19 genes encoding different cytochrome P450 aromatase isozymes in brain and ovary. Endocrinology 139:2179-2189.
Terashima M, Toda K, Kawamoto T, Kuribayashi I, Ogawa Y, Maeda T, Shizuta Y (1991). Isolation of a full-length cDNA encoding mouse aromatase P450. Arch. Bioch. Biophys. 285:231-237.
Tessema M, Mьller-Belecke A, Hцrstgen-Schwark G (2006). Effect of rearing temperatures on the sex ratios of Oreochromis niloticus populations. Aquaculture 258:270-277.
Thorgaard GH (1977). Heteromorphic sex chromosomes in male rainbow trout. Science 196:900-902.
Toft D, Gorski J (1966). A receptor molecule for estrogens: isolation from the rat uterus and preliminary characterization. Proc. Natl. Acad. Sci. USA 55:1574-1581.
Tong SK, Chiang EF, Hsiao PH, Chung B (2001). Phylogeny, expression and enzyme activity of zebraWsh cyp19 (P450 aromatase) genes. J. Steroid Bioch. Mol. Biol. 79:299-303.
Trant JM (1994). Isolation and characterization of the cDNA encoding the channel catfish (*Ictalurus punctatus*) form of cytochrome P450 arom. Gen. Comp. Endocrinol. 95:155-168.
Trant JM, Gavasso S, Ackers J, Chung BC, Place AR (2001). Developmental expression of cytochrome P450 aromatase genes (CYP19a and CYP19b) in zebrafish fry (Danio rerio). J. Exp. Zool. 290:475-483.
Tsai CL, Chang SL, Wang LH, Chao TY (2003). Temperature influences the ontogenetic expression of aromatase and oestrogen receptor mRNA in the developing tilapia (Oreochromis mossambicus) brain. J. Neuroendocrinol. 15:97-102.
Uchida D, Yamashita M, Kitano T, Iguchi T (2004). An aromatase inhibitor or high water temperature induce oocyte apoptosis and depletion of P450 aromatase activity in the gonads of genetic female zebrafish during sex-reversal. Comp Bioch Physiol Parte A: Mol. Integr. Physiol. 137:11-20.
Van Den Hurk R, Lambert JGD (1982). Temperature and steroid effects on gonadal sex differentiation in rainbow trout. In: Richter, C.J.J.; Goos, H.J.T. (Eds.), Proceeding of the International Symposium on the Reproductive Physiology of Fish. pp. 69-72. Pudoc: Wageningen.
Wang DS, Kobayashi T, Zhou L, Nagahama Y (2004). Molecular cloning and gene expression of Foxl2 in the Nile Tilapia Oreochromis niloticus. Biochem. Biophys. 320:83-89.
Wang DS, Kobayashi T, Zhou LY, Paul-Prasanth B, Ijiri S, Sakai F, Okubo K, Morohashi K, Nagahama Y (2007). Foxl2 up-regulates aromatase gene transcription in a female-specific manner by binding to the promoter as well as interacting with ad4 binding protein/steroidogenic factor 1. Mol. Endocrinol. 21:712-725.
Wang DS, Zhou LY, Kobayashi T, Matsuda M, Yasushi S, Sakai F, Nagahama Y (2010). Doublesex-and Mab-3 Related Transcription Factor-1 Repression of Aromatase Transcription, a possible mechanism favoring the male pathway in tilapia. Endocrinology 151:1331-1340.
Wang LH, Tsai CL (2000). Temperature affects the development of central neurotransmitter systems of tilapia, Oreochromis mossambicus. Neurosci. Lett. 285:95-99.
Warner DA, Shine R (2008). The adaptive significance of temperature-dependent sex determination in a reptile. Nature 451:566-568.
Watanabe M, Tanaka M, Kobayashi D, Yoshiura Y, Oba Y, Nagahama Y (1999). Medaka (Oryzias latipes) FTZ-F1 potentially regulates the transcription of P450 aromatase in ovarian follicles: cDNA cloning and functional characterization. Mol. Cell Endocrinol. 149:221-228.
Wessels S, Horstgen-Schwark G (2007) Selection experiments to increase the proportion of males in Nile tilapia (Oreochromis niloticus) by means of temperature treatment. Aquaculture 272:80-87.
Western P, Harry J, Marshal-Graves J, Sinclair A. (2000). Temperature-dependent sex determination in the American alligator: Expression of SF1, WT1, and DAX1 during gonadogenesis. Gene 241:223-232.
Yamamoto E (1999). Studies on sex-manipulation and production of cloned populations in hirame, Paralichthys olivaceus (Temminck et Schlegel). Aquaculture 173:235-246.
Yamamoto T (1969). Sex differentiation. In: Hoar, W.; Randall, D. (Eds.), Academic Press. Fish Physiol. pp. 117-175
Yan YL, Miller CT, Nissen RM, Singer A, Liu D, Kirn A, Draper B, Willoughby J, Morcos PA, Amsterdam A, Chung BC, Westerfield M, Haffter P, Hopkins N, Kimmel C, Postlethwait JH, Nissen R (2002). A zebrafish sox9 gene required for cartilage morphogenesis.

Development 129:5065-5079.
Yao HH, Capel B (2005). Temperature, genes, and sex: a comparative view of sex determination in *Trachemys scripta* and *Mus musculus*. Biochem. J. 138:5-12.
Yoshiura Y, Senthilkumaran B, Watanabe M, Oba Y, Kobayashi T, Nagahama Y (2003). Synergistic expression of Ad4BP/SF-1 and cytochrome P-450 aromatase (ovarian type) in the ovary of Nile tilapia, *Oreochromis niloticus*, during vitellogenesis suggests transcriptional interaction. Biol. Reprod. 68:1545-1553.
Zanaria E, Muscatelli F, Bardoni B, Strom T, Guioli S, Guo W, Lalli E, Moser C, Walker A, McCabe E, Meitinger T, Monaco A, Sassone-Corsi P, Camerino G (1994). An unusual member of the nuclear hormone receptor superfamily responsible for X-linked adrenal hypoplasia congenita. Nature 372:635-641.
Zhou RJ, Cheng HH, Zhang QY, Guo YQ, Cooper RK, Tiersch TR (2002). Sry-Related Genes in the Genome of the Rice Field Eel (Monopterus Albus). Genet. Selec. Evol. 34:129-137.

Genetic differentiation between the black skinned and white skinned snails *(Archachatina marginata)* using random amplified polymorphic DNAs

Okon, B., Ibom, L. A. and Njume, G. N.

Department of Animal Science, University of Calabar, Calabar, Nigeria.

The study investigates the genetic differentiation between the black skinned and white skinned ectotypes of the giant African land snails *(Archachatina marginata)* from Cross River State in Niger Delta region of Nigeria. The random amplified polymorphic DNA (RAPD) technique was employed in this study. Five (5) oligonucleotide primers (OPAD-09, OPAE-04, OPAE-05, OPAF-07 and OPAF-09) were used to amplify DNA from five samples of *A. marginata* constituting black skinned and white skinned ectotypes. A total of 58 RAPD bands with 24 polymorphic bands (40.18%) with size range of 150 to 5,500 bp, were scored from the populations. The black skinned ectotype had mean percentage polymorphism of 34.50%, while the white skinned ectotype recorded mean percentage polymorphism of 41.40%. Genetic similarity coefficient ranged from 60 to 63%, while the genetic distance ranged from 0.37 to 0.40. The genetic similarity between the two ectotypes of *A. marginata* from Cross River State is high and depicted low genetic differences. This reveals that the genetic variability of the species *(A. marginata)* from Cross River State is gradually eroding. Efforts should be made to conserve the genetic pool of this species, as the erosion of genetic variability is dangerous and could continue unnoticed till extinction is imminent.

Key words: Genetic, differentiation, snail, ectotypes, Niger Delta.

INTRODUCTION

Archachatina marginata is one of the species of African land snails known as Giant African land snails (GALS) (Akinnusi, 2002; 2004; Okon et al., 2008; Okon and Ibom, 2012). It belongs to the phylum Mollusca and the class gastropoda, it could be black- or white- skinned (Ejidike, 2002; Akinnusi, 2002, 2004; Okon et al., 2008; Ibom, 2009; Okon and Ibom, 2012). The snail is popular and highly appreciated as a valuable source of animal protein in many countries of Africa and beyond (Akinnusi, 2002, 2004; Ibom, 2009; Okon and Ibom, 2012). In Nigeria, *A. marginata* constitutes an important component of the food of numerous rural dwellers, especially in the rain-forest zones (Akinnusi, 2004).

A. marginata offers an opportunity to complement the supply of animal protein from conventional and regular sources such as beef, pork, chevon, mutton, among others. This will to a great extent reduce the problem of animal protein inadequacy, as the per capita protein consumption of an average Nigerian is still low (World Bank, 2001). However, several workers (Ojating, 2001; Ojating and Ogar, 2002; Okon et al., 2008; Okon et al., 2009) have reported a drastic depletion of populations of GALS owing to excessive and indiscriminate gathering of snails (both juvenile and adult) from the wild; coupled with deforestation, bush burning and other human activities. Consequently, the genetic diversity of GALS

Table 1. Primers and size of fragments generated.

Primer	Size of fragment (bp)
OPAD-09	200 - 1500
OPAE-04	250 - 2500
OPAE-05	150 - 2500
OPAF-07	250 - 3000
OPAF-09	200 - 5,500

(*A. marginata)* is gradually eroding thus threatened with extinction. Puurtinen et al. (2004) also noted that genetic threats associated with small population size may contribute to the elevated extinction risk of small and isolated populations. This necessitates a swift study to provide genetic informations on GALS (*A. marginata*) as there is presently little or no information on genetics differentiation and breeding of GALS.

The study of the structure and function of genes at molecular level in a breeding population can help to determine the genetic differentiation and similarity of such population (Lee et al., 2011). One of the modern marker techniques for such study is random amplified polymorphic DNAs (RAPD) (Williams et al., 1998). The technique is simple and does not require prior knowledge of the genome. Also, it requires small amount of DNA (Hadrys et al., 1992). The technique also enables detection of polymorphism in closely related organisms. The objective of this study was to genetically differentiate between the black skinned and the white skinned ectotypes of *A. marginata* using RAPD.

MATERIALS AND METHODS

Collection of samples

A total of 60 snails (*A. marginata*) comprising of 40 black skinned and 20 white skinned ectotypes were randomly picked from the three agro-ecological zones (Southern agro-ecological zone, Central agro-ecological zone and Northern agro-ecological zone) of Cross River State, Nigeria.

DNA extraction and RAPD

Live snails were transferred to the molecular laboratory of Federal University of Agriculture, Abeokuta (FUNAAB) for genotyping. DNA was extracted following a modified phenol chloroform extraction and ethanol precipitation as described by Sambrook et al. (1989). About 10 µl of each DNA was taken into eppendorf tube with 90 µl sterile distilled water to obtain 20 - 50 ng/µl. The total reaction volume of 20 µl was used with final concentration containing 20 ng of genomic DNA, 10x reaction buffer (2.0 µl), 25 mM magnesium chloride (1.6 µl), 5% Tween 20 (2.0 µl) water (8.2 µl). The amplification was programmed at 45 cycles for 20 s of denaturation at 94°C, 20 s of annealing temperature at 37°C, 40 s of primers extension at 72°C and final extension of 7 min at 72°C. This process was repeated four times to ensure accuracy. Five RAPD primers (Table 1) from Operon Kit A (with 60% G + C content) were used. PCR product was electrophoresed on 1.4% agarose gel in 1 x TAE buffer at 150 V and 0.5 miniamp for 2 h. The gel was stained in 2.5 mg/L ethidium bromide for 3 min and further destained for 10 min in sterile water and then viewed under ultra violet light and the picture was taken.

Data analysis

The RAPDs banding profiles were visually scored for all the DNA samples and for each primer. The recording of the data was according to the presence/absence criterion. Clear and visible bands were marked as present (1) or absent (0). Bands that were faintly stained were not considered in the data. These bands were considered as polymorphic when they were absent in some samples in frequency greater than 1% (Jorde, 1995) and change in band intensity was not considered as polymorphism. Similarity coefficients were calculated across all the possible pair wise comparisons of snail samples among populations, using the formula below:

$$Sxy = \frac{2nxy}{nx + ny}$$

Where, nxy is the number of common bands shown in both individuals x and y, and nx and ny are the total numbers of bands observed in individuals x and y respectively (Nei and Li, 1979). As means of providing a visual representation of genetic relationships, a dendrogram was constructed based on the similarity coefficient values (1 - Sxy) between pairs of snail samples. The dendrogram was constructed using the unweighted pair group method of Arithmetic averging (UPGMA) employing the sequential, agglomerative, hierarchical and nested clustering (SAHN) from NTSYS-PC program (Rohlf and Marcus, 1993).

RESULTS AND DISCUSSION

Generation of RAPD pattern of Genomic DNA

Five primers from the Operon Kit A (OPAD-09, OPAE-04, OPAE-05, OPAF-07, OPAF-09) with 60% G+C content were used in this study. These primers were used to generate RAPD pattern of genomic DNA for samples of snails (*A. marginata*) from Cross River State, Nigeria. The results showed that the largest fragment was 5,500 bp, while the smallest fragment was about 150 bp in size (Table 1).

The results of the RAPD DNA fragment showed that different primers generated different numbers of fragments and lengths of DNA amplification products (Table 2). From the five samples, a total of 58 bands were generated using the five primers. Primers OPAD-09 and OPAE-04 each generated 13 fragments; primer OPAE-05 generated 10 fragments, OPAF-07 generated seven fragments, while OPA-09 generated 15 fragments (Figure 1).

The numbers of RAPDs bands generated by the five primers and in each of the two ectotypes of *A. marginata* studied are shown in Table 3. The total numberof polymorphic bands were 24 out of 58 reproducible bands

Table 2. Code sequence, nucleotide lengths and G+C content of primers used in the RAPD analysis.

Primer name	Primer sequence (5' to 3')	Nucleotide length	G+C content (%)
OPAD-09	TCGCTTCTCC	10-mers	60
OPAE-04	CCAGCACTTC	10-mers	60
OPAE-05	CCTGTCAGTG	10-mers	60
OPAF-07	GGAAAAGCGTC	10-mers	60
OPAF-09	CCCCTCAGAA	10-mers	60

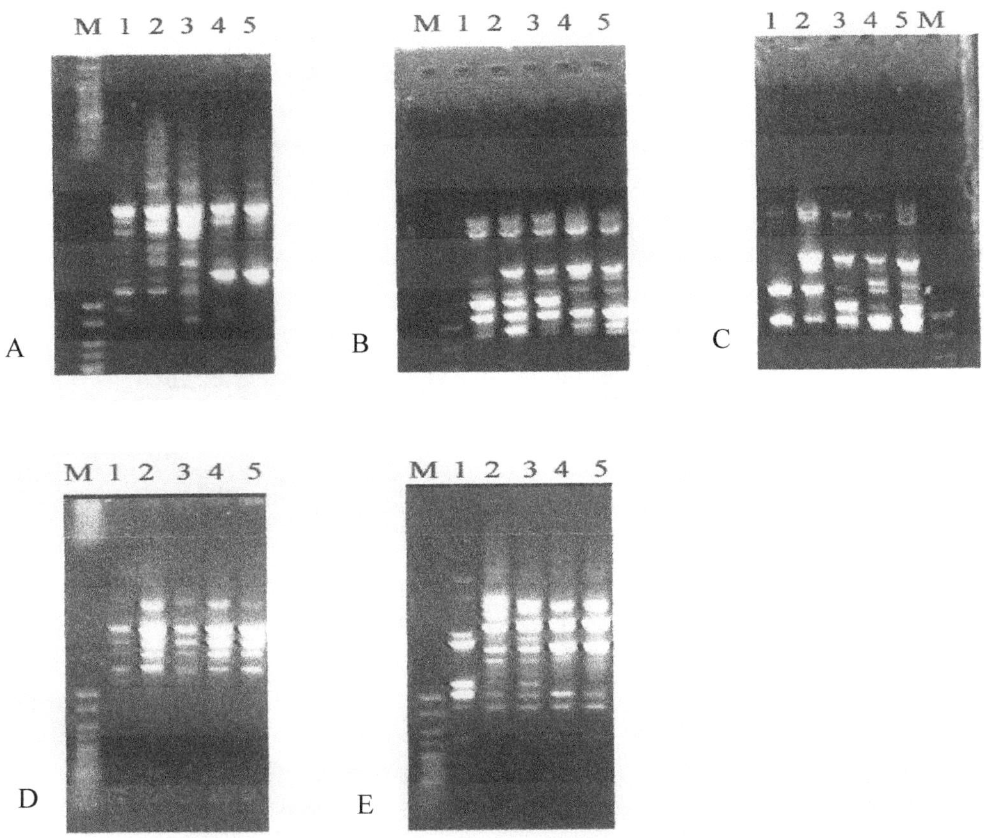

Figure 1. Electrophoresis gel pictures for the RAPD primers. A, OPAD-09; B, OPAE-04; C, OPAE-05; OPAF-07; E, OPAF-09.

Table 3. The number of RAPD fragments and polymorphic fragments generated by five primers in black skinned and white skinned *A. marginata*.

Primer	Total reproducible band	Total polymorphic band	Number of polymorphic bands according to ectotypes		Percentage polymorphic bands (%)
			*BSS	*WSS	
OPAD-09	13	8	7	8	61.54
OPAE-04	13	4	4	4	30.77
OPAE-05	10	4	3	4	40.00
OPAF-07	7	2	1	2	28.57
OPAF-09	15	6	5	6	40.00
Total	58	24	20	24	40.18
Percentage polymorphism (%)			34.50	41.40	

*BSS, Black skinned ectotype; *WSS, white skinned ectotype.

Table 4. Similarity coefficient between black and white skinned snails (*A. marginata*).

Sample	1	2	3	4	5
1	-				
2	60	-			
3	72	60	-		
4	60	63	60	-	
5	60	63	60	83	-

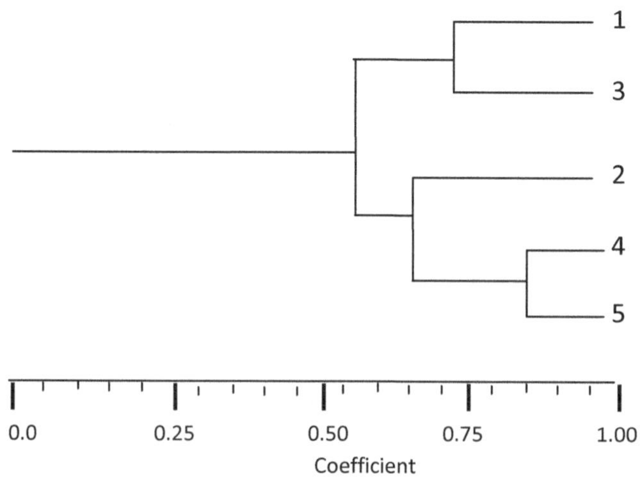

Figure 2. Neigbouring-joining dendrogram showing the genetic relationship between populations of white skinned and black skinned *A. marginata* in Cross River State of Nigeria.

generated. Primer OPAD-09 generated the highest percentage of polymorphic bands (61.54%), followed by primers OPAE-05 and OPAF-09 (40.00%), while primer OPA-07 recorded the least percentage of polymorphic bands (28.57%). Meanwhile the white skinned snail recorded the highest percentage of polymorphic bands of 41.40% against the black skinned snails with 34.50% (Table 3).

This high polymorphism in white skinned ectotype of *A. marginata* could point towards a moderately higher population of white skinned ectotype and thus higher diversity than black skinned ectotype in the state (Puurtinen et al., 2004). According to Uboh et al. (2010), there is a cultural discrimination against consumption of white skinned snails in some parts of Cross River State, Nigeria. This implies that more of the black skinned ectotype is removed from the wild than the white skinned ectotype, thereby explaining the higher population of white skinned snails in the state.

The similarity coefficient between the white skinned and black skinned snails ranged from 60 to 83% (Table 4) which are quite high. This showed that the genetic differences between the two snail ectotypes is moderately low. This is in agreement with the findings of Wan et al. (2009) who noted that high genetic similarity indicates low genetic differences or variability between snail species. These low genetic differences between the white skinned and black skinned *A. marginata* could be attributed to a relatively small population size of these snails in the state of study, which might have resulted in high level of inbreeding.

This confirms the findings of Ojating (2001), Ojating and Ogar (2002), Okon et al. (2008) and Okon and Ibom (2012) that the excessive and indiscriminate gathering of GALS from the wild, couple with deforestation, bush burning have led to depletion of GALS population, thus subjecting the genetic diversity of the snail to threat of extinction. The low genetic variability or high genetic similarity could also be as a result of high rate of self-fertilization in these snails (Charlesworth et al., 1993; Worth et al., 1997). Furthermore, low genetic variability is reported to arise from reduction in effective population size which then allow further accumulation of deleterious mutations, thus reducing the fitness of the population (Puurtinen et al., 2004).

The dendrogram (Figure 2) revealed that the snail samples were not just grouped based on spatial distribution but also based on the genetic closeness. This is indicated in the second cluster where the white skinned sample (5) snail is grouped along the black skinned snails' samples (2) and (4). This implies that the black skinned and white skinned *A. marginata* are of the same genetic background, otherwise there would have been distinct cluster for the white skinned snail.

Thus, the genetic differences here could be due to environmental factors such as random drift, mutation and possibly, selection (Weiner, 1994; Akinnusi, 2004; Puurtinen et al., 2004; Okon et al., 2008; Okon and Ibom, 2012). The genetic distance estimates between black skinned and white skinned *A. marginata* ranged from 0.37 to 0.40.

This small distance estimate may indicate population substructure like subpopulations in which there is random mating, but between which there is reduced amount of gene flow (Brent, 1996; Wan, 2009).

Conclusion

The genetic differentiation between samples of black skinned and white skinned snails *(A. marginata)* from Cross River State is low. This is depicted in the high genetic similarity (60 to 83%). Thus, it can be inferred that the genetic differences between these snails is mostly based on the influence of the environment (random drift, mutation and selection), since the snails are of the same genetic background. The results also demonstrated that loss of genetic variability or low genetic differences between these snails is associated with drastic decrease in population size, high level of inbreeding, high level of self-fertilization and thus limited gene flow. The gradual erosion of genetic variability is dangerous, because it may go unnoticed, especially since the population is regulated by ecological factors, until the reproduction rate of the population is so low that extinction is unavoidable. Hence, more research should be carried out towards the conservation of genetic pool of GALS.

REFERENCES

Akinnusi FAO (2004). Introduction to snails and snail farming, 2nd Edition. Abeokuta, Nigeria. Triolas Exquisite Ventures. 1-90.

Akinnusi O (2002). Introduction to snails and snail farming. Abeokuta, Nigeria. Triolas Exquisite Ventures. 34-37.

Charlesworth B, Morgan MT, Charleworth D (1993). The effect of deleterious mutations on neutral molecular variation. Genetics 134:1289-1303.

Ejidike BN (2002). Snail rearing practices in Southern Nigeria. Proceedings of the 27th Annual Conference of Nigeria Society for Animal Production (NSAP) March 17-21, 2002. Akure, Nigeria. pp. 307-308.

Hadrys H, Balick M, Shierwater B (1992). Applications of Random Amplified Polymorphic DNA (RAPD) in molecules. Ecology 1:55-63.

Ibom LA (2009). Variations in reproductive and growth performance traits of white-skinned x black skinned African giant snail hatchlings [*Archachatina marginata* (Swainson)] in Obubra, Nigeria. Ph.D. Thesis. Animal Science Department, University of Calabar, Calabar.

Lee HK, Chung CH, Cho SW, Na HJ, Jin HS, Jang KW, Lee YM, Han J, Ohh BK (2011). Analysis of Genetic characteristics of Koream native chicken using biochemical genetic markers. Genetics 115:78 - 92.

Nei M, Li WH (1979). Mathematical model for studying genetic variation in terms of restriction endonuclease. Proc. Natl Acad. Sci. USA. 7:5269-5273.

Ojating I (2001). Rearing, food predilection and reproductive performance of African land snail, *A. marginata* in kitchen farm at Okuku and Umudike. J. Sustain. Agric. Environ. 3(1):135-140.

Ojating I, Ogar GE (2002). Effect of feed types on the growth performance of African giant snails fattened at room temperature. J. Sustain. Agric. Environ. 4(2):303-308.

Okon B, Ibom LA (2012). Snail breeding and snailery management. Freshdew Productions. Calabar, Cross River State, Nigeria. 1-90.

Okon B, Ibom LA, Etuk NE, Akpan EW (2008). Variations in growth patterns and conformation of snails: Influence of strain and locations on isometry of growth in Cross River State. J. Agric. For. Soc. Sci. 6(2):218-225.

Okon B, Ibom, LA, Ekpo IA, Ewa EC (2009). Evaluation of reproductive performance and some egg quality parameters of Albino snails (*Archachatina marginata saturalis* (Swainson). J. Appl. Sci. 12(1):8234-8240.

Puurtinen K, Knott E, Suenpaa S, Oik T, Kaitala V (2004). Genetic variability and drift load in populations of an aquatic snail. Evolution 58(4):749-756.

Rohlf FJ, Marcus LF (1993). A revolution in morphometrics. Trends Ecol. Evol. 8:129-132.

Sambrook J, Fritsch EF, Mariatis T (1989). Molecular cloning: A laboratory manual, 2nd Edition, Cold Spring Harbour Laboratory Press. 10-60.

Uboh FE, Ebong PE, Mbi E (2010). Cultural Discrimination in consumption of black snail *(Archachatina marginata)* and white snail *(Achatina achatina)*: Any scientific justification? Int. Res. J. Microbiol. 1(1):013-017.

Weiner G (1994). Animal Breeding. The Tropical Agriculturist, Macmillan Education between Towns Road, Oxford. p. 43.

Williams JGK, Kubelik AR, Livak KJ, Rafaski JA, Tingey SV (1998). DNA Polymorphism amplified by arbitrary primers are useful as genetic markers. Nucleic Acid Res. 18:6531-6535.

Worth T, Baur A, Baur B (1997). Mating system and genetic variability in the simultaneously hermaphroditic terrestrial gastropod *Balea perversa* on the Baltic island of Oland, Sweden. Hereditas 126:199-209.

Genetic analysis of *japonica* x *indica* recombinant inbred lines and characterization of major fragrance gene by microsatellite markers

Vikram Singh[1] , R. K. Jain[2] and Mukesh Kumar[1]

[1]Department of Genetics and Plant Breeding, Chaudhary Charan Singh Haryana Agricultural University, Hisar (Haryana) 125004, India.
[2]Department of Molecular Biology and Biotechnology, Chaudhary Charan Singh Haryana Agricultural University, Hisar (Haryana) 125004, India.

Traditional basmati rice varieties are very low yielding due to their tendency to lodging and increasing susceptibility to diseases. To improve the characters of basmati rice variety and study the inheritance of various physio-morphological and quality characters, F_5 population comprising of 204 lines from the cross between NPT II (non-aromatic, *japonica*) and Taraori Basmati or HBC19 (aromatic, *indica*), were evaluated. Ample amount of genetic variability was observed for the characters plant height, tillers per plant, kernel length, kernel breath and L/B ratio. The grain yield/plant showed positive correlation with productive tiller/plant and test weight. Path coefficient analysis showed that the productive tiller/plant and test weight contribute to grain yield/ plant through direct effect. The parent off-spring regression was high for all the characters under study suggesting improvement of these characters by mere selection. Based on divergence study, 204 lines were categorized in seven clusters whereas parents were grouped in different clusters. Molecular restricted selection using specific SSR markers with depicting high correlation with aroma could offer great promise to select high yielding rice among high aroma lines. A total of 54 randomly selected F_5 plants were subjected to SSR marker analysis using SSR markers. The F_5 plants had an allele from either of the two parental lines (homozygous condition) or alleles from both the parental rice varieties (heterozygous condition). At some SSR loci, new/recombinant alleles were observed, which indicate the active recombination between genomes of two rice varieties and can be used for linkage mapping once complete homozygosity is achieved. SSR allelic profile based on two dimensional principal component analysis demonstrated high level of diversity among parents and F_5 plants spread between them.

Key words: *Oryza sativa* L., basmati, microsatellite, phenotyping, rice, recombinant inbred lines (RILs).

INTRODUCTION

Based on a number of morphological, physiological, biochemical and molecular traits, Asian cultivated rices are organized in two major subspecies, that is, *Oryza sativa japonica* and *Oryza sativa indica* (Oka, 1988). These two subspecies are commonly associated with differences in growth habitat and are the products of independent domestication events from ancestral *Oryza rufipogon* populations in different locations and at different times (Sang and Ge, 2007). Basmati rices are known to originate in undivided India and Pakistani Punjab. Traditional basmati varieties have unique quality in the world. However, these are tall, photoperiod sensitive low yielder, weak stemmed non-responsive to increased nitrogenous fertilizer dosages. Basmati rice varieties with exquisite aroma, superfine grain characteristics and excellent cooking qualities have gain greater attention due to greater preference in domestic and international markets. According to its grain shape, rice is primarily classified into long, medium and short categories. The cooking quality of rice grain is associated with grain shape (Yang et al., 2001; Luo et al., 2004). In most cases, long grain rice has a high grain amylose content and after cooking, it is often firm and fluffy (not sticky); medium grain rice has a low amylose content and after cooking, it is often soft, moist and sticky in texture. The cooking quality and amylose content in short grain rice are similar to those of rice in the medium grain category. Grain shape has attracted significant attention in rice breeding programs due to its contributions to rice yield and quality.

Grain shape has been widely accepted as a complex trait controlled by multiple genes with small effects. Grain size is usually evaluated by test weight, which is one of the three key components, along with grain length, grain width and panicle number per plant. The most important component of fragrance in basmati varieties is 2-acetyl-1-pyrroline (2-ACP) (Buttery et al., 1983; Paule and Powers, 1989; Petrov et al., 1996). The detailed biosynthesis pathway of this compound has not yet been completely elucidated (Lorieux et al., 1996; Bradbury et al., 2005). The studies have established that in most of the varieties, a single recessive gene is responsible for fragrance (Lorieux et al., 1996). Identification of ACP by smelling or chewing seeds is subjective and non-reliable as analyst sense become saturated or physical damage occurs from chewing hard grain. Further, identification of ACP using gas chromatography require large sample of tissue and also time consuming. Initial genetic studies performed by Tanksley's group (Ahn et al., 1992), localized a gene controlling aroma or fragrance (*frg* gene) in Della (Jasmine-derived aromatic variety) on the long arm of chromosome 8. Later, Lorieux et al. (1996) tagged this gene as a major and recessive quantitative trait locus (QTL) in the same region, but limited to a 12 cm genetic interval and in a (IR64 × Azucena) DH population where the traditional upland variety Azucena was the donor of aroma. Application of molecular marker technology in linkage mapping and molecular dissection of the complex traits such as fragrance can greatly enhance the efficiency and accuracy of breeding process. It is difficult to intogress desirable traits into basmati rice by conventional breeding methods due to complex nature of basmati rice grain quality and its poor combining ability with other rice genotypes (Singh et al., 2000). Glaszmann (1987) identified basmati rice as a genetically distinct cluster, distinct from that of typical *indica* and *japonica*. The genetics of basmati rice quality traits are not well understood, and finding desirable segregants possessing all the desirable basmati characteristics has been difficult.

In the present investigation, we describe the development and phenotyping of RIL's derived from cross NPT II and HBC 19 with respect to yield, yield attribute and quality and also determines association between Rg-28 locus and basmati rice aroma.

MATERIALS AND METHODS

Mapping populations of RILs were developed from a cross between NPT II (IR 68552-100-1-2-2) and HBC 19 (pureline selection from Taraori basmati). NPT-II is a *japonica*, non-scented high yielding plant type. It was crossed as female to HBC 19, a commercially important premium basmati variety having long slender grain, intermediate amylose content with high alkali spreading value and elongation ratio. The two parental rice varieties, NPT II and HBC 19 and 204 F_5 recombinant inbreed lines were grown in Augmented design at CCSHAU. Each progenies / lines was planted in a row of 2.5 m length.

Phenotyping of population

RILs in F_5 generation were evaluated for agronomic traits namely, plant height, effective tillers/plant, test weight and grain yield/plant. Mature F_6 grains of the RILs and parental lines were dried at 37°C for 15 days and hulled in laboratory mill. The 10 representative unbroken rice grains were measured for Kernel length and Kernel breath using photographic enlarger. The correlation coefficient was calculated according to the method suggested by Johnson et al. (1955) and path coefficients were obtained according to Dewey and Lu (1959). The degree of resemblance between relatives indicates the proportionate amount of additive genetic variance in the population. For measuring the resemblance between offspring and parents, the grouping of observation was done in a different way. The sum of cross products is used to calculate the covariance of offspring with parents and the regression of offspring on parents is calculated to measure the degree of resemblance. The regression of parents b_{op} was calculated following Falconer method (1983). In order to quantify the genetic diversity between parental lines and RILs, Euclidean Cluster Analysis was employed. Constellation of genotypes into cluster was done following Ward's minimum variance method (Ward, 1963). The aroma of F_6 grains was determined by placing ten milled rice grains in a Petri plate containing 1.7% KOH and incubated at ambient temperature for 10 min as suggested by Sood and Siddiq (1978) method. The RILs were scored as aromatic, moderately aromatic and non-aromatic.

Genotyping of RILs

A total of 54 plants were randomly selected from the NPT II × HBC 19 F_5 population for microsatellite marker analysis. Genomic DNA was isolated from one month-old plant leaf samples (~ 100 mg each) using CTAB method (22N). DNA was checked for its quality and quantity by 1% agarose gel electrophoresis using a standard containing 100 ng/µl genomic λ DNA. Thirty five (35) markers (obtained from Research Genetics, Huntsville, AL, U.S.A.) were evaluated for polymorphism between NPT II and HBC 19 (data not

Table 1. Mean, range, phenotypic and genotypic coefficients of variations, heritability and genetic advance for various traits in F_5 population derived from cross NPT II x HBC 19.

Character	Mean± SE(m)	Range	PCV	GCV	Heritability	Genetic advance (%mean)
Plant height (cm)	121.58 ± 0.81	90.00 - 157.00	10.32	9.59	86.3	23.51
Productive tillers/plant	8.29 ± 0.14	5.00 - 23.50	47.29	25.70	29.5	36.86
Yield/plant (g)	8.39 ± 0.25	1.90 - 32.76	46.53	44.52	92.3	112.9
1000 grain weight (g)	18.67 ± 0.19	12.10 - 28.50	18.17	15.09	69	33.08
Kernel length(mm)	7.09 ± 0.05	5.00 - 10.50	24.11	10.80	20.1	12.78
Kernel breath(mm)	1.95 ± 0.02	1.38 - 3.00	33.69	15.48	21.1	18.77
L/B ratio	3.71 ± 0.05	2.09 - 6.00	22.14	19.73	79.4	46.43

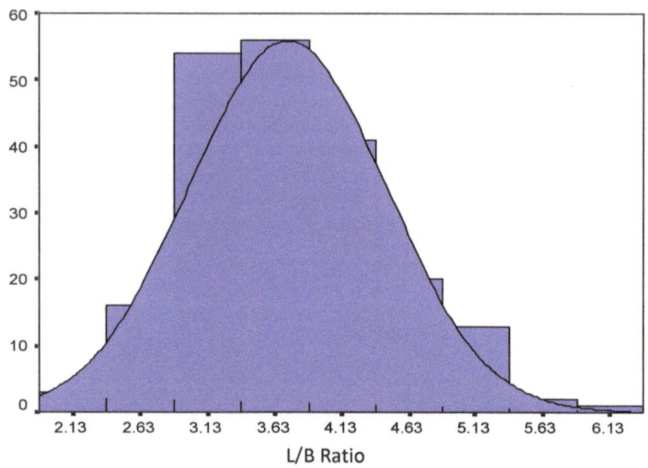

Figure 1. Frequency distribution graph of the RIL population for L/B ratio.

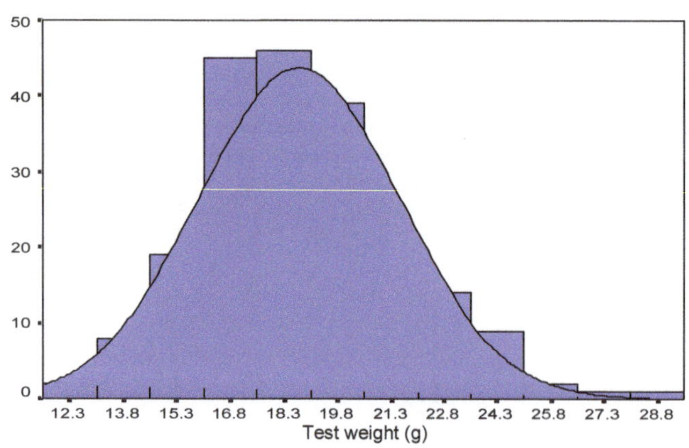

Figure 2. Frequency distribution graph of the RIL population for test weight (g).

shown) and six of these showing polymorphism and clear discrete banding profiles were used for analysis of NPT II x HBC 19 F_5 plants. The PCR reaction was conducted in a reaction volume of 20 µl containing 1x PCR buffer, 100 µM dNTPs, 0.4 µM of each primer, 1.2 mM $MgCL_2$, 1 unit Taq DNA polymerase and 50 ng template DNA. PCR amplification was performed with initial denaturation at

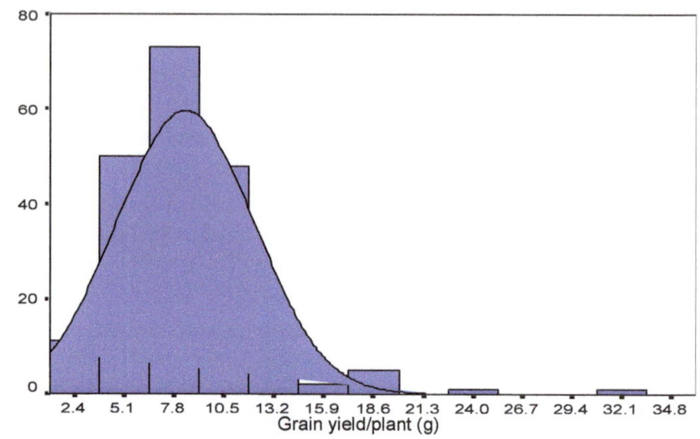

Figure 3. Frequency distribution graph of the RIL population for grain yield per plant (g).

94°C for 5 min followed by 35 cycles of 94°C for 1 min, 55°C for 1 min, 72°C for 2 min and final extension at 72°C for 7 min before cooling at 4°C. Amplification products were denatured and resolved at 4% denatured polyacrylamide gel, as described by Saini et al. (2004). The size (in nucleotides base pairs) of the most intensely amplified band of stutter for each microsatellite marker was determined based on its migration relative to molecular weight size marker (10 bp DNA ladder from Gibco BRL, Md.).

RESULTS AND DISCUSSION

Phenotyping variation among RILs

F_5 population showed significant variation among 204 RILs for all the characters. Ample amount of variation was observed for almost all the characters (Table 1) showing these traits are responsible for variation in grain yield in various lines. The coefficient of skewness for grain yield/plant, plant height, test weight and L/B ratio were 2.248, 0.297, 0.582 and 0.446; was less than the standard deviation 3.714, 11.66, 2.817 and 0.7338, respectively which indicates normal distribution of the population (Figures 1 to 3) and that these characters are governed by polygene. Long slender grain is a defining characteristic of basmati rice varieties. The grain length in

Table 2. Phenotypic correlation coefficient in F_5 population derived from cross NPT II x HBC 19.

Character	Plant height (cm)	Productive tillers/plant	Yield/plant (g)	1000 grain weight (g)	Kernel length (mm)	Kernel breath (mm)	L/B ratio
Plant height (cm)	-	0.0760	0.0054	0.0530	0.0541	0.0037	0.0148
Productive tillers/plant			0.3222**	0.0880	0.0778	0.0174	0.0658
Yield/plant (g)				0.2008**	-0.0886	0.0515	-0.0416
1000 grain weight (g)					0.0911	0.1387	-0.0537
Kernel length(mm)						-0.1052	0.6116**
Kernel breath(mm)							-0.8283**
L/B ratio							

**Significant at 1%; *Significant at 5%.

Table 3. Direct (diagonal) and indirect effects of component traits on grain yield/plant in F_5 population derived from cross NPT II x HBC 19.

Character	Plant height (cm)	Productive tillers/plant	1000 grain weight (g)	Kernel length (mm)	Kernel breath (mm)	r with grain yield/plant
Plant height (cm)	-0.0216	-0.0016	-0.0011	-0.0012	-0.0001	0.0054
Productive tillers/plant	0.0241	0.3175	0.0279	0.0247	0.0055	0.3222**
1000 grain weight (g)	0.0098	0.0163	0.1848	0.0168	0.0256	0.2008**
Kernel length (mm)	-0.0069	-0.0100	-0.0117	-0.1283	0.0135	-0.0886
Kernel breath (mm)	0.0000	0.0001	0.0010	-0.0007	0.0070	0.0515

RILs ranged from 5 to 10.50 mm. The two parental variety HBC 19 and NPT II are quiet similar in test weight measuring 24 and 24.5, respectively. However, test weight in the RILs ranged from 12 to 28.5 g with a population mean 18.67 g. The transgressive segregants were noticed for almost all the characters. High heritability together with genetic advance was recorded for grain yield/plant, plant height, test weight and L/B ratio indicate the presence of additive gene effect and ample scope of improving these characters through simple selection. High genotypic and phenotypic coefficient of variance for grain yield and its component were observed indicating that variability among the genotype is genetic and selection for these characters will be effective. Characters with low GCV such as plant height and test weight showed high heritability which indicate that variation for these characters is heritable.

Phenotypic correlation coefficient analysis was carried out to assess the association between various traits in NPT II x HBC19 F_5 population (Table 2). Yield per plant had positive correlation with productive tillers per plant and test weight (0.3222** and 0.2008**, respectively). Kernel length had positive correlation with L/B ratio (0.6116**) and negative correlation with kernel breath (-0.8283**). Correlation between the traits could be due to linkage or pleiotropy. Path coefficient analysis provides a clean and more realistic picture of a complex situation that exists at the correlation level. Portioning of genotypes correlation between yield/plant and its component characters indicated that all the characters contribute to grain yield /plant through direct effect, not by indirect effect (Table 3). Productive tillers/plant (0.3175) and test weight (0.1848) had high positive direct effect on grain yield/plant. The degree of resemblance between relative indicates the proportionate amount of additive genetic variance in the population. To phrase it otherwise, the resemblance between relatives which is one of the basic genetic properties of the population is determined by the relative magnitude of the additive and non-additive components of genetic variation. Since additive component is fixable part of the genetic variance, its relative magnitude plays a key role in deciding the breeding methodology for improving the character. More, the resemblance between the relatives, higher the magnitude of additive genetic variance and thus more easy is the improvement in the character.

The parent offspring regression (b) were 0.783, 0.553, 0.431, 0.328, 0.293, 0.246 and 0.213 for 1000 grain weight, grain yield per plant, plant height, kernel breath, L/B ratio, kernel length and tillers per plant, respectively. It was high for most of the characters suggesting that these characters can be improved mere by selection.

Cluster analysis

Based on divergence studies, parents and 204 RILs were categorized into nine clusters (Table 4) with highest

Table 4. Mean values of different clusters for 7 traits in F_5 population derived from cross NPT II x HBC 19.

Cluster	Number of RILs	Plant height (cm)	Productive tillers/plant	Yield/plant (g)	1000 grain weight (g)	Kernel length (mm)	Kernel breath (mm)	L/B ratio
1	43	120.23	8.15	8.77	17.32	3.91	1.60	4.36
2	42	111.91	8.11	7.66	18.85	7.11	2.13	3.37
3	33	120.45	7.01	8.28	17.00	6.50	2.08	3.17
4	32	127.06	8.06	9.13	22.77	7.15	1.96	3.70
5	22	129.65	10.86	9.10	18.35	6.91	2.30	3.04
6	18	130.48	7.83	5.87	18.98	7.69	1.91	4.04
7	14	123.80	7.95	6.54	16.14	8.46	1.80	4.74
8	HBC 19	136.00	23.50	24.69	24.00	8.25	1.38	6.00
9	NPT II	96.00	12.50	32.76	24.55	5.75	2.75	2.09

Table 5. Microsatellite alleles for the polymorohic marker on aromatic and non-aromatic rice.

S/N	Primer	HBC 19	NPT II
		Allele size (bp)	Allele size (bp)
1	RM 223	151	158
2	RM 149	240	259
3	RM 210	143	157
4	RM 42	159	165
5	RM 308	131	146
6	RM 195	302	308

number of genotypes in cluster I (43). Parents and RILs come in different group; this means that recombination occurred in segregating population and the RILs differ from parents. Test weight of cluster IV (22.77 g) is comparable to the parents whereas cluster VII has more kernel length (8.46 mm) than that of HBC 19(8.25).

Aroma

Aroma is one of the most important quality trait for basmati rice consumers. Petrov et al. (1996) reported more than 100 volatile compounds in the rice grain, of which 2-acetyl-1-pyrroline was the most predominant component. The two parental lines differ in grain aroma; HBC 19 highly aromatic whereas, NPT II non-aromatic. Out of 204 lines, very few lines have almost same aroma and yield (20 g) as comparable to HBC 19. Out of 54 agronomically superior lines which are subjected to SSR analysis, 32 lines are aromatic and 22 lines are non-aromatic. Among these RILs line, no. 111 and 143 has highest kernel length (10 and 9 mm) and L/B ratio (5.25 and 4.5) which is much higher than better Better parent HBC 19. No intra line variation for aroma was observed. These lines also differ for grain shades ranging from red to white (data not shown). Few RILs (5.55%) reconstitute the original aroma of HBC 19 suggesting involvement of 3 or more genes as their expected proportion in RILs will be ½ n with n number of genes (Loriex, 1996).

Microsatellite marker analysis

Micosatellite marker analysis was carried out on 54 (32 aromatic) randomly selected F_5 plants (NPT II × HBC 19) and two parental varieties using six polymorphic markers (Table 5) to assess allelic distribution and percent homozygosity achieved so far and association of molecular marker with aroma. Eleven (11) RILs (7, 8, 10, 12, 13, 16, 17, 24, 28 and 30) were homozygous for all the locus except RM 195; whereas nine (Line no. 23, 26, 34, 35, 39, 44, 46, 48 and 50) out of 54 RILs were giving banding pattern similar to HBC 19 except at locus RM 195 which shows NPT II type banding and biochemical assay showed the presence of aroma which might be due to recombination and consequently Rg 28 locus was transferred to NPT II background or vice-versa. RM 342 was significantly positively correlated to aroma while all other markers are positively correlated to aroma except RM 149 (Table 6). This may be due to the fact that aroma in generally is recessive or possibly because of failure to recognize the effect of xenia and the fact that aroma is an endosperm characteristic product of sexual fertilization is a one generation advance from that of the plant on which

Table 6. Correlation coefficient between molecular markers and aroma in F_5 population derived from NPT II x HBC 19.

Primer	RM 223	RM 149	RM 210	RM 308	RM 42	RM 195	Aroma
RM 223	--	0.399**	0.183	0.431**	0.094	0.048	0.235
RM 149			0.382**	0.486**	0.128	0.121	-0.100
RM 210				0.569**	0.543**	0.217	0.211
RM 308					0.389**	0.122	0.156
RM 42						0.205	0.283**
RM 195							0.174

**Significant at 1%; *significant at 5%.

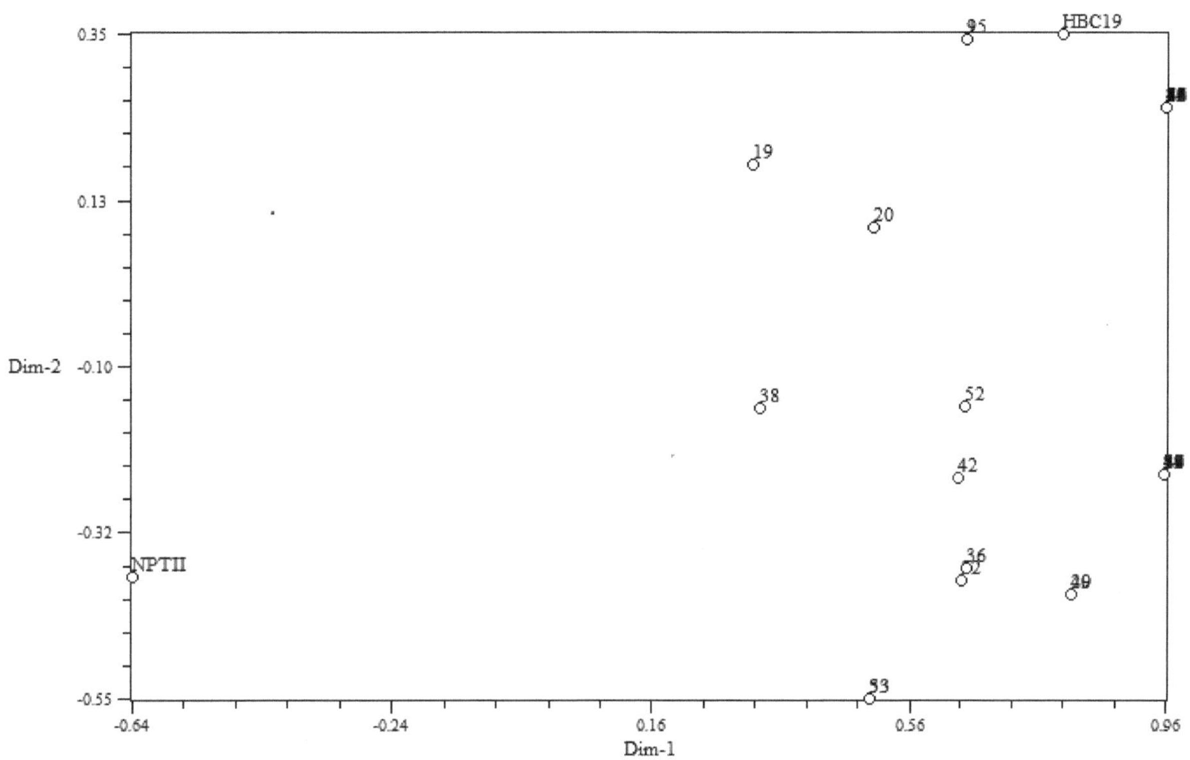

Figure 4. Two dimensional principal component analysis (PCA) scaling of 54 NPT II x HBC 19 F_5 genotypes using Jaccard similarity coefficient data of polymorphic markers.

it develops (Bollic, 1992).

Molecular analysis showed that there was still some residual heterozygosity in RILs probably due to insufficient number of selfing cycles at the F_5 generation. Percent homozygosity varied from 52 to 100% with an average of 72%. This value is lower than the expected value of 96.8% in F_5 generation. Deviation of observed frequency of the two segregating alleles of individual markers from the expected 1:1 Mendelian ratio may be due to segregating distortion which can seriously affect the QTL mapping results (Xu et al., 1997). At one loci, RM 210 one (line no. 19) out of 54 F_5 plants showed the presence of new or recombinant alleles different from those present in either of the two parental rice varieties. Occurrence of such alleles may result from crossing over, and also some microsatellite loci have been reported to be more hot spots, where mutation occurs up to 100 times more frequently than the normal mutation rate (Jain et al., 2004). The two-dimensional PCA (Rohlf, 1993) generated using SSR allelic diversity data showed interspersing of $54F_5$ plants between the two diverse parental genotypes (Figure 4). Some of the aromatic F_5 plants were close to parents whereas some lines shows variability beyond the parents which may be consequence of recombination between the genomes of two parental lines and the subsequent segregation that leads to the occurrence of unique genotypes.

In self pollinated crops, the germplasm is available in

the form of multitude of pure lines. The breeding endeavor revolves around recombining the genes of interest spread over different pure lines in one or few backgrounds. By way of recombination breeding which involves crossing of pure lines followed by selection in segregating generations till the selected lines become homozygous. In this context, the gene constellation governing aroma, yield, kernel length, test weight can be standardized by crossing. Such an attempt was expected to generate a population with much expended variability toward concurrent tandem selection for yield and grain quality including aroma. Molecular restricted selection using specific SSR markers with depicting high correlation with aroma could offer great promise to select high yielding rice among high aroma lines. The present study thus can be rated as an integrated approach for rice improvement using conventional recombination breeding compliment by molecular marker assisted selection.

REFERENCES

Ahn SN, Bollisch CN, Tanksley SD (1992). RFLP tagging of a gene for aroma in rice. Theor. Appl. Genet. 84:825-828.

Bollic CN, Rutger JN, Webb BD (1992). Development in rice research in United States. Int. Rice Commission Newsl. 41:32-34.

Bradbury LM, Fitgerald TL, Henry RJ, Jin Q, Waters DLE (2005). The gene for fragrance in rice. Plant Biotech. J. 3:363-370.

Buttery RG, Ling LC, Juliano BO, Turnbaugh JG (1983). Cooked rice aroma and 2-acetyl-1-pyroline in rice. J. Agric. Food Chem. 31:823-826.

Dewey DR, Lu KH (1959). A correlation and path coefficient analysis of crested wheat-grass. Agron. J. 51:515-518.

Falconer DS (1983). Introduction to quantative genetics (2nd edition). Longman Group Lyd.

Glaszmann JC (1987). Isozymes and classification of Asian rice varieties. Theor. Appl. Genet. 74:21-30.

Johnson HW, Robinson HF, Comstock RE (1955). Estimates of genetic and environmental variability in soyabean. Agron. J.47:314-318

Lorieux M, Petrov M, Huang N, Guiderdoni E, Ghesquière A (1996). Aroma in rice: genetic analysis of a quantitative trait. Theo. Appl. Genet. 93:1145-1151.

Luo YK, Zhu ZW, Chen N, Duan BW, Zhang LP (2004). Grain types and related quality characteristics of rice in China. Chinese J. Rice Sci. 18:135-139.

Oka HI (1988). Weedy forms of rice. In: Origin of cultivated rice. Elsevier, Amsterdam. pp 107-114.

Paule CM, Powers JJ (1989). Sensory and chemical examination of aromatic and non aromatic rices. J. Food Sci. 54:343-346.

Petrov M, Danzart M, Giampaoli P, Faure J, Richard H (1996). Rice aroma analysis Discrimination between a scented and a non scented rice. Sci. Aliments 16:347-360.

Rohlf FJ (1993). NTSYS-PC:Numerical taxonomy and multivariate analysis system. Version 1.8 Exeter Sowftware, New York.

Saini N, Jain N, Jain S, Jain RK (2004). Assessment of genetic diversity within and among basmati and non-basmati rice varieties using AFLP, ISSR and SSR markers. Euphytica 140:133-146.

Sang T, Ge S (2007) Genetics and phylogenetics of rice domestication. Curr. Opin. Genet. Dev. 17:1-6.

Singh RK, Gautam L, Saxena S, Singh S (2000) Scented rice germplasm: conservation, evaluation and utilization. *In:* Aromatic Rices (ed. R.K. Singh, U.S. Singh and G.S. Khush), Oxford & IBH Pub., New Delhi: 107-133.

Sood BC, Siddiq EA (1978). A rapid technique for scent determination in rice. Indian J. Genet. Plant Breed. 38:268-271.

Ward H J (1963). Hierarchical grouping to optimize an objective function. J. Am. Stat. Assoc. 58:236-244.

Xu Y, Zhu L, Xiao J, Huang N, McCouch S (1997). Chromosomal regions associated with segregation distortion of molecular markers in F2, backcross, doubled haploid, and recombinant inbred populations in rice (Oryza sativa L.). Mol. Gen. Genet. 253:535-545.

Yang LS, Bai YS, Zhang PJ, Xu CW, Hu XM, Wang WM, She DH, Chen GZ (2001). Studies on the correlation between grain shape and grain quality in rice. Hybrid Rice 16:48-50 [in Chinese].

A critical scientific review on South African governance of genetically modified organisms (GMOs)

F. W. Jansen van Rijssen[1], E. Jane Morris[2] and J. N. Eloff[1]*

[1]Phytomedicine Programme, Department of Paraclinical Sciences, Faculty of Veterinary Science,University of Pretoria, Private Bag X04, Onderstepoort, 0110, South Africa.
[2]Department of Biochemistry, Faculty of Natural and Agricultural Sciences, University of Pretoria and African Centre for Gene Technologies, PO Box 75011, Lynnwood Ridge 0040, South Africa.

Credible governance of genetically modified organisms (GMOs) is essential because of public concerns in South Africa (SA) and internationally. In this preliminary study, the opinions of a number of scientists with experience and/or interest in GMO governance were determined by means of two questionnaires to determine their perceptions on the credibility of risk governance of GMOs in SA. The respondents felt 'some improvement' was required in criteria related to good governance. Excellence (quality of risk assessment) and effectiveness, such as protracted regulatory processes needed 'some to much improvement'. The responses were evaluated against an analysis of the South African GMO Act, regulations, policy guidelines and available information. The Act provides a pro-active basis for good governance comparable to internationally described risk governance models, but implementation seemed to follow the less advanced technocratic model. A number of reasons were identified such as unclear roles of decision makers. Some of the causes for protracted decision-making identified by respondents were: a) excessive precaution in decision making, and b) different mandates resulting in no unanimity among government departments. Proposals for improvement in credibility included communication as a critical component of risk governance and continued training of reviewers and decision makers.

Key words: Genetically modified organisms, risk assessment, risk governance, South Africa.

INTRODUCTION

A recent study, confirming the need for further investigation of the South Africa's (SA) risk governance of genetically modified organisms (GMOs) (Jaffe, 2008), points to various deficiencies in governance that could impact on credibility and could cause delays in processing of permits, resulting in increasing costs of registering new GMO products. Incidents that raised concern to ensure that risks are addressed in a scientific way (Paarlberg, 2000) and that illustrated conflicting approaches between departments that could affect credibility, were (DAFF, http://www.daff.gov.za):
a) protracted decision making such as the embargo on commodity clearances (import of GM grains) since 2005 until 2011(DAFF, http://www.daff.gov.za); b) delays in stacked trait cotton approvals and several appeals against decisions (DAFF, http://www.daff.gov.za); c) poorly justified socio-economic reasons for refusal of permits (DAFF, http://www.daff.gov.za); and, d) the

*Corresponding author. E-mail: kobus.eloff@up.ac.za.

Abbreviations: AC, GMO Advisory Committee; **CAC or Codex,** Codex Alimentarius Commission; **DAFF,** Department of Agriculture, Fisheries and Forestry; **DEA,** Department of Environmental Affairs; **DOH,** Department of Health; **DST,** Department of Science and Technology; **the DTI,** the Department of Trade and Industry; **EC,** GMO Executive Committee.

Department of Trade and Industry's (the DTI) mandatory requirement for labelling of genetically modified ingredients and components (not defined) in the Consumer Protection Act (South Africa, 2008) to enable consumer food choices, despite existing labelling regulations by the Department of Health (DOH) (South Africa, 1972; 2004).

The current study addressed the question: How credible is governance of GMOs in SA as perceived by scientists with knowledge of the system?

Much has been said on improvement of trust in governance in general, such as the need for more transparency in decision making (FAO/WHO, 2006). Health hazards such as the bovine spongiform encephalitis (BSE), a transmissible, neurodegenerative, fatal brain disease of cattle, and debates on food GMOs in Europe resulted in changes in government food safety systems (Dreyer et al., 2006; Atkins and Norman, 2009). The establishment of the European Food Safety Authority (EFSA) in 2002 as an independent and transparent organization to advise the Commission of the European Communities on food safety was one of the major reforms in Europe to restore public trust (CEC, 2002).

The concept of risk governance has received considerable attention lately (Dreyer et al., 2006, Dreyer and Renn, 2009; CAC, 2010). The Codex Alimentarius Commission (CAC), an International Commission addressing food standards, has pro-actively described the process of risk analysis (risk assessment, risk management and risk communication) and the principles that could be applied by governments to improve open and transparent decision making (CAC, 2010). A comprehensive scope of food safety/risk governance that could equally be applied to environmental safety /risks has been defined as: the totality of actors, rules, conventions, processes and mechanisms concerned with how relevant food risk information is collected, analysed and communicated, and decisions on food safety management are made.

Food risk governance includes, but extends beyond the three components of risk analysis (risk assessment, risk management, risk communication). It also involves co-ordination between public bodies, commercial and civil society actors, and wider contextual factors such as institutional arrangements, legislative procedures and political culture (Dreyer, et al., 2006).

Millstone (2007) identified three successive models of governance as a) technocratic, b) decisionist, and c) co-evolutionary, demonstrating increasing interaction with stakeholders and including matters of socio-economic importance. In technocratic governance models, the roles of risk assessors and risk managers (decision makers) are not well separated. This is found in older governance structures and may result in poor decisions, such as the BSE scandal in the United Kingdom (Dreyer et al., 2006).

The decisionist model strongly favours functional separation (Millstone, 2007). The CAC supports functional separation to ensure the scientific integrity of the risk assessment, to avoid confusion over the functions to be performed by risk assessor and risk managers and to reduce conflict of interest' (CAC, 2011).

The co-evolutionary or democratic model (transparent model according to Renn (2008), includes additional structured reciprocal links between science (risk assessment) and policy (risk assessment framing, management and decision making) as well as an evaluation step to evaluate risks versus benefits (Morris, 2011). This requires stakeholder participation (Millstone, 2007; Dreyer et al., 2009). Presently, more research is being conducted on the concept of participation. The role of democratic decision making depends on 'opening up the governance process' (Stirling et al., 2009). Consultation can occur through advisory committees, *ad hoc* consultations public hearings and regulatory instruments. The Commission of the European Communities (CEC) identified the following criteria of good governance: independency in risk assessment, ethical conduct, transparency, openness, participation, accountability, excellence in conduct and effectiveness (CEC, 2000, 2001). Models of good governance were used to compare development of SA governance of GMOs with international examples of GMO governance.

The SA GMO Act (South Africa, 1997) legislates all GMOs including GM seed and grain, GM micro-organisms and GM animals. All activities are included from importation, contained use, trials, general release, monitoring and exportation. The SA area planted according to 2011 statistics (James, 2011) are a combined 2.3 million hectares of GM maize, GM soybean and GM cotton (maize 72% or 1.873 million hectares of total maize planted; soybean 85 % or 1,873million hectares of total; cotton 100 % or 15000hectares). No microorganisms or GM animals have been approved for general release. GM vaccines are only in clinical trials.

In SA, the GMO Act (Act 15 of 1997) is often held as an example for future food safety governance models (Chanda et al., 2010). This model partly resembles the framework for risk governance in Europe. With the exception of the labelling of GMO foods, governance in the SA of GMOs is legislated in terms of the GMO Act as amended (South Africa, 1997) and administered by the Department of Agriculture, Forestry and Fisheries (DAFF). The Act establishes a decision making body, the GMO Executive Council (EC), constituted in practice of one representative from each of six government departments plus the chairperson of the GMO Advisory Committee (AC). The government departments have mandates in terms of their own legislation except for Department of Science and Technology (DST) whose mandate also derives from the National Biotechnology Strategy for South Africa (South Africa, 2001). DOH reactively regulates the safety of food (South Africa, 1972). The Department of Environmental Affairs (DEA) has additional responsibility for environmental safety of

GMOs in terms of the National Environmental Management: Biodiversity Act (South Africa, 2004). The AC consists of independent scientists from academia, research institutes and the private sector, and submits recommendations on permit applications to the EC.

The risk assessment step in risk governance is of necessity and partly non-transparent because of the confidentiality of certain company information. Government policy may reduce transparency. This was exemplified by the fact that the EC discouraged government GMO regulatory scientists from responding to the survey described in this paper, by stipulating that these scientists must first obtain written ministerial approval.

Therefore, determination of credibility could not be based on first-hand information but only on perce-ptions by those who viewed the system from outside. These perceptions do however, indicate problem areas. This study should be regarded as preliminary because policy makers and scientists directly involved in the regulatory process were not allowed to participate.

MATERIALS AND METHODS

The study includes an analysis of applicable legislation, analysis of available official guidelines, interviews with the SA government officials and two questionnaires distributed to the SA scientists with knowledge of GMO legislation and risk assessments. The first questionnaire, targeting a limited number of scientists from industry, public research and academic institutes, was designed to: a) qualitatively describe a credibility profile of the SA governance of GMOs from responses to criteria and sub-criteria described in Table 1a, b) probe perspectives in general on criteria of good governance based on the statements in Table 1b. These statements and results were grouped under three categories of good governance, namely policies/procedures (statements 10.1 to 10.15), excellence scientific (statements 11.1 to 11.8) and transparency (statements 12.1 to 12.9).

For the first questionnaire, 24 responses (response rate of 10.2%including possible responses from regulatory authorities and members of the AC) were obtained and considered as a fair number in this field, considering the constraints encountered. The low response rate could be ascribed to: i) Some potential respondents were unfamiliar with the subject; ii) Government officials as well as advisors to government (a possible 54 responses) did not participate and neither did anti- GM lobby groups; iii) in a number of cases, a single response was received per biotechnology seed company or a research institute, instead of responses from individual persons and iv) The internet approach for questionnaires presented several technical problems.

The participants could be considered as a homogenous group, representing applicants or potential applicants for permits and having in common marketing (or general release) of GMO products or an academic interest.

Statements in the three categories of good governance of the questionnaire indicate a reasonable general understanding of 'good governance' among respondents. Although the response numbers were relatively low, valuable information was obtained that should lead to a more in-depth future study of risk governance. The current study focused on assessment of priority needs to improve legislation, policy and implementation.

In a second questionnaire, a few key scientists from the agricultural biotech seed industry in the SA and scientists responsible for preparation of permit applications were questioned regarding their views on reasons for delays in issuing of permits and proposed remedial actions (Figure 2a, b). Six responses were obtained. The participants were existing or potential permit-holders who had submitted new applications in the last five years. Importers of GM grain were not included as they were not directly involved in new submissions. They were also involved in a legal dispute with government at the time of the study.

RESULTS AND DISCUSSION

The majority view of the scientists was that some improvements were required to ensure credible governance. In particular, results to responses 'transparency', 'openness', legislation' and 'participation' needed 'some improvement'. 'Effectiveness', 'excellence' 'scientific' and 'accountability'needed 'some to much improvement', whereas, 'ethical'conduct and 'independency' in risk assessments needed 'no to some improvement' (Figure 1). The results are discussed in more detail in terms of three categories that underpin credible risk governance: i) functional separation between risk assessment and risk management (policies and procedures); ii) excellence in performance (risk assessors and reviewers, review procedures); and iii) transparency in governance (communication, participation).

Functional separation between risk assessment and risk management

The legal framework for governance of GMOs in the SA provides for functional separation between risk assessment and risk management (South Africa, 1997).This brings the SA in line with Codex guidelines (CAC, 2011) but also with the most developed examples of independency in risk governance such as the EFSA (EFSA, http://www.efsa.europa.eu).

Almost all responding SA scientists agreed on the importance of separate roles for the EC (Table 1b, 10.5) and AC (Table 1b, 10.3), but felt that some improvement was needed in clarifying those roles (Figure 1, Table 1a, 3). The statement 'EC members do not have a role as reviewers of risk assessment data/information' (Table 1b, 10.4), created some disagreement (21% disagree, 67% agree, 12% unsure). Some felt that government by virtue of its own legislation has a responsibility to conduct independent reviews. This is indeed the case with pro-active SA environmental legislation (South Africa, 2004) and is now also included in the amendment to the GMO Act (South Africa, 1997). One respondent felt that 'the legislation needs some work; the DEA calls for 'coordinated regulations of GMOs, but undertakes its own risk assessments'. The respondent suggested: 'the AC ...will undertake a complete risk assessment of all safety issues. The EC members must address non-safety issues, such as cultural impact, loss of traditional knowledge, impact on trade and labour etc.'

Table 1a. South African GMO risk governance: Scores for credibility criteria and sub-criteria.

Criteria	Sub-criteria	NI	SI	MI	DNK	Total
1. Legislation		4	11	3	1	19
1.1	Following the latest international model of risk analysis	4	12	3	3	
2. Effectiveness		0	10	8	1	19
2.1	Clear roles in the legislative processes	7	11	5	1	
2.2	Evaluation of future impact	2	13	7	2	
2.3	Past experience, where possible	3	14	4	3	
2.4	Clear guidelines	3	13	7	1	
3. Accountability		3	8	6	3	20
3.1	Clear roles in the legislative processes	4	12	4	3	
3.2	Clear roles in the executive processes	3	12	5	3	
3.3	Clear roles for risk assessors	4	10	5	3	
3.4	Parties to assume responsibility for their roles	1	8	9	5	
4. Independency		7	4	4	4	19
4.1	No pressure on risk assessors from policy makers	10	5	3	5	
4.2	Risk assessors acceptable to all parties	7	8	3	5	
4.3	No pressure from stakeholders on risk assessors	9	7	3	4	
5. Scientific excellence		2	9	7	1	19
5.1	Enough suitably qualified specialist risk assessors	3	10	8	3	
5.2	Peer-reviewed assessments of scientific information	4	13	4	3	
5.3	Best use of available information systems	1	13	6	4	
5.4	Consulting with international organizations	0	13	3	8	
5.5	Networking with national food safety authorities	1	9	6	8	
5.6	Consulting with independent experts	1	13	4	6	
5.7	Risk assessments: international standards/guidelines	5	11	3	5	
5.8	SA accredited laboratories	2	11	7	4	
5.9	International standards	4	12	3	5	
6. Ethical conduct		7	5	2	5	7
6.1	Risk assessors do not have a conflict of interest	9	8	2	5	
6.2	Risk assessors have confidentiality clearance	8	4	1	11	
7. Openness		1	12	5	1	19
7.1	Interaction with stakeholders	4	11	8	1	
7.2	Decision making	3	8	12	1	
8. Participation		3	10	4	2	19
8.1	Inclusive approach	3	10	9	2	
9. Transparency		1	12	5	1	19
9.1	Clear procedures	3	16	4	1	
9.2	Communicating uncertainty in risk assessment	2	10	8	4	
9.3	Risk assessors' names and qualifications known	4	7	5	8	

NI = No improvement; SI = some improvement; MI = much improvement; DNK = do not know

Independency as a criterion for credibility was regarded as important. The majority of respondents (92%) agreed that risk assessment should be conducted independently from risk management (Table 1b, 10.2). Respondents also agreed that risk assessors were independent and not subject to pressure from policy makers or stakeholders. The members of the AC were acceptable to most respondents (Table 1a, 4). One area of future concern, though not identified by respondents to the questionnaire, is that the representation on the AC in terms of the amended GMO Act may include two officials representing the 'public sector' --undefined (South Africa, 1997). Presumably, they could be from semi-state institutes or government departments. The concern is that government officials are bound by political policy and may not be regarded as independent.

A specific characteristic of 'independency' is that the risk assessment should be 'a purely scientific activity' (CAC, 2011) and the majority of respondents agreed that this was the case (Table 1b, 10.1). This is the requirement in Section 3.3.a of the GMO Regulations which states that applicants must submit a 'scientifically-based risk assessment', again repeated in the heading of Section 4 (South Africa, 1997). Some sociologists contest this view as not completely possible, as risk assessors (or reviewers) would to some extent approach the assessment subjectively (Meghani, 2009). This argument is held in favour of more transparency in assessment of risks that need to be considered in new governance models.

Ethical conduct relates to the criterion 'independency from pressure of applicants' and includes 'confidential clearance' and lack of 'conflict of interest'. Some respondents were uncertain of the degree of lack of

Table 1b. South African GMO risk governance: Scores for statements on credibility (%) (AC, GMO Advisory Committee; EC, GMO Executive Council).

Number	Statement	Total response	Agree	Disagree	Unsure
10	**Policies and Procedures**				
10.1	RISK ASSESSMENT of GMO's is deemed to be a purely scientific activity	24	84	16	0
10.2	RISK ASSESSMENT should be conducted independently from risk management (including from political influences)	24	92	8	0
10.3	The role of the AC is to review food/feed safety data of new GMOs and make recommendations to the EC	23	100	0	0
10.4	EC members do not have a role as reviewer of risk assessment data/information	24	67	20	13
10.5	The role of the risk managers, including the EC, is to consider managerial risk options regarding risk assessments as proposed by the AC	24	92	4	4
10.6	Any uncertainty / disagreement / need for more information regarding scientific issues should be referred back to the AC for advice before final decisions are made at EC level	24	100	0	0
10.7	The AC should be more pro-active by not only advising on proposals for specific activities or projects but also make proposals in this respect	22	91	9	2
10.8	Lack of policies on some regulatory matters	24	63	8	29
10.9	The AC should initiate new policies pertaining to assessments, e.g. , guidelines for risk assessment requirements	24	92	4	4
10.10.	The function of the AC should include the development of guidelines for regulatory risk assessment requirements	24	96	4	0
10.11	Case-by-case assessments could result in uncertainty in regulatory requirements	24	50	46	4
10.12	Guidelines for regulatory requirements of GM food and feed safety should have more detail	24	71	13	17
10.13	Too many separate application forms	24	29	33	38
10.14	An EVALUATION step to consider socio-economic effects and benefits, should be included in the pre-regulatory assessment, before the managerial decision making step	24	75	17	8
10.15	The AC should receive more legal status as an independent advisory body	24	42	25	33
11	**Excellence**				
11.1	Peer-reviewing of information is very important	24	100	0	0
11.2	A single multi-disciplinarian cannot replace a team of specialists	24	100	0	0
11.3	Risk assessors should be involved in research to remain in touch with science	24	79	17	4
11.4	The AC specialists should be included in the government team to international meetings/conferences	24	92	4	4
11.5	South Africa should keep a roster of all details of potential risk assessors for GMOs	24	88	4	8
11.6	PhD-degree qualification with at least 3 years experience in the relevant discipline	23	87	13	0
11.7	MSc-degree qualification with at least 5 years experience in the relevant discipline	24	75	21	4
11.8	To be nominated as a member of the AC, the specialist should have conducted at least 10 GM food/feed safety assessments	23	39	39	22
12	**Transparency**				
12.1	RISK ANALYSIS policies should be developed in collaboration with stakeholders	24	100	0	0

Table 2. Contd.

12.2	Stakeholders communication in risk analysis of GMOs in important	24	100	0	0
12.3	Stakeholder participation in the scientific reviewing of company information is not acceptable	24	42	29	29
12.4	Stakeholder participation regarding the evaluation of the potential risk in the context of socio-economic impact is invaluable	24	71	17	12
12.5	Commencement of risk assessment of new GMOs should be announced in the media	24	38	54	8
12.6	The risk assessment report of the AC should be made available to the applicant for comments to be considered by the EC	24	92	8	0
12.7	The risk assessment report of the AC should be published for information	24	62	25	13
12.8	The final approved report of the EC, including risk analysis decisions as well as socio-economic and benefit considerations, should be published on the internet/media for public information	24	75	17	8
12.9	There should be an opportunity for objections/comments to decisions of the GMO Council	24	84	8	8

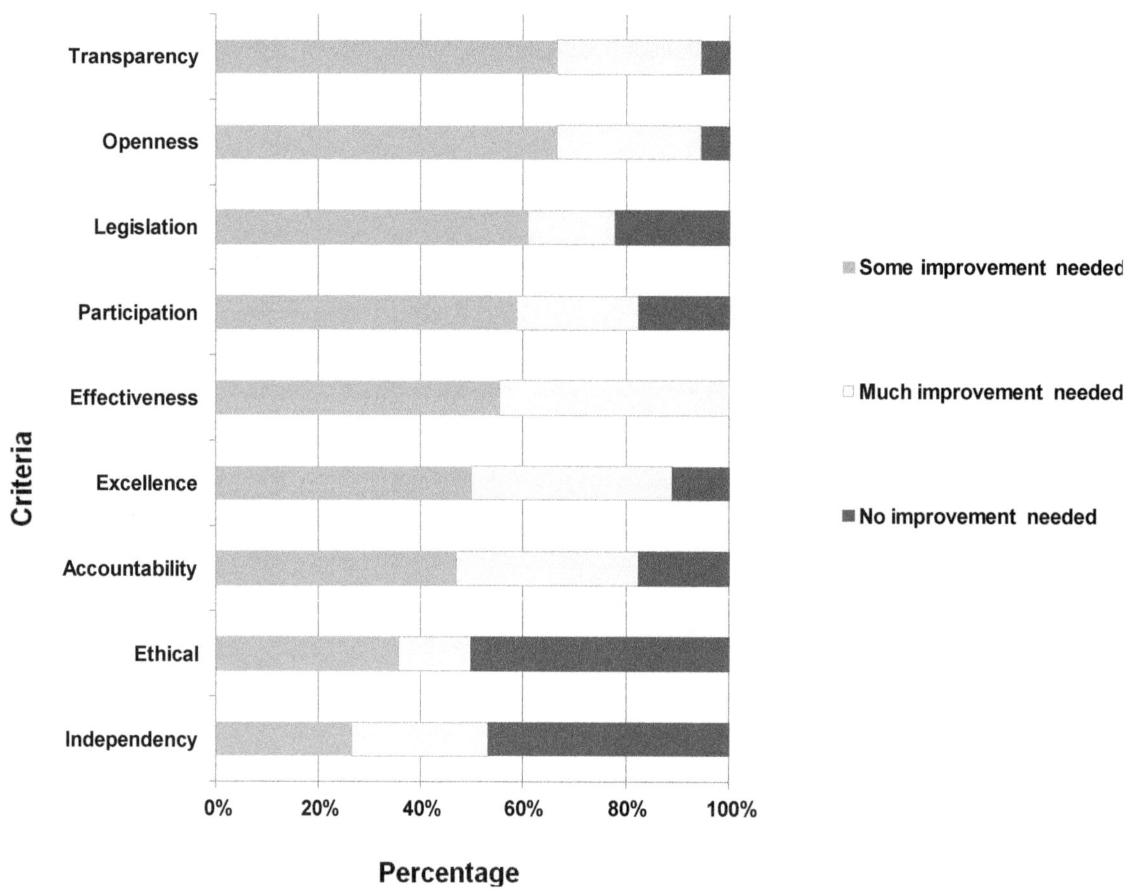

Figure 1. Responses to status of GMO risk governance sorted on 'some improvement needed'.

conflict, probably because names of the sub-committee and most of the AC members were not publically available (Figure 1; Table 1a). One respondent remarked that it was impossible for scientists not to have vested interests.

The AC has a general advisory function but also specific functions prescribed by the GMO AC (South Africa, 1997). Respondents gave strong support for specific functions in addition to reviewing of dossiers. These included advising on uncertainties and disagreements at EC level (Table 1b, 10.6, 100%), being pro-active by pro-

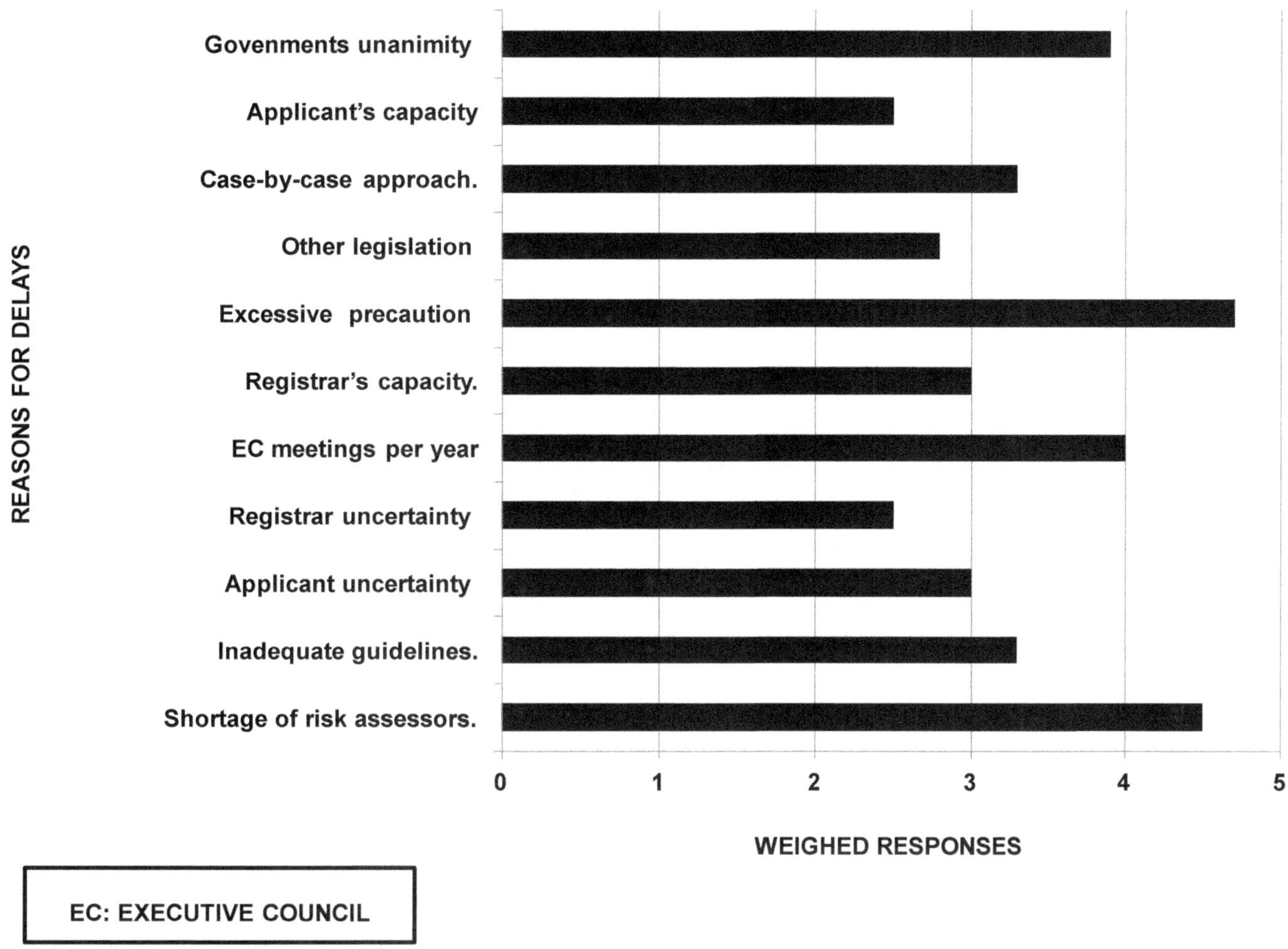

Figure 2a. Responses from the agricultural biotechnology industry / applicants on the reasons for delays in approval of GMO permits. (Weighed responses, ranking groups: 1 = unimportant to 5 = extremely important).

proposing specific activities and projects (Table 1b, 10.7, 91%), and initiating new policies pertaining to assessments such as guidelines (Table 1b, 10.9, 92%). Such pro-activity exists in the self-tasking functions of EFSA, (http://www.efsa.europa.eu). Availability of competent scientists could be a limiting factor for members of the AC as they are mostly employed full-time by universities or research institutes and though appointment to the AC is a prestigious position, their available time is limited. There seems to be a need for more risk assessors (reviewers) (Figure 2a).

Respondents highlighted the need for improved effectiveness in the regulatory system. Only a few procedures were touched on in this study e.g. managerial responsibility such as planning (Table 1a, 2) and decision making (Table 1b, 10). Effectiveness was reflected by the time taken to reach a decision on the issuing of a permit; approvals for commodity clearances were delayed from 2005 until 2011. A number of reasons for delays were given (Table 2a). There seemed to be inconsistency in requests for more information from different panels or similar information repeatedly requested, and requests for additional information not requested before. Excessive precaution in decision making, shortage of risk assessors, infrequent EC meetings, no unanimity between government departments on national policy; and inadequate guidelines were cited as main causes. This study did not consider for example the need for improvement of quality of dossiers and more training of applicants as often experienced in regulatory situations by one of us as a former regulator. An example of poor quality and incompleteness of a dossier was the case of a banana application (ACB, 2011). Effectiveness of the regulatory system needs addressing in much detail.

Scientific excellence

The quality of risk assessments appeared to require substantial improvement (Table 1a, 5). Codex guidelines were generally followed and the opinion was that international guidelines were being adhered to (Table 1a, 5.7, 5.9). It is possible to conclude that a reviewer with

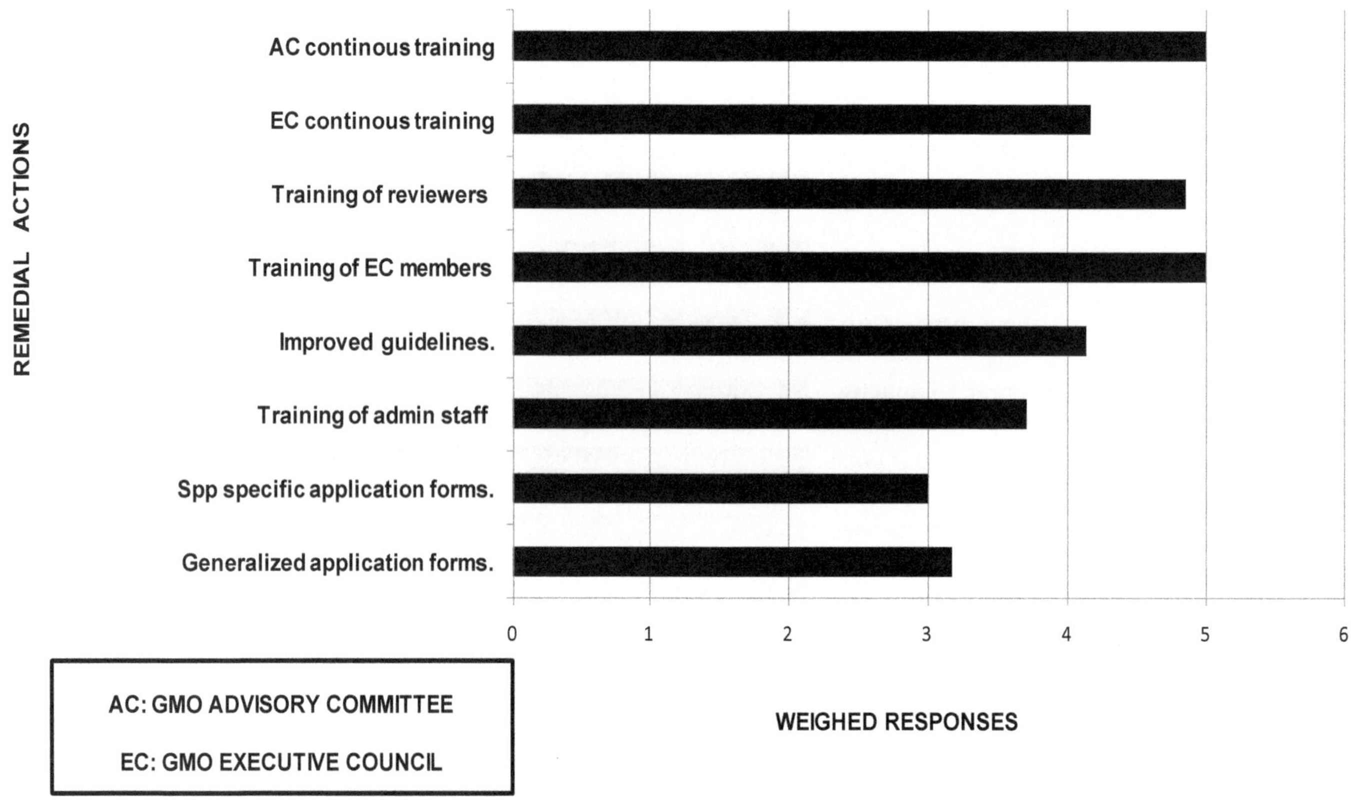

Figure 2b. Respondents from agricultural biotechnology industries/applicants on proposed remedial actions for delays in GMO permit approvals. (Weighed responses from ranking groups: 1 = unimportant to 5 = extremely important).

less experience tends to request much more additional information than an experienced reviewer. Reviewers undertake their tasks independently and therefore may form a variety of opinions. There is little opportunity for reviewers to discuss amongst themselves or for inexperienced reviewers to learn from their more experienced colleagues. In these circumstances, Codex guidelines may also be interpreted differently by different reviewers. It is also unclear what approaches to risk assessments are followed, the comparative risk analysis approach (CAC, 2009; Kuiper et al., 2001) or a toxicological risk assessment (Millstone et al., 1999). Excessive caution in decision making was identified as the main reason for delays in approvals (Figure 2a). Training of risk assessors (reviewers) and decision makers was considered as an important remedial action (Figure 2b).

All respondents agreed on the importance of peer review of risk assessment reports (Table 1b, 11.1) and that a single multi-disciplinarian could not replace a team of specialists (Table 1b, 11.2). The current procedure (South Africa, 2008) is to appoint, on a case-by-case basis, a panel of three reviewers from the AC subcommittee with a member of the AC as chairperson. Each reviewer's report and recommendations are included in a final report to the EC. Additional information may be requested from the applicant by the reviewers. It is not clear from the Guidelines (South Africa, 2008; DAFF, http://www.daff.gov.za) whether the full AC meeting would discuss or review the final report. The procedure seems to emphasize 'reporting' rather than 'reviewing'. The EC would consider the AC reports, public inputs, as well as reviews from the different government departments. This transferred final reviewing to the EC rather than the AC level. The current procedure reduces time taken for finalizing initial reviews but could also lead to inconsistent recommendations and requests for more information, causing delays, as identified in the second questionnaire (Table 2a). Adequacy of expertise within small panels to cover a number of disciplines such as both food and environmental issues in general release permits, may be a constraint. These aspects may be too broad for the small panel to cover although additional expertise could be co-opted. An option would be to conduct peer reviewing at the level of the full AC meeting where a greater diversity of expertise is present. Another option would be to conduct reviewing by focused sub-panels. Against this approach, one respondent remarked that 'there is also a need for overall multi-disciplinary understanding as opposed to over-specialized'. Peer reviewing needs to be critically investigated for optimization of this process.

Respondents agreed that a PhD-degree with at least three years of experience or an MSc-degree with at least five years of experience in relevant disciplines seemed adequate for eligibility for the AC. There was no consensus on any particular number of reviews that should have been handled before being nominated (Table 1b,11.6 to 11.8). Countries such as Brazil require a PhD-degree as a minimum qualification. That may not be a practical suggestion because of lack of capacity at this stage in SA.

The respondents identified the lack of sufficient adequately trained risk assessors as a problem (Table 1b, 5.1; Figure. 2a). One reason could be the poor remuneration which is considerably lower than that stipulated in the guidelines of the South African Council for Natural Scientific Professions (South Africa, 2003). Reasons could also include exposure to criticism or lack of interest or inadequate time allocated. Respondents recommended that a detailed roster of potential risk assessors should be kept (Table 1b, 11.5). Such a roster of expertise is a requirement of the Biosafety Clearing House (BCH,) established under the Cartagena Protocol on Biosafety to the Convention on Biological Diversity (CPB, http://www.cbd.int/biosafety/protocol.shtml). At present, no experts from SA are listed on the BCH website, despite many names having been submitted, making it difficult to source expertise. In addition, the latest Protocol records show that the Ministry of Environmental Affairs is the national focal point with only one communication record received, both this department and the Ministry of Agriculture are listed as competent authorities, and four records and 13 decisions (permit approvals) received from DAFF, though more than 350 permits are granted annually (DAFF, http://www.daff.gov.za). DAFF's clearing house link seems to be inoperative at times.

As a remedial option to address delays, respondents agreed on the need for AC members to have continued exposure to new information, to attend international meetings and conferences, and to be involved in research and new developments in relevant disciplines including risk assessment approaches (Table 1b, 11.3 and 11.4). Training of EC and AC members was regarded as essential to help reduce delays in the regulatory process (Figure 2b).

One respondent felt that assessors (reviewers) also needed 'contact and input from industry experts and practices'. Industry, especially the multi-national industry, has the means to employ world class specialists to conduct research in specialized areas'. Knowledge and experience of scientists from the industry may be considered in specific cases according to the Food and Agricultural Organization and World Health Organization's expert group (FAO/WHO) (2006).

International outreach programmes often focus only on regulatory authorities unless the need of risk assessors (reviewers) is brought to their attention. Where skills and experience are scarce, capturing of institutional memory collectively is important. This has not been well exploited in the SA risk analysis framework. Coordination in this respect is important. The majority of respondents felt that scientists who were members of the AC should be included in government delegations to international meetings/conferences. There seemed to be a general awareness of the importance of access to international scientific exposure and the need for improvement in this regard (Table 2a, 5.3 to 5.5). In a strategic review of Food Standards Australia New Zealand (FSANZ, 2003-2004), it was stated that credibility 'was founded on the quality of its science and its scientific risk assessments'. One of the recommendations was that the FSANZ continued to forge strong linkages with international experts and other regulatory agencies. FAO/WHO (2006) confirmed continuous exposure to international developments was important.

Transparency

The majority of respondents rated transparency as needing some improvement. The questionnaire tested a few pertinent issues, namely clear procedures; communication of uncertainty; and identification of the risk assessors (Table 1a, 9.1-3). The need was identified for some improvement in all three areas. Procedures, guidelines, application forms, and permits are adequately published on the website of the DAFF, but there appears to be a lack of transparency in EC decision making. The minutes of the EC meetings are too cryptic, suffer delayed listing and risk assessment recommendations are not available. The Promotion of Access to Information Act (South Africa, 2000a) that aims to 'foster a culture of transparency and accountability', provides at a nominal fee a legal opportunity to demand additional information. Compared with the extent of accessibility of reports in Australia (FSANZ, http://www.foodstandards.gov.au; http://www.ogtr.gov.au) and EFSA (EFSA 2006, 2009), the SA website information (DAFF, http://www.daff.gov.za) has limitations.

Participation

The majority of respondents agreed that stakeholder participation regarding the evaluation of the potential risk in the context of socio-economic impact was invaluable (Table 1b, 12.4). Participatory procedures are not new to the SA decision making but represent a new concept in the risk analysis process. In general, respondents indicated that significant improvement was needed for participation in the SA system, including participation in policy development and risk assessment (Table1a, 8). All respondents agreed that risk analysis policies should be developed in collaboration with stakeholders (Table 1b, 12.1) and that communication was important (Table 1b, 12.2). Opportunity for public input is included in the GMO

Act (South Africa, 1997) and should be exploited in a structured way.

In line with Article 26 of the Cartagena Protocol on Biosafety (Secretariat of the Convention on Biological Diversity, 2000), the GMO Act includes provision for socio-economic assessment as part of the risk analysis. This may include cost-benefit and risk-benefit comparisons but Section 5.1 in the regulations is extremely vague on required, but undefined consideration of the socio-economic impact on biodiversity, access to natural resources, cultural traditions, knowledge, and practices. The majority (75%) of respondents agreed that socio-economic analysis should take place prior to decision making (Table 1b, 10.14). At this point, 'opening up the governance process through public participation' (CEC, 2000) would be important. Stakeholders may contribute to 'democratic' decisions of the risk managers at some stage in the process. One respondent stated that, 'non-safety issues should be considered by the EC (mandate to review) and not the AC (focus on safety and science). Non-safety issues are relevant only to general release, not to contained or confined activities, which are experimental and short term'. Another respondent commented that 'socio-economic effects should have a minor influence on risk decisions'. Although socio-economic issues may be addressed as a requirement of the GMO Act (South Africa, 1997), little information regarding the procedures and requirements for socio-economic analysis is available.

Limitations of the study

(1) Statistical analysis of the survey, althoughthe ideal means of analysis, is not possible with small target groups. This is unfortunately the case with a very specialized field such as risk governance of GMOs in a country with limited resources. In an almost similar study by Wentholt et al., (2009) on the risk analysis of GMOs, only 33 out of 106 invited European respondents participated and 19 out of 60 international non-EU participants. They were from a range of professions and occupations. Government officials with intimate knowledge of the system would most likely not be inclined to participate. Their opinion, should they be officially requested, would most likely not be spontaneous. However, their experience in the process of risk assessment is of great importance to identify needs of applicants.

(2) One deficiency identified in the study, is that we did not obtain the opinion of regulators on the quality of applications and provision of required information.

(3) The responses should largely be considered as 'perceptions'. Experience of the respondents are real, perceived, or from hearsay. Perceptions are valid observations that can be changed by improving the system and need be taken into consideration in participative policy development, particularly on sociological matters.

(4) This study illustrates the opinions of some stakeholders at the time of the study. This may change with time for the same group. A group (different segment of the population) with priorities other than marketing of their products may have a different perception of credibility.

Consumers, in general, are not familiar with the technical side of regulatory matters and, therefore, a more simplified questionnaire will have to be designed for them.

(5) The study should be considered as preliminary as a more balanced opinion would include the experience of those intimately involved in SA risk governance of GMOs according to the mentioned criteria. They should include regulatory authorities, advisors to government as well as members of the AC.

Conclusion

This study tried to determine: How credible is governance of GMOs in SA as perceived by scientists with knowledge of the system?"

In general, respondents felt that 'some to much' improvement is necessary to ensure credibility. An analysis of the GMO Act, policy guidelines and delays in issuing of permits, as well as available information on the implementation of the Act, confirmed the perceptions. Based on these results, an indication is given where improvements are most needed to increase credibility of the system.

Although functional separation of risk assessment (AC) and decision making (EC) has been established in the SA legislation, the scope of decision makers' responsibilities needs be clearly defined as it seems as if the EC still functions according to intentions of the 'technocratic' model that may over-ride recommendations by the advisory body. This could cause increase in the workload and may be outside the expertise and mandate of some EC members, resulting in further delays in decision making.

A clearly identified deficiency was excellence in performance. This implied the need for improved review processes, and elimination of delays. The degree to which potential poor preparation of application forms contributed to delays could not be assessed due to lack of participation by regulators. Continued constructive exposure to new information, research and/or development as well as training of both decision makers and risk assessors were identified as priorities. Improved guidelines that address new challenges such as assessment of stacked traits need urgent attention.

Many aspects of transparency really need some improvement. 'Participation', in particular, as a democratic principle, has not yet been clearly developed in the SA governance of GMOs. This includes the contribution of stakeholders and scientists to framing of the assessments. Framing includes scientific inputs, policy development on various matters such as scientific approaches to risk assessments and socio-economic impacts and cultural considerations. New research on how to address participation by stakeholders at different stages of risk analysis needs attention.

Government has to consider development of integrated strategies and policies for good governance to achieve greater credibility of risk assessment of GMOs, not only of the risk assessment component itself, but also for all other components of governance. These should include:

(1) Harmonization between government departments in the approach to risk analysis, taking into account the National Strategy for Biotechnology for SA(South Africa, 2001) and including policy issues such as the approach to precaution.

(2) Policy on training, recruitment and remuneration of risk assessors (South Africa. 2003).

(3) An additional step could be consideredin the iterative process of risk analysis as described by Codex Alimentarius Commission for development of risk assessment policy that could give direction according to national strategy and good governance policies.

(4) The majority of the criteria of good governance could be grouped as communication related, transparency, participation, and openness. Therefore, improved communication is critical.

Although the credibility of the SA governance of GMOs could be improved, it has certain strengths that could be applied to other risk analysis systems. Functional separation of the risk assessment and decision making as contained in GMO legislation could be followed as an example to ensure scientific integrity; however, implementation of new legislation with respect to risk assessment such as for pesticides, should be critically considered with the necessary awareness of good risk governance to ensure credibility of decisions.

The availability of experienced risk assessors is a limiting factor in credible governance. The use of regional or sub-regional independent experts could be considered while in the meantime, educating and training of local candidate risk assessors should be a priority.

In general, the robustness of the governance of GMOs is reflected in the track record of safety since 1990, as no significant impact from accidents or adverse effects on humans or animals have been recorded. Trust and confidence will depend largely on the introduction of more democratic governance for both GMOs and other food safety matters.

ACKNOWLEDGEMENTS

We thank The National Research Foundation for providing funds, The Department of Statistics, University of Pretoria, a private biometrics consultant, Marie Smith, for the advice on statistical analysis and Cora Bezuidenhout for the reference management. We also wish to thank the respondents who shared their knowledge and perceptions with us and the reviewers of this manuscript.

REFERENCES

ACB - African Centre for Biosafety (2011).GM banana slips in South Africa: key issues and concerns.http://www.acbio.org.za/images/stories/dmdocuments/GM-banabas.pdf. Downloaded April 2012.

Atkins D, Norman JA (2009). A risk management perspective on the governance framework, in Food Safety Governance: integrating science, precaution and public involvement, Springer-Verlag: 207-238. See also UK-FSA -United Kingdom, Food Standards Agency (2006). FSA chief scientist annual report.www.food.gov.uk. Downloaded April 2012.

BCH - Biosafety Clearing House.http://www.cbd.int/biosafety/protocol/shtml. Downloaded March 2010.

CAC - Codex Alimentarius Commission (2011).Procedural manual.20th ed. Secretariat of the Codex Alimentarius Commission. Joint FAO/WHO Food Standards Programme. Rome: 105-113.http://www.codexalimentarius.org/procedures-strategies/procedural-manual/en/. Downloaded May 2012.

CAC - Codex Alimentarius Commission) (2009). Foods derived from modern biotechnology. Guideline for the conduct of food safety assessment of foods derived from recombinant DNA plants, CAC/GL 45-2003 as amended in 2008, FAO/WHO Food Standards Programme.2nd ed. Rome.ftp://ftp.fao.org/docrep/fao/011/a1554e/a1554e00.pdf. Downloaded May 2012.

CEC - Commission of the European Communities (2001). European governance: a white paper. COM (2001): 428. http://ec.europa.eu/governance/white_paper/en.pdf. Downloaded March 2010.

CEC - Commission of the European Communities (2000). White paper on food safety.COM(1999): 719. http://ec.europa.eu/dgs/health_consumer/library/pub/pub06_en.pdf. Downloaded March 2010.

CEC - Commission of the European Communities (2002). Laying down the general principles and requirements of food law, establishing the European Food Safety Authority and laying down procedures in matters of food safety.Regulation (EC) No 178/2002 of the European Parliament and the Council, 28 January 2002, OJEC.L31(1): 1-24.

Chanda R, Fincham R, Venter P (2010). A review of the South African food control system: Challenges of fragmentation. Food Contrl. 21(6):816-824.

DAFF (2006-2012). Minutes of the GMO Executive Council.South African Department of Agriculture, Forestry and Fisheries, Pretoria.http://www.daff.gov.za. Downloaded May 2012.

DAFF (2008).Guideline document for subcommittees to assist the advisory committee in terms of Section 11(2) of the Genetically Modified Organisms Act of 1997.http://www.daff.agric.za. Downloaded April 2012.

Dreyer M, Renn O (2009). Food safety governance: integrating science, precaution and public involvement, Springer Verlag:3-245.

Dreyer M, Renn O, Borkhart K, Ortleb J (2006). Institutional re-arrangements in European food safety governance: a comparative analysis, in Food Safety Regulation in Europe: A Comparative Institutional Analysis, Series Ius Commune, Intersentia Publishing, Antwerp. pp. 9-64.

EFSA - European Food Safety Authority (2006). Transparency in risk assessment carried out by EFSA: guidance document on procedural aspects. EFSA J. 353:1-16.

EFSA - European Food Safety Authority (2009).Transparency in risk assessment - scientific aspects. Guidance of the Scientific Committee on transparency in the scientific aspects of risk assessment carried out by EFSA, pt 2: General principles. EFSA J. 1051:1-22.

EFSA - European Food Safety Authority.http://www.efsa.europa.eu. Downloaded April 2012.

EU - European Union (2004). European Union 6th Framework Project: SAFE FOODS.http://www.safefoods.wur.nl. Downloaded April 2012.

FAO/WHO - Food and Agricultural Organization and World Health Organization of the United Nations (2006). Food safety risk analysis: a guide for national food safety authorities, FAO Nutrition paper no. 87, Rome.

FSANZ - Food Safety Australia New-Zealand (2003-2004). Annual report and FSANZ Science strategy.http://www.foodstandards.gov.au/scienceand_education_/.Downloaded April 2012. See also Bansemer A (2003). Food standards Australia New Zealand, a strategic review.http://www.foodstandards.gov.au/aboutus/strategicreview.cfm. Downloaded February 2005.

Jaffe G (2008). Ensuring Biosafety at the National Level: Suggestions to Improve the Operation of the South African Biosafety Regulatory System, Brief number 11, PBS, Intl Food Policy Inst. http://www.ifpri.org/publication/ensuring-biosafety-national-level. Downloaded March 2010.

James C (2011) Global status of commercialised biotech/GM crops: 2011. ISAAA Brief number 43, ISAAA: Ithaca, NY: 129-35.

Kuiper HA, Kleter GA, Noteborn HPJM, Kok EJ (2001). Assessment of the food safety issues related to genetically modified foods. Plant J, 27(6):503-528.

Meghani Z (2009). The US Food and Drug Administration, Normativity of risk assessment, GMOs and American democracy. J. Agr. Environ. Ethic.22: 125-139.

Millstone E (2007). Can food safety policy-making be both scientifically and democratically legitimated? If so, how? J. Agr. Environ. Ethic. 20(5): 483-508.

Millstone E, Brunner E, Mayer S (1999). Beyond substantial equivalence.Nature.401(6753): 525-526.

Millstone E, van Zwanenberg P, Levidow L, Spök A, Hirakawa H, Matsuo M (2008).Risk-assessment policies: differences across jurisdictions, EUR 23259 EN. Joint Research Centre, European Commission. http://publications.jrc.ec.europe.eu/. Downloaded March 2010.

Morris EJ (2011). A semi-quantitative approach to GMO risk-benefit analysis. Transgenic Res. 20:1055-1071.

OGTR - Office of the Gene Technology Regulator, Department of Health and Ageing, Australia. http://www.ogtr.gov.au. Downloaded April 2012. See also FSANZ.

Paarlberg RL, Gruhn P, Goletti F, Yudelman M (2000). Governing the GM crop revolution: policy choices for developing countries.Intl.Food Policy Res. Inst. http://www.ifpri.org/sites/default/files/pubs/2020/dp/2020dp33.pdf.Downloaded April 2012.

Renn O (2008). Risk governance: coping with uncertainty in a complex world. Earthscan/James & James:1-443.

Secretariat of the Convention on Biological Diversity (2000).Cartagena Protocol on Biosafety to the Convention on Biological Diversity: text and annexes,Montreal. http://www.cbd.int/doc/legal/cartagena-protocol-en.pdf. Downloaded May 2012.

South Africa (1972).Foodstuffs, Cosmetics and Disinfectants Act of 1972, updated by Act No 39, 2007, Government Gazette 32012, 20 March 2009. http://www.daff.gov.za. Downloaded April 2012.

South Africa (1997).Genetically Modified Organisms (Act No. 15, 1997), Government Gazette 18029: 323, May 1997; as amended (Act No 23, 2007), Government Gazette 29803: 502, April 2007; and Regulations (2010), Government Gazette 32966: 120, 26 February 2010. http://www.daff.gov.za. Downloaded April 2012.

South Africa (2000a). Promotion of Access to information (Act no. 2, 2000), Government Gazette 20852/416, 3 February 2000.http://www.justice.gov.za. Downloaded April 2012.

South Africa (2000b). Promotion of Administrative Justice (Act No.3, 2000). Government Gazette 20853/416, 3 February 2000. http://www.justice.gov.za. Downloaded April 2012.

South Africa (2001).A national Biotechnology strategy for South Africa, Department of Science and Technology.http://www.pub.ac.za/pdfs/biotechstrategy_2002.pdf. Downloaded April 2012.

South Africa (2003).Natural Scientific Professions Act (Act No. 27, 2003). Government Gazette 25774/461, 28 November 2003; and, recommended consultation fees, Board Notice 43, 2011. Government Gazette 34072, 2011.http://www.sacnasp.org.za. Downloaded April 2012.

South Africa (2004).National Environmental Management Biodiversity Act (Act No. 10, 2004), ch5 pt3 s78, Government Gazette 26436: 467, 7 June 2004: 62-64. http://www.environment.gov.za. Downloaded April 2012.

South Africa (2008).Consumer Protection Act (Act No. 68 of 2008), Government Gazette 32186/526, 29 April 2009; and Regulations (R 293, 1 April 2011). Government Gazette 34180/550 of 1 April 2011. http://www:thedit.gov.za. Downloaded April 2012.

Stirling EA, Dreyer M, Renn O, Vos E, Wendler F (2009). The need for change, in Food safety governance: integrating science, precaution and public involvement, Springer-Verlag, Berlin.

Wentholt MTA, Rowe G, König A, Marvin HJP, Frewer LJ (2009). The views of key stakeholders on an evolving food risk governance framework: results from a Delphi study. Food Policy 34:539-548.

Human immunodeficiency virus type-1 (HIV-1) genetic diversity and prevalence of antiretroviral drug resistance mutations in treatment-naïve adults in Jos, North Central Nigeria

Anejo-Okopi J. A.[1], Agbaji O. O.[1,2], Agaba P. A.[1,3], Ugoagwu P. O.[1], Were K.[4], Onywera H.[4], Owiti P.[4], Isa S. E[1,2], Otecko N.[4], Okwori A. E. J[5], Musa J.[1,6], Oguche S.[1,7], Sagay A. S.[1,6], Idoko J. A[1,2], Nimzing L.[1,8], Jatau E. D.[9] and Olonitola, O. S.[9]

[1]AIDS Prevention Initiative in Nigeria (APIN), Jos University Teaching Hospital, Nigeria.
[2]Department of Medicine, University of Jos/Jos University Teaching Hospital, Nigeria.
[3]Department of Family Medicine, University of Jos/Jos University Teaching Hospital, Nigeria.
[4]HIV-R Laboratory, Kenya Medical Research Institute, Kisian Kisumu, Kenya.
[5]National Veterinary Research Institute, Vom, Nigeria.
[6]Department of Obstetrics and Gynaecology, University of Jos/Jos University Teaching Hospital, Nigeria.
[7]Department of Paediatrics, University of Jos/Jos University Teaching Hospital, Nigeria.
[8]Department of Medical Microbiology, University of Jos, Nigeria.
[9]Department of Microbiology, Ahmadu Bello University Zaria, Nigeria.

The presence of human immunodeficiency virus (HIV) type-1 diversity has an impact on vaccine efficacy and drug resistance. It is important to know the circulating genetic variants and associated drug-resistance mutations in the context of scale up of antiretroviral therapy (ART) in Nigeria. The objective of this study was to determine the genetic diversity of HIV-1 and the prevalence of antiretroviral (ARV) drug resistance mutations among antiretroviral treatment-naïve HIV-1 infected patients in Jos, North Central Nigeria. Plasma samples were collected from 105 ARV drug-naïve patients enrolled for HIV care at the Jos University Teaching Hospital (JUTH) HIV Treatment Center between October 2010 and April 2011. One hundred (100) samples were successfully amplified. Viral subtyping was done using REGA subtyping tool and by phylogenetic analysis using PAUP software. The drug resistance mutations were determined using the Stanford University HIVdb sequence interpretation algorithm. HIV-1 subtypes identified were; CRF02_AG (48.0%), G (41.0%), CRF06_cpx (6.0%) and A1 (5.0%). 8% of the patients' isolates had at least one major resistance mutation in the *RT* gene: Nucleoside reverse transcriptase inhibitors: M41L (1%), K65KR (1%), M184IM (1%), M184V (2%) and T215ADNT (1%), non-nucleoside reverse transcriptase inhibitors: K103N (2%), K101E (1%), G190A (1%), P225HP (1%), Y181I (1%), Y188L (1%), and Y181C (1%). Among antiretroviral (ARV) naïve patients in Jos, North Central Nigeria, the common HIV-1 subtypes was CRF_02 and G. And the prevalence of drug resistance mutations was found to be high (8%). Further study and national surveillance will be critically important to understand the clinical impact of transmitted resistance mutations on ART naïve individuals in resource limited settings.

Key words: HIV-1 subtypes, antiretroviral (ARV), treatment-naïve, drug-resistance, mutation, accessory and polymorphisms, Nigeria.

INTRODUCTION

Most human immunodeficiency virus type-1 (HIV-1) strains infecting individuals in the Americas, Europe and Australia belong to subtype B, whereas the majority of the remaining viral subtypes are found among persons in sub-Saharan Africa (Robertson et al., 2000). The genetic diversity within HIV-1 presents a challenge for global management of HIV infection even as antiretroviral (ARV) drugs become increasingly available. The use of highly active antiretroviral therapy (HAART) in developed countries has led to a marked reduction in the mortality rate among HIV infected patients. Until recently, access to life-saving ARVs was limited in low and middle-income countries. However, the introduction of affordable ARV combinations through local and international efforts has led to a significant increase in access to ARVs. Treatment failure is mostly related to emergence of drug-resistant variants and its transmission to uninfected individuals raises serious clinical concerns for initiation of therapy. The prevalence of resistant mutations in newly infected individuals range between 10 to 25% in Europe and the Americans (Wensing et al., 2005), while in developing countries like Nigeria, where access to ARVs is increasing, resistance in the treatment-naïve HIV-1 infected is rarely reported (Vergne et al., 2006). Drug resistance arises from mutations in the *pol* (protease-PR and reverse transcriptase-RT) genes that encode the molecular targets for the drugs. The viral polymorphisms are due to the high rate of HIV-1 replication, low fidelity of RT (Wei et al., 1995), recombination (Nukoolkarn et al., 2004) and variants accrual during the infectivity period (Hirsch et al., 1998). The emergence of amino acid substitutions associated with resistance can be classified as major (primary) and accessory (secondary) mutations. Some major mutations may lead to several fold decrease in sensitivity to one or more antiretroviral drugs by single (mutation) appearance. Accessory mutations may not result in a significant decrease in sensitivity but are associated with an increase in viral fitness. Thus, the appearance of major mutation in a genome already containing accessory mutations could influence the speed with which highly resistant viruses are selected during therapy. Studies have revealed that HIV-1 subtypes A1, CRF02-AG and G predominate the epidemic in Nigeria (Chaplin et al., 2011; Ajoge et al., 2012). Given the increasing genetic heterogeneity of HIV-1 in West Africa and the increasing use of antiretroviral drugs, it is important to characterize *pol* gene sequences from HIV-1 subtypes occurring among drug-naive populations. Recent studies have documented that Nigeria has a prevalence of 1.6% in untreated patients (Hamers et al., 2011).

Nigeria has the second highest burden of HIV infection in the world, second only to South Africa. The report of the 2010 HIV survey showed that the North Central Zone had the highest HIV prevalence (7.5%) of all the six geo-political zones in the country (GARPR, 2012). The current number of Nigerians on ARVs has been estimated to be 620,000 (personal communication). This figure has not been disaggregated by geo-political zone; hence the number of persons on ARVs in north central zone in currently not known. However, with the rapid scale up of ARV use and limited infrastructure for care and monitoring, Nigeria may face an increase in the emergence of HIV-1 resistant strains among the drug-naïve population. In this study, we characterized HIV-1 subtypes and determined prevalence of drug resistance mutations in treatment-naïve patients in Jos, North Central Nigeria.

MATERIALS AND METHODS

Patients

This study was carried out at the AIDS Prevention Initiative in Nigeria (APIN) supported HIV clinic at the Jos University Teaching Hospital (JUTH), Jos. The entry point for all patients was either through HIV counseling and testing (HCT) or referred HIV positive patients from other services within the hospital and the surrounding community. One hundred and five (105) HIV-1 infected treatment-naive patients were recruited sequentially after obtaining informed consent between October 2010 and April 2011. The Jos University Teaching Hospital Ethics Committee approved the study protocol. A questionnaire was used to collect basic demographic data from each study participant. Criteria for inclusion in this study was patients who had no previous ARV exposure and were aged 18 years and above. Blood samples were collected in athylenediaminetetraacetic acid (EDTA) lined containers and plasma was extracted and cryopreserved. The samples were subsequently shipped in ice parked containers to the Kenya Medical Research Institute HIV-Resistance Laboratory, Kisian Kisumu, where genotypic testing using In-house Genotyping System was done. Out of 105 samples tested, 100 were successfully amplified for genotypic drug resistance testing

HIV-1 genotyping

Isolation and amplification of HIV-1 partial pol gene by RT-PCR and nested PCR

Two amplification protocols for HIV *pol* gene were used as previously described (McNulty et al., 2007). Adhering to a unidirectional flow, the HIV viral RNA was extracted from plasma using the QIAamp Viral RNA mini kit (Qiagen, Hilden, Germany) followed by reverse-transcription of the RNA into cDNA using the outer primers Prt-F1-forward (2253-2275 nucleotides, nt) and RT-R1 reverse (nt 3370-3348) for RT-PCR. The cDNA were amplified by nested-PCR with primer Prt-F2 (forward, 2265-2288 nt) and RT-R2 (reverse, 3326-3304 nt). The amplified DNA fragments from nested-PCR were verified by visually comparing the intensity of each sample's band to that of the DNA mass ladder's bands of known DNA quantity for expected size by electrophoresis in 1.0% agarose gel stained with 0.5 µg/ml ethidium bromide and photographed under ultraviolet illumination. The fragments were purified using the QIAquick PCR purification kit (Qiagen, Hilden Germany) in spin columns, and direct population-based sequencing performed on both strands using BigDye® Terminator v3.1 Cycle Sequencing kit on an automated ABI 3130X1 Genetic Analyzer (Applied Biosystems).

Sequence analyses and determination of drug resistance mutation

The generated nucleotide sequences were viewed using the Sequence Analysis Software v3.7, and aligned and edited using Sequencher version 5.0, which assembles the six overlapping sequence segments for the six primers to form a contiguous sequence. Sequences with frame shifts or stop codons were excluded from analysis. The quality of the generated sequences was checked using Sequence Quality Assessment Tool (SQUAT). The sequences in fasta format were then subjected to the Stanford HIVdb algorithm (http://hivdb.stanford.edu/) for subtyping and determination of mutations conferring various antiretroviral drugs resistances. Mutations in the sequences were defined as differences from the consensus B reference sequence and were further characterized as RTI and PI associated resistance mutations.

Subtyping and phylogenetic analyses

HIV-1 subtyping was performed using REGA HIV-1 subtyping tool V2.0 from Stanford HIV drug resistance database (http://hivdb.stanford.edu/), a worldwide subtype references were obtained from Los Alamos database (http://hiv-web.lanl.gov), and sequences were aligned against the known reference strains. Phylogenetic inferences were performed by the neighbor-joining (NJ) method as implemented in PAUP* version 4.0 (Swofford et al., 2002).The reliability of the tree topology was tested with 1000 bootstrap analyses and bootstraps of ≥70% were considered significant. The bootscanning method was used to detect and study recombination, as implemented in the SIMPLOT software, version 2.5 (Saminen et al., 1995). Figure 2 shows a rooted tree of the 84 selected pol sequences for clarity with group O sequence O.CM.91 as an out group. The recombination was further confirmed with Recombination Identification Program (RIP) version 3.0 available online (http://www.hiv.lanl.gov/content/sequence/RIP/RIP.html). Subtypes assignments by this method are shown to the right of the tree with isolates clustered around the various references.

RESULTS

Baseline characteristics of the study population

A total of 105 patients were sequentially recruited and categorized. Samples for five patients could not be amplified and they were excluded from analysis. The mean age (years) was 37±9.5. The proportion of women and men were 55 and 45%, respectively. 98% of the patients reported a heterosexual mode of transmission while 2% was via blood transfusion. None of the patients were men who had sex with men (MSM) or injection drug users (IDU). 66% of the participants resided in Plateau state, while 34% resided in other states. 69% were married, 19% were single (never married), while 8% were separated/divorced and 5% were widowed. 11, 22, 36 and 32% of the subjects had no formal education, elementary, secondary and tertiary education respectively. WHO disease staging of the subjects was: stage I, 24%; II, 21%; III, 36%, and IV, 19%. 50% of the spouses of the participants were HIV positive, while 50% were HIV negative. The participants whose spouses were on ARV were 14%, and those whose spouses were not on ARV were 86%. The median baseline CD4 cell count was 186 cells/mm^3 (IQR: 12-737).

Phylogenetic analyses of the PR and RT Sequences using PAUP v. 4.0

Figure 1 summarizes the distribution of different genetic subtypes in the *pol* region using 21 references from Los Alamos HIV Sequence database. Figure 2 shows results of the phylogenetic analyses of the sequences and illustrates the different clusters within the isolates. The subtypes AG and G clustered around references from Nigeria and other West and Central African neighboring regions. Phylogenetic analysis of the partial pol sequences showed that 48/100 (48%) of the isolates were CRF02_AG, 41/100 (41%) subtype G, 6/100 (6%) CRF06_ cpx and 5/100 (5%) subtype A1 (Figure 1).

Reverse transcriptase sequence variability of isolates

After the analysis of PR and RT sequences, 8% of the patients (8/100) had at least one resistance mutation (95% CI: 3.9 to 16.2%). The distributions of the subtypes and profiles of drug-resistance mutations according to drug classes (Table 1), and the mutations conferring to resistance to NRTIs included M41L (1%), K65KR (1%), M184IM (1%), M184V (2%) and T215ADNT (1%), while mutations that conferred resistance to NNRTIs comprised K101E1%), G190A1%), P225HP (1%), Y181I (1%), Y188L- (1%, K103N (2%) and Y181C (2%) (Table 2). Of the eight patients, seven females harbored viruses with drug resistance mutations to both RT inhibitors. Of four individuals that harbored multiple mutations, three of them were females (Table 1). The frequency of the accessory mutations in the RT region are as shown in Figure 3 and were as follows; R211K (65%), V179IE (8%), G196EG- (8%), V118I (5%), V106I (5%), A98SG (2%), T69N-(2%), L228HL (2%), and E44DE (1%), V90I (1%), E138A (1%) and L210M (1%).

Protease sequence variability in isolates

There was very low prevalence of major mutations detected in the non-B strains to protease inhibitors from

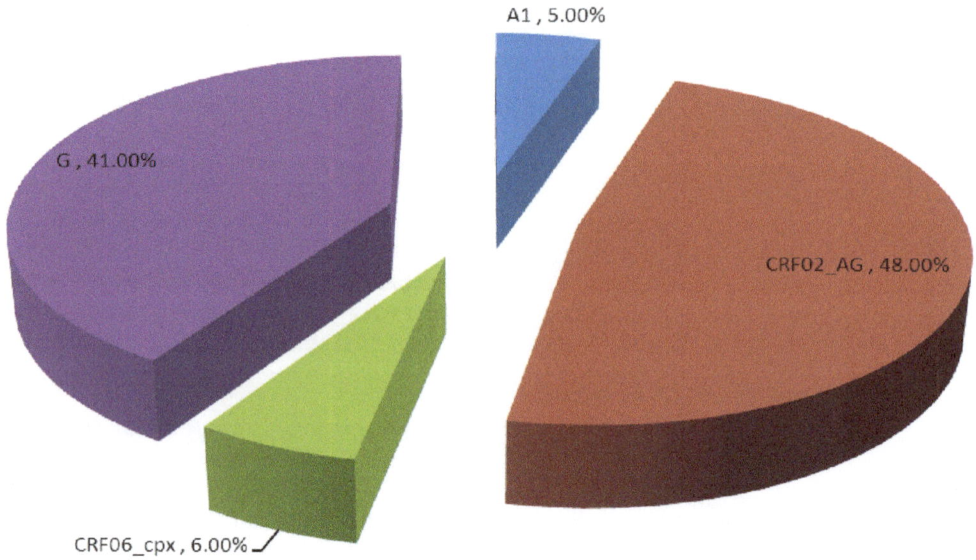

Figure 1. Analysis of HIV-1 genetic diversity of the *pol* gene. Phylogenetic analysis of the PR and RT sequences revealed distribution of HIV-1 subtype's diversity circulating among ARV treatment-naïve patients. The prevalence of observed subtypes are: 48% of the sequences were subtype CRF02_AG, 41% were subtype G, 6% subtype were CRF06_cpx, and 5% subtype A1. Subtypes AG and G were predominant in circulation than subtypes CRF06_cpx and A1.

the 100 samples. In contrast, many minor mutations were found at the following positions, in order of decreasing frequency (Figure 3); K20VMI (97%), M36I (94 %), E35DEGQ (42%), V82IV (40%), L63P (25%), L10I/V (20%), V77I (5%), I62V (3%), L10M (1%), V11LV (1%), L33F (1%), Q58E (1%) and L90V (1%). The single mutations observed were; L33F, L90V, Q58E, V11LV, and L10M. The substitution of valine at codon 90 and methionine at codon 10 against the known amino acid with subtype B was rare to non-B subtype. The study observed that each sequence had at least one known accessory PI drug resistance mutation (Figure 4).

DISCUSSION

Primary HIV drug resistance represents a challenge for the treatment of HIV infection because it can reduce the efficacy of first-line ART and has impact clinical outcomes. It is established that primary drug-resistance will emerge in the region where ART has been widely available for years (Shekelle et al., 2007), and that HIV-1 subtypes have been associated with drug resistance mutations. We analyzed the PR and RT gene-coding regions of 100 HIV-1 isolates from treatment-naive patients. This is the largest number of isolates among treatment naïve patients studied so far in North Central Nigeria. Sixty-three (63%) of our patients had CD4 count of ≤200 cell/mm^3, which may suggest long-standing infection and were therefore eligible to commence antiretroviral treatment using the existing national treatment guidelines at the time of the study.

The WHO disease staging was also evidence of advanced infection as majority of the patients were classified in stages III and IV. The frequency of those whose spouses were on ARV was 14% and this may be explained by the fact that patients enrolled in the ARV Centre upon disclosure of their HIV serostatus invite their spouses for provider-initiative HIV counseling and testing in the program if their HIV status turns out to be positive.

Interestingly, the most predominant strains circulating in the zone were subtype CRF02_AG with proportion of 48% followed by subtype G-41%, CRF06_cpx 6% and subtype A-5%, which were comparable with previous studies in Nigeria, Chad, Niger and Cameroon (Peeters et al., 2000; Vidal et al., 2003; Chaplin et al., 2011). The subtype AG clustered around the isolates from Ibadan, Nigeria and Cameroon. The high prevalence of CRF02_AG and G may explain differences in infectivity, elevated viral replication suggesting a higher rate of transmission. The fact that HIV-1 subtype G was observed with multiple drug resistance mutations could suggest higher rates of mutation and replicative capacity within the studied population. The diversity of HIV-1 viruses in Nigeria is becoming more evident, although the prevalence of subtype AG and G appears to be on the rise.

The presence of resistance mutations is a central

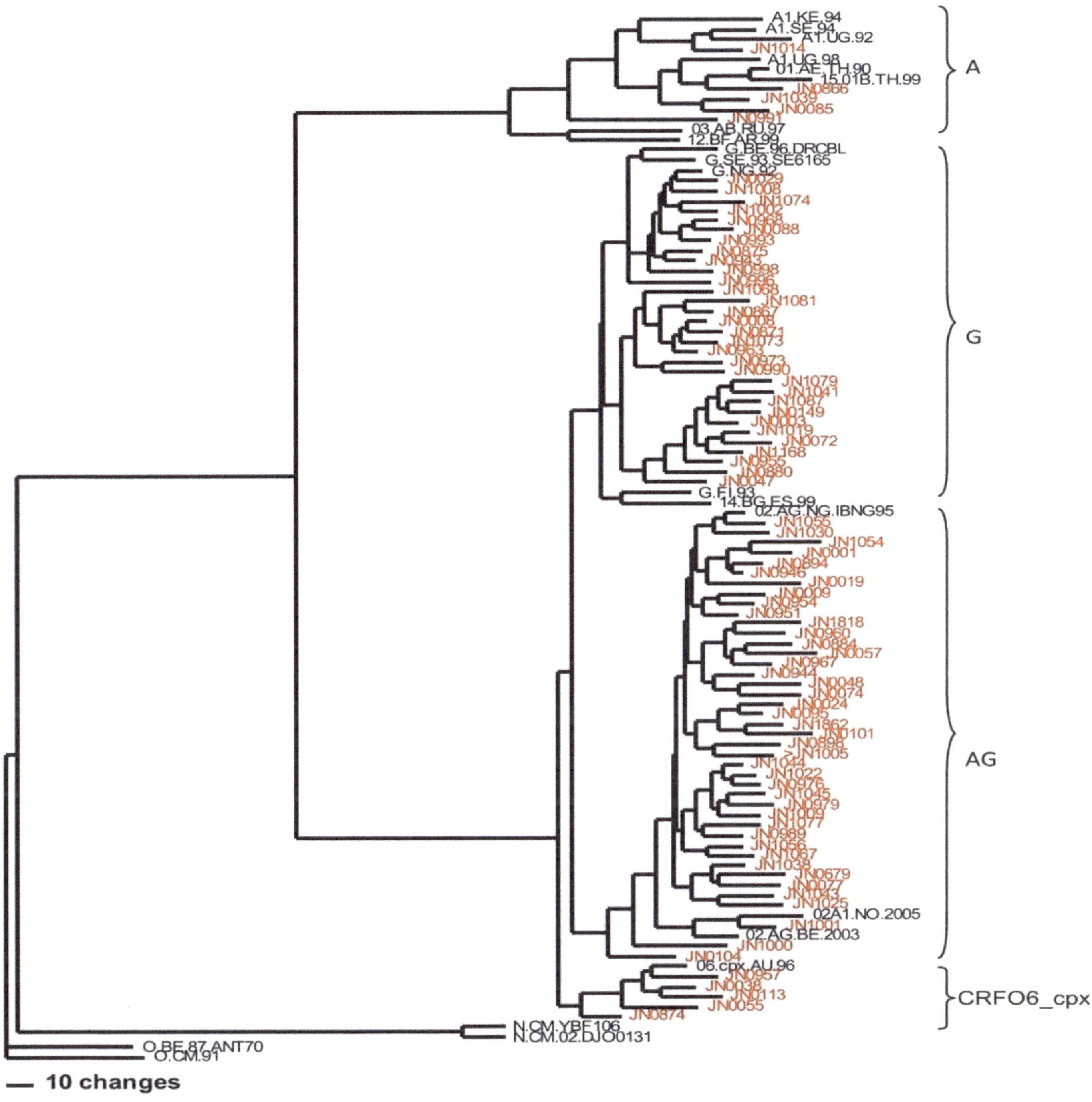

Figure 2. Phylogenetic analysis of Isolate from antiretroviral treatment-naïve patients. Neighbor– Joining (NJ) Rooted Tree of selected 84 HIV-1 isolates colored red by prefix "JN" with 21 aligned sequences (coloured black) from Los Alamos HIV database using PAUP version 4. Clustering of HIV-1 isolates among ARV treatment-naïve patients: the HIV-1 pol gene sequences were compared including Nigerian sequences obtained from Los Alamos HIV sequence data base. The subtype CRF02_AG and G clustered significantly around sequences from Nigeria (02.AG.NG.IBNG95 Ibadan isolates), Belgium (02.AG.BE.2003) and Jos Isolate (G.NG.92) respectively. Subtypes A1 and CRF06_cpx clustered around references from Thailand (15.01B.TH.99), Uganda (A1.UG.98) and 06.cpx.AU.96 respectively. Isolates JN1000 and JN0104 did have an earlier branching compared to other AG isolates suggesting an early earlier existing transmitted isolates. An additional pruning analysis, which consists of removing respective reference sequences from an alignment and rerunning the phylogenetic analysis, revealed that our CRF02_AG, subtype G sequences all maintained their distinct clustering patterns even when their respective reference sequences were absent. This further confirms the stability of the tree topology and subtype assignments. In addition, the subtype A and CRF06_cpx sequences clustered separately, and among themselves from the crown groups of AG and G viruses. Taken together, these data imply that CRF02_AG, G subtype A1, and CRF06_cpx have been in circulation for some time.

Table 1. Distribution of HIV-1 subtypes and mutations conferring resistance to NRTIs and NNRTIs among treatment-naïve patients.

PTID	Age	Sex	CD4 Cell count	Duration of HIV-1 Infection (AI)	HIV-1 Subtype	Spouse on ARV	RT major mutation	
							NRTI	NNRTI
JN1041	36	F	18	LI	G	Yes	M184IM	K101E, Y181I
JN0055	31	F	31	Recent	CRF06_cpx	No	-	Y188L
JN0679	34	M	93	LI	CRF02_AG	No	K65KR	Y181C
JN0866	27	F	21	LI	A1	Yes	T215ADNT	-
JN0990	39	F	272	LI	G	Yes	M184V	K103N, P225HP
JN0088	45	F	389	LI	G	Yes	M184V	Y181C, G190A
JN0110	29	F	237	LI	CRF02_AG	No	M41L	-
JN0114	43	F	21	LI	CRF02_AG	No	-	K103N

PTID, Patient identity; AI, avidity index; LI, long infection; ARV, antiretroviral; RT, reverse transcriptase; NRTI, nucleoside reverse transcriptase inhibitor; NNRTI, non-nucleoside reverse transcriptase inhibitor.

Table 2. Frequency of NRTI and NNRTI drug resistance mutations among treatment-naïve patients.

NRTI mutation	Number	%	NNRTI	Number	%
M41L	1	1.00	K101E	1	1.00
K65KR	1	1.00	K103N	2	1.00
M184IM	1	1.00	Y181I	1	1.00
M184V	2	2.00	Y181C	2	2.00
T215ADNT	1	1.00	Y188L	1	1.00
			G190A	1	1.00
			P225HP	1	1.00

NRTI, nucleoside reverse transcriptase inhibitor; NNRTI, non nucleoside reverse transcriptase inhibitor. Frequency of major NRTI and NNRTI mutations are reported in the total study population of 100 patients

problem to treatment efficacy. Of 100 sequenced samples, 8% of the subjects harbored at least one primary mutation that confers resistance to either NRTI or NNRTI (Table 1). The high prevalence of NNRTIs resistance among cases with transmitted resistance is a potential problem as the national guidelines recommends efavirenz based HAART as a preferred first-line therapy. NNRTI is known to have low genetic barriers, with which a single mutation may confer high level resistance to more than one drug in the class (K103N, Y18C/I, Y188L). Multiple major RT mutations were observed in four patients. Among the major drug-resistance mutations, the most common multiple mutations observed were NNRTI resistance mutations as observed in three three female patients (Table 1); this may suggest transmitted resistance since their spouses were known to be on ARV drug. The presences of T215ADNT mutation which is a revertant, suggests a previous infection with HIV-1 strains containing T215Y/F or represents a transition on the reversion from T215F/Y.

The observed prevalence of 8% ARV drug-resistance mutations in drug naïve patients is in contrast to 1.6% prevalence previously reported (Hamers et al., 2011). This study reveals that most individuals had long established infection (89% data not shown), although the exact time of infection is unknown. Therefore, the observed prevalence of resistance may have been influenced by the long duration of HIV-1 infection and another possibility, is that high TDR is expected in patients whose spouses have poor adherence to ART or spouses who have multiple sexual contacts other than their spouse. Theoretically, newly infected patients are more likely to harbor more resistant viruses than chronic or long established infection because transmitted resistance may have reversed after a while without drug pressure (Perelson et al., 1996). This might be explained by the fact that wild type virus may become predominant in the absence of selective drug pressure when individuals are infected with both wild type and mutant viruses. However, several studies have demonstrated the persistence of mutations of major resistance for periods as long as five to seven years (Barbour et al., 2004). The observed multiple primary resistance mutations were common in subtype G suggesting low genetic restrictions. Also identified was the replacement of leucine (L) with methionine (M) instead of tryptophan (W) at drug

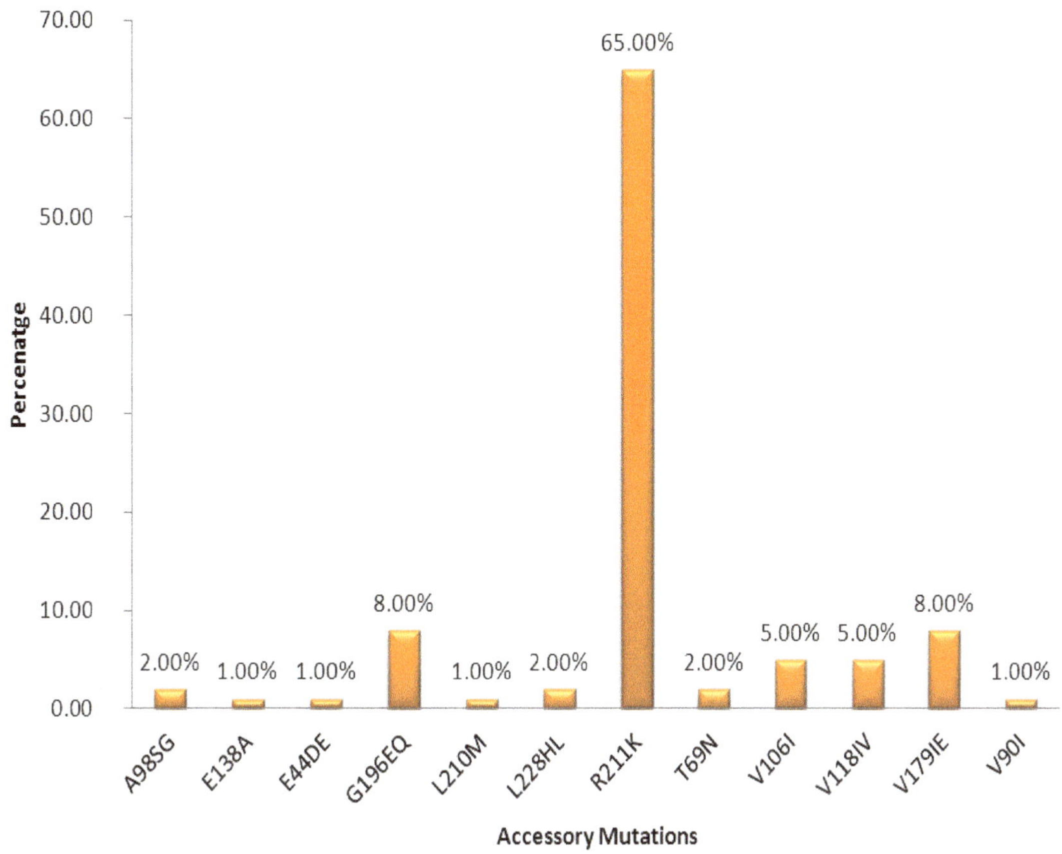

Figure 3. Prevalence of RT (NRTI and NNRTI) accessory mutations among the ARV treatment naïve patients. This figure shows the result of the sequence analysis blasted into the Stanford HIV consensus drug resistance database reveals some accessory (minor) mutations that confer some form of resistance to antiretroviral drugs. In NRTI the highest prevalence of amino acid substitution was observed in position R211K (65%), followed by Q196EQ (8%), L228HL (2%), A98S (2%), and E138, E44DE and L210M as 1% each. The frequency of amino acid position in NNRTI are in order of decrease; V179IE (8%), V106 and V1185 (5%), T69N (2%), and V9011 (1%).

resistance position 210W; which is a rare finding. This mutation could be a reversion from thymidine analogue mutation-II (TAM) L210W mutation which confers high level resistance to some NNRTIs (stavudine and zidovudine).

The frequency of (TAM) M41L was low with only 1% of patients having virus with any TAM, and is associated with most nucleoside analogues except lamivudine. Overall, the protease gene region was less conserved than the RT gene because of its genome flexibility. Again, protease can afford to retain its altered dynamics via the compensatory mutations (Arts and Haduza, 2012). However, accessory mutations in the protease gene are known to reduce "genetic barrier" to drug resistance by increasing the selection of the resistant strains in the presence of other mutations (Wainberg and Brenner, 2012). Several other accessory mutations that are not classically associated with high level resistance were observed in *RT* gene; E44DE (1%), V90I (1%), A98SG (1%), E138K (1%), L210M (1%), L228HL (2%), T69N (2%), V118I (5%), V106I (5%), G196EG (8%), 179IE (8%) and R211K (65%). These mutations by themselves do not have a substantial effect on phenotype but have contributory role in the development of resistance mutations by modifying the effects of drug selected mutations, possibly by compensatory changes in the enzyme backbone. It has been reported that some of these RT positions (NRTI-44, 69, 211, 228; NNRTI-90, 98, 106,118,179) actually confer intermediate resistance to some available nucleoside and non-nucleoside analogues with diverse compensatory roles (Johnson et al., 2011). The prevalence of these mutations in the HIV-1 isolates infecting an individual suggests that the outcome of drug treatment may adversely affect the study population.

There were high frequencies of accessory mutations observed in protease gene, and the predominant mutations were at positions K20VMI, M36I, E35DEGQ

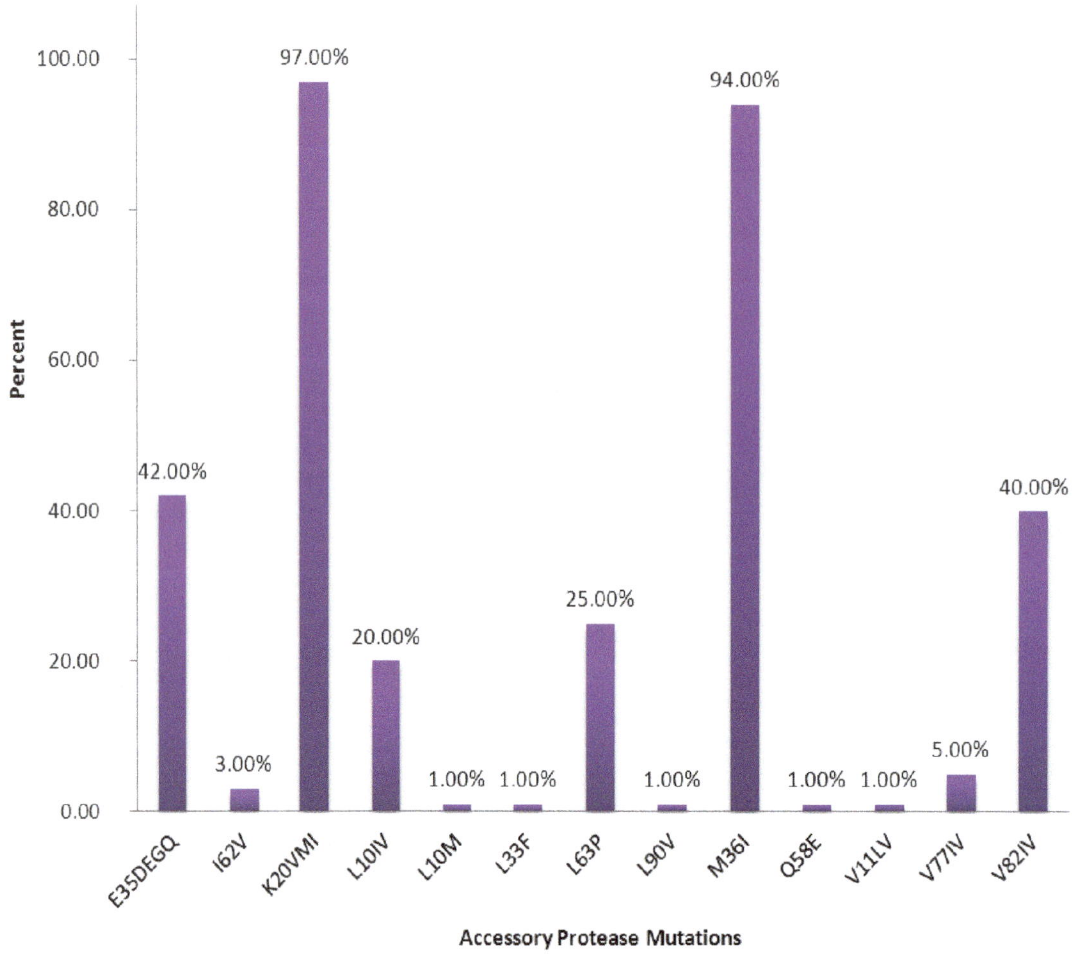

Figure 4. Prevalence of Protease Accessory mutations among the ARV treatment naïve patients. The figure shows the prevalence of accessory mutations associated with protease inhibitors in decreasing order; K20VMI (97%), M36I (94%), E35DEGQ (42%), V82VI (40%), L63P (25%), L10IV (20%), V77IV (5%), I62V (3%), and L10M, V11LV, L33F, Q58E, and L90V as 1% each. The most common substitutions were K20VMI, M36I, E35DEGQ and V82I. The single mutations observed were; L33F, L90V, Q58E, V11LV, and L10M. The substitution of valine at codon 90 and methionine at codon 10 against the known amino acid with subtype B is rare to non-B subtype.

and V82IV. The presence of these minor mutations do not lead to high level resistance when occurring alone but they either improve viral fitness or increases the drug resistance level in the presence of major PI mutations (Nijhuis et al., 1999; van Maarseveen et al., 2006; Scherrer et al., 2012).

From comparative analyses, the study also demonstrates that in HIV-1 subtypes CRF02_AG and subtype G, the specific codon positions are often utilized differently, either in contrast with one another or with subtype B in encoding amino acids, hence leading to distinct functional motifs that may influence divergent pathways to drug resistance. These findings emphasize the significance of subtype specific determinants of susceptibility to antiretroviral therapy and immune response (Wainberg and Brenner, 2012). However, it is important to keep in mind that late detection of HIV disease and accessing of ARVs are the characteristics of most patients attending the tertiary treatment center in Nigeria. Over 60% of individuals enrolled in this study were diagnosed with <200 cells/mm^3, and were therefore late presenters. The observed result reflects transmitted drug resistance (TDR) levels characteristics of individuals with established infection or advanced HIV disease. In Nigeria, the population of treated patients in the scale-up exercise is not well monitored, and poor adherence to ARV medication could lead to emergence of drug resistance and drug failure. This may suggest that the transmission of resistance mutations from treated individuals to treatment-naïve patients is likely to occur more frequently.

In conclusion, this study indicates that ARV resistant

variants and drug-resistance mutations exist among HIV-1 infected treatment naïve patients in Jos, North Central Nigeria. The diversity of subtypes has implications for treatment and vaccine design strategies. Population surveillance for ARV drug resistance is important and should be included in all implementation programs. The findings from this study also underscores the need for targeted genotyping the HIV-1 viral genome before beginning of ARV treatment and recommendations of appropriate choice of first-line regimens in resource limited countries.

ACKNOWLEDGMENTS

We are deeply indebted to the patients who agreed to participate in this study. We thank the Director and staff of HIV-1 Drug Resistance Laboratory, Kenya Medical Research Institute Kisian Kisumu, and Dr. Clement Zeh who permitted my training and analyses of the samples. We are also grateful for the support of the following people: Prof I. Ujah mni, Director General Nigeria Institute of Medical Research Lagos, Mr J. Adetunji of Medicom Laboratories Jos, and Dr. P. Okonkwo of the AIDS Prevention Initiative in Nigeria. Ramyil. S, F. Moulton and T. Obadiah were instrumental in overseeing sample collection, transport and processing.

REFERENCES

Ajoge HO, Gordon ML, Ibrahim S, Shittu SO, Ndungu T, Olonitola SO (2012). Drug resistance pattern of HIV type 1 isolates sampled in 2007 from therapy-naive pregnant women in North-Central Nigeria. AIDS Res. Hum. Retrov. 28(1):115-118.

Arts JE, Hazuda JD (2012). HIV-1 Antiretroviral Drug Therapy. Cold Spring Harb Perspect Med. 2(4):a007161.

Barbour JD, Hecht FM, Wrin T, Liegler TJ, Ramstead CA, Busch MP, Segal MR, Petropoulos CJ, Grant RM (2004). Persistence of primary drug resistance among recently HIV-1 infected adults. AIDS 18:1683-1693.

Chaplin B, EisenG, Idoko J, Onwujekwe D, Idigbe E, Adewole I, Gashau W, Meloni S, Sarr DA, Sankalé LJ, Ekong E, Murphy LR, and Kanki P (2011). Impact of HIV Type 1 Subtype on Drug Resistance Mutations in Nigerian Patients Failing First-Line Therapy. AIDS Res. Hum. Retrov. 27(1):71-80.

Global AIDS Response Progress Report (GARPR), (2012). National Agency for the control of AIDS Federal Republic of Nigeria: January-December 2011 reporting period. pp. 15-22.

Hamers RL, Wallis LC, Kityo C, Siwale M, Mandaliya K, Conradie F, Botes EM, Wellington M, Osibogun A, Sigaloff CEK, Nankya I, Schuurman R, Wit WF, Stevens SW, Vugt VM, Wit TFR (2011). HIV-1 Drug-resistance in antiretroviral-naive individuals in Sub-Saharan Africa after rollout of antiretroviral Therapy: A multi-center observational Study. Lancet 10:1473-1499.

Hirsch MS, Conway B, D'Aquila RT, Johnson VA, Brun-Vézinet F, Clotet B, Demeter LM, Hammer SM, Jacobsen DM, Kuritzkes DR, Loveday C, Mellors JW, Vella S, Richman DD (1998). Antiretroviral drug resistance testing in adults with HIV infection. JAMA 279(24):1984-1991.

Johnnson VA, Clavez V, Gunthard HF, Paredes R, Pillay D, ShaferR, Wensing AM, Richman DD (2011). Update of the drug resistance mutations in HIV-1. Top. Antivir. Med. 4:156-164. Los Alamose HIV dataset, http://www.hiv.lanl.gov.

McNulty A, Jennings C, Bennett D, Fitzgibbon J, Bremer WJ, Ussery M, Kalish LM, Heneine W, García-Lerma JG (2007). Evaluation of Dried Blood Spots for HIV-1 1 Drug Resistance Testing. J. Clin. Microbiol. 45(2):517-521.

Nukoolkarn S, Pongthapisith V, Panyim S, Leelamanit W (2004). Sequence Variability of the HIV Type 1 Protease Gene in Thai Patients Experienced with Antiretroviral Therapy. AIDS Res. Hum. Retrov. 20(12):1368-13721.

Nijhuis M, Schuurman R, de Jong D, Erickson J, Gustchina E, Albert J, Schipper P, Gulnik S, Boucher CA. (1999). Increased fitness of drug resistant HIV-1 protease as a result of acquisition of compensatory mutations during suboptimal therapy. AIDS 13(17):2349-2359.

Peeters M, Esu-Williams E, Vergne L, Montavon C, Mulanga-Kabeya C, Harry T, Ibironke A, Lesage D, Patrel D, Delaporte E. (2000). Predominance of subtype A and G HIV type 1 in Nigeria, with geographical differences in their distribution. AIDS Res. Hum. Retrov. 16(4):315-325.

Perelson AS, Neumann AU, Markowitz M, Leonard J M, Ho DD (1996). HIV-1 Dynamics *in vivo*: Virion Clearance Rate, Infected Cell Life-Span, and Viral Generation Time. Science 271:1582-1586.

Robertson DL, Anderson JP, Bradac JA, Carr JK, Foley B, Funkhouser RK, Gao F, Hahn BH, Kalish ML, Kuiken C, Learn GH, Leithner T, McCutchan F, Osmanov S, Peeters M, Pieniazek D, Salminen M, Sharp PM, Wolinsky S, Korber B (2000). HIV-1 Nomenclature proposal. Science 288:55-56.

Saminen MO, Carr JK, Burke DS, McCutchan FE (1995). Identification of breakpoints in intergenotypic recombinants of HIV type 1 by bootscanning AIDS Res. Hum. Retrov. 11:1423-1425.

Scherrer UA, Ledergerber B, von Wyl V, Böni J, Yerly S, Klimkait T, Cellerai C, Furrer H, Calmy A, Cavassini A, Elzi L, Vernazza LP, Bernasconi E (2012). Minor Protease Inhibitor Mutations at Baseline Do Not Increase the Risk for a Virological Failure in HIV-1 Subtype B Infected Patients. PLoS One. 7(6):e37983

Shekelle P, Maglione M, GeotzMB, Wagner G, Wang Z, Hilton L, Carter J, ChenS, Tringle C, Mojica W, New-Berry S (2007). Antiretroviral (ARV) drug resistance in the developing world. Evid. Rep. Technol. Ass. (Full Rep.):1-74.

Swofford DL (2002). PAUP*. Phylogenetic Analysis Using Parsimony (*and other methods). Version 4. Sunderland, MA: Sinauer Associates.

van Maarseveen NM, de Jong D, Boucher CA, Nijhuis M (2006). An increase in viral replicative capacity drives the evolution of protease inhibitor-resistant human immunodeficiency virus type 1 in the absence of drugs. J. Acquir. Immune Defic. Syndr. 42(2):162-168.

Vergne L, Diagbouga S, Kouanfack C, Aghokeng A, Butel C, Laurent C, Noussi N, Tordy M, Sawadogo A, Drabo J, Hien H, Zekeng L, Delaporte E, Peeters M (2006). HIV-1 drug-resistance mutations among newly diagnosed patients before scaling-up programmes in Burkina Faso and Cameroon. Antivir. Ther. 11:575-579.

Vidal N, Koyalta D, Richard V, Lechiche C, NdinaromtanT, Djimasgar A, Delaporte E, Peeters M (2003). High genetic diversity of HIV-1 strains in Chad, West Central Africa. AIDS J. Acquir. Immune Def. Syndr. 33(2):239-246.

Wainberg AM, Brenner GB (2012). The Impact of HIV Genetic Polymorphisms and Subtype Differences on the Occurrence of Resistance to Antiretroviral Drugs. Mol. Biol. Int. 6; 10.1155/2012/256982.

Wei XS, Ghosh SK, Taylor EM, Johnson AV, Emini AE, Deutsch P, Lifson JD, Bonhoefler S, Nowak MA, Hahn BH, Saag MS, Shaw GM. (1995). Viral Dynamics in HIV-1 Infection. Nature 373:117-122.

Wensing AMJ, van de Vijver AD, Angarano G, Asjo B, Balotta C, Boeri E, Camacho R, Chaix, M, Costagliola D, De Luca A, Derdelinckx I, Grossman Z, Hamouda O, Hatzakis,A, Hemmer R, Hoepelman A, Horban A, Korn K, Ku¨cherer C, Leitner T, Loveday C, MacRae E, Maljkovic I, de Mendoza C, Meyer L, Nielsen C, Op de Coul, EL, Ormaasen V, Paraskevis D Perrin L, Puchhammer-Stock E, Ruiz L, Salminen M, Schmit J, Schneider F, Schuurman R, Soriano V, Stanczak G, Stanojevic M, Vandamme A, Van Laethem K, Violin M, Wilbe K, Yerly S, Zazzi M, Boucher AC (2005). Prevalence of Drug-Resistant HIV-1 Variants in Untreated Individuals in Europe: Implications for Clinical Management. J. Infect. Dis. 192(6):958-966.

Simple sequence repeat (SSR) markers for assessing genetic diversity among the parental lines of hybrid rice (*Oryza sativa* L.)

Mahalingam. A[1], R. Saraswathi[2] and J. Ramalingam[3]

[1]Centre for Plant Breeding and Genetics, Tamil Nadu Agricultural University, Coimbatore, India.
[2]Tamil Nadu Rice Research Institute, Tamil Nadu Agricultural University, Coimbatore, India.
[3]Centre for Plant Molecular Biology, Tamil Nadu Agricultural University, Coimbatore, India.

The present scientific study was carried out at Tamil Nadu Rice Research Institute, Aduthurai with the objective of studying the genetic divergence among 51 restorer and five maintainer lines in terms of grain quality and fertility restorer genes (*Rf_3, Rf_4* and *Rf_7*) using 55 primers of simple sequence repeat (SSR) markers. Among them, 37 SSR markers were found to be polymorphic and the number of amplified fragments ranged from one to five. The highest polymorphic information content (PIC) value (more than 0.60) was observed for eight primers *viz.,* AB 443, RM 3, RM 29, RM 226, RM 228, RM 304, RM 1812 and RM 3873 and average PIC value was 0.444. Cluster analysis using NTSYS generated dendrogram divided all the 56 parental lines into two distinct groups *viz.,* maintainer line (Group I) and restorer line (Group II) at 76% coefficient of similarity. Maintainer line group consisted of five genotypes (COMS 23B, COMS 24B, COMS 25B, CRMS 31B and CRMS 32B) and restorer line group had 51 genotypes. Further, at 80% similarity, all fifty one restorer lines were again grouped into nine clusters. With higher polymorphism revealed by SSR markers, parental lines having the similar genetic background from pedigree information were grouped into different clusters. The combination of pedigree analysis and SSR markers could be a more reliable method to study the diversity and grouping of parental lines of hybrid rice. Hybridization between diverse restorer and CMS groups identified from this study would be expected to yield hybrid combinations with premium grain quality and good fertility restoration.

Key words: Simple sequence repeat (SSR), maintainer and restorer genetic diversity, grain quality, fertility restoration.

INTRODUCTION

Rice (*Oryza sativa* L.) is the major staple cereal food crop fulfilling about 60% dietary requirement, 20% calorie and 14% protein requirement of the world's population. In the present decade, the rate of increase in rice production is lower (1.5% per year) than the increase in population (1.8% per year). The present world population of 6.3 billion is likely to reach 8.5 billion by 2030. Out of this, 5 billion people will be rice consumers and there is a need of 38% more rice by 2030. To meet this challenge there is a need to develop rice varieties with higher yield potential and greater stability (Khush, 2006). Hybrid rice technology is one of the strategy to meet this immense challenge imposed by ever growing populations. Hybrid rice varieties have clearly shown a yield advantage of 1.0

to 1.5 tonnes per hectare (20 to 30%) over conventionally bred modern varieties (Virmani et al., 2003). India is the second country after China to develop and release the first rice hybrid during 1994 (Janaiah and Hussain, 2003). Breeding hybrid rice requires the development of elite parental lines in either a three-line or two-line system.

The heterotic potential of hybrids is associated with genetic divergence between their parental lines. Empirical evidence has indicated that establishment of divergent maintainer and restorer groups are the key for success of hybrid breeding programs using the CMS system. In order to increase hybrid rice breeding efficiency, the relationship between genetic diversity and heterosis can be investigated by using phenotypic traits and molecular markers (Julfiquar, 1985). Selection of diverse parents with desirable traits can be done based on the analysis using morphological, biochemical and or DNA markers. Though morphological characterization is an important methodology for studying diversity, it is cumbersome, whereas there are limitations in using biochemical markers as they are limited in number and are predisposed by environmental factors or the developmental stage of the plant (Winter and Kahl, 1995). DNA markers are the most widely used and are predominant due to their abundance and repeatability and they remain unaffected across different stages, seasons, locations and agronomic practices (McCouch et al., 1997). Of the several classes of DNA markers available, microsatellite or simple sequence repeat (SSR) markers are considered as most amenable for several applications including genetic diversity studies due to their multiallelic nature, high reproducibility, co-dominant inheritance, abundance and extensive genome coverage (McCouch et al., 2002). A carefully chosen set of SSR markers providing genome-wide coverage will facilitate an unbiased assay of genetic diversity, thus giving a robust, unambiguous molecular description of rice cultivars. SSR markers have been effectively used to study genetic diversity among closely related rice cultivars (Spada et al., 2004) and distantly related genotypes (Giarrocco et al., 2007). Hence the present investigation was taken up with the objective of studying the genetic divergence among the parental lines (Maintainer or 'B'lines and Restorer or `R` lines) of hybrid rice using Simple Sequence Repeat (SSR) markers which are distributed throughout the rice genome.

MATERIALS AND METHODS

Five maintainer or B lines (COMS 23A, COMS 24A, COMS 25A, CRMS 31A and CRMS 32A) and fifty one tester or Restorer lines (18 AICRIP parental lines, 13 recently stabilized breeding lines and 20 Advanced Cultures / lines under evaluation) formed the basic genetic materials for this study. The pure seeds of male sterile and maintainer lines were obtained from Paddy Breeding Station, Centre for Plant Breeding and Genetics, Coimbatore and the restorer lines were obtained from Tamil Nadu Rice Research Institute, Aduthurai. The parents were raised in a Randomized block design with three replications during September 2009 at Tamil Nadu Rice Research Institute, Aduthurai. The details of the chosen parental genotypes are given in Table 1.

Isolation of genomic DNA

Total genomic DNA was isolated from 56 parental lines at tillering stage (45 DAS) from fresh leaves of each genotype following the CTAB (cetyltrimethylammonium bromide) procedure as described by Dellapota et al. (1983). Fifty five SSR markers (Table 2) representing the entire rice genome associated with grain quality parameters (grain size, amylose content, alkali spreading value, cooking traits, protein content, plant height, panicle length, grains per panicle) and fertility restorer genes (*Rf*$_3$, *Rf*$_4$ and *Rf*$_7$) were selected from the published framework map (Akagi et al., 1996; Temnykh et al., 2000). Information regarding chromosomal localizations and repeat types of the SSRs can be found in the Web database (www.gramene.org). The isolated genomic DNA was quantified by NanoDrop^{TM} 1000 Spectrophotometer. Further, the isolated DNA was checked for intactness, homogeneity and purity by electrophoresis in 0.8% agarose gel.

PCR amplification

PCR reactions were carried out in Programmable Thermal Cycler (PTC) MJ research Inc. USA. The reaction volume was 15 µl containing 2 µl of genomic DNA, 1X assay buffer, 200 µM of dNTPs, 2 µM $MgCl_2$, 0.2 µM each primer (Forward and Reverse) and 1 unit of *Taq* polymerase (Banglore Genei). The temperature cycles were programmed as 95°C for 2 min, 94°C for 45 s, 55°C for 1 min, 72°C for 1.30 min for 35 cycles and additional temperature of 72°C for 10 min for extension and 4°C for cooling.

The amplified PCR products were separated in 3% agarose gel prepared in 1X TBE buffer stained with Ethidium bromide. The gel was run in 1X TBE buffer at a voltage of 90 V for a period of 45 min to 1 h. The gel was visualized in UV transilluminator and photographs taken using Alpha Digidoc gel documentation instrument. Clearly resolved, unambiguous bands were scored visually for their presence or absence with each primer. The scores were obtained in the form of matrix with '1' and '0', which indicate the presence and absence of bands in each variety respectively.

Data analysis

SSR markers scoring data were used to estimate the similarity on the basis of number of shared bands. Similarity was calculated with SIMQUAL function of NTSYSpc2.0 version that computes a variety of similarity and dissimilarity coefficients for qualitative data. The similarity matrix was used to generate dendrogram based on UPGMA. In order to estimate the congruence among dendrograms, cophenetic matrices for which marker and index type were computed and compared using the Mantel test. The PIC value of each locus was calculated as

$$PIC_j = 1 - \sum_{i-1 \text{ to } L} P_{ij}^2$$

Where, P_{ij} is the relative frequency of the ith allele for the locus j and is summed across all the alleles (L) over all lines. PIC provides an estimate of the discriminatory power of a locus by taking into account, not only the number of alleles that are expressed, but also the relative frequencies of those alleles. PIC values may range from 0 (monomorphic) to 1 (very highly discriminative), with many alleles in equal frequencies.

Table 1. Details of the parents used in the study.

S/N	Tester lines		Source
	Name of the entry	Parentage	
	AICRIP parental lines		
1	IET 19863	GR 11 / Pusa Basmati 1 / 6-12-1-4-1-3	AICRIP
2	IET 19922	IET 12356 / IR 64	
3	IET 20601	OR 1530 – 8 / IR 68181 – B- 49	
4	IET 20605	SYE – 1 / Tadukan	
5	IET 20881	Abhaya / SKL – 8	
6	IET 20885	Improved White Ponni / Tetep	
7	IET 20888	ADT 38 / CR 1014	
8	IET 20890	GR – 11 / Pusa Basmati 1 / 6- 12- 1- 4 – 1 – 4 -3 -2	
9	IET 20895	Erramallelu / RNR – M7	
10	IET 20896	Erramallelu / Kavya	
11	IET 20897	NLR – 145 / Kavya	
12	IET 20898	NLR – 145 / Kavya	
13	IET 20899	Tella Hamsa / Erramallelu	
14	IET 20904	Kavya / RNR – M7	
15	IET 20932	Keshava / NLR – 33655	
16	IET 20937	OR – 1530 – 8 / IR 68181 – B – 49	
17	IET 20944	Jaya / Mahsuri	
18	IET 20945	Kavya / BPT 5204 // Vijetha	
		Recently stabilized breeding lines	
19	AD 09522	Imp Rascadam / ASD 19 – 2-1-7-4-1	TRRI, Aduthurai
20	AD 09523	I.W.Ponni / Ratna – 1-1-2-3-1	
21	AD 09524	ASD 19 / WGL 32100 – 1-3-3-11-5	
22	AD 09525	ASD 19 / WGL 32100 – 13-2-2-1-1	
23	AD 09526	WGL 32100 / Swarna – 7-2-3-3-1	
24	AD 09527	BPT 5204 / Imp Rascadam – 4-7-1-2-1	
25	AD 09528	BPT 5204 / Imp Rascadam – 2-4-5-6-6	
26	AD 09529	I.W.Ponni / Kalajoha – 5 -1 – 5 - 4	
27	AD 09530	I.W.Ponni / Kalajoha – 21-3-3-3-1	
28	AD 09531	BPT 5204 / Azucena – 6-2-2-3-1	
29	AD 09532	ADT 43 / WGL 32100 – 7-3-3-2-2	
30	AD 09533	ADT 43 / WGL 32100 – 12-3-4-3-1	
31	AD 09534	ADT 43 / WGL 32100 – 3-3-7-10-1	
		Advanced Cultures / lines under evaluation	
32	AD 04072	Improved White Ponni / P 1110 – 5 – 1 - 2	TRRI, Aduthurai
33	AD 06084	ADT 42 / JGL 384	
34	AD 07073	ADT 43 / JGL 384	
35	AD 07076	ADT 43 / JGL 384	
36	AD 07083	ADT 43 / I. W. Ponni	
37	AD 07158	BPT 5204 / AD 02235	
38	AD 07309	Ajaya / IET 11613 – 22 – 5 – 2	
39	AD 08005	ADT 43 / ACK 03002	
40	AD 08009	ADT (R) 45 / AD 01236	
41	AD 08010	ADT (R) 45 / AD 01236	
42	AD 08013	ADT (R) 45 / ADT (R) 47	
43	AD 09193	ADT 43 / JGL 384	
44	AD 09194	ADT (R) 47 / Karnataka Ponni	

Table 1. Contd.

45	AD 09203	ADT 43 / ADT 37	
46	AD 09206	ADT 43 / ADT 37	
47	AD 09216	ADT (R)45 / AD 01236	
48	AD 09222	ADT (R)45 / IET 18208	
49	AD 09223	ADT (R)45 / IET 18208	
50	AD 09231	PY 3 / ADT (R) 47	
51	AD 09241	AD 01236 / AD 01205	
Male sterile and Maintainer Lines			
52	COMS 23B		Coimbatore
53	COMS 24B		Coimbatore
54	COMS 25B		Coimbatore
55	CRMS 31B		Cuttack
56	CRMS 32B		Cuttack

RESULTS AND DISCUSSION

In the amplification of genomic DNA of 56 hybrid rice parental genotypes using 54 primers of SSR markers, 37 were found to be polymorphic. The number of amplified fragments ranged from one to five. Of the total amplified bands, the average polymorphic fragment per primer was 2.46. Among the 54 primers, RM 228, RM 304, RM 336 and RM 1812 produced the maximum number of alleles. Polymorphism revealed by RM 29 co segregating with width expansion ratio of cooked rice and RM3873 which is co segregating with fertility restoration gene (*Rf*$_3$) are presented in Figures 1 and 2. The PIC values representing allelic diversity and frequency among genotypes varied from one locus to another. The highest PIC value (more than 0.60) was observed for eight primers *viz*., AB 443, RM 3, RM 29, RM 226, RM 228, RM 304, RM 1812 and RM 3873 while it was the lowest for RM 339 (0.035). The average PIC value was 0.444 (Table 3). The results of this study, that is, the number of alleles and PIC value are comparable to those reported by Cho et al. (2000), Yu et al. (2003), Thomson et al. (2007), Lapitan et al. (2007) and Seetharam et al. (2009). The above results provided an overview of the genetic diversity in all the genotypes.

The cluster analysis using NTSYS generated dendrogram (Figure 3) throws light on genetic similarity among the genotypes. It divided all the 56 parental lines into two distinct groups viz., maintainer group (Group I) and restorer group (Group II) at 76% coefficient of similarity (Table 4). Maintainer group consisted of five genotypes (COMS 23B, COMS 24B, COMS 25B, CRMS 31B and CRMS 32B) and restorer group had 51 genotypes. Further, at 80% similarity all fifty one restorer lines were again grouped into nine clusters. Fifteen (15) genotypes viz., AD 07073, AD 07076, AD 07083, AD 07309, AD 07158, AD 08005, AD 08009, AD 08010, AD 08013, AD 09193, AD 09194, AD 09203, AD 09206, AD 09216, and AD 09231 were grouped in cluster VI and all belonged to advanced cultures originated from TRRI, Aduthurai. Out of the 15 genotypes, seven and four had the common origin from ADT 43 and ADT (R) 45 as the female parents respectively. Out of the 13 recently stabilized breeding lines, most of them were grouped in cluster IV (AD 09526, AD 09531, AD 09532, AD 09533, AD 09534, AD 09528, AD 09530), followed by cluster III (AD 09522, AD 09524, AD 09525). The clusters II, V and VIII had one genotype each. The genotype AD 09529 (I.W.Ponni × Kalajoha) alone fell in a single cluster V. The genotypes from AICRIP were grouped exclusively in cluster I (6 genotypes) and VII (3 genotypes). Another set of seven genotypes viz., IET 20897, IET 20898, IET 20899, IET 20932, IET 20937, IET 20944 and IET 20945 were grouped in cluster VIII. One genotype each was grouped in cluster II (IET 20885) and III (IET 20904). In cluster III, the genotypes viz., AD 09522, AD 09524 and AD 09525 had ASD 19 as one of the parents. In cluster II, the entries IET 20885 and AD 09523 had I.W.Ponni as one of the parent. But another genotype AD 09523 also with I.W.Ponni as one of the parents fell into cluster VIII.

In contrast to the morphological or biometrical traits, the molecular markers reveal polymorphism at the DNA level, suggesting a very powerful tool for characterization of genotype and estimation of genetic diversity. Among them, the microsatellite or SSR (simple sequence repeats) markers show a high potential for identification and estimation of genetic diversity (Hashimoto et al., 2004). In this study, newly stabilized breeding lines viz., AD 09527 and AD 09528 with the same genetic background (AD 09527: BPT 5204 × Imp Rascadam – 4-7-1-2-1, AD 09528: BPT 5204 × Imp Rascadam – 2-4-5-6-6) were grouped into separate clusters *viz.,* II and IV. Likewise, AD 09529 and AD 09530 had the same genetic background (AD 09529: I.W.Ponni × Kalajoha – 5 -1 – 5 – 4, AD 09530: I.W.Ponni × Kalajoha – 21-3-3-3-1) but were also grouped into separate clusters viz., IV and V.

Table 2. List of SSR primers used for diversity study.

Primer	Sequence	Annealing Temperature (°C)	Chromosomal location
RM431	F - TCCTGCGAACTGAAGAGTTG R - AGAGCAAAACCCTGGTTCAC	55.5	1
RM11	F - TCTCCTCTTCCCCCGATC R - ATAGCGGGCGAGGCTTAG	56.5	7
RM336	F - CTTACAGAGAAACGGCATCG R - GCTGGTTTGTTTCAGGTTCG	54.5	7
RM505	F - AGAGTTATGAGCCGGGTGTG R - GATTTGGCGATCTTAGCAGC	56.0	7
RM1812	F - CAGCTAGTGAGCTCCTAGTG R - GCTAACCCACCAACTTATTC	53	11
RM209	F - ATATGAGTTGCTGTCGTGCG R - CAACTTGCATCCTCCCCTCC	57	11
RM3	F - ACACTGTAGCGGCCACTG R - CCTCCACTGCTCCACATCTT	57.5	6
RM217	F -ATCGCAGCAATGCCTCGT R - GGGTGTGAACAAAGACAC	55	6
RM225	F - TGCCCATATGGTCTGGATG R - GAAAGTGGATCAGGAAGGC	54	6
RM247	F - TAGTGCCGATCGATGTAACG R - CATATGGTTTTGACAAAGCG	53	12
RM81	F - GAGTGCTTGTGCAAGATCCA R - CTTCTTCACTCATGCAGTTC	53.5	3
Wx (GBSS)	F - CTTTGTCTATCTCAAGACAC R - TTGCAGATGTTCTTCCTGAT	50.0	6
RM10	F - TTGTCAAGAGGAGGCATCG R - CAGAATGGGAAATGGGTCC	54	7
RM6	F - GTCCCCTCCACCCAATTC R - TCGTCTACTGTTGGCTGCAC	56.5	2
RM238	F - GATGGAAAGCACGTGCACTA R - ACAGGCAATCCGTAGACTCG	56.5	6
RM44	F - ACGGGCAATCCGAACAACC R - TCGGGAAAACCTACCCTACC	57.5	8
RM137	F - GACATCGCCACCAGCCCACCAC R - CGGGTGGTCCCCGAGGATCTTG	65.0	8
RM325A	F - GACGATGAATCAGGAGAACG R - GGCATGCATCTGAGTAATGG	53.5	8
RM331	F - GAACCAGAGGACAAAAATGC R - CATCATACATTTGCAGCCAG	52.0	8
RM339	F - GTAATCGATGCTGTGGGAAG R - GAGTCATGTGATAGCCGATATG	53.0	8
RM342	F - CCATCCTCCTACTTCAATGAAG R - ACTATGCAGTGGTGTCACCC	55.0	8
RM224	F - ATCGATCGATCTTCACGAGG R - TGCTATAAAAGGCATTCGGG	53.5	11
RM211	F - CCGATCTCATCAACCAACTG R - CTTCACGAGGATCTCAAAGG	53.5	2

Table 2. Contd.

Primer	Sequence	Annealing Temperature (°C)	Chromosomal location
RM29	F - CAGGGACCCACCTGTCATAC R - AACGTTGGTCATATCGGTGG	56.5	2
RM232	F - CCGGTATCCTTCGATATTGC R - CCGACTTTTCCTCCTGACG	54.5	3
RM215	F - CAAAATGGAGCAGCAAGAGC R - TGAGCACCTCCTTCTCTGTAG	55.5	9
RM190	F- CTTTGTCTATCTCAAGACAC R - TTGCAGATGTTCTTCCTGATG	50.5	6
RM204	F - GTGACTGACTTGGTCATAGGG R - GCTAGCCATGCTCTCGTACC	56.5	6
RM226	F- AGCTAAGGTCTGGGAGAAACC R - AAGTAGGATGGGGCACAAGCTC	58.0	1
RM140	F - TGCCTCTTCCCTGGCTCCCCTG R - GGCATGCCGAATGAAATGCATG	62.0	1
RM411	F - ACACCAACTCTTGCCTGCAT R- TGAAGCAAAAACATGGCTAGG	56.0	3
RM507	F - CTTAAGCTCCAGCCGAAATG R - CTCACCCTCATCATCGCC	55.0	5
RM1	F - GCGAAAACACAATGCAAAAA R - GCGTTGGTTGGACCTGAC	54.0	1
RM171	F - AACGCGAGGACACGTACTTAC R - ACGAGATACGTACGCCTTTG	56.0	10
RM228	F - CTGGCCATTAGTCCTTGG R - GCTTGCGGCTCTGCTTAC	55.0	10
RM6737	F - CATTGGGGGTGGATAAAGAG R - TATCCTCTACTCCCTCGGCC	55.5	10
AB443	F - CACTCGTATAGTAACTTGAGACGAT R - CCAAACAAGTTCCCACAATAGA	53.4	10
RM304	F - TCAAACCGGCACATATAAGAC R - GATAGGGAGCTGAAGGAGATG	53.5	10
RM3873	F - GCTAGCTAGGACCGACATGC R - CCTCCTCCTTATCCTCCCTG	56.0	1
RM3233	F - GTGGTGAGTAAACAGTGGTGG R - GAGAGCAGAGCAGAGGCAAC	57.0	1
RM6100	F - TCCTCTACCAGTACCGCACC R- GCTGGATCACAGATCATTGC	56.5	10
RM210	F - TCACATTCGGTGGCATTG R - CGAGGATGGTTGTTCACTTG	54.0	8
RM315	F - GAGGTACTTCCTCCGTTTCAC R - AGTCAGCTCACTGTGCAGTG	56.5	1
RM282	F - CTGTGTCGAAAGGCTGCAC R - CAGTCCTGTGTTGCAGCAAG	56.8	3
RM333	F - GTACGACTACGAGTGTCACCAA R- GTCTTCGCGATCACTCGC	56.5	6
RM234	F - ACAGTATCCAAGGCCCTGG R - CACGTGAGACAAAGACGGAG	56.5	7
RM250	F - GGTTCAAACCAAGCTGATCA R - GATGAAGGCCTTCCACGCAG	56.0	2

Table 2. Contd.

S/N	Primer	Sequences	Annealing temperature (°C)	Chromosomal Location
48	RM208	F - TCTGCAAGCCTTGTCTGATG R - TAAGTCGATCATTGTGTGGACC	55.0	2
49	RM5	F - TGCAACTTCTAGCTGCTCGA R - GCATCCGATCTTGATGGG	55.0	1
50	RM490	F - ATCTGCACACTGCAAACACC R - AGCAAGCAGTGCTTTCAGAG	56.0	1
51	RM1108	F - GCTCGCGAATCAATCCAC R - CTGGATCCTGGACAGACGAG	55.5	10
52	RM5841	F - CCTCTCTCTCTCTCTCCCCC R - TGTTATTGGCACGTGGTGTG	57.0	10
53	RM6344	F - ACACGCCATGGATGATGAC R - TGGCATCATCACTTCCTCAC	55.5	7
54	RM443	F - GATGGTTTTCATCGGCTACG R - AGTCCCAGAATGTCGTTTCG	54.5	1
55	RM7003	F - GGCAGACATACAGCTTATAGGC R - TGCAAATGAACCCCTCTAGC	55.5	12

These results indicated that SSR analysis could be a better method to study the diversity of parental lines of hybrid rice.

Presently, molecular marker techniques have been widely used in studying genetic variation and diversity among populations, species and varieties (Hashimoto et al., 2004; Yu et al., 2005). The use of SSR markers for assessing genetic diversity in the present study indicated that tester lines had wider genetic bases than the CMS maintainer lines. In addition, there were clear and wider genetic differences between the CMS lines and the tester lines. The original two parental groups *viz.,* CMS lines and testers have become more complex, through the emergence of new patterns of heterotic combinations. Therefore, the partitioning of parental lines to new parental groups and subgroups (AICRIP cultures, stabilized breeding lines and advanced cultures) is more important than the recognition of two parental groups (that is, maintainer and tester/restorer group). Nghia et al. (1999) reported genetic diversity existing among thermo-sen-sitive genic male sterile (TGMS) lines was much lower than that among cytoplasmic genic male sterile (CMS) lines. The cluster analysis based on RAPD markers was able to reveal close genetic relationships between different rice genotypes used in the hybrid rice breeding program.

Also SSR markers are an indispensable complementation to pedigree analysis in the identification of parental groups. In general, the pedigree analysis is considered to have no effect on selection and mutation. Therefore, pedigree analysis cannot reveal the relationship between progeny and their parents exactly. On the contrary, SSR markers can detect genetic variation at DNA level. SSR analysis showed that CRMS 31B was more similar to CRMS 32B than to CMS lines from TNAU (Figure 3). A combination of pedigree analysis and SSR markers will be helpful in more reliable grouping.

Conclusion

In the present study, the maintainer lines and tester/ restorer lines of rice were clearly discriminated on the basis of SSR analysis. Furthermore, with higher polymorphism revealed by SSR markers, some of the parental lines having similar genetic background from pedigree information were also grouped into different clusters. The identi-

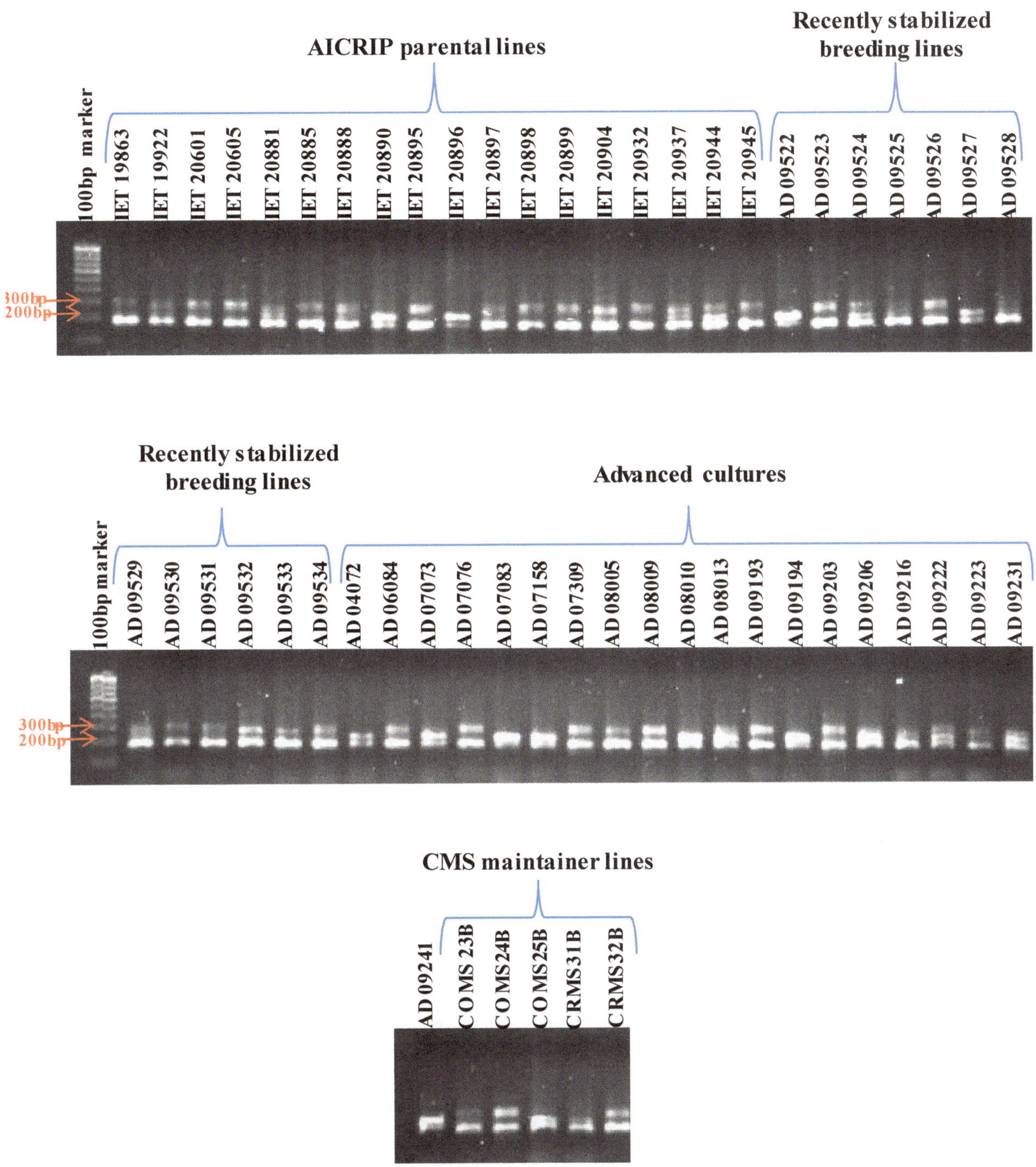

Figure 1. SSR polymorphism revealed by RM29 co segragating with width expansion of cooked rice.

fication of heterotic groups and patterns among breeding populations and lines provides fundamental information in order to help the plant breeders to gain more information on heterosis.

Systematic studies on classifying the breeding lines into heterotic groups are limited. Therefore, the grouping result based on SSR analysis might be helpful to identify heterotic groups for hybrid rice breeding.

Rice grain quality is mainly determined by the combination of many physical as well as chemical characters. Physical quality characters include kernel size, shape, hulling, milling percentage and head rice recovery. Chemical quality is mainly determined by amylose content, gelatinization temperature, gel consistency. High volume expansion and greater length wise expansion of kernel during cooking decides the consumer preference. Rice

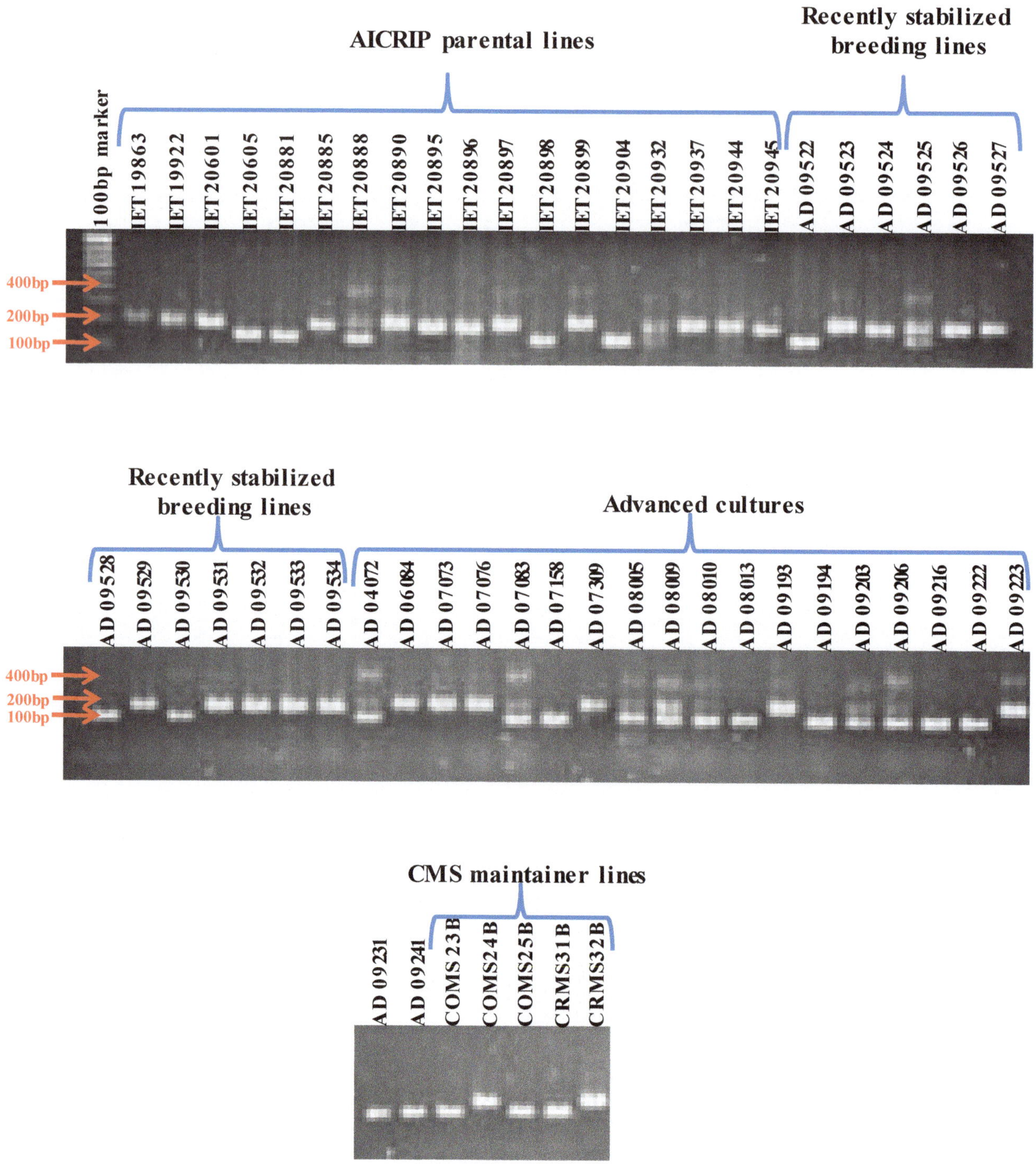

Figure 2. SSR polymorphism revealed by RM3873 co segragating with *Rf3* gene.

with soft to medium gel consistency, intermediate amylose content and gelatinization temperature is a preferred level for the consumers which determines the eating and cooking quality of rice grains (Bao et al., 2002). Amylose content is governed by the *waxy* (*Wx*) locus which is located on chromosome 6 (Septiningsih et al., 2003) and it is linked to the gene for alkali spreading score, which is also an indicator of temperature at which rice grain becomes gelatinous during cooking (Sano, 1984). SSRs used in this study were co-segregating with grain quality

Table 3. Number of alleles, percentage of polymorphism and PIC value of 37 polymorphic SSR markers.

S/N	Marker	Number of alleles	Number of polymorphic bands	Percentage of polymorphism	Polymorphic information content (PIC)
1	AB 443	2	2	100.00	0.716
2	RM 1	2	2	100.00	0.311
3	RM 3	3	2	66.67	0.602
4	RM 6	2	1	50.00	0.500
5	RM 29	3	3	100.00	0.641
6	RM 44	2	2	100.00	0.222
7	RM 140	2	1	50.00	0.337
8	RM 171	3	1	33.33	0.462
9	RM 204	3	2	66.67	0.528
10	RM 208	2	2	100.00	0.222
11	RM 209	2	1	50.00	0.494
12	RM 211	2	1	50.00	0.477
13	RM 224	2	1	50.00	0.469
14	RM 225	2	2	100.00	0.487
15	RM 226	3	3	100.00	0.603
16	RM 228	4	4	100.00	0.664
17	RM 232	2	1	50.00	0.469
18	RM 247	2	1	50.00	0.036
19	RM 304	**5**	4	80.00	0.708
20	RM 336	4	4	100.00	0.579
21	RM 339	2	1	50.00	0.035
22	RM 490	2	1	50.00	0.316
23	RM 505	3	2	66.67	0.525
24	RM 1108	2	1	50.00	0.500
25	RM1812	4	3	75.00	0.726
26	RM 3873	3	3	100.00	0.642
27	RM 6100	2	1	50.00	0.103
28	RM 137	2	1	50.00	0.494
29	RM 331	2	2	100.00	0.495
30	RM 190	2	1	50.00	0.320
31	RM 210	2	1	50.00	0.433
32	RM 250	2	1	50.00	0.397
33	RM 333	3	2	66.67	0.283
34	RM 411	2	1	50.00	0.274
35	Wx (GBSS)	2	2	100.00	0.380
36	RM 217	2	1	50.00	0.494
37	RM 234	2	2	100.00	0.497
	Mean	2.46	1.78	70.41	0.444

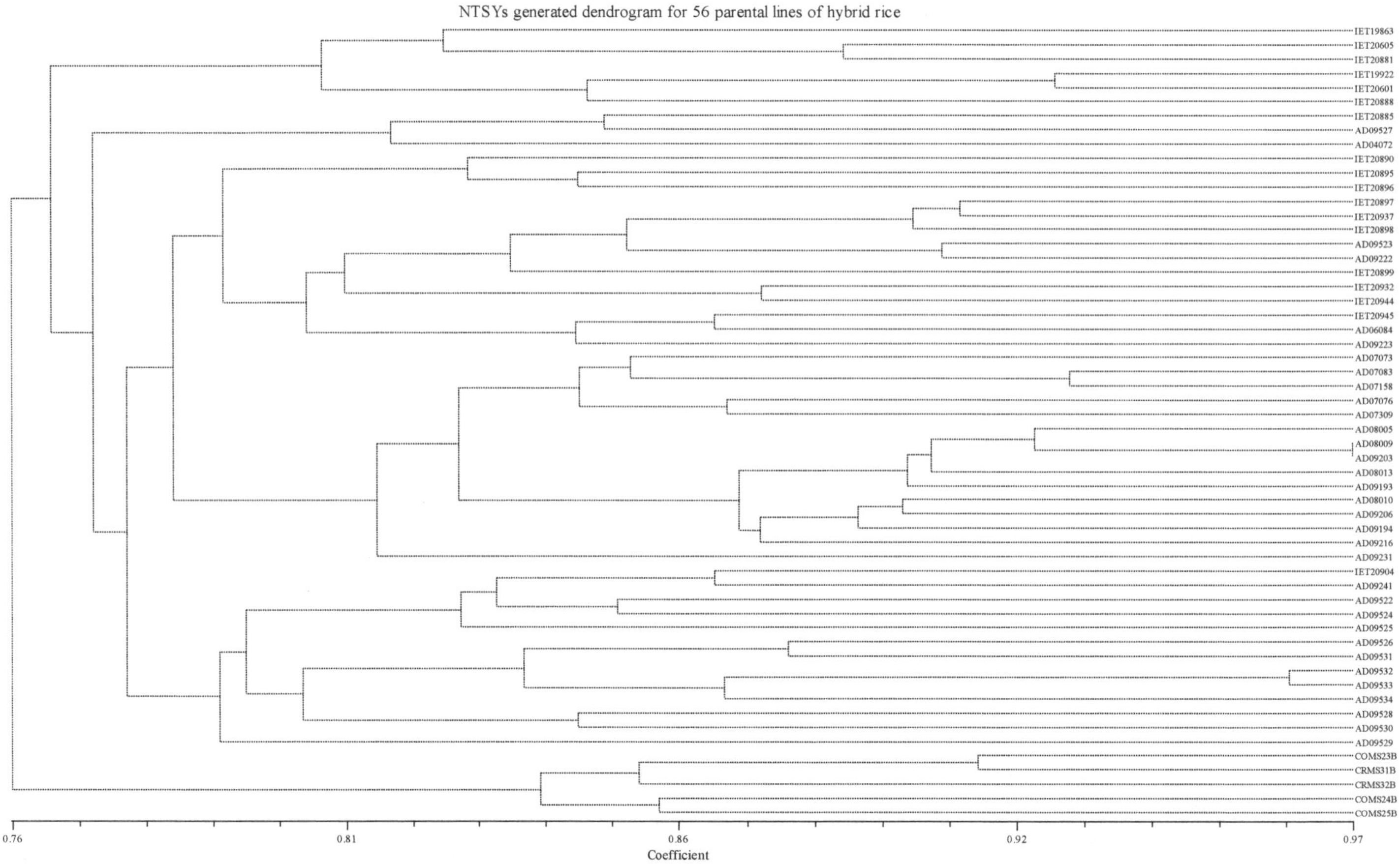

Figure 3. NTSYs generated dendrogram for hybrid rice parental lines from TNRRI.

Table 4. Clustering pattern of 56 rice genotypes based on SSR markers information.

Cluster	Genotype	Number of genotype
I	IET 19922, IET 19863, IET 20601, IET 20605, IET 20881, IET 20888	6
II	IET 20885, AD 09527, AD 04072	3
III	IET 20904, AD 09522, AD 09524, AD 09525, AD 09241	5
IV	AD 09526, AD 09531, AD 09532, AD 09533, AD 09534, AD 09528, AD 09530	7
V	AD 09529	1
VI	AD 07073, AD 07076, AD 07083, AD 07309, AD 07158, AD 08005, AD 08009, AD 08010, AD 08013, AD 09193, AD 09194, AD 09203, AD 09206, AD 09216, AD 09231	15
VII	IET 20890, IET 20895, IET 20896	3
VIII	IET 20897, IET 20898, IET 20899, IET 20932, IET 20937, IET 20944, IET 20945, AD 09523, AD 06084, AD 09222, AD 09223	11
IX	COMS 23B, COMS 24B, COMS 25B, CRMS 31B, CRMS 32B	5
	Total	56

and fertility restorer genes (Rf_3, Rf_4 and Rf_7). The diverse restorer lines identified from this SSR based diversity study will be hybridized with diverse group of CMS lines (COMS23A, COMS24A, COMS25A, COMS31A, and COMS32A) to develop hybrid combinations with superior grain quality and greater restoration ability. It could maximize opportunities to obtain superior hybrid combinations because unrelated parents would be expected to contribute unique desirable alleles at different loci. It is generally accepted that molecular markers represent genetic variation at DNA level, providing more accurate measures of relationships between individuals without the influence of environmental variation.

REFERENCES

Akagi H, Yokozeki Y, Inagaki A, Fujimura T (1996). Microsatellite DNA markers for rice chromosomes. Theor. Appl. Genet. 93: 1071-1077.

Bao JS, Wu YR, Hu B, Wu P, Cui HR and Shu QY (2002). QTL for rice grain quality based on a DH population derived from parents with similar apparent amylose content. Euphytica. 128:317-324.

Cho YG, Ishii T, Temnykh S, Chen X, Lipovich L, Park WD, Ayres N, Cartinhour S, McCouch SR (2000). Diversity of microsatellites derived from genomic libraries and Gen Bank sequences in rice (*Oryza sativa* L.). Theor. Appl. Genet., 100: 713 -722.

Dellapota SL, Woode J, Hicks JB (1983). A plant DNA minipreperation : Version 2, plant mol. Biol. rep. 1: 10-22.

Giarrocco LE, Marassi MA, Salerno GL (2007). Assessment of the genetic diversity in Argentine rice cultivars with SSR Markers. Crop Sci. 47(2): 853-860.

Hashimoto Z, Mori N, Kawamura M, Ishii T, Yoshida S, Ikegami M, Takumi S, Nakamura C (2004). Genetic diversity and phylogeny of Japanese sake-brewing rice as revealed by AFLP and nuclear and chloroplast SSR markers. Theor Appl Genet. 109(8): 1586-1596.

Janaiah A, Hussain M (2003). Can hybrid rice technology help productivity growth in asian tropics Farmers experiences. Econ. Polit. Weekly 38(25: 2492 – 2501).

Julfiquar AW (1985). Genetic Divergence Among Some Maintainer and Restorer Lines in Relation to Hybrid Rice Breeding. MS thesis, UPLB.

Khush GS (2006). Rice Research challenge for the 21st century. J. Rice Res. 1(1)

Lapitan VC, Brar DS, Abe T, Redona ED (2007). Assesment of Genetic diversity of Philippines rice cultivators carrying good quality traits using SSR markers. Breed. Sci. 57(2): 263-270.

McCouch SR, Leonid T, Xu YB, Katarzyna BL, Karen C, Mark W, Fu BY, Reycel M, Li ZK, Xing YZ, Zhang QF, Izumi K, Masahiro Y, Robert F, Genevieve D, David S, Samuel C, Doreen W, Lincoln S (2002). Development and mapping of 2240 new SSR markers for rice (*Oryza sativa* L). DNA Res. 9(6): 199–207

McCouch SR, Chen X, Panaud O, Temnykh S, Xu Y, Cho YG, Huang N, Ishii T, Blair M (1997). Microsatellite marker development, mapping and applications in rice genetics and breeding. Plant Mol. Biol. 35(1-2): 89–99.

Nghia PT, Malik JPS, Pandey MP, Singh NK (1999). Genetic distance analysis of hybrid rice parental lines based on morphological traits and RAPD markers. Omonrice, 7: PP 57-69.

Sano Y (1984). Differential regulation of waxy gene expression in rice endosperm. Theor. Appl. Genet. 68:467–473.

Seetharam K, Thirumeni S and Paramasivam K (2009). Estimation of genetic diversity in rice (*Oryza sativa L.*) genotypes using SSR markers and morphological characters. Afr. J. Biotechnol. 8(10): 2050-2059.

Septiningsih EM, Trijatmiko KR, Moeljopawiro S, McCouch SR (2003). Identification of quantitative trait loci for grain quality in an advanced backcross population derived from the *Oryza sativa* variety IR64 and the wild relative *O. rufipogon*. Theor. Appl. Genet .107:1433–1441.

Spada A, Mantegazza R, Biloni M, Capoali E, Sala F (2004). Italian rice varieties: historical data, molecular markers, and pedigrees to reveal their genetic relationships. Plant Breed. 123: 105–111.

Temnykh S, Park WD, Ayers N, Cartinhour S, Hauck N, Lipovich L, Cho YG, Ishii T, McCough SR (2000). Mapping and genome organization of microsatellite sequences in rice (*Oryza sativa* L.). Theor. Appl. Genet. 100: 697 - 712.

Thomson J. Michael Septiningsihn, Endang M, Suwardjo, Fatimah Santoso, Tri J, Silitonga, Tiur S, McCouch SR (2007). Genetic diversity analysis of traditional and improved Indonesian rice (*Oryza sativa* L.) germplasm using microsatellite markers. Theor. Appl. Genet. 114(3): Pp. 559-568.

Virmani SS, Mao CX, Hardy B (2003). Hybrid Rice for Food Security, Poverty Alleviation, and Environmental Protection. Proceedings of the 4th International Symposium on Hybrid Rice, Hanoi, Vietnam, 14-17 May 2002. Los Baños (Philippines), IRRI. 407 p.

Winter P, Kahl G (1995). Molecular marker technologies for plant improvement. World J. Microbiol. Biotechnol., 11, 438–448.

Yu SB, Xu WJ, Vijayakumar CHM, Ali J, Fu BY, Xu JL, Jilang YZ, Maghirang R, Domingo JR., Aquino C, Virmani SS, Li ZK (2003). Molecular diversity and multilocus organization of parental lines used in the International Rice Molecular Breeding Program. Theor. Appl. Genet., 108(2): 131-140.

Yu GQ, Bao Y, Shi CH, CQ Dong, Ge S (2005). Genetic diversity and population differentiation of Liaoning weedy rice detected by RAPD and SSR markers. Biochem. Genet. 43(5-6): 261-270.

Rapid approach for cloning bacterial single-genes directly from soils

Chenguang Zhu, Peipei Wang, Chunlei Shen, Wei Wang, Yuanping Tang, Bing Mei, Zhengkai Xu and Rentao Song

Shanghai Key Laboratory of Bio-energy Crops, School of Life Sciences, Shanghai University, Shanghai 200444, P. R. China.

Obtaining functional genes of bacteria from environmental samples usually depends on library-based approach which is not favored as its large amount of work with small possibility of positive clones. A kind of bacterial single-gene encoding glutamine synthetase (GS) was selected as example to detect the efficiency of cloning strategy in this study. Five *GS* genes were directly cloned from soils using degenerate primers with two steps of nested polymerase chains reactions. The genes showed 94 to 99% amino acid identities to the homologs in the known database, and encoded proteins affiliated to GS I and GS II families, respectively. All the five genes could rescue the growth of *Escherichia coli* glutamine auxotroph mutant ET6017 in minimum medium (ammonium chloride was sole nitrogen source in this medium). This study develops one rapid approach for cloning bacterial single-genes directly from soils. Comparing with the conventional strategies for gene cloning from complex environmental samples, this method did not need making genomic library and isolating target genes from large amount of library clones. This approach distinctively demonstrates its advantages of rapidity and effectiveness particularly when it aims at cloning short single-genes that had known homologs in all kinds of nucleic acid databases.

Key words: Gene cloning, soil, glutamine synthetase, nested PCR, single-gene.

INTRODUCTION

The fast development of modern biological science brings the urgent needs of cloning functional genes from various prokaryotic organisms in complex environments, as prokaryotes (mainly bacteria) exist in every part of our planet and demonstrate the huge diversity of gene resources that has far not yet to be developed and utilized by human societies (Cowan, 2000). Metagenome represents mixed genomes of multiple microorganisms from an environmental sample, and is a large gene resource that has been paid a lot of attention in recent years (Handelsman et al., 1998; Lorenz et al., 2002; Rajendhran and Gunasekaran, 2008; Cretoiu et al., 2012; Ekkers et al., 2012). Soil is one of the complex environments, which appears to be a major habitat of microbes. Many works have reported that novel genes and proteins are discovered from the soil metagenomes by library-based enzyme activity screening or nucleotide sequence screening (Henne et al., 1999; Majernik et al., 2001; Brady et al., 2002; Voget et al., 2003; Daniel et al., 2004; Riesenfeld et al., 2004; Treusch et al., 2004; Tringe et al., 2005). But for cloning the bacterial single-genes which have known homologs in the public databases from soils, the library-based approach is not favored because of its large amount of work with small probability of positive clones. Therefore, the more convenient approach for cloning such genes from environmental samples is keen to be developed.

Glutamine synthetase (GS), which produces glutamine from glutamate and ammonia, is the key enzyme in the system of regulating nitrogen assimilation of all the microbes (Shapiro and Stadtman, 1970). The reported coding region of *GS* genes from various bacteria is usually 1 to 2.3 kb (Patriarca et al., 1992; Kumada et al., 1993). The widely existence and suitable open reading frame (ORF) fragment length make the bacterial *GS* gene appropriate example for detecting the efficiency of various cloning methods. In this study, a convenient and effective approach to clone bacterial single-genes directly from the soils was developed, using *GS* gene as the target gene. This strategy was free of performing metagenomic library and screening target genes from the library. The advantage and potentially applicative field of this approach were also discussed in this work.

MATERIALS AND METHODS

Soil samples and bacterial strains

Six samples of soils were collected at different places (grassland, roots of tree, riverside, etc.) in the campus of Shanghai Univeristy. After collection, the soil samples were immediately performed with soil DNA extraction kit to isolate the metagenomic DNAs from these samples. *Escherichia coli* Top10 (stored in this laboratory) and ET6017 (provided by *E. coli* Genetic Stock Center, Yale University, USA) were respectively used for conventional DNA manipulation and functional analysis of cloned *GS* genes. *E. coli* was grown in Luria-Bertani (LB) medium or M9 medium (Sambrook and Russell, 2001) at 37°C. Ampicillin was used to screen *E. coli* transformants at concentration of 100 μg ml^{-1}.

Conventional genetic techniques

Extraction of bacterial plasmid DNA, restriction enzyme (Fermentas Life Sciences) digestion, DNA ligation, and transformation of *E. coli* cells were performed as described in the standard procedures (Sambrook and Russell, 2001). Soil metagenomic DNAs were extracted using the E.Z.N.A.™ (easy nucleic acid isolation) soil DNA kit (Omega Bio-tek, Inc.) according to the protocol provided by the kit manufacturer.

Cloning the bacterial *GS* genes

The degenerate primers specific for *GS* genes from *Bacillus cereus*, *Rhizobium* spp. and *Pseudomonas* spp. (these bacteria are known to be widely distributed in various soil samples) were designed to amplify the positive gene fragments. For designing the primer pairs of *GS* genes from each bacterial species, the most abundant DNA sequences of known homologous *GS* genes from each bacterial species were chosen to do conservative analysis. The degenerate primers were designed based on the conservative analysis of these genes. The sequences of diverse primer pairs were listed as follows:
For amplification of *GS* genes from *Pseudomonas* spp.:

Ps4-elF: 5'-GCATCACCCAAATTCAAGGG-3'
Ps4-F: 5'-ATGTCGAAGTCGGTTCAACT-3'
Ps4-R: 5'-TCAGCAGCTGTAGTAMAGCT-3'

For amplification of GS genes from *B. cereus*:

Bc1-elF: 5'-ACTGATTCTGAAGGTGTTTA-3'
Bc1-F: 5'-ATGGCTAGGTACACAAAAGA-3'
Bc1-R: 5'-TTAGTAHAGAGACATATATT-3'

For amplification of *GS* genes from *Rhizobium* spp.:

Rz1-elF: 5'-CGCAAAGTCCGWTACCGTCA-3'
Rz1-F: 5'-ATGGCGACCGCAAGCGAAAT-3'
Rz1-R: 5'-TCAGGCCGAATARTACATGT-3'
Rz2-elF: 5'-CTCCGGRTTTGCTCACCTTC-3'
Rz2-F: 5'-ATGACAAAATATAAGCTCGA-3'
Rz2-R: 5'-TYARGCKGCAGCGGAAGCCG-3'

The nested PCR protocol was applied in this study. This protocol includes two steps of PCR amplification procedures. The primer pairs titled "-elF" and "-R" were used in the first amplification procedure to amplify the whole ORF and part region (40 to 65 bps) upstream the initial codon of the ORF of the various *GS* genes. The amplified products were purified by AxyPrep PCR purification Kit (Axygen Biosciences) and were used as the template DNA for the second PCR amplification. The primer pairs titled "-F" and "-R" were used in the second amplification procedure to amplify the intact ORF of the various *GS* genes. The Ex Taq DNA polymerase (TaKaRa Co., Ltd) was used for the whole PCR procedures. The following amplification condition was applied as reference: 5 min at 94°C, 30 cycles of 50 s at 94°C, 55 s at 50°C, 90 s at 72°C, finally, 10 min at 72°C. The positive DNA fragments amplified from the second PCR procedure were purified by agarose gel purification using AxyPrep Gel Purification Kit (Axygen Biosciences), then were ligated with pMD18-T vector (TaKaRa Co., Ltd) to obtain the recombinant plasmids.

The positive clones were picked out by PCR detection using primer pairs titled "-F" and "-R". The nucleotide sequences of the positive clones were determined by double direction sequencing using M13 primer pair (forward primer 5'-CAGGAAACAGCTATGAC-3' and revise primer 5'-GTTTTCCCAGTCACGAC -3') in a MegaBACE 4500 DNA Analysis System (Amersham Biosciences).

Functional complement of *GS* genes in *E. coli* GS mutant ET6017

The *E. coli* glutamine auxotroph mutant ET6017 *[[araD139]$_{B/r}$ ⊿(argF-lac)169 flhD5301 ⊿(fruK-yeiR)725(fruA25) relA1 rpsL150(strR) ⊿(glnG-glnA)229 rha-10 deoC1]* (provided by *E. coli* Genetic Stock Center, Yale University, USA) was used for the GS genes complement study. For expression in *E. coli*, the positive clones in which the inserted *GS* gene has the identical transcriptional direction with the *lac* promoter located at the pMD18-T, were picked out and sequenced to confirm their correctness of gene sequences. Then, the plasmids of these positive clones were isolated and transformed into *E. coli* strain ET6017 by electrotransformation. The recombinant *E. coli* clones were grown in M9 minimal medium as described by Sambrook and Russell (2001), supplemented with 100 μg ml^{-1} of ampicillin. The OD_{600} values of recombinant ET6017 cells were monitored by Cary 50 UV-visible spectrophotometer (Varian Inc., USA) to quantitatively analyze their growth status.

Phylogenetic analysis of the cloned GSs

Sequences were compared with NCBI GenBank entries (http://www.ncbi.nlm.nih.gov/) using the protein-protein BLAST. For

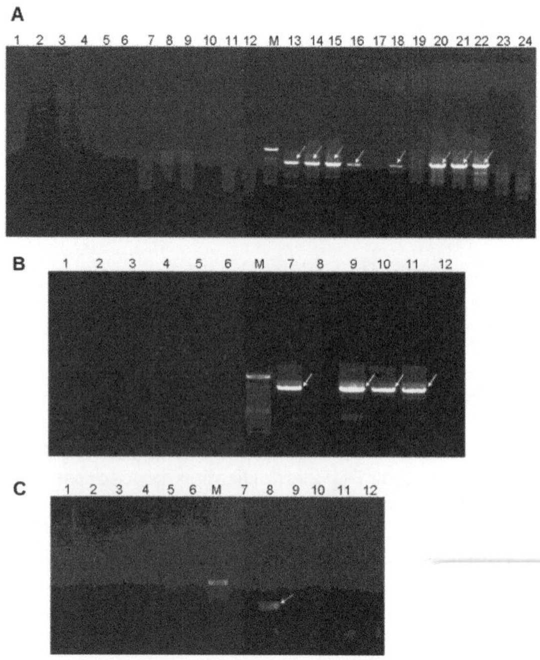

Figure 1. The agarose gel electrophoresis of nested PCR products from six samples of soils. (A) Lanes 1 to 6, the first PCR product amplified by primer pairs Bc1-elF and Bc1-R; lanes 7 to 12, the first PCR product amplified by primer pairs Rz1-elF and Rz1-R; lanes 13 to 18, the second PCR product amplified by primer pairs Bc1-F and Bc1-R; lanes 19 to 24, the second PCR product amplified by primer pairs Rz1-F and Rz1-R; M, 100 bp DNA ladder. (B) Lanes 1 to 6, the first PCR product amplified by primer pairs Ps4-elF and Ps4-R; lanes 7 to 12, the second PCR product amplified by primer pairs Ps4-F and Ps4-R; M, 100 bp DNA Ladder. (C) lanes 1 to 6, the first PCR product amplified by primer pairs Rz2-elF and Rz2-R; lanes 7 to 12, the second PCR product amplified by primer pairs Rz2-F and Rz2-R; M, 100 bp DNA Ladder. The white arrows point the specific *GS* gene fragments.

sequence alignment of GSs, the software Clustal X 1.81 (Thompson et al., 1997) was used with the following alignment parameters: gap opening penalty 10, gap extension penalty 0.2, delay divergent sequences 30%, DNA transition weight 0.5, protein weight matrix Gonnet series and DNA weight matrix IUB. Phylogenetic tree was constructed by the neighbour-joining method of the MEGA 4 program (Tamura et al., 2007) using the bootstrap test based on 1000 replicates.

Nucleotide sequence accession numbers

The nucleotide sequences of *GS* genes cloned from soil samples were deposited in the GenBank data library under accession number JX017366 ~ JX017370.

RESULTS AND DISCUSSION

Optimization of PCR procedures, cloning and sequence analysis of GS genes

At the beginning of this study, we tried to use the primer pairs '-F" and "-R" of various *GS* genes to rapidly clone the ORF of these genes from soils, but the results showed that it was hard to obtain the clear and optimal PCR fragments directly from soil DNAs (data not shown). This was probably due to the existence of humic compounds in the soil DNA templates together with the low abundance of target bacterial genome templates among the soil metagenomes, which resulted in the poor amplification of target genes. As PCR was a kind of sensitive enzymatic reaction, the purity and specialty of template DNA seemed to be very important. Therefore, we designed a series of nested primers termed as "-elF". Utilization of primer pairs "-elF" and '-R" could ensure that the integrate ORF of various *GS* genes had been amplified from the soil, although the concentration of the amplified specific DNA fragments was too low to directly do the subcloning experiments. The target ORF of the *GS* genes was obviously amplified in the second PCR reaction (Figure 1) because after purification of the amplified products of the first PCR reaction, the contaminants including humic compounds, nucleases and soil genomes had been almost removed, and subsequently, the amplified DNA fragments from the first step of PCR became to be the predominant template DNA in the reactionary system of the second PCR. This could be the major reason why the clear and abundant DNA fragments were abundantly amplified in the second PCR process.

The amplified fragments in the second PCR procedure were purified and sequenced. Altogether five *GS* genes were cloned by nested PCR. These genes were respectively termed as *grassB4*, *raodR6*, *grassP1*, *riverP6* and *grassRz2*. Amino acid homology analysis revealed that the *grassB4*, *raodR6*, *grassP1* and *riverP6* encoded proteins with highest amino acid identities respectively to the glutamine synthetases type I from *Bacillus cereus* Rock4-18 (accession no. EEL59774, 98% identity), *Rhizobium leguminosarum* bv. *viciae* 3841 (accession no. CAK07882, 99% identity), *Pseudomonas fluorescens* SBW25 (accession no. CAY46625, 96% identity) and *Pseudomonas putida* GB-1 (accession no. ABZ00983, 94% identity). The *grassRz2* encoded protein with highest amino acid identity to the glutamine synthetase type II from *Rhizobium etli* CFN 42 (accession no. ABC91870, 99% identity) (Table 1). These results indicate that all five genes cloned in this study belonged to bacterial GS gene families.

Phylogenetic analysis of cloned GS proteins

After homology analysis, phylogenetic tree of diverse GS proteins from bacteria, green alga, higher plants and animals, together with five GS proteins obtained in this study was performed to investigate the evolutionary relationship of theirs. The result indicate that GrassB4, RoadR6, GrassP1 and RiverP6 were closely affiliated to the GS I family which was widely distributed among bacteria (Figure 2). The GrassRz2 was affiliated to the GS

Table 1. Protein homologs of five GS genes cloned from the soil metagenomes.

Gene	ORF length (bps)	Protein length	Function	% protein identity (number of residues)	Organism	Accession number
grassB4	1335	444	Glutamine synthetase	98(436)	*Bacillus cereus* Rock4-18	EEL59774
				98(435)	*Bacillus cereus* Rock1-3	EEL21814
				98(435)	*Bacillus cereus* Rock3-28	EEL33362
raodR6	1410	469	Glutamine synthetase	99(465)	*Rhizobium leguminosarum* bv. *viciae* 3841	CAK07882
				99(465)	*Rhizobium leguminosarum* bv. *trifolii* WSM1325	ACS56202
				99(464)	*Rhizobium leguminosarum* bv. *trifolii* WSM2304	ACI55023
grassP1	1407	468	Glutamine synthetase	96(451)	*Pseudomonas fluorescens* SBW25	CAY46625
				96(450)	*Pseudomonas fluorescens* WH6	EFQ65563
				95(445)	*Pseudomonas brassicacearum* subsp. *brassicacearum* NFM421	AEA66503
riverP6	1407	468	Glutamine synthetase	94(440)	*Pseudomonas putida* GB-1	ABZ00983
				94(440)	*Pseudomonas* sp. TJI-51	EGB97092
				94(438)	*Pseudomonas fluorescens* Pf-5	AAY95795
grassRz2	1041	346	Glutamine synthetase	99(342)	*Rhizobium etli* CFN 42	ABC91870
				98(340)	*Rhizobium leguminosarum* bv. *trifolii* WSM2304	ACI56112
				98(340)	*Rhizobium etli* CNPAF512	EGE56659

II family that is widely distributed among various eukaryotes and several soil bacteria. All the five GS proteins cloned in this study were obviously different with the GS III proteins judging from the evolutional distances in the tree (Figure 2).

Functional complement of five GS genes in *E. coli* GS mutant ET6017

Strain ET6017 lacked the ability of synthesizing glutamine via ammonium. When this strain was inoculated into the M9 medium which contained the ammonium chloride as the sole nitrogen source, it could not grow. In this study, in order to confirm the exact function of five GS genes, the *E. coli* GS mutant ET6017 was chosen to be the receptor strain. After the GS genes were respectively transformed into strain ET6017, the recombinant cells all grew well in the M9 medium according to their OD_{600} values, however, the ET6017 cells bearing the pMD18-T vector failed to grow continuously in the M9 medium (Figure 3). This result confirms that all the GS genes could be expressed in the ET6017 cells and demonstrated the enzymatic functions of glutamine synthetase.

Significance and applicative potential of this approach

Soil is a complex environment which is a major resource of microbial diversity (Gans et al., 2005). Current opinion indicates that more than 99% of the microbes present in many natural environments such as soil are not culturable. Therefore, analysis of nucleic acids directly extracted from environmental samples facilitates the process of mining unknown genes (proteins) without the need for microbial isolation and cultivation (Schloss and Handelsman, 2003). Presently, the conventional approach of cloning genes from soil metagenomes is to construct the clone library (using vectors such as BAC, cosmid, etc.) and isolate genes from the library (Rondon et al., 2000; Rajendhran and Gunasekaran, 2008). However, the existence of contaminants including

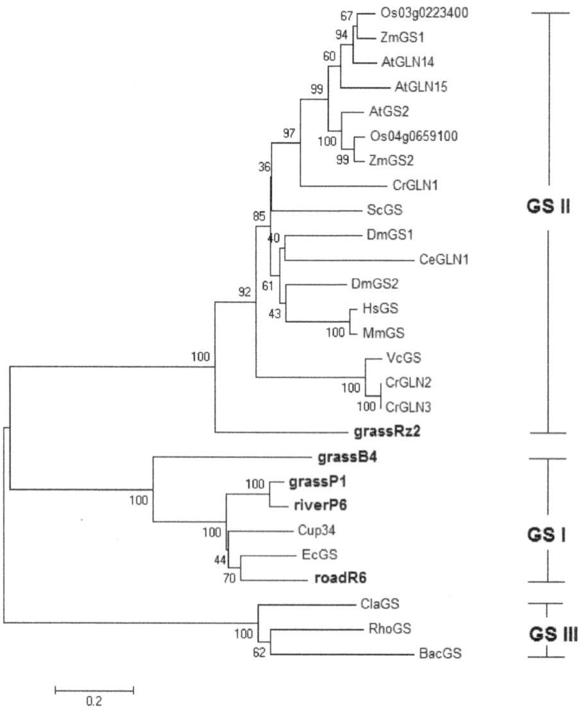

Figure 2. Phylogenetic relationship between the five cloned GS proteins and partial homologs of known organisms. The scale bar represents the substitution frequencies per amino acid residue. The values of bootstrap for the clade in 1000 replicates are indicated by numbers at nodes. Os04g0659100 and Os03g0223400, GS from *Oryza sativa Japonica* (NP_001054133 and NP_001049424); ZmGS1 and ZmGS2, GS from *Zea mays* (BAA03433, NP_001105725); AtGLN14, AtGLN15 and AtGS2, GS from *Arabidopsis thaliana* (NP_568335, NP_175280, NP_198413); CrGLN1, CrGLN2 and CrGLN3, GS from *Chlamydomonas reinhardtii* (EDP07598, Q42689, XP_001692927); DmGS1 and DmGS2, GS from *Drosophila melanogaster* (NP_722606, NP_727525); EcGS, GS from *Escherichia coli* str. K-12 substr. MG1655 (NP_418306); HsGS, GS from *Homo sapiens* (NP_001028216); ScGS, GS from *Saccharomyces cerevisiae* S288c (NP_015360); MmGS, GS from *Mus musculus* (NP_032157); CeGLN1, GS from *Caenorhabditis elegan* (NP_509012); VcGS, GS from *Volvox carteri* f. *nagariensis* (XP_002947417); Cup34, GS from *Cupriavidus metallidurans* CH34 (ABF08941); ClaGS, GS from *Clostridium acetobutylicum* ATCC824 (AAK80605); RhoGS, GS from *Rhodococcus erythropolis* PR4 (BAH34552); BacGS, GS from *Bacteroides fragilis* YCH46 (BAD47790).

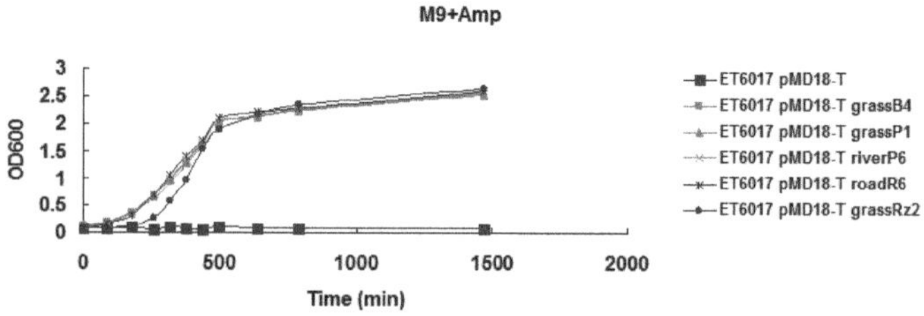

Figure 3. The growth curve of the pMD18-T based recombinant plasmids in *Escherichia coli* ET6017. The mutant ET6017 was transformed with pMD18-T, pMD18-T grassB4, pMD18-T grassP1, pMD18-T riverP6, pMD18-T roadR6, pMD18-T grassRz2. The transformants were cultivated in liquid M9 medium (100 μg ml^{-1} of ampicillin was supplemented). The growth curve was measured in the 1500 min.

humic compounds, nucleases,together with the difficulty of screening functional genes from the library, makes the approach based on genomic library analysis hard work and low efficient. In this study, a strategy for cloning bacterial single-genes directly from soil samples was developed without microbial isolation and cultivation. Based on this method, the target genes could be easily obtained to meet the subsequent needs for functional research. The practical premise of this method was to know which species of microbes living in the target soil sample and the homologs of target genes could be exactly found in the known gene database to conduct the sequence comparison and design the degenerate primers.

Different with the genomic library approach which was usually used for mining unknown genes and studying the function of large gene clusters, our method was favorable to be utilized in cloning short single-genes (no longer than 2 kbs) having known information of their homologs. Compared with the library screening approach, this method demonstrated the advantages of rapidity, effectiveness and convenience.

Compared with the conventional clone strategy from the bacterial strains, this protocol did not need to isolate the bacterial strains. It was postulated that this approach could attribute to provide rich bacterial genes for many genetic engineering improvements. Moreover, the primers designed in this work were also inferred to be used to clone the same GS genes not only from soils, but also from other complex environmental samples containing identical bacterial species.

ACKNOWLEDGEMENTS

This work was supported by the grants from the National Natural Sciences Foundation of China (30871278, 30970242), the Ministry of Agriculture of China (2013ZX08003005-002), and Shanghai Municipal Science and Technology Committee (10DZ2271800).

REFERENCES

Brady SF, Chao CJ, Clardy J (2002). New natural product families from an environmental DNA (eDNA) gene cluster. J. Am. Chem. Soc. 124:9968-9969.

Cowan DA (2000). Microbial genomes-the untapped resource. Trends Biotechnol. 18:14-16.

Cretoiu MS, Kielak AM, Al-Soud WA, Sørensen SJ, van Elsas JD (2012). Mining of unexplored habitats for novel chitinases-*chiA* as a helper gene proxy in metagenomics. Appl. Microbiol. Biotechnol. 94:1347-1358.

Daniel R (2004). The soil metagenome - a rich resource for the discovery of novel natural products. Curr. Opin. Biotechnol. 15:199-204.
Ekkers DM, Cretoiu MS, Kielak AM, Elsas JD (2012). The great screen anomaly - a new frontier in product discovery through functional metagenomics. Appl. Microbiol. Biotechnol. 93:1005-1020.
Gans J, Wolinsky M, Dunbar J (2005). Computational improvements reveal great bacterial diversity and high metal toxicity in soil. Sci. 309:1387-1390.
Handelsman J, Rondon MR, Brady SF, Clardy J, Goodman RM (1998). Molecular biological access to the chemistry of unknown soil microbes: a new frontier for natural products. Chem. Biol. 5: R245-R249.
Henne A, Daniel R, Schmitz RA, Gottschalk G (1999). Construction of environmental DNA libraries in *Escherichia coli* and screening for the presence of genes conferring utilization of 4-hydroxybutyrate. Appl. Environ. Microbiol. 65: 3901-3907.
Kumada Y, Benson DR, Hillemann D, Hosted TJ, Rochefort DA, Thompson CJ, Wohlleben W, Tateno Y (1993). Evolution of the glutamine synthetase gene, one of the oldest existing and functioning genes. Proc. Natl. Acad. Sci. USA. 90:3009-3013.
Lorenz P, Liebeton K, Niehaus F, Eck J (2002). Screening for novel enzymes for biocatalytic processes: accessing the metagenome as a resource of novel functional sequence space. Curr. Opin. Biotechnol. 13:572-577.
Majernik A, Gottschalk G, Daniel R (2001). Screening of environmental DNA libraries for the presence of genes conferring $Na^+(Li^+)/H^+$ antiporter activity on *Escherichia coli*: characterization of the recovered genes and the corresponding gene products. J. Bacteriol. 183:6645-6653.
Patriarca EJ, Chiurazzi M, Manco G, Riccio A, Lamberti A, De Paolis A, Rossi M, Defez R, Iaccarino M (1992). Activation of the *Rhizobium leguminosarum glnII* gene by NtrC is dependent on upstream DNA sequences. Mol. Gen. Genet. 234:337-345.
Rajendhran J, Gunasekaran P (2008). Strategies for accessing soil metagenome for desired applications. Biotechnol. Adv. 26: 576-590.
Riesenfeld CS, Goodman RM, Handelsman J (2004). Uncultured soil bacteria are a reservoir of new antibiotic resistance genes. Environ. Microbiol. 6:981-989.
Rondon MR, August PR, Bettermann AD, Brady SF, Grossman TH, Liles MR, Loiacono KA, Lynch BA, MacNeil IA, Minor C, Tiong CL, Gilman M, Osburne MS, Clardy J, Handelsman J, Goodman RM (2000). Cloning the soil metagenome: a strategy for accessing the genetic and functional diversity of uncultured microorganisms. Appl. Environ. Microbiol. 66:2541-2547.
Sambrook J, Russell JW (2001). Molecular cloning, a laboratory manual. Cold Spring Harbor Laboratory Press, New York, USA.
Schloss PD, Handelsman J (2003). Biotechnological prospects from metagenomics. Curr. Opin. Biotechnol. 14:303-310.
Shapiro BM, Stadtman ER (1970). The regulation of glutamine synthesis in microorganisms. Annu. Rev. Microbiol. 24:501-524.
Tamura K, Dudley J, Nei M, Kumar S (2007). MEGA4: Molecular Evolutionary Genetics Analysis (MEGA) software version 4.0. Mol. Biol. Evol. 24:1596-1599.
Thompson JD, Gibson TJ, Plewniak F, Jeanmougin F, Higgins DG (1997). The CLUSTAL_X windows interface: flexible strategies for multiple sequence alignment aided by quality analysis tools. Nucleic Acids Res. 25:4876-4882.
Treusch AH, Kletzin A, Raddatz G, Ochsenreiter T, Quaiser A, Meurer G, Schuster SC, Schleper C (2004). Characterization of large-insert DNA libraries from soil for environmental genomic studies of archaea. Environ. Microbiol. 6:970-980.
Tringe SG, Von Mering C, Kobayashi A, Salamov AA, Chen K, Chang HW, Podar M, Short JM, Mathur EJ, Detter JC, Bork P, Hugenholtz P, Rubin EM (2005). Comparative metagenomics of microbial communities. Sci. 308:554-557.
Voget S, Leggewie C, Uesbeck A, Raasch C, Jaeger KE, Streit WR (2003). Prospecting for novel biocatalysts in a soil metagenome. Appl. Environ. Microbiol. 69:6235-6242.

Genetic diversity, taxonomy and legumins implications of seed storage protein profiling in Fabaceae

Conrad A. Omonhinmin and Oluwaseun O. Ogunbodede

Applied Biology and Biotechnology Unit, Department of Biological Sciences, College of Science and Technology, Covenant University, Ota, Ogun State, Nigeria.

Proteomic evidences can be pivotal to the discovery of new plant proteins and plant relationships, due to the diversity of form it can reveal. Seed storage protein profiles of 20 Fabaceae species: 4 grain-legumes and 16 non-pulses; of 16 genera and 10 tribes were analysed by sodium dodecyl sulphate polyacrylamide gel electrophoresis (SDS-PAGE) to estimate protein content diversity and the possible genetic relatedness. 28.3% similarity and 71.7% proteomic polymorphism was scored for the species. The high variability expressed by the lot reflects the genetic diversity amongst Fabaceae population. Dendrogram based on the proteomic data clustered the species into four groups. Aside two species, *Albizia lebbeck* and *Albizia zygia* belonging to the tribe *Ingeae* and those of the tribe *Caesalpinieae*, the other species clustered with several other non-traditional cohorts resulting in a rearrangement that showed least semblance with phylogenetic relationships based on traditional morphology taxonomic delimitation. The similarity in profiles can be preliminarily forensic for proteins of importance whether for nutritional, industrial or for improvement of existing crops or for entirely new plants as crops. The protein mix, and the resultant relationship based on seed storage proteins instigates a review of erstwhile taxonomic, agricultural and research perspectives for the Fabaceae.

Key words: Fabaceae, seed protein, polymorphism, genetic diversity, taxonomy, single protein.

INTRODUCTION

Legumes vary from annual and perennial herbs to shrubs, trees, vines/lianas, and even a few aquatics; in size from some of the smallest plants of deserts and arctic or alpine regions to the tallest of rain forest trees as well as constitute conspicuous, and often dominant component of most of the vegetation types distributed throughout temperate and tropical regions of the world (Rundel, 1989). Legumes are particularly diverse in tropical forests and temperate shrub lands with a seasonally dry or arid climate. This preference for semi-arid to arid habitats is related to a nitrogen-demanding metabolism (Sprent and McKey, 1994; Sprent, 2001).

Taxonomically, the family Fabaceae is traditionally divided into three subfamilies, the Caesalpinioideae, Mimosoideae and Papilionoideae; a recognition that is based mainly on floral characteristics with 39 tribes and some 670 genera recognized (Polhill and Raven, 1981; Polhill, 1994). However, recent update of the tribal and generic re-evaluation of the classification of Fabaceae, have resulted from more than 10 years of intensive molecular phylogenetic studies; recognizes 36 tribes, 727 genera and 19,327 species (Lewis et al., 2005).

Table 1. Fabaceae taxa used in seed-protein electrophoresis analysis.

S/N	Sub family	Species	Voucher number	Status	Geographical distribution/sources
1	Mimosoideae	*Acacia auriculiformis*	CU0025	Cultivated	South west, Nigeria
2		*Adenanthera pavonina*	CU0026/13	Cultivated	BGBM, Germany*
3		*Acacia scorpioides*	CU0081/15	Cultivated/Wild	BGBM, Germany*
4		*Pentaclethra macrophylla*	CU0056	Wild/Cultivated	Southern Nigeria
5		*Albizia lebbeck*	CU0030	Wild	Southern Nigeria
6		*Albizia zygia*	CU0031	Wild	Southern Nigeria
7	Caesalpinioideae	*Bauhinia tomentosa*	CU0027	Cultivated	Southern Nigeria
8		*Bauhinia purpurea*	CU0028/16	Cultivated	Southern Nigeria.
9		*Delonix regia*	CU0032	Cultivated/Wild	Southern Nigeria
10		*Dialium guineense*	CU0033	Wild	Southern Nigeria
11		*Cassia fistula*	CU0018	Cultivated	South West
12		*Cassia javanica*	CU0019/08	Cultivated/Wild	BGBM, Germany*
13		*Caesalpinia pulcherrima*	CU0020	Cultivated	Southern Nigeria
14		*Afzelia africana*	CU0079	Wild	Southern Nigeria
15		*Peltophorum pterocarpum*	CU0055	Cultivated/Wild	Southern Nigeria
16		*Senna semia*	CU0064	Wild	South West, South South, Nigeria
17	Papilionoideae /Faboideae	*Phaselous vulgaris*	CU0077	Cultivated	Southern Nigeria
18		*Arachis hypogaea*	CU0080	Cultivated	Southern Nigeria
19		*Vigna unguiculata*	CU0076	Cultivated	Southern Nigeria
20		*Amorpha fruiticosa*	CU0078/108	Cultivated	BGBM, Germany*

* Boanisscher Garten, Botanischer Museum, Berlin Germany.

These genomic or proteomic studies are considered more reliable than morphological or cytological evidences, particularly, sodium dodecylsulphate polyacrylamide gel electrophoresis (SDS-PAGE) of total seed proteins widely applied in resolving systematic relationships and for inter and intra specific studies (Karihaloo et al., 2002; Yousaf et al., 2008).

Seed proteins are physiologically stable, easy to handle and they operate at the level of gene product where the environment has very little influence (Javaid et al., 2004; Iqbal et al., 2005). These proteins are expressed form of genome and can be used as biomarkers and were properly traced between specimens, they can lead to the identification of proteins of industrial, medicinal or nutritional prospective applications.

Exploring plants' genetic potential is becoming a frontline study, because the information it provides on depleted gene pool of cultivated plants and the genetic erosion that have accompanied human developmental tendencies, are integral to making informed decisions about plant protection, conservation and improvement programs.

The objective of the present study was to evaluate the genetic diversity, taxonomical relationships, and possible exposition of important proteins among 20 Fabaceae species, using SDS-PAGE analysis of seed storage proteins.

MATERIALS AND METHODS

Samples and protein extraction

Seed samples of 20 species of the plant family Fabaceae were received from field surveys collections of eight states in southern Nigeria (8.5°N – 6.45°N; 3.38°E – 7.5°E); and seed lot from Boanisscher Garten, Botanischer Museum, Berlin Germany (Table 1). Seed whole protein extraction was carried out with Norgen All-in-One purification kit® Norgen Biotek Corporation. The flow-through proteins content were stored at -200°C until separation analysis. Gel was fixed with a 500 mL of USP-grade 95% (v/v) ethanol in water and stained in 0.1% (w/v) Coomassie blue R350, 20% (v/v) methanol and 10% (v/v) acetic acid and afterward resultant banding were capture with a digital transilluminator.

Analysis

The numbers of monomorphic and polymorphic protein bands were scored for each sample based on staining intensity as: low, medium and high, which is an index of the subunit constituents of the protein bands. A similarity matrix based on Jaccard's similarity coefficient was generated from protein bands scored as 0 (absent) and 1 (present) and followed by a distance matrix and analysed using

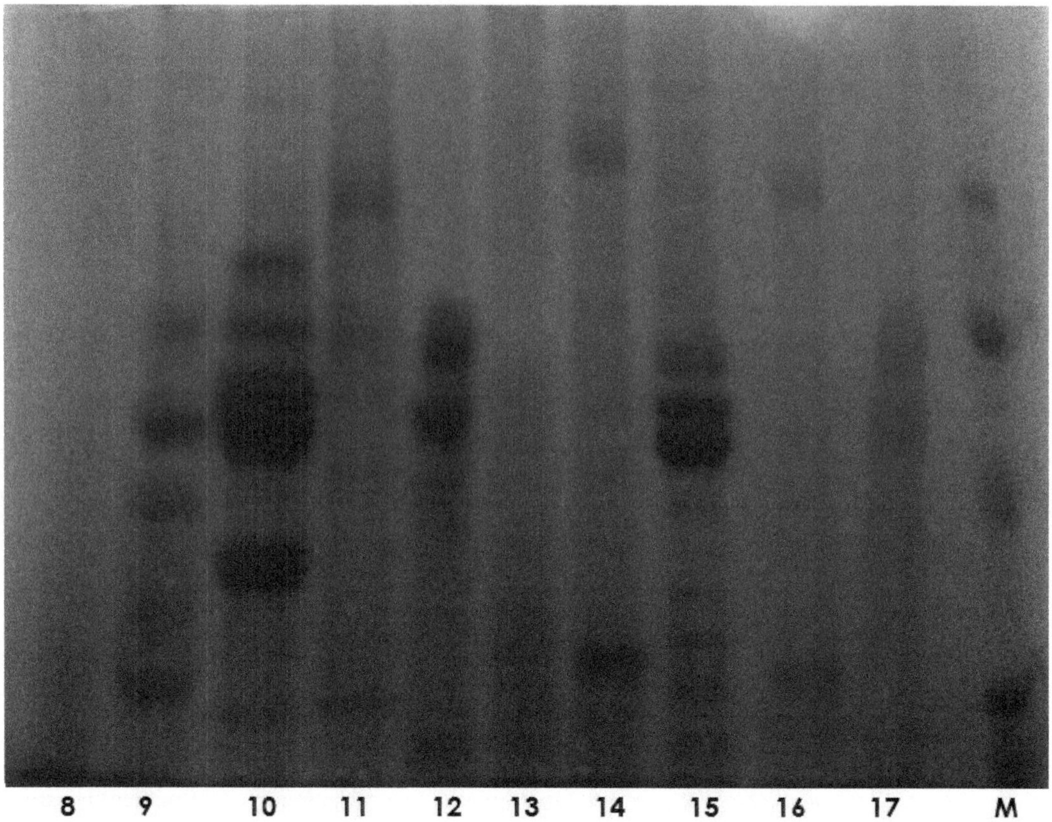

Figure 1. The banding pattern obtained by SDS-PAGE with 12% separating gel from some of the Fabaceae species of interest. Lane M = protein standard; lazne 8 = *C. pulcherrima*; lane 9 = *D. regia*; lane 10 = *D. guineense*; lane 11 = *P. pterocarpum*; lane 12= *P. vulgaris*; lane 13 = *S. semia*; lane 14 = *P. macrophylla;* lane 15 = *A. hypogeal*; lane 16 = *A. lebbeck*; lane 17 =*A. zygia*.

SPSS 15.0 for Windows. A hierarchical cluster was generated from the similarity matrix and compared with a previous similarity cluster generated from morphological data according to Polhill (1994).

RESULTS

The SDS gel electrophoresis of reproducible storage proteins for the 20 Fabaceae species resulted in bands that ranged from molecular weights of 14 to >100 kDa. Eighty nine (89) bands were detected (Figure 1).

The total seed storage proteins segregated into six distinct groups based on molecular weight of the proteins, ranging from 14 to 24, 25 to 30, 31 to 40, 41 to 62, 63to 100 and >100 kDa proteins. Each group of protein phenotypes consist of several subunits, from which specific proteins can be identified (Kottapalli et al., 2008). Higher polymorphism was recorded for proteins ranging from 31 to 62 kDa, accounting for 43.5% of the total protein bands. The least polymorphism and numbers of protein bands (9.6%) were recorded for the lower molecular weight (<24 kDa) proteins (Table 2). This indicates that the Fabaceae may predominantly express medium molecular weight genome products.

Protein profile similarity index

Analysis of the Jaccard similarity coefficient and distance for the protein profile resulted in a mean similarity of 28.3% and thus 71.7% (approximately 72%) dissimilarity. The pairwise analysis of the species (20) against the protein groups (6) recorded the least (5%) similarity and the highest (50%) similarity. Likewise, a high mean Jaccard distance of 0.668 was recorded for the taxa proteins profiled, an indication of the degree of dissimilarity between the taxa.

Cluster analysis

A dendogram of the protein polymorphism generated resulted in a phylogenetic tree construct that highlighted the clumping of the taxa into related groups based on the protein polymorphism data. Using average linkage distance, the hierarchical cluster was grouped into four distinct clusters (Figure 2) defining the phylogenetic relationship at intra-subsectional for the taxa studied.

The dendogram at 40% cluster distance revealed four clusters. The first cluster (Group A) with eight members:

Table 2. Characterization of six protein-banding patterns (groups) generated from 20 Fabaceae species.

Well lane	Taxa	Protein groups by molecular weight (kDa)						
		14-24	25-30	31-40	41-62	63-100	>100	Number of band
1	*A. auriculiformis*	Abs	Low	Abs	Low	Abs	Low	3
2	*A. pavonina*	Abs	Low	Low	Low	Abs	Low	4
3	*B. tomentosa*	Abs	Low	Abs	Abs	Abs	Abs	1
4	*B. purpurea*	Low	Low	Med	Abs	Low	Abs	5
5	*A. scorpioides*	Abs	Low	Med	Abs	Low	Abs	4
6	*C. fistula*	Abs	Abs	Med	Abs	Low	Med	5
7	*C. javanica*	Low	Abs	Low	Med	Abs	Low	5
8	*C. pulcherrima*	Low	Abs	Abs	Abs	Abs	Abs	1
9	*D. regia*	Abs	Low	Low	High	Low	Abs	6
10	*D. guineense*	Abs	Low	Med	High	Low	Low	9
11	*P. pterocarpum*	Abs	Low	Abs	Med	Med	Abs	5
12	*P. vulgaris*	Low	Abs	Abs	High	Abs	Abs	4
13	*S. semia*	Abs	Abs	Low	Abs	Abs	Abs	1
14	*P. macrophylla*	Abs	Abs	Low	Low	Low	Low	4
15	*A. hypogaea*	Low	Med	Med	High	Med	Med	11
16	*A. lebbeck*	Abs	Abs	Low	Med	Low	Abs	4
17	*A. zygia*	Abs	Abs	Low	Med	Abs	Abs	3
18	*V. unguiculata*	Low	Abs	Med	Abs	Low	Abs	4
19	*A. africana*	Abs	Abs	Med	Abs	Low	Med	5
20	*A. fruiticosa*	Low	Abs	Low	Med	Abs	Low	5

Abs = Absence; Med = medium.

Delonix regia, *Amorpha fruiticosa*, *Albizia lebbeck*, *Albizia zygia*, *Pentaclethra macrophylla*, *Phaselous vulgaris*, *Cassia javanica* and*Peltophorum pterocarpum* from the 3 sub families linking 6 tribes: *Cassieae, Caesalpinieae, Ingeae, Phaseoleae* and *Amorphheae.* The second cluster (Group B) with five members: *Caesalpinia pulcherrima*, *Senna semia*, *Bauhinia tomentosa*, *Acacia auriculiformis* and *Adenanthera pavonina* from 2 sub-families; Caesalpinoideae and Mimosoideae, linking five tribes: *Cassieae, Caesalpinieae, Circideae, Acacieae* and *Mimoseae.* The third cluster (Group C) with five members: *Cassia fistula, Afzelia africana, Bauhinia purpurea, Vigna unguiculata* and *Acacia scorpioides* from the three sub families and five tribes: *Cassieae, Circideae, Detarieae, Acacieae* and *Phaseoleae.* The forth cluster (Group D) with two members: *Dialum guineense, Arachis hypogaea* from two subfamilies and two tribes; *Cassieae* and *Aeschynomeneae.*

DISCUSSION

Genetic diversity

Protein homology (28.3%) was observed across the taxa studied, which represent the measure of similarity and thus the degree of inter-specific closeness amongst the species. 71.7% polymorphism was recorded for the seed lot. These results confirm as expected, some degree of closeness between the taxa as well as reveals the level of variations that can exist among species members of the same family (Carmona et al., 2010).

The taxa studied presented three categories of high, moderate and low molecular weight polypeptides. Recording 89 bands and a similarity of 28.3 and 71.7% protein polymorphism, the protocol applied could be useful for species/cultivar identification and protein markers generation for the family, particularly with majority of the protein homology within the 30 to 40 and 41 to 62 kDa molecular weight protein range (Figure 1 and Table 2). SDS-PAGE studies recording protein similarity and differences across different species, accessions have been highlighted by Valizadeh (2001) for grain legumes in Iran, Ghafoor and Ahmad (2005) for *Vigna mungo*, Ishtiaq et al. (2010) for Ranunculaceae and De Britto et al. (2012) for Apocynaceae. These studies including the present one, establishes that profiling seed storage protein regarded as independent of environmental fluctuations using SDS-PAGE is a reliable tool for investigating intra and inter-specific variations in plants (Hames, 1990; Carmona et al., 2010; Tchiagam et al., 2011).

The degree of polymorphism recorded shows the diversity of DNA products (proteins) within the Fabaceae and offers potentials for research, industrial and economic applications of such proteins when deciphered.

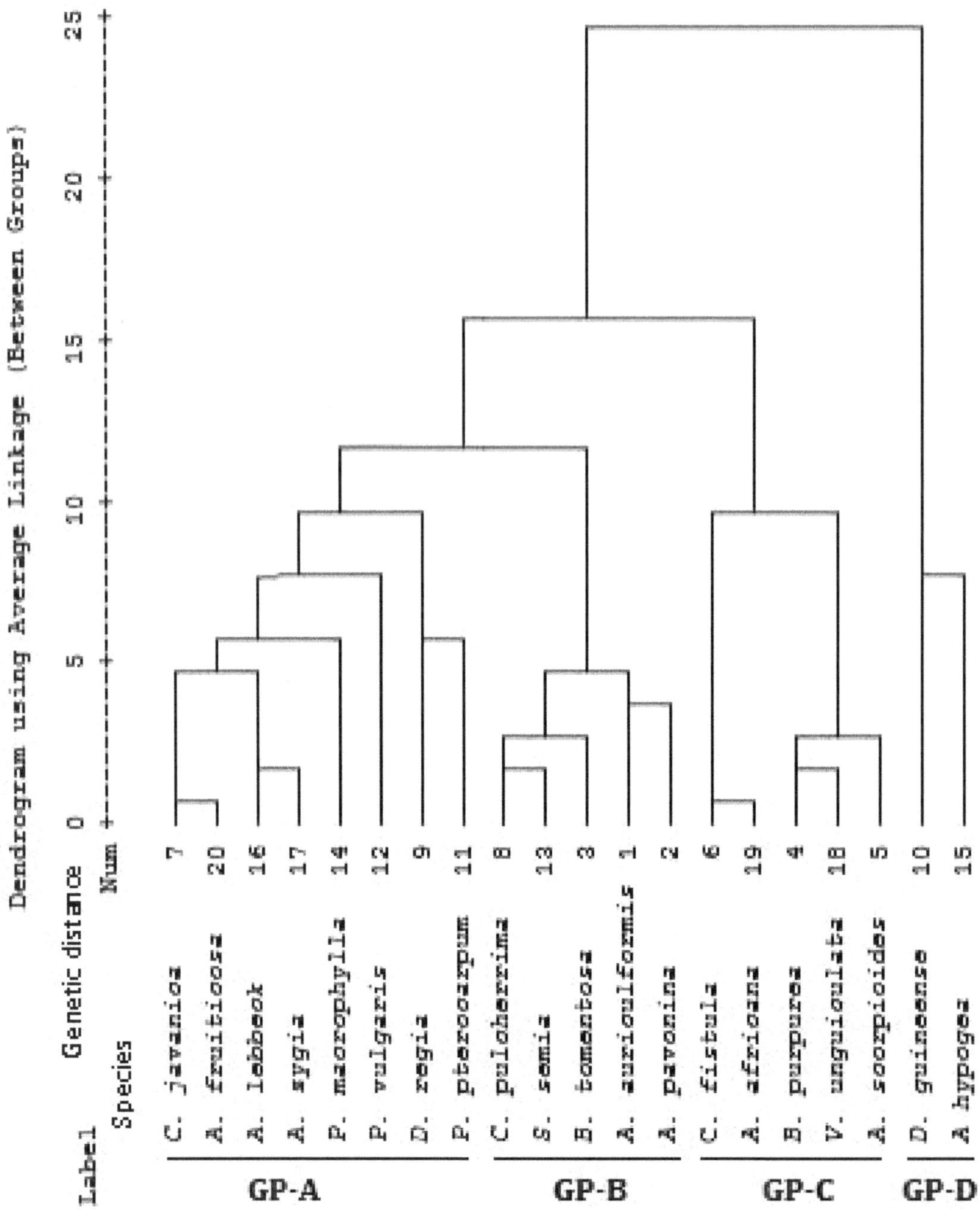

Figure 2. Dendrogram of 20 Fabaceae species based on distances produced from Jaccard's similarity coefficients of seed storage protein; with four clusters: GP-A (8 members); GP-B (5 members); GP-C (5 members); GP-D (2 members).

Taxonomic characterization

Taxonomically, the family Fabaceae have been traditionally divided into three subfamilies; Caesalpinioideae, Mimosoideae and Papilionoideae; and according to last formal classification by Polhill (1994), prior to the advent of molecular evidences, 39 tribes and 670 genera were determined for the family.

In the present study, emerging groups (A to D) in the dendogram from the Jaccard's similarity matrix revealed a divergence from the traditional taxonomic status of the Fabaceae by majority of the species studied. Comparatively, only two (*A. lebbeck* and *A. zygia*) of the 20 species maintained parallel positions in the phylogenetic trees (Figure 3) from the morphological (Polhill, 1994) and the proteomic evidences (Figure 2). In

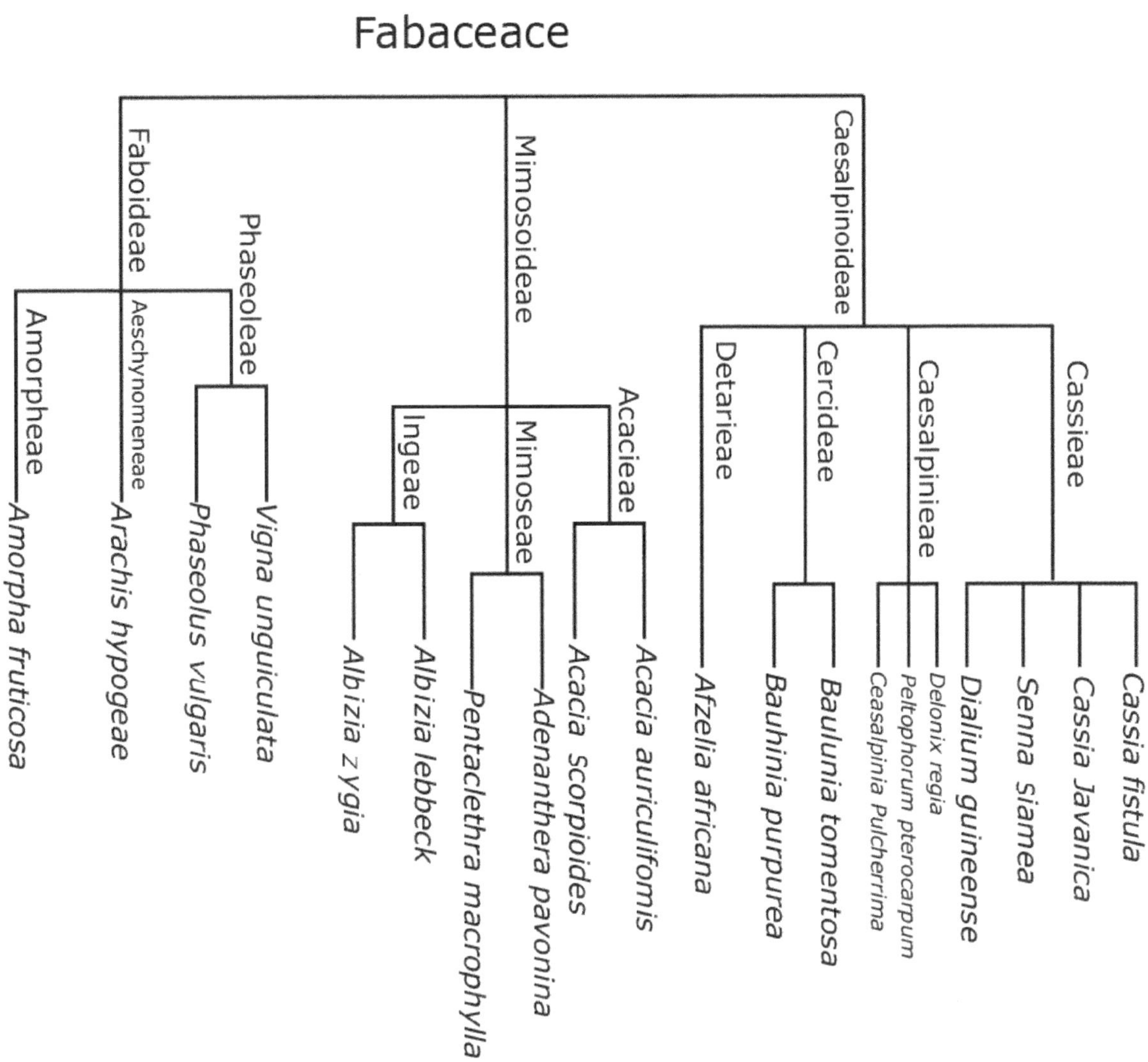

Figure 3. Traditional delimitation of the 20 Fabaceae species into three sub-families and ten tribes based on morphological characteristics according to Polhill (1994).

addition, the species *A. hypogaea* that is traditionally grouped in the sub-family Papilionoideae; increasing segregates from the group, forming an out-group (Group D) with *D. guineense*. In the seed protein profile of 11 grain legumes, Valizadeh (2001), recorded similar (single-member) out-group for *A. hypogaea*.

The species studied shared morphological characteristics that warranted their groupings, such as the tree habit for the tribe *Cassieae;* long cylindrical indehiscent pods for the genus *Cassia* and members of the tribe *Phaseoleae* are grain legumes with edible seeds. However, the present protein profile reflects a considerable re-alignment of the taxa across sub-family and tribal boundaries, which suggest that SDS-PAGE seed storage proteins profiling is a reliable tool for taxonomic studies and the evidences from such studies can lead to the identification of new groups and in some cases the delimitation into a new clade for species like *A. hypogaea*. When added to the growing pool of molecular (genomic and proteomic) evidences, these data will allow for better clarification and taxonomic review of the family, Fabaceae.

Single proteins and leaf protein concentrate applications

Comparative analysis of the study taxa profile highlighted possible legumins of importance. In the present study, *D. guineense*, clustered with *A hypogaea* and earlier profiling and identification of *A hypogaea* proteins (Kottapalli et al., 2008) recorded *Arachi*-specific as well as universal single polypeptides ranging from 18.0 to 96.6 kDa (Javaid et al., 2004). Two of such protein spots aligned with similar spots in *D. guineense* with molecular weight of 27.8 and 32.14 kDa corresponding to the proteins glycinin, and Gly1 (Catsimpoolas et al., 1971; Kottapalli et al., 2008). These 11S-globulin proteins have been implicated in reduction of cardiovascular, pro-atherogenic factors and incidence of chronic diseases (Duranti, 2006; Fassini et al., 2011). While genomic mapping may be required to mark the precise genes coding for such proteins, quantitative proofs using techniques like protein profiling are fundamental to such molecular studies.

Fabaceae species like *Cicer arietinum, Desmodium distortum, Lens culinaris, Phaseolus calcaratus, Phaseolus mungo, Pisum sativum, Psophocarpus tetragonolobus, Vigna radiate* and *Vigna unguiculata*, have shown excellent amino acid contents comparable to *Medicago sativa* (alfalfa) and *Glycine max* meal (OTA, 1983; Zia-Ul-Haq et al., 2012; Khalid et al., 2012). In the present study, *V. unguiculata* clustered with four other species like *C. fistula* and *A. africana.* Similarly, *P. vulgaris* clustered with seven other species such as *P. macrophylla.* These species posses seasonal variability and other characters that qualify them as good candidates for commercial level leaf protein concentrate (LPC) production in tropical Africa, alongside other non-leguminous plants like *Telferia occidentalis* (Agbede et al., 2008; Adeyeye et al., 2011; Aletor and Adebayo, 2012).

Storage protein profiling of Fabaceae seeds from the study highlighted a high degree of genetic diversity within the Fabaceae. The level of inter-specific variations may warrants a review of the traditional taxonomic delimitations in the family. The comparative protein banding underscores possible important species for single proteins like glycinin as well as for LPC production.

ACKNOWLEDGEMENTS

We acknowledge Covenant University, Ota and the technical assistance of Dr. Rotimi S. O.

REFERENCES

Adeyeye EI, Omolayo FO (2011). Chemical composition and functional properties of leaf protein concentrates of *Amaranthus hybridus* and *Telfairia occidentalis.* Agric. Biol. J. N. Am. 3:499-511.

Agbede JO, Adegbenro M, Onibi GE, Oboh C, Aletor VA (2008). Nutritive evaluation of *Telfairia occidentalis* leaf protein concentrates in infant foods. Afr. J. Biotechnol. 7(15):2721-2727.

Aletor O, Adebayo AO (2012). Nutritive and Physico-Chemical Characteristics of the Leaves and Leaf Protein Concentrates from Two Edible Vegetables: A Comparative Evaluation. W. J. Dairy Food Sci. 7(1):54-58.

Carmona S, Alvarez JB, Caballero L (2010). Genetic diversity for morphological traits and seed storage proteins in Spanish rivet wheat. Biol. Planta. 54(1):69-75.

Catsimpoolas N, Kenney JA, Meyer EW, Szuhaj BF (1971). Molecular weight and amino acid composition of glycinin subunits. J. Sci. Food Agric. 22(9):448-450.

De Britto AJ, Kumar PBJR, Gracelin DHS (2012). Studies on protein profile of some medicinally important species of Apocyanaceae family using SDS-PAGE. Chem. Bio. Phy. Sci. 2(2):792-796.

Duranti M (2006). Grain legume proteins and nutraceutical properties. Fitoterapia 77:67-82.

Fassini PG, Noda RB, Ferreira ES, Silva MA, Neves VA, Demonte A (2011). Soybean glycinin improves HDL-C and suppresses the effects of rosuvastatin on hypercholesterolemic rats. Lip. Health Dis. 10:165-172.

Ghafoor A, Ahmad Z (2005). Diversity of agronomic traits and total seed protein in black gram *Vigna mungo* (L.) Hepper. Acta Biol. Cra. Ser. Bot. 47(2):69-75.

Hames BD (1990). One-dimensional polyacrylamide gel electrophoresis. In: Hames, B.D. and Rickwood, D., Eds., *Gel Electrophoresis of Proteins*, 2nd Edition, Oxford University Press, New York, 382.

Iqbal SH, Ghafoor A, Ayub N (2005). Relationship between SDS-PAGE markers and *Ascochyta* blight in chickpea. Pak. J. Bot. 37:87-96.

Ishtiaq Ch M, Mumtaz SA, Wang Y, Yiyu C, Mehmood T, Ashraf M (2010). Proteins as Biomarkers for Taxonomic Identification of Traditional Chinese Medicines (TCMs) from Subsection *Rectae* Genus *Clematis* from China. World Appl. Sci. J. (Special Issue of Biotechnology & Genetic Engineering) 8:62-70.

Javaid A, Ghafoor A, Anwar R (2004).Seed storage protein electrophoresis in groundnut for evaluating genetic diversity. Pak. J. Bot. 36:25-29.

Karihaloo JL, Kaur M, Singh S (2002). Seed protein diversity in *Solanum melongena* L., and its wild and weedy relatives. Gent. Res. Crop Eval. 49(6):533-539.

Khalid II, Elhardallou SB, Elkhalifa EA (2012). Composition and Functional Properties of Cowpea (*Vigna ungiculata* L. Walp) Flour and Protein Isolates. Am. J. Food Technol. 7:113-122. doi: 10.3923/ajft.2012.113.122.

Kottapalli KR, Payton P, Rakwal R, Agrawal GK, Shibato J, Burow M, Puppala N (2008). Proteomics analysis of mature seed of four peanut cultivars using two-dimensional gel electrophoresis reveals distinct differential expression of storage, anti-nutritional, and allergenic proteins. Plant Sci. 175:321-329.

Lewis G, Schrire B, MacKinder B, Lock M (2005). Legumes of the world. Royal Botanic Gardens, Kew, UK.

OTA (1983). Plants: The Potentials for Extracting Protein, Medicines, and Other Useful Chemicals–Workshop Proceedings (Washington, D. C.: U.S. Congress, Office of Technology Assessment, OTA-BP-F-23, September 1983).

Polhill RM (1994). Classification of the Leguminosae. pp. xxxv–xlviii in Phytochemical Dictionary of the Leguminosae (FA, Bisby, J, Buckingham, JB, Harborne). Chapman and Hall, New York, NY.

Polhill RM, Raven PH (1981). Advances in legume systematics, parts 1 and 2. Royal Botanic Gardens, Kew, UK.

Rundel PW (1989). Ecological success in relation to plant form and function in the woody legumes. **In**: CH, Stirton, JL, Zarucchi,). Advances in legume biology, Monog. Syst. Bot. Missou. Bot. Gard. 29:377-398.

Sprent JI, McKey D (1994). Advances in legume systematics, Part 5, The nitrogen factor. Royal Botanic Gardens, Kew, UK.

Sprent JI (2001). Nodulation in legumes. Royal Botanic Gardens, Kew, UK.

Tchiagam JN, Bell JM, Nassourou AM, Njintang NY, Youmbi E (2011). Genetic analysis of seed proteins contents in cowpea (*Vigna unguiculata* L. Walp.). Afr. J. Biotechnol. 10(16):3077-3086.

Valizadeh M (2001). Seed Storage Protein Profile of Grain Legumes Grown in Iran, Using SDS-PAGE. J. Agric. Sci. Technol. 3:29.37-292.

Yousaf Z, Masood S, Shinwari ZK, Khan MA, Rabani A (2008). Evaluation of taxonomic status of medicinal Species of the genus *Hyoscyamous, Withania, Atropa* and *Datura* based on poly acrylamide gel electrophoresis. Pak. J. Bot. 40(6):229.39-2297.

Zia-ul-haq M, Shahid SA, Ahmad S, Qayum M, Rasool N (2012). Mineral contents and antioxidant potential of selected legumes of Pakistan. J. Med. Plant Res. 6(32):4735-4740. doi: 10.5897/JMPR12.255.

Morphometric characteristics of *Lotus corniculatus* L. genotypes

Đorđe Gatarić[1], Vojo Radić[1], Branko Đurić[1], Milenko Šarić[2], Zora Čolović[2] and Boro Petković[2]

[1]Agriculture Faculty, University of Banja Luka, Banja Luka, Republic of Srpska.
[2]Center for the Advancement and Rural Development, Banja Luka, Republic of Srpska.

The aim of this study was to examine the degree of variability in morphological and agronomic characteristics of 20 *Lotus corniculatus* L. local genotypes, and also to set aside germplasm that will be used as a source of genetic basis for improvement of the studied properties. In poor quality soils, *L. corniculatus* L. plays an important role providing protein component in animal feeds. The importance of *L. corniculatus* L. in livestock development at hilly-mountainous areas is enormous. This study presents productive characteristics of 20 *L. corniculatus* L. genotypes selected from local populations in the previous period of selection. Significant differences were observed for most of the monitored parameters. The studied genotypes showed the greatest variability in green matter yield (CV 49, 58), number of pods per plant (CV 26, 27) and seed yield (CV 34, 31). To determine the significance of differences between the genotypes and their ranking for the level of significance, Duncan's multiple range test was used. Based on the results of Duncan's multiple range test and analysis of the significance of differences between genotypes, it was ascertained that genotypes G1, G3, G15 and G17 are the basis for the creation of synthetic variety with good production properties.

Key words: *Lotus corniculatus* L., genotype, variability.

INTRODUCTION

Birdsfoot trefoil (*Lotus corniculatus* L.) is a perennial legume, important for forage production. Genus *Lotus* is highly polymorphic and widespread (Buselinck and Grant, 1995; Balan et al., 2002). It comprises about 200 annual and perennial species (Grant and Niizeki, 1999). From the agricultural standpoint, birdsfoot trefoil is the most important species in this genus, with high nutritive value. The greatest divergence of *L. corniculatus* L. occurs in the Mediterranean region (Grant and Niizeki, 1999), and includes regions of Serbia, Bosnia and Herzegovina (Vojin and Jelačić Slavica, 2007). Large collections of variability for a great number of agronomically important traits are local populations that are well adapted to local ecological conditions. Large number of *L. corniculatus* L. populations exists in climatic and soil conditions of Republic of Srpska and possess great genetic variability (Gatarić et al., 2010). Variability for the great number of agronomically traits is very important in order to identify the most productive genotypes, and to introduce them into the further breeding process. Genetic improvement of the population cannot be achieved beyond the boundaries that are determined by genes. Therefore, the selection of germplasm that will be included in the breeding is a very sensitive stage of breeding process (Rumbaugh et al., 1988). In order to achieve the expected progress, sources of parental germplasm, population size and intensity of selection are very important. Most breeders (Prosperi et al., 2006) reported a higher breeding value of locally adapted populations in relation to germplasm from other geographic areas.

The aim of this study was to determine the divergence degree of morphological and agronomical characteristics among 20 *L. corniculatus* L. genotypes, and to set aside germplasm that will be used as a source of genetic basis for improvement of the studied properties. Parental combinations with the most positive features for the new cultivars creation should be allocated on the basis of morphological and biological variability.

MATERIALS AND METHODS

From different localities throughout Bosnia and Herzegovina, 20 *L. corniculatus* L. genotypes were obtained in the previous cycle of selection from natural populations. All together, 20 *L. corniculatus* L. genotypes of national origin were examined. The experiment was established during three growing seasons (2006 to 2008) at the site of Bumbara (N 44°44' 59" E, 17° 02' 52", 431 m above sea level), close to the city of Banja Luka. Into a Jiffy pots filled with substrate, sowing of seeds was carried out on 17th March 2006. A greenhouse was used for the production of transplants. The seed was planted in the substrate and after planting, containers were placed in the greenhouse. After thirty days, plants were set out in the open due to hardening. When the plants were about 10 cm high, transplantation was performed on experimental plots. Transplanting of young plants was completed on 10th May, 2006. From each genotype, 10 plants were transplanted at the distance of 50 cm between plants and 80 cm between rows. There was no isolation between the plants, and pollination was free. Morphometric analysis of genotypes was carried out during 2006 to 2007 on five randomly selected plants of each genotype. The following parameters were measured: green mass yield (g), plant height (cm), number of stems (pieces), number of pods (pieces), thickness of the stems (mm), yield of air-dried hay (g), seed yield (g), weight of 1000 seeds (g) and number of seeds per pod (pieces). These parameters were measured on 100 plants in one growing season, and at the same plants these parameters were measured in the following year.

Analyses were performed on first cut plants, and the second year the same parameters were analyzed on second cut plants. In the first year of the research, manual harvesting of individual plants was done when 65% of the pods got dark brown color (the technological maturity of the seeds). The harvesting was done at that stage of maturity in order to avoid dispersal and the loss of seed. In the second year of the research, the first mowing was conducted when 50% of flowers were in bloom. The second mowing was conducted in August 2007 when 65% of pods got golden brown color, that is, when the shooting of the most mature pods began. The green mass yield was measured on an electronic scale after manual harvesting, directly at the experimental plot. The yield of air-dried hay was determined from the samples that were dried in the greenhouse, and then weighed on electronic scale. Plant height was measured immediately prior to mowing, on the five stems of each genotype and expressed as the average plant height (cm). Stem thickness was measured immediately after harvest (between first and second nodes above the cut), on the five stems and expressed as the average stem thickness (mm). Number of stems and number of pods per plant was analyzed by counting on each plant in the first and second year after harvest. Separation of seeds was done manually from pods on each plant. After cleaning, seed yield per plant was determined by weighing (g). Mass of 1000 seeds (g) was determined in the laboratory. The number of seeds per pod was calculated as the average from the ten pods from one plant.

The results of biometric measurements were processed by PC applications for Windows (SAS Institute Inc., 2003). The results of the studied traits were analyzed by analysis of variance (ANOVA) using SAS 9.1. program with the GLM procedure. Duncan's multiple range test (DMRT) was used to determine the significance of differences between genotypes and plants within genotypes and their ranking for the level of significance P = 0.05.

RESULTS A ND DISCUSSION

The success of selection depends on the existence of characteristics variability. Coefficient of variation is a very usable and useful measure in the selection work. It allows a direct comparison of the properties expressed in different units of measurement.

For the studied genotypes, high coefficients of variability were determined for most measured traits in both years of research. The coefficient of variation ranged from 6.51 for the mass of 1000 seeds (2006) up to 49.58 for the green mass yield (2006). The obtained values indicate high level of differences between analyzed traits. In 2006, the green matter yield had the highest coefficient of variation (CV 49.58), and in 2007, it was lower (CV 24.77) (Table 1). The coefficient of variation of air-dried hay yield was in the proportion of green mass coefficient. Similar results were obtained by other researchers (Radovic et al., 2007). Examined genotypes had higher coefficient of variation for the number of stems per plant, number of pods per plant and seed yield per plant than for plant height, stem thickness, mass of 1000 seeds and number of seeds per pod. Number of stems per plant had a great coefficient of variation, but it can be explained by the fact that observed genotypes had a large number of stems. Based on the maximum and minimum values, it can be concluded that there is wide amplitude of examined characteristics values in the studied material. Observed plants had wide variation within the ecological valence, and the aim was to identify and isolate individuals with more positive features for the further selection (Garcia de Santos et al., 2001). Based on the results of Duncan's test, it can be concluded that within observed genotypes, there are three significantly different groups in green mass yield, number of stems per plant and plant height.

The first group comprises G1, G3 and G17; although, between them, there was statistically significant difference. Thus, genotype G1, had the highest green mass yield and differed from all other genotypes, while between the genotypes G3 and G17, there was no statistically significant difference. In the second group that comprises genotypes G2, G4, G7, G9, G11 and G13, there were no statistically significant differences. Genotypes of this group had more or less uniform green mass yield and they were the closest to the average values of all investigated genotypes. The third group included the G6, G8 and G16 and those were the genotypes that had modest yield. Genotype G8 was statistically different from all other genotypes, and between G6 and G16, there was no statistically significant difference. Statistically, other genotypes did not differ from second group (Table 2). Regarding

Table 1. Descriptive statistical parameters of the tested genotype properties.

Variable	N	μ	MIN	MAX	STD	CV (%)
GMY*	20	70.45	34.72	181.02	34.93	49.58
DMY*	20	26.85	12.91	64.62	12.45	46.37
PH*	20	32.83	28.08	40.48	3.87	11.80
SW*	20	1.78	1.31	2.40	0.27	15.18
NSP*	20	33.20	24.60	56.20	7.95	23.96
NPP*	20	344.24	257.80	566.60	90.44	26.27
SYP*	20	3.28	1.92	5.77	1.13	34.31
TSW*	20	1.12	1.01	1.26	0.07	6.51
NSPo*	20	17.53	12.43	20.30	1.64	9.35
GMY**	20	189.18	107.68	276.98	46.86	24.77
DMY**	20	68.08	45.72	97.07	14.37	21.11
PH**	20	36.18	29.22	40.66	2.88	7.96
SW**	20	1.46	1.18	1.80	0.17	11.61
NSP**	20	105.08	65.40	164.00	19.84	18.88
NPP**	20	837.88	520.00	1113.20	179.08	21.37
SYP**	20	8.12	3.97	12.50	2.03	24.98
TSW**	20	1.03	0.88	1.14	0.07	6.92
NSPo**	20	12.87	9.83	16.30	1.84	14.26

*2006th year, **2007th year; The SAS system, the means procedure; GMY, Green mass yield (g); DMY, air-dry hay yield (g); PH, plant height (cm); SW, stem thickness (mm); NSP, number of stems per plant; NPP, number of pods / plant, SYP, seed yield per plant (g); TSW, mass of 1000 seeds (g); NSPo, the number of seeds per pod; N, repetition; μ, average; MIN, minimum; MAX, maximum; STD, standard deviation; CV, coeficient of variation; Values of examined morphological and agronomic traits, obtained by analysis of variance, illustrate statistically significant differences between both genotypes and years of use. Interaction effects were not determined, which means that conclusions can be based on the values of basic parameters. Duncan's multiple range test was used to determine the significance of differences between genotypes.

the number of pods per plant, it can be concluded that there were three groups of genotypes which are significantly different. The first group comprised G1, G3 and G15; although, between them, there was a statistically significant difference. Genotypes G1 and G3 had the highest number of pods per plant and they were significantly different from all other genotypes in the number of pods per plant and seed yield. In subsequent selection cycles, G1, G3 and G15 will represent a significant source of germplasm for the characteristics number of pods per plant and seeds production (Sindhu, 2004; Vuckovic et al., 2005). Based on the Duncan's test, results and analysis of the significance of differences, it can be concluded that genotype G7 had the highest, while the genotypes G1 and G3 had the lowest number of seeds per pod.

In subsequent selection cycles, G7 could be important source of germplasm for the characteristic number of seeds per pod. Thus, the final assessment should be made based on correlations and not on the basis of partial properties (Radic et al., 2011). Crossing between genotypes that showed good results can give synthetic variety with high yield components (Mc Graw et al., 1986; Smith et al., 2009).

Conclusions

Research of biological, morphological and agronomic characteristics of 20 *L. corniculatus* L. genotypes showed a number of variable traits in the studied genotypes. High coefficients of variability for most measured traits were determined on the studied genotypes in both years of research. The highest variability was found for the green matter yield (CV 49, 58). In subsequent selection cycles, genotypes G1, G3 and G17 represent a significant germplasm source for high green mass yield, while the genotypes G1, G3 and G15 had the highest production of pods per plant and seed yield. The results obtained in this study can be used for a variety of selection processes, and represent a good base for the selection activities of the domestic *L. corniculatus* L. cultivars with good production properties. The erosion of diversity of most cultivated plant species indicates the need for collection, study and evaluation of genetic resources, and for creation of new germplasm as a genetic resource in future breeding programs.

Table 2. Average values of examined traits (DMRT).

Genotype	GMY	DMY	PH	SW	NSP	NPP	SYP	TSW	NSPo
G1	221.76^{A}	80.85^{A}	39.92^{A}	1.94^{AB}	89.00^{AB}	839.90^{A}	8.29^{AB}	1.013^{DE}	11.90^{C}
G2	116.13^{BCDEF}	43.02^{BCD}	33.42^{ABC}	1.44^{CD}	68.90^{ABC}	542.90^{CDEF}	5.72^{ABCD}	1.003^{DE}	14.80^{ABC}
G3	175.12^{AB}	66.05^{AB}	38.35^{AB}	1.63^{BC}	78.20^{ABC}	818.00^{AB}	9.12^{A}	1.005^{E}	14.00^{BC}
G4	135.11^{BCDEF}	50.31^{BCD}	37.85^{AB}	1.69^{BC}	71.70^{ABC}	639.60^{ABCDE}	6.74^{ABCD}	1.157^{ABCD}	16.40^{AB}
G5	107.01^{CDEF}	44.18^{BCD}	32.81^{ABC}	1.50^{CD}	65.60^{ABC}	685.60^{ABCD}	6.60^{ABCD}	1.131^{ABCDE}	15.70^{ABC}
G6	83.38^{EF}	33.85^{CD}	29.57^{C}	1.45^{CD}	45.00^{C}	393.40^{F}	3.25^{D}	1.028^{E}	14.40^{ABC}
G7	124.20^{BCDEF}	49.68^{BCD}	36.00^{ABC}	1.76^{BC}	62.00^{ABC}	598.20^{CDEF}	6.64^{ABCD}	1.189^{AB}	17.80^{A}
G8	71.20^{F}	29.32^{D}	29.86^{C}	1.26^{D}	59.70^{BC}	453.00^{EF}	4.00^{CD}	1.025^{CDE}	14.50^{ABC}
G9	110.47^{BCDEF}	42.93^{BCD}	35.12^{ABC}	1.57^{CD}	61.70^{ABC}	503.50^{CDEF}	5.03^{BCD}	1.196^{A}	17.10^{AB}
G10	107.41^{CDEF}	40.93^{CD}	34.09^{ABC}	1.44^{CD}	62.00^{ABC}	476.40^{DEF}	5.50^{BCD}	1.034^{BCDE}	17.90^{AB}
G11	115.80^{BCDEF}	44.27^{BCD}	35.05^{ABC}	1.67^{BC}	82.60^{AB}	685.10^{ABCD}	5.51^{BCD}	1.126^{ABCDE}	14.50^{ABC}
G12	140.03^{BCDE}	50.02^{BCD}	36.10^{ABC}	1.70^{BC}	73.10^{ABC}	553.20^{CDEF}	5.52^{BCD}	1.133^{ABCDE}	15.70^{ABC}
G13	118.05^{BCDEF}	43.52^{BCD}	35.86^{ABC}	1.75^{BC}	58.70^{BC}	607.70^{BCDEF}	5.36^{BCD}	1.153^{ABCDE}	15.00^{ABC}
G14	99.29^{DEF}	35.40^{CD}	32.27^{BC}	1.64^{BC}	56.90^{BC}	413.20^{F}	4.27^{CD}	1.093^{ABCDE}	16.20^{AB}
G15	142.01^{BCDE}	52.08^{BCD}	34.08^{ABC}	1.66^{BC}	71.30^{ABC}	706.40^{ABC}	7.19^{ABC}	1.104^{ABCDE}	14.80^{ABC}
G16	92.28^{EF}	32.23^{CD}	33.66^{ABC}	1.61^{BCD}	71.10^{ABC}	529.10^{CDEF}	4.14^{CD}	1.055^{ABCDE}	13.80^{BC}
G17	169.34^{ABC}	55.50^{BC}	34.16^{ABC}	1.48^{CD}	96.40^{A}	655.20^{ABCDE}	5.73^{ABCD}	1.134^{ABCDE}	15.20^{ABC}
G18	147.10^{BCDE}	49.72^{BCD}	31.31^{BC}	1.48^{CD}	72.10^{ABC}	546.70^{CDEF}	5.66^{ABCD}	1.102^{ABCDE}	15.40^{ABC}
G19	158.35^{BCD}	53.36^{BC}	32.69^{ABC}	1.55^{CD}	65.20^{ABC}	568.00^{CDEF}	5.28^{BCD}	1.093^{ABCDE}	14.80^{ABC}
G20	162.24^{BCD}	52.14^{BCD}	37.94^{AB}	2.10^{A}	71.60^{ABC}	606.10^{BCDEF}	4.52^{CD}	1.206^{ABC}	15.70^{ABC}

The SAS system, The GLM procedure. Values marked with same letter are not significantly different at the level of r = 0.05 (DMRT); GMY, Green mass yield (g); DMY, air-dry hay yield (g); PH, plant height (cm); SW, stem thickness (mm); NSP, number of stems per plant; NPP, number of pods/plant; SYP, seed yield per plant (g); TSW, mass of 1000 seeds (g); NSPo, the number of seeds per pod.

REFERENCES

Balan M, Breazu I, Oprea G, Neagu M (2002). Genetic diversity among accessions of perennial grasses and *Lotus corniculatus* L. varieties. EGF. 19 (7):400-401.

Buselinck PR, Grant WF, Barnes RF, Miller DA, Nelson CJ (1995). Forages, An Introduction to Grassland Agriculture. 1(5): 237-248.

Garcia de los Santos G, Steiner JJ, Beuselinck PR (2001). Adaptive ecology of *Lotus corniculatus* L. genotypes. Crop Sci. 41:564-570.

Gatarić Đ, Kovačević Z, Đurić B, Radić V, Lakić Ž (2010). Genetic resources of legumes and grasses in Republic of Srpska, XII symposium on fodder crops in Republic of Serbia. Proceeding, Kruševac. 26(1-7). (in Serbian).

Grant WF, Niizeki M (1999). Birdsfoot Trefoil (*Lotus corniculatus* L.), in RJ Singh, Ed., Genetic Resources, Chromosome Engineering, and Crop Improvement. Forage Crops 5:153-204.

McGraw RL, Beuselinck PR, Smith PR (1986). Effect of latitude on genotype x environment interactions for seed yield of birdsfoot trefoil. Crop Sci. 26: 603-605.

Prosperi JM, Jenczewski E, Angevain M, Ronfort J (2006). Morphologic and agronomic diversity of wild genetic resources of *Medicago sativa* L. collected in Spain. Genet. Resour. Crop Evol. 53(4): 843-856.

Radić V, Gatarić Đ, Đurić B (2011). Correlation between the yield and the yield components of *Lotus corniculatus* L. genotypes, 46th Croatian & 6th International Symposium on Agriculture, Opatija, Croatia. 473-476.

Radović Jasmina, Lugić Z, Sokolović D, Štrbanović R, Marković J (2007). Variability of productive characteristics and fodder quality of selected Birdsfoot Trefoil (*Lotus corniculatus* L.) genotypes. *A Periodical of Scientific Research on Field and Vegetable Crops*, Novi Sad, Serbia. 44 (1):45-50.

Rumbaugh MD, Caddel JL, Rowe DE(1988). Breeding and Quantitative Genetics. *In* Hanson AA, Barnes DK, and Hill RR (1988) Alfalfa and alfalfa improvement. Agron. Monogr. 29. pp. 777–808.

SAS Institute Inc. (2003). SAS/STAT Software, Version 9.1, SAS Institute, Cary, NC.

Sindhu S (2004). Seed production potentia in birdsfoot trefoil *Lotus corniculatus* L.. Lotus Newsletter 34:5-11.

Smith M, Barbara, Diaz Anita, Daniels R, Winder L, Holland JM (2009). Regional and Ecotype Traits in *Lotus corniculatus* L., Restor. Ecol. 17:12-23.

Vojin S, Jelačić Slavica (2007). Morphological and nutritional propertis of birdsfoot trefoil *Lotus corniculatus* L. autochthonous populations in Serbia and Bosnia and Hercegovina. Crop Evolution 54:421-428.

Vučković S, Krstanović S, Ćupina B, Simić A, Stojanović I, Stanisavljević R, Vučković M (2005). Technology of seed production of Birdsfoot trefoil. Proceeding of scientific papers, Institute PKB Agroeconomic, Belgrade. 11 (1-2):125-132 (in Serbian).

Ribosomal DNA variation in landraces of bambara groundnut

Peter M Waziri[1] , Festo J Massawe[2], Samson B Wayah[1] and Jaafar M Sani[1]

[1]Department of Biochemistry, Kaduna State University, Main Campus, P.M.B. 2339, Kaduna, Nigeria.
[2]Department of Crop biotechnology, School of Biosciences, The University of Nottingham, Malaysia Campus, Jalan Broga, 43500 Semenyih, Selangor, Malaysia.

The ribosomal DNA (rDNA) is a basic component of all eukaryotic chromosomes which codes for rRNA. The rDNA is made of tandemly repeated units of 18S, 5.8S, 28S, ITS and IGS. The 18S, 5.8S and 28S are coding regions while the ITS and IGS are non coding. The rDNA evolves in a concerted manner such that rDNA units within an individual are almost completely identical making it a useful tool in taxonomic studies. The rDNA marker is used to discriminate between species in the same or different genus. Bambara groundnut is an under-utilized legume that is highly nutritious and adapted to low input farming systems. Despite its potential, Bambara groundnut is still cultivated in the form of landraces, and no true varieties have been developed. Precise estimate of genetic variation and genetic relationship is *sine qua non* for the development of improved cultivars. About 48 individual seeds (3 each) from seven landraces of Bambara groundnut, 9 wild/cultivated species of other legumes (cowpea, mungbean and groundnut) were used in this study. Analysis of the rDNA gene unit shows genetic variation in the species of the genus *Vigna* and *Arachis*. The UPGMA dendrogram constructed separates the plant species in this study into three (3) main clusters. The landraces of Bambara groundnut were placed in two different clusters indicating variation of rDNA gene units between the landraces. The analysis of rDNA gene in this study reveals the presence of genetic variation between the landraces of Bambara groundnut which could be used by scientists and breeders in the production of improved cultivars.

Key words: Ribosomal DNA, Bambara groundnut, ITS, IGS, DNA variation.

INTRODUCTION

Bambara groundnut (*Vigna subterranea* (L) Verdc.) is an important leguminous crop of African origin. Bambara is a name of a tribe and district in Mali, which is apparently where this legume derives its name from (Rassel, 1960). Because it is widely cultivated in the whole of sub-Saharan Africa, the specific centre of origin of Bambara groundnut is still unclear. There are two botanical varieties of Bambara groundnut: *Vigna subterranea* var *subterranea* which is the cultivated type and *Vigna subterranea* var *spontanea* which is the wild type (Vijaykumar et al., 2011). Bambara groundnut is locally referred to as 'gurjiya' or 'kwaruru' by the Hausas' in Nigeria although each dialect has a unique local name for Bambara groundnut in Africa. The International Institute of Tropical Agriculture (IITA) in Ibadan, Nigeria, is the major repository of the genetic resources of Bambara groundnut in Africa. Bambara groundnut is the most important legume after cowpea (*Vigna unguiculata*) and groundnut (*Arachis hypogea*) (Howell et al., 1994). As a preferred food crop to many in Africa, it is a good

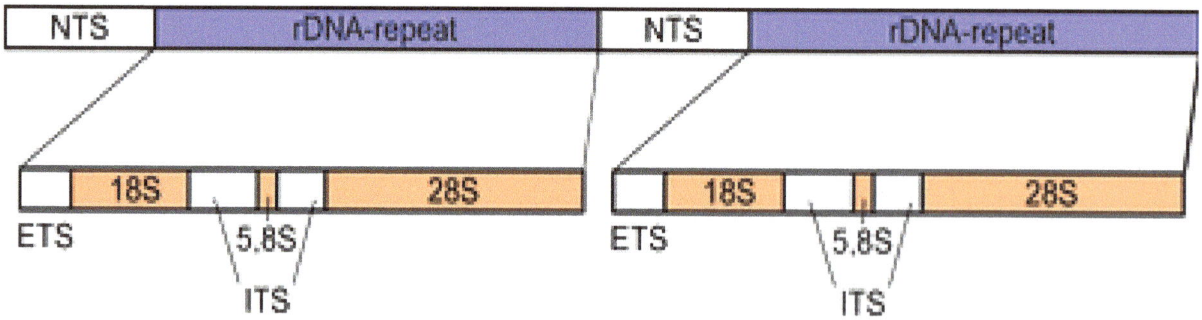

Figure 1. Ribosomal DNA cluster. Source: www.en.wikipedia.org/wiki/file:eucaryot_rdna.png.

supplement to cereal based diet and nutritionally superior to other grain legumes (Azam-Ali et al., 2001). Bambara groundnut is commonly consumed together with cereals because it contains sufficient amount of lysine which is lacking in cereals.

Ribosomal DNA (rDNA) is a component of all eukaryotic DNA. It is a conserved region that codes for ribosomal RNA (rRNA) which is involved in intra cellular protein synthesis (Kollipara et al., 1997). The ribosomal DNA consists of tandemly repeated units of 18S, 5.8S, 28S, intragenic spacer region (IGS) or non transcribed spacer (NTS) and internally transcribed spacer (ITS) as shown in Figure 1. The 18S, 5.8S and 28S regions are highly conserved in all eukaryotes unlike the ITS and IGS regions which are highly variable and non coding.

The high variability of the non coding regions (ITS and IGS) has been attributed to insertion, deletion and gene mutations. And it is this variation that is used to discriminate between individuals of the same or different species (Kollipara et al., 1997). For more than two decades now, the rDNA has been used as a powerful tool in comparative evolutionary and phylogenetic studies (Hillis and Davis, 1986; Mindell and Honeycutt, 1990; Wesson et al., 1992; Schlotterer et al., 1994; Vogler and DeSalle, 1994; Tang et al., 1996).

At present, there are no improved varieties of Bambara groundnut, as such the crop is still cultivated in the form of landraces in semi-arid regions where rainfall is limiting and environmental conditions for sustainable growth are harsh and unsuitable. A landrace is a combination of different genotypes with the ability to withstand biotic and abiotic stress and still produce good yields under low-input farming system (Zeven, 1998). Ever since Bambara groundnut was discovered in native West Africa, so many attempts have been made to characterize its complete germplasm in order to provide useful information that would facilitate the development of improved cultivars. Up till date, the complete germplasm of Bambara groundnut is yet to be fully exploited and no true variety has been produced.

The use of rDNA marker in this particular study is aimed at resolving molecular phylogeny between landraces of Bambara groundnut and some of the wild and cultivated species within the genus *Vigna* and *Arachis*. A similar study on tomato revealed the crop to be a close relative of potato and so taxonomists have placed both tomato and potato in the same genus *Solanum*, hence the change of name of tomato from *Lycopersicon esculentum* to *Solanum lycopersicon* (Doyle and Gaut, 2000). The information obtained from model crops through molecular phylogeny can thus be applied to the under-utilized crops in a bid to improve them.

The proposed study aims to evaluate genetic variation based on the occurrence of distinct rDNA within the landraces of Bambara groundnut which will thus, facilitate the appropriate selection of superior germplasms/genotypes for the breeding of true varieties of Bambara groundnut.

MATERIALS AND METHODS

About 48 individual seeds (3 each) from 7 landraces of Bambara groundnut and 9 wild/cultivated species of other legumes (cowpea, groundnut and mungbean) were obtained from the AAR-UNMC biotechnology research centre. The seeds were originally collected from Nigeria, Namibia, Botswana and Malaysia. The list of the seed material used is shown in Table 1. Finn peat soil contained in plastic pots (20 cm diameter) was used as a planting medium.

DNA extraction

Leaves were harvested from 3 weeks old plants (Table 1) and used for DNA extraction. The harvested leaves were ground to fine powder in liquid nitrogen. The fine powder was poured into 50 ml falcon tubes and DNA was then extracted using an optimized protocol developed at the AAR-UNMC biotechnology research centre. The protocol used was modified from previous protocol described by Dellaporta et al. (1983). To further confirm the integrity of the DNA extracted, 0.8% agarose gel electrophoresis was carried out. The DNA extracted was used as template for the PCR reactions.

Polymerase chain reaction

In order to optimize the primer conditions, a gradient PCR was carried out where all the primers were subjected to different annealing temperatures to select the best (optimum) primer annealing temperature.

The ranges of temperature used were between 50 – 60°C for the

Table 1. The list of plant accessions used, their status, origin and lanes they appeared on the gel pictures.

Plant	Status	Origin	Lane
V subterranea (Kano I)	Landrace	Nigeria	1
V subterranea (Kano II)	Landrace	Nigeria	2
V subterranea (Kano III)	Landrace	Nigeria	3
V subterranea (Malaysia Red)	Landrace	Malaysia	4
V subterranea (Getso)	Landrace	Nigeria	5
V subterranea (Namibia I)	Landrace	Namibia	6
V subterranea (Botswana I)	Landrace	Botswana	7
V radiata	Wild	Ghana	8
V mungo	Wild	India	9
V membranacea	Wild	Ethiopia	10
V lasiocarpa	Wild	Brazil	11
V racemosa	Wild	Ghana	12
V trilobata	Wild	Belgium	13
V unguiculata (Cowpea)	Cultivated	Nigeria	14
Arachis hypogea (Groundnut)	Cultivated	Malaysia	15
V radiata (Mungbean)	Cultivated	Malaysia	16

18S, IGS, ITS I and ITS II primers. The best annealing temperatures observed were selected and included in the main PCR amplification conditions. Similarly, for every reaction set (for each primer), a negative control (containing all PCR components except DNA template) was included. PCR amplification was carried out in a total reaction volume of 25 µl. The concentration of primer used was 0.2 µM and the amount of DNA added is 25 ng. The PCR amplification conditions used were: 94°C for 4 min (initial denaturation); 35 cycles of 94°C (denaturation) for 1 min, 55°C (primer annealing) for 1 min, 72°C (for amplification) for 1 min, and a final extension period of 10 min at 72°C.

The 4 primer pairs used to analyze the rDNA segments of Bambara groundnut and other species in the genus *Vigna* and *Arachis* in this study were synthesized by Bioline Marker Corporations, North Korea. The primer pair sequences are as follows: 18S (Forward: NS7 - 5' GAG GCA ATA ACA GGT CTG TGA TGC 3', Reverse : R635 - 5' GGT CCG TGT TTC AAG ACG G 3') (Diaz et al., 2000), 5S-IGS (Forward: VR5SL - 5' CCATCAGAACTCCGCAGTTA 3', Reverse: VR5SR - 5' GGATCCGGTGCATTAGTGCT 3') (Vijaykumar et al., 2011), ITS I: Forward - 5' AAGTCGTAACAAGGTTTCCGTAG 3', Reverse - 5' AAAGACTCGATGGTTCACG 3' (Souframanien et al., 2003) and ITS II: Forward – 5' TAGCGAAATGCGATA CTTGGT 3', Reverse - 5' GTTAGTTTCTTTTCCTCC 3') (Souframanien et al., 2003).

On completing the amplification of rDNA segments of the isolated genomic DNA, all the products were separated on a 1.5% agarose gel. One microliter (1µl) of SYBR safe (Invitrogen Oregon, USA) was added to the gel prior to casting. The PCR product was mixed with 6x loading dye (Fermentas life science, Canada) in a ratio of 5:1. Tris-borate-EDTA (TBE) buffer was used to cast as well as run the gels. The gels were run at 80V for 90 min.

Data analysis

Gel results from agarose electrophoresis were scored as (1) for the presence or (0) for the absence of bands. Each band size was treated independently. The scored data was prepared on a Microsoft excel spreadsheet and imported into the multivariate statistical (software) program (MSSP). The data consists of seven variables and sixteen samples and each size of amplified band fragment was treated as a variable. The data was then analyzed using average genetic distance and Bray-Curtis similarity coefficient. The similarity matrix obtained was used to construct the dendrogram based on UPGMA (Un-weighted Pair Group method with Arithmetic Averaging) clustering method. The diversity indices were also computed based on Shannon's diversity index ($\log_{10}$) without data transposition using the MSSP. Evolution in plant species is the basis on which the UPGMA algorithm analyses molecular data (Olukolu et al., 2011).

The % polymorphism of the rDNA markers were also calculated using the formula:

$$\frac{\text{Total number of polymorphic bands X 100}}{\text{Total number of bands}}$$

Table 2. Percentage polymorphism of the four primer pairs.

Primer	Number of bands	Polymorphism (%)
18S	12	16%
ITS I	14	14.2%
ITS II	9	33.3%
IGS	34	58%
Total bands	69	

RESULTS

The percentage polymorphisms of all the four primer pairs used are shown in Table 2. Of all the four primers used, IGS had the highest polymorphism (58%) followed by ITS II (33%). The dendrogram of the seven landraces of Bambara groundnut is shown in Figure 6 while the dendrogram of the 15 plant species in the genus *Vigna*

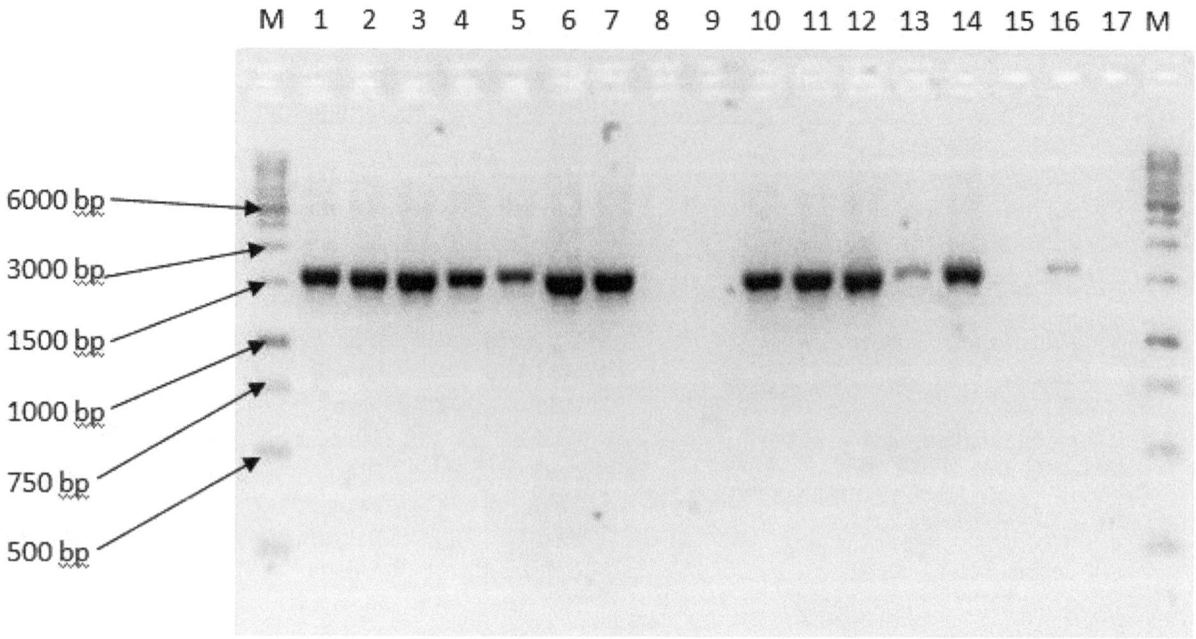

Figure 2. Gel picture of the PCR amplified 18S fragments. Lanes 1 – 7 correspond to the landraces of Bambara groundnut while lanes 8 – 13 correspond to the wild species listed in Table 1. Lane 14 – *V unguiculata*, lane 15 – *A hypogea*, lane 16 – *V radiata*, lane 17 – negative control while M represents the DNA ladder.

and groundnut (genus *Arachis*) is shown in Figure 7.

The dendrogram in Figure 6 separates the landraces of Bambara groundnut into two main clusters. Botswana I, Namibia I, Malaysia red, getso and Kano III formed the first cluster. The first cluster has two subclusters with Namibia I, Botswana I and Malaysian red forming the first subcluster while Kano III and getso formed the second subcluster. The second cluster has only Kano I and Kano II. Kano II formed the first subcluster while Kano I formed the second cluster.

The dendrogram, as shown in Figure 7 separates the landraces of Bambara groundnut used in this study into three (3) main clusters. The first cluster consists of cultivated mungbean, wild mungbean and wild *Vigna mungo* with an average genetic distance of (0.60). The cultivated mungbean formed the first subcluster while the wild mungbean and *Vigna mungo* formed the second subcluster. Similarly, all the cultivated landraces of Bambara groundnut except Kano I and Kano II formed the second cluster with an average genetic distance of (0.38). The Bambara groundnut landraces Botswana I, Namibia I and Malaysia red were very similar (bands produced by these landraces were consistent) and in addition to wild *Vigna membranaceae* they formed the first subcluster. Getso and Kano III formed the second subcluster.

The third cluster was formed by cowpea, groundnut, Kano I, Kano II, wild species of *Vigna racemosa*, *Vigna lasiocarpa* and *Vigna trilobata* with an average genetic distance of (0.58). In the third cluster, cowpea alone formed the first subcluster while the wild species of *Vigna racemosa* and *Vigna lasiocarpa* formed the second subcluster. The third subcluster was formed by groundnut and wild *Vigna trilobata* while the fourth subcluster was formed by Kano I and Kano II.

DISCUSSION

There is no sequence length variation in the 18S region of all landraces of Bambara groundnut (lanes 1 – 7) which produced only 1500 bp fragments (Figure 2). This confirms the fact that the 18S region of rDNA is highly conserved since it codes for 18S rRNA.

The result obtained in Figure 3 suggests that the ITS regions of Bambara groundnut landraces (lanes 1 – 7) are highly conserved because they have the same sequence length of ITS I region (350 bp). However, the result in Figure 4 suggests the ITS II region to be more diverged than the ITS I region as revealed by the varieties and pattern of band sizes produced in the landraces of Bambara groundnut. All landraces produced bands (400 bp) except Kano II (lane 2) and Kano III (lane 3) from Nigeria. This difference between the landraces of Bambara groundnut could be due to geographic origin as reported in previous studies (Massawe et al., 2002).

Similarly, the results in Figure 5 indicates that the sequence length of IGS varies between landraces of Bambara groundnut, wild species of *Vigna*, cultivated species of cowpea, mungbean and peanut which makes it a powerful tool in phylogenetic studies. This justifies its wide use in resolving taxonomic differences (Singh et al., 2008).

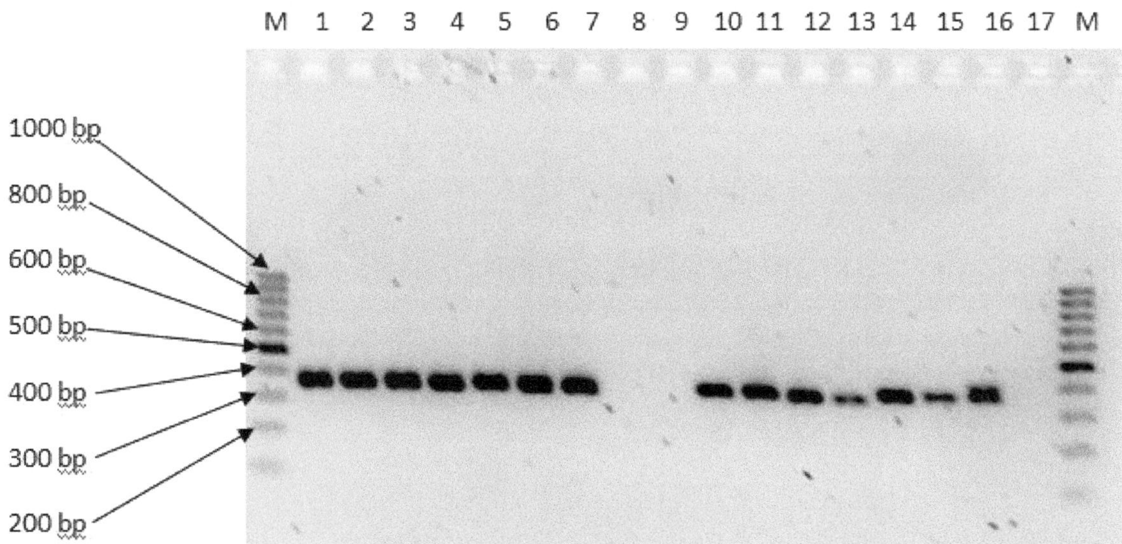

Figure 3. Gel picture of the PCR amplified ITS I fragments. Lanes 1 – 7 correspond to the landraces while lanes 8 – 13 correspond to the wild species listed in Table 1. Lane 14 – *V unguiculata*, lane 15 – *A hypogea*, lane 16 – *V radiata*, lane 17 – negative control while M represents the DNA ladder.

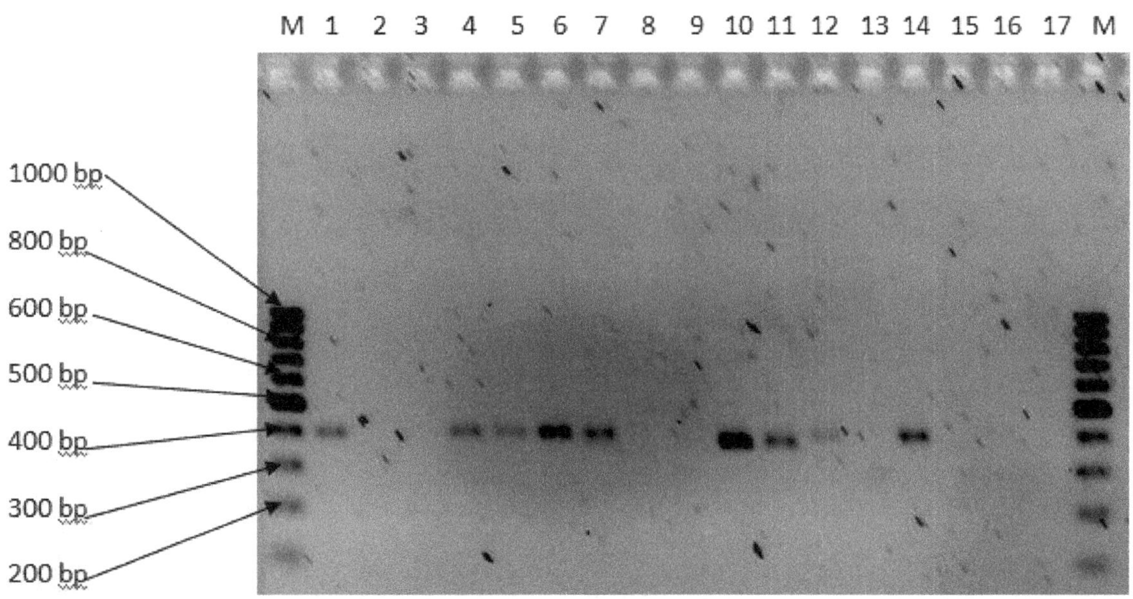

Figure 4. Gel picture of the PCR amplified ITS II fragments. Lanes 1 – 7 correspond to the landraces while lanes 8 – 13 correspond to the wild species listed in Table 1. Lane 14 – *V unguiculata*, lane 15 – *A hypogea*, lane 16 – *V radiata*, lane 17 – negative control while M represents the DNA ladder.

Phylogenetic analysis

The taxonomy (classification) of plant species including those within the genus *Vigna* has been a subject of controversy over the years. So many wild crop species are yet to be identified and this can be a limitation to breeding since wild species of plants are known to contain large repositories of genetic information. Similarly the genetic improvement of under-utilized legumes such as Bambara groundnut also relies on the identification and characterization of wild species within the genus *Vigna*. In a bid to prevent the wild type species from going extinct, phylogenetic studies enables the identification and preservation of the wild plant species. Crosses made between wild and cultivated species could produce new and improved cultivars of Bambara groundnut.

In this study, the average genetic distance within single cluster of the dendrogram (Figure 7) was uniform because

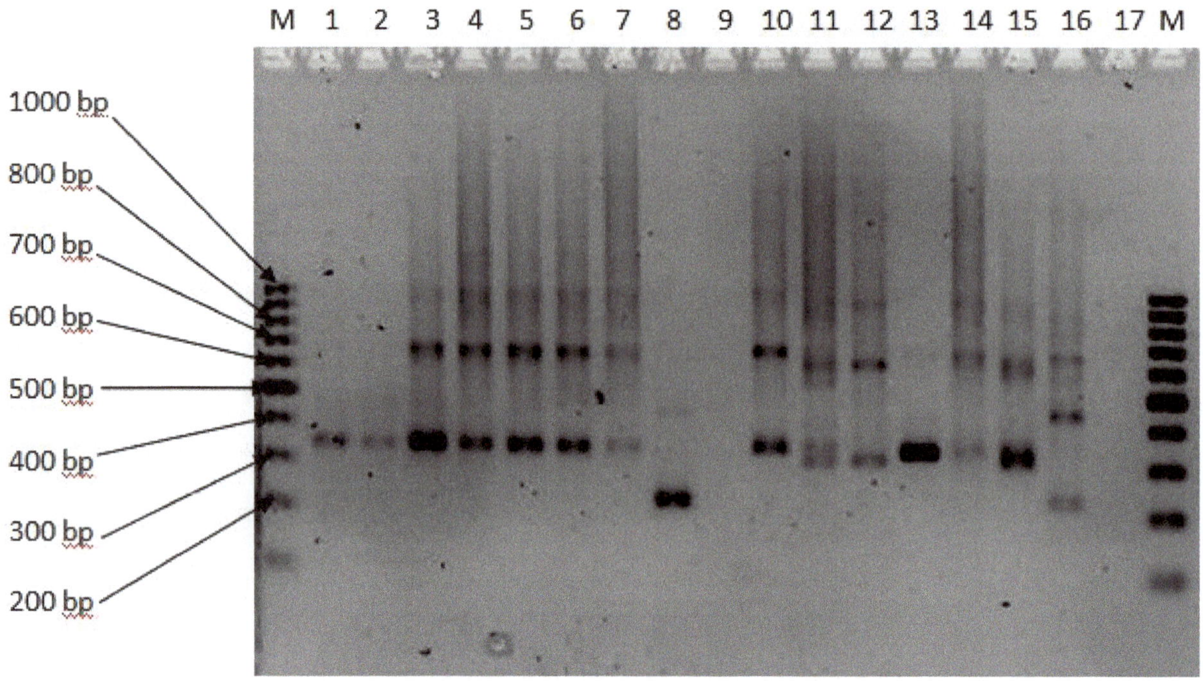

Figure 5. Gel picture of the PCR amplified IGS fragments. Lanes 1 – 7 correspond to the landraces while lanes 8 – 13 correspond to the wild species listed in Table 1. Lane 14 – *V unguiculata*, lane 15 – *A hypogea*, lane 16 – *V radiata*, lane 17 – negative control while M represents the DNA ladder.

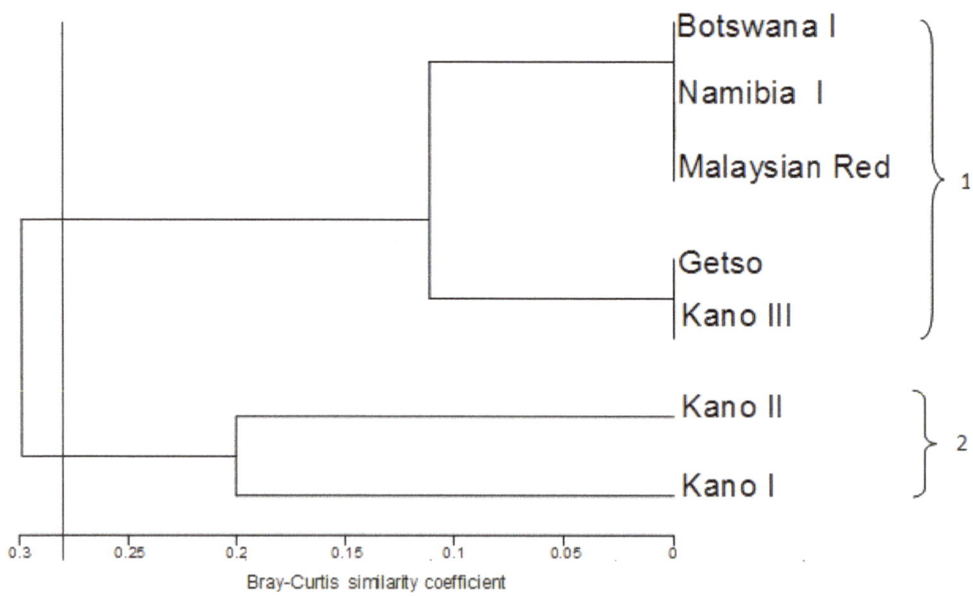

Figure 6. UPGMA dendrogram of the seven landraces of Bambara groundnut based on Bray-Curtis similarity coefficient.

the plants in the cluster have similar sequence length of rDNA genes based on the (UPGMA) clustering analysis. The classification of organisms through clustering algorithms based on average genetic distance has been described as a reliable tool in inferring phylogenetic relationship (Oosthuizen et al., 2006). Usually, low genetic distances signify close relationship while high genetic distances signify distant relationship (Oosthuizen et al., 2006). The average genetic distance of the 15 species in the genus *Vigna* is 0.52. This value is slightly more than the average genetic distance expected (0.40) for species within the same genus (Grant et al., 1988).

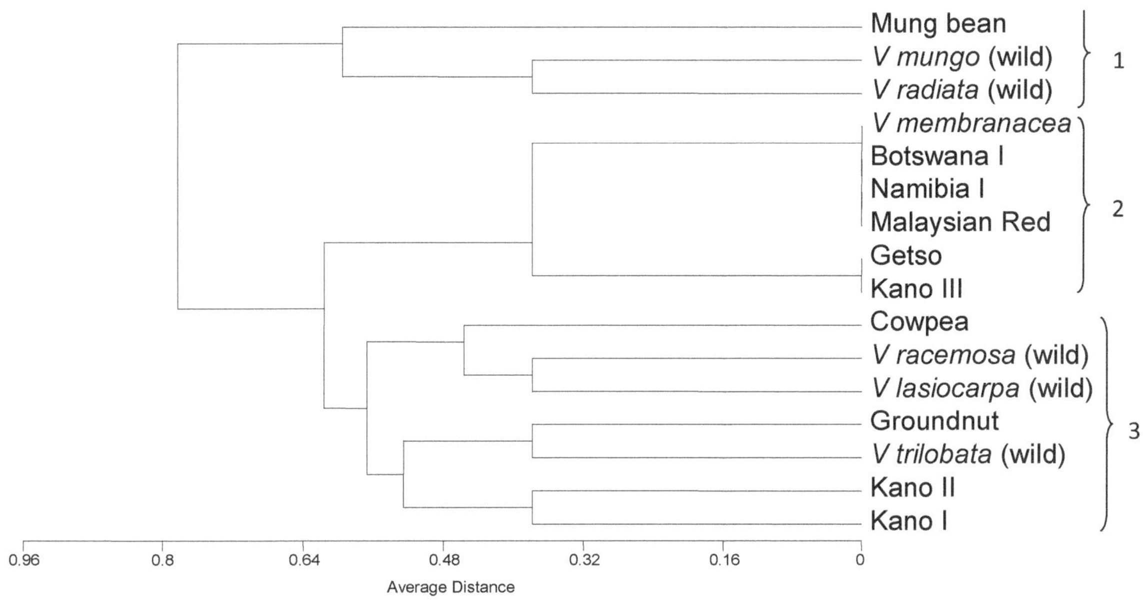

Figure 7. UPGMA dendrogram of the 15 species of the genus *Vigna* and groundnut (genus *Arachis*) based on their average genetic distance.

One important aspect of this study is that the seven (7) landraces of Bambara groundnut were placed in two separate clusters (Figures 6 and 7) implying that there is variation of rDNA gene units between the landraces. Geographic origin may be responsible for the non inclusion of the Kano I and Kano II in the second cluster (Figure 7) because they were originally collected from Kano state, Nigeria. Previous studies using RAPD marker indicates that landraces of Bambara groundnut are related to their place of origin (Massawe et al., 2002).

The landrace variation could be responsible for the inconsistent yields in Bambara groundnut. One of the goals of plant breeding is to develop pure lines of a crop that would produce uniform yield under low input farming systems. A recent study by Singrun and Sheckel (2004) indicates that landraces of Bambara groundnut from the Southern parts of Africa have between 3 – 8 different genotypes. The different genotypes are believed to confer agronomic traits such as the ability of Bambara groundnut to withstand low input farming and still produce significant yields. The task ahead lies in identifying the high yielding genotypes of Bambara groundnut and using them to make crosses that would produce improved and uniform yields in Bambara groundnut. Even though creation of pure lines in Bambara groundnut could have its own limitations, it will ensure consistency in yields obtained. One of the drawbacks of pure line development is that it could be wiped out in the case of disease outbreak because of its genetic uniformity. However, development of multiple in-bred lines could help integrate many traits in a single cultivar and overcome the limitations associated with the production of pure lines in harsh environments. Notwithstanding, achieving this goal largely depends on the identification of sufficient variation within the germplasms of Bambara groundnut.

The fact that the sequence length of rDNA genes in landraces of Bambara groundnut from Nigeria varied considerably could be due to the enormous genetic diversity of the landraces in West Africa. It is a common knowledge that the evaluation of genetic diversity is *sine qua non* for the improvement of under-utilized crop species because rich germplasms (high genetic diversity) provides surplus raw material for the production of superior varieties of crops.

Cowpea which is also believed to have originated from West Africa was also grouped in the same cluster as Kano I and Kano II supporting the classification of both cowpea and Bambara groundnut in the subgenera *Vigna*. This is due to the presence of paralogous genes in both cowpea and Bambara groundnut. However, none of the landraces were grouped together with mungbean, another legume from the genus *Vigna*. But interestingly, the cultivated and wild mungbean (*V radiata*) were clustered together thus corroborating previous claim that the wild mungbean is the progenitor of the present day cultivated mungbean. Only both of them (wild and cultivated mungbean) produced 250 bp fragments within the IGS region and were grouped together in the first main cluster. This also implies that the cultivated mungbean still contain genes homologous to its wild counterpart even after going through evolution. Mungbean is an important legume consumed and cultivated majorly in Asia. It can also be improved by making intra-specific crosses with its wild counterpart.

The dendrogram (Figure 7) also reveals a close relationship between *Vigna radiata* and *Vigna mungo* which corroborates similar findings of Kaga et al. (1996) and Goel et al. (2001). Both *V radiata* and *V mungo* are in the subgenera *Ceratotropis* also known as Asian genus *Vigna* and their germplasms are useful for making crosses within the species of *Vigna*.

The spacer regions (IGS and ITS) exhibited more variability in this study and has helped in analyzing the phylogenetic relationship in different species of the genus *Vigna*. This variability is an important asset to the taxonomist where it is used in resolving taxonomic differences in plants (Penteado et al., 1996). Phylogenetic relationship does not refer to the similarity between species but the relative times in the past when species shared common ancestors. However, the variability within the spacer regions could be due to gene flow where genes are exchanged between species through natural and artificial means (Nepolo et al., 2010). This makes gene flow to be an invaluable source of genetic variation within a particular species as seen in the landraces of Bambara groundnut and the other species of *Vigna* used in this study. As such, the rDNA marker can be associated with specific traits of interest, and thus facilitate breeding true varieties of Bambara groundnut.

Conclusion

One of the goals of this study is to explore the level of genetic variation in Bambara groundnut available to plant breeders for genetic improvement purposes. Analysis of the non coding regions (ITS and IGS) reveal the presence of genetic variation between the landraces of Bambara groundnut as well as in other species of the genus *Vigna* and *Arachis* in this study. This study supports the classification of Bambara groundnut and cowpea in the same subgenera *Vigna* in contrast to previous report by Phansak et al. (2005) where Bambara groundnut and cowpea were placed in the subgenera *ceratotropis* and *Vigna* respectively. The close relationship between Bambara groundnut and other *Vigna* species observed in this study justifies the change of name of Bambara groundnut from *Voandzeia subterranea* to *Vigna subterranea*. The average genetic distance (0.52) obtained is also within the range expected for species within the same genus and family. Analysis of the rDNA genes support claims that wild cowpea and mungbean are progenytors of the present day cultivated cowpea and mungbean respectively. Similarly, the analysis of the rDNA genes reveals the presence of homologous genes in *Vigna radiata* and *Vigna mungo* corroborating previous studies by Kaga et al. (1996) and Goel et al. (2001).

The variation revealed between the landraces of Bambara groundnut in this study makes the rDNA gene an invaluable tool in molecular phylogeny. The genetic variation obtained in this study adds to and further our understanding of Bambara groundnut germplasm and could help facilitate the production of ideal Bambara groundnut types with useful agronomic traits that would meet consumer preferences. Estimation of genetic variability based on the rDNA genes in plants as done in this study is very imperative because, species in the genus *Vigna* belong to the family of legumes known for their ability to fix nitrogen. As such, they have tremendous roles to play in maintaining soil fertility and sustaining global agriculture.

Future studies

This particular study adds to the list of studies that confirms the usefulness of rDNA in molecular phylogeny. The rDNA gene analysis should be extended to other crops whose origin and genetic relationship is still unclear. Because of the enormous potential of Bambara groundnut, future studies should be focused on assessing genetic variability in all of the existing landraces of Bambara groundnut distributed across Africa. This would help in providing sufficient genetic variation that is required for the production of improved cultivars of Bambara groundnut.

Legumes play significant roles in agriculture especially in low input farming systems where they are intercropped with cereals. Being cheap source of proteins, they complement cereals (and other foods) in feeding a large population of over 7 billion people that live in the world today. The main focus should be on improving the production of these cheap food products and also unraveling the mystery behind the genes that help in feeding the world.

Biotechnology offers one of the best tools of improving under-utilized crop species and so requires a concerted effort by researchers, local farmers and funding agencies. Promotion is equally important in making these underutilized (including Bambara groundnut) crops reach their potentials.

ACKNOWLEDGEMENT

I want to thank God Almighty for making this research work come to fruition. I also want show my sincere gratitude to Dr Festo Massawe for the guidance he provided me with all through the course of this research. And to the entire staff of biotechnology research centre, University of Nottingham-Malaysia Campus, thank you for your support.

REFERENCES

Azam-Ali SN, Sesay A, Karikari SK, Massawe FJ, Anguilar-Manjarrez J, Bannayan M, Hampson KJ (2001). Assessing the potential of under-utilized crop – A case study using Bambara groundnut. Exp. Agric. 37: 433-472.

Dellaporta SL, Wood J, Hicks JB (1983). A plant DNA minipreparation: Version II. Plant Mol. Biol. Rep. 1: 19–21.

Diaz MR, Fell JW (2000). Molecular analyses of the IGS & ITS regions of rDNA of the psychrophilic yeasts in the genus *Mrakia*. Antonie van Leeuwenh 77: 7–12.

Doyle JJ, Gaut SB (2000). Plant Molecular Evolution. Kluwer Academic publications, Dordrecht, Netherlands.

Goel S, Raina SN, Ogihara Y (2001). Molecular Evolution and Phylogenetic Implications of Internal Transcribed Spacer Sequences of Nuclear Ribosomal DNA in the *Phaseolus–Vigna* Complex. Mol. Phylogenet. Evol. 22: 1–19.

Grant WS, Dempster YL, Da Silva FM (1988). Use of protein electrophoresis in evolutionary systematic. Trans R. Soc. S. Afr. 46(4): 295–311.

Hillis DM, Davis SK (1986). Evolution of ribosomal DNA: fifty million years of recorded history in the frog genus *Rana*. Evolution 40: 1275–1288.

Howell JA, Eshbaugh WH, Guttman S, Rabakonandrianina E (1994). Common names given to Bambara groundnut (*Vigna subterranea*) in Madagascar. Econ. Bot. 48: 217–221.

Kaga A, Tomooka N, Egawa Y, Hosaka K, Kamijima O (1996). Species relationships in subgenus *Ceratotropis* (genus *Vigna*) as revealed by RAPD analysis. Euphytica 88: 17–24.

Kollipara KP, Singh RJ, Hymowitz T (1997). Phylogenetic and genomic relationships in the genus *Glycine* Willd. based on sequences from the ITS region of nuclear DNA. Genome 40: 57–68.

Massawe FJ, Dickinson M, Roberts JA, Azam-Ali SN (2002). Genetic diversity in Bambara groundnut landraces (*Vigna subterranea* (L.) Verdc) revealed by AFLP markers. Genome 45: 1175–1180.

Mindell DP, Honeycutt RL (1990). Ribosomal RNA in vertebrates: evolution and phylogenetic applications. Ann. Rev. Ecol. Syst. 21: 541–566.

Nepolo E, Takundwa M, Chimwamurombe PM, Cullis CA, Kunert K (2010). A review of geographical distribution of Marama bean [*Tylosema esculentum* (Burchell) Schreiber] and genetic diversity in the Namibian germplasm. Afr. J. Biotechnol. 8(10): 2088–2093.

Oosthuizen S, Van der Bank FH, Ferreira JT (2006). Genetic distances between and evolutionary relationships for *Tilapia guinasana, T. rendali T. sparrmani*. Biochem. Syst. Ecol. 21(3): 351–362.

Penteado MI, Garcia P, Pérez de la Vega M (1996). Genetic variability and mating system in three species of the genus Centrosema. J. Hered. 87: 124–130.

Phansak P, Taylor PWJ, Mongkolporn O (2005). Genetic diversity in yardlong bean (*Vigna unguiculata* ssp. *sesquipedalis*) and related *Vigna species* using sequence tagged microsatellite site analysis. Sci. Hortic. 106: 137–146.

Rassel A (1960). Le voandzou *Voandzeia subterranea* Thou. et sa culture au Kwango. Bull. agric. du Congo Belge et du Ruanda-Urundi 51: 1–26.

Schlotterer C, Hauser MT, Von Haeseler A, Tautz D (1994). Comparative evolutionary analysis of rDNA ITS regions in *Drosophila*. Mol. Biol. Evol. 11: 513–522.

Singh A, Deverumath RM, RamaRao S, Singh VP, Raina SN (2008). Assessment of genetic diversity, and phylogenetic relationships based on ribosomal DNA repeat unit length variation and Internal Transcribed Spacer (ITS) sequences in chickpea (*Cicer arietinum*) cultivars and its wild species. Genet. Resour. Crop Evol. 55: 65–79.

Singrun C, Schenkel W (2004). Fingerprinting of Bambara groundnut germplasm with molecular markers. Proceedings of the Int. Symposium on Bambara Groundnut, Botswana College of Agriculture 8 – 12 September, 2003.

Souframanien J, Joshi A, Gopalakrishna T (2003). Intraspecific variation in the internal transcribed spacer region of rDNA in black gram (*Vigna mungo* (L.) Hepper). Curr. Sci. 85(6): 798–802.

Tang J, Toe L, Back C, Unnasch TR (1996). Intra-specific heterogeneity of the rDNA internal transcribed spacer in the *Simulium damnosum* (Diptera: *Simuliidae*) complex. Mol. Biol. Evol. 13: 244–252.

Vijaykumar A, Saini A, Jawali N (2011). Molecular characterization of intergenic spacer region of 5S ribosomal RNA genes in subgenus *Vigna*: extensive hybridization among *V. unguiculata* subspecies. Plant Syst. Evol. 294: 39–55.

Vogler AP, DeSalle R (1994). Evolution and phylogenetic information content of the ITS-1 region in the tiger beetle *Cicindela dorsalis*. Mol. Biol. Evol. 11: 393–405.

Wesson DM, Porter CH, Collins FH (1992). Sequence and secondary structure comparisons of ITS rDNA in mosquitoes (Diptera: *Culicidae*). Mol. Phylogenet. Evol. 1: 253-269.

Zeven AC (1998). Landraces: A review of definitions and classifications. Euphytica 104: 127–139.

An evaluation of some mutant cotton (*Gossypium hirsutum* L.) varieties from Azerbaijan in Southeast Anatolian region of Turkey

Lale EFE, Fatih KILLI and A. Sefer MUSTAFAYEV

Kahramanmaras Sutcu Imam University, Faculty of Agriculture, Field Crops Department, Kahramanmaras, Turkey.

The aim of this study was to determine seed cotton yield, yield components and fiber technological properties of four mutant cotton varieties from Azerbaijan (Agdas-3, Agdas-6, Agdas-7 and Agdas-17) and three local standard varieties (Maras-92, Sayar-314 and Stoneville-453) belonging to *Gossypium hirsutum* L. The study was carried out at the experimental field of Koruklu Agricultural Research Station in Sanliurfa Province (Akcakale-Harran Plain) in the South-east Anatolian Region of Turkey. The trial was established in randomized block design with four replications in four years (2001 to 2004). In the study, plant height, monopodia, number of sympodia and boll, weight of seed cotton per boll, ginning outturn, 100 seed weight, seed cotton yield, earliness ratio, fiber length, fiber fineness, strength and uniformity were investigated. According to the results of four years' trials, the varieties, in all investigated characters except fiber fineness, strength and uniformity and year x variety interactions for plant height, number of bolls per plant, seed cotton yield and first harvest ratio, were found to be significant. The highest sympodial number was taken from Agdas-3 variety (13.79), the highest number of boll per plant was taken from Agdas-17 (15.41) and the highest seed cotton weight per boll was taken from Agdas-7 (5.54 g). All Agdas varieties had less ginning outturn than local standard varieties. The highest seed cotton yield was obtained from Agdas-17 (3654.2 kg ha^{-1}), followed by Agdas-3 (3593.8 kg ha^{-1}). All Agdas varieties had more seed cotton yield than standards except Agdas-7. Among Azerbaijan varieties, Agdas-3 was the earliest maturing variety with 81.3% of first harvest ratio. All investigated varieties were similar to each other in fiber technological properties. In conclusion, it can be said that among Azerbaijan varieties, Agdas-17 and Agdas-3 had higher seed cotton yield than local standard varieties, and they are the most hopeful varieties for Southeast Anatolian Region.

Key words: Cotton, *Gossypium hirsutum* L., seed cotton yield and fiber technological properties.

INTRODUCTION

Increase in world population and high living standards are raising demand of natural fibers, particularly cotton lint. Cotton is an important fiber crop in the world. Its seed is an important raw material in oil and forage industries due to the high percentage of the seed oil and protein.

Cotton sowing area is about 33.9 million ha., and cotton lint production is 25.5 million tonnes in the world (Anonymous, 2012). Approximately, 80% of the world production is supplied by 8 countries such as China, India, USA, Pakistan, Brazil, Uzbekistan, CFA and Turkey. Turkey has 542 000 ha. sowing area, 2 580 000 tonnes of cotton seed, 954 600 tonnes of cotton lint and 1760 kg ha^{-1} lint yield (Anonymous, 2011).

In Turkey, cotton is grown on a large scale in Southeast Anatolian, Mediterranean and Aegean Regions. Cotton sowing areas were 313.9 thousand ha in Southeast

Anatolian Region, 129.9 thousand ha in Mediterranean Region and 96.7 thousand ha in Aegean Region (Anonymous, 2011). In recent years, South east Anatolian Region has become the most important cotton production region of Turkey in terms of sowing extent and production quantity. 53.6% of total domestic cotton production has been supplied from this region. It has been estimated that cotton sowing areas will increase much in parallel to increasing of irrigated fields.

Production increase in Southeast Anatolian Region resulted from incease in sowing area and yield per unit area. In recent years, lint yield in this region has risen to the lint yield level of the other cotton growing regions due to use of higher yield, quality cultivars and increasing of irrigated area.

Yield, yield components and fiber technological traits in cotton can vary depending on the genotypic structure of cultivar, environmental conditions and growing techniques (Murray and Verhalen, 1970; Meredith and Bridge, 1973; Meredith, 1990; Brown, 1991).

Begum and Hossain (2011) evaluated twelve advanced cotton genotypes with a standard variety at Central Cotton Research, Training and Seed Multiplication Farm, Sreepur, Gazipur, Bangladesh during 2006 to 2007. They reported that significant differences among the genotypes were observed for different characters. The genotypes, SR-05 and SR-12, produced the highest cotton seed yield (2634.73 and 2632.59 kg ha^{-1}). The genotypes, SR-05 (56.03 g) and CB-9 (56.57 g), produced the highest average weight of cotton seed per ten bolls.

Ashokkumar and Ravikesavan (2011) conducted a field experiment to find out the morphological diversity and performance per se of yield component traits, yield, and a fiber quality traits of eleven cotton genotypes (*Gossypium hirsutum* L.) during 2005 to 2006. Significant differences were recorded for all the fourteen traits observed. Researchers reported that the cultivar, MCU 12, produced maximum yield (95.33 g $plant^{-1}$) due to more number of sympodia (24.11 $plant^{-1}$), number of bolls (25.17 $plant^{-1}$), boll weight (4.20 g), ginning outturn (35.43%), lint index (6.12 g) and seed index (10.44 g).

Long et al. (2010) conducted two field experiments to asses the fiber quality and yarn performance of Australian bred cotton (five *G. hirsutum* L. and one *G. barbadense* L.) genotypes. The strongest yarns were produced using genotypes with the longest and finest fiber. They noticed that *G. hirsutum* L. breeding lines CHQX12B and CHQX377 each had micronaire values of 4.4; but CHQX377 spun stronger yarns due to its finer and more mature fiber.

Karademir et al. (2010) carried out the study from 2007 to 2008 on cotton growing season at Southeastern Anatolia Research Institute in Diyarbakir, Turkey, in order to determine the relationship between yield, fiber length and other fiber related traits in advanced cotton strains, derived from a cotton breeding program. In the study, nine advanced cotton strains and two check varieties were used as plant materials. Researchers reported that the advanced cotton strains had significant differences in terms of the investigated characteristics when compared to the check varieties. The highest cotton seed yield (4087.0 kg ha^{-1}) and fiber yield (1632.2 kg ha^{-1}) were obtained from 'SET-34' cotton strain which had acceptable fiber quality properties.

Singh et al. (2008) conducted a field experiment in Bathinda, Punjab, India, during the spring- summer season of 2005 to 2006 to evaluate the feasibility of growing cotton genotypes in the spring-summer season. Researchers revealed that the highest number of monopodial branches (2.8 per plant) was observed in PIL-8, whereas the highest number of sympodial branches (29.4) and total numbers of boll per plant (33.2) were found in LH-2074.

Mahmood et al. (2007) tested new candidate varieties/strains of cotton (*G. hirsutum* L.) from different research stations/institutes for their yield performance. They reported that cotton seed yield of VH-142 was the highest with 5417 kg ha^{-1} having 52.1 bolls per plant with maximum value, as compared to the other lines. DNH-57 and NIAB-999 remained second and third with 5234 and 5095 kg ha^{-1} cotton seed yield, respectively. BH-160 and CRIS-467 were found to be second and third highest boll bearing varieties with an average of 51.0 and 40.2 bolls $plant^{-1}$, respectively.

Copur (2006) carried out a study during 2000 and 2001 growing season in Sanliurfa Province, Southeastern Anatolia Region of Turkey, to determine the yield, yield components and fiber technological traits of some cotton (*G. hirsutum* L.) cultivars. Fifteen cotton cultivars were used to determine the higher yielding cultivars. They noticed that cotton seed yields varied between 1884 and 4322 kg ha^{-1} and statistical significant differences were determined among cultivars in all the observed characters. Also, they reported that Stoneville-453 was the highest yielding cultivar just ahead of Sayar-314 under irrigated conditions in Southeastern Anatolia Region of Turkey.

Begum et al. (2005) investigated performances of one advanced line encoded SR-05 with three other existing Cotton Development Board (CDB) varieties namely CB-1, CB-5 and CB-9 at four SDB research stations of Bangladesh during 2004 to 2005. They reported that significant variations among the line and varieties were observed for number of vegetative branches per plant, nod number of first fruiting branches, days to first flowering, days to first boll split and cotton seed weight per boll. Although, they also noticed that primary fruiting branches, secondary fruiting branches, number of bolls per plant, plant height at harvest, and cotton seed yield did not show any significant differences.

Hanif et al. (2005) evaluated the performance of 15 candidate cotton genotypes against that of three commercial cultivars in Multan, Pakistan, during 2003 to 2004. CIM-496 (4629 kg ha^{-1}), CIM-506 (4622 kg ha^{-1}),

IR-CIM-448 (4294 kg ha^{-1}), BH-160 (4184 kg ha^{-1}) and MNH-700 (4072 kg ha^{-1}) had higher yields than commercial cultivars [CIM-473 (3782 kg ha^{-1}), CIM-499 (3687 kg ha^{-1}) and FH-1000 (3352 kg ha^{-1})]. CIM-707 produced the tallest plants (159 cm). CIM-506, BH-160 and CIM-496 recorded the highest number of bolls per plant (59, 58 and 54 respectively).The greatest boll weight was recorded for CIM-497 (4.6 g). Researchers noticed that CIM-496, CIM-506 and BH-160 were superior among the evaluated genotypes and were recommended for cultivation in Multan.

Soomro et al. (2005) compared the yield and yield components of six commercial cotton cultivars in an experiment conducted in Multan, Pakistan during 2004 cotton season. They revealed that the cultivars were statistically similar for boll weight and cotton yield per plant of second pick at 120 days after planting (DAP). The cultivars statistically showed significant differences for cotton yield per plant of first pick at 90 DAP, cotton yield per plant of third pick at 150 DAP, number of bolls per plant and cotton yield per hectare. The mean performance of genotypes ranged from 11.22 to 25.85 for bolls per plant, from 3.07 to 3.30 g for boll weight, from 1.50 to 8.80, 7.91 to 13.57 and 19.59 to 63.89 g for cotton yield per plant of first pick at 90 DAP, second pick at 120 DAP and third pick at 150 DAP, respectively; while the range recorded for cotton yield per hectare was 1424 to 3393 kg ha^{-1}.

Basbag and Temiz (2004) conducted a study in Diyarbakir, Turkey, in 2001 to 2002, involving cotton lines and cultivars. They reported that Ogosta 644 had the greatest plant height (90.44 cm) followed by green fibered line (90.28 cm); and Beliizvor 432 (89.46 cm) and Sayar-314 had the highest number of bolls (22.00 per plant), ginning percentage (39.45%) and cotton seed yield (3907.5 kg ha^{-1}). They also noticed that Beliizvor, green fibered line and Sayar-314 recorded the best results for fibre length (30.02 mm), fiber fineness (2.87 mic.) and fiber strength (33.41 g tex^{-1}) compared with the other cultivars and lines.

Arshad et al. (2003) evaluated the performance of new cotton strains, along with a commercial cultivar in Multan, Pakistan, during 1997 to 1998 and 1998 to 1999. They reported that the highest cotton seed yield (3026 kg ha^{-1}) was recorded with CIM-433 followed by CIM-435 (2896 kg ha^{-1}). The longest stapel (28.13 mm) was recorded in CIM-435 followed by CIM-443 (26.43 mm). Researchers also reported that CIM-443 was significantly the best in ginning outturn percentage (37.98%) and number of bolls per plant (36.00) followed by CIM-435. Boll weight was highest (4.34 g) in CIM-435.

Wondimu (2000) tested a set of 12 cotton (*G. hirsutum* L.) genotypes during the main growing season, from 1985 to 1990 at Abobo, Ethiopia, which is characterized by low elevation (530 m above sea level) and a sub-humid climate. They noticed that combined analysis of yield data indicated significant differences among genotypes and environments. Genotype-environment interaction was highly significant, indicating differential performance of the cultivars in an array of environments.

In this study, cultivars of Agdas-3, Agdas-6, Agdas-7 and Agdas-17 (*G. hirsutum* L.) brought from Azerbaijan were compared with local standard cultivars of Sayar-314, Maras-92 and Stoneville-453 and their adaptations were evaluated taking into consideration cotton seed yield, yield components and fiber technological properties.

MATERIALS AND METHODS

In this study, seven cotton (*G. hirsutum* L.) varieties were used. Four of these (Agdas-3, Agdas-6, Agdas-7 and Agdas-17) were early maturing mutant varieties created in Azerbaijan by using mutation method of 1975 to 1990. The other three (Sayar-314, Maras-92 and Stoneville-453) were local standard varieties (std) in the Sanliurfa Province in Southeast Anatolian Region of Turkey. Some agronomical and technological characteristics of the varieties tested are given in Table 1.

The study was conducted in 2001 to 2004 at the Koruklu Agricultural Research Station's experimental field in Akcakale-Sanliurfa Province, based on randomized block design with four replications.

Soil properties of the experiment location

The experiment location is flat or nearly flat, generally deep and has much lime, heavy texture and clay structure. Organic matter (0.09 to 0.4%) and salt content are too low, cation exchange capacity is high, pH of the soil varies from 7.4 to 7.6 (Dinc et al., 1988). Physical and chemical properties of the experiment location soil are given in Table 2.

Climatic properties of the experiment location

Sanliurfa Province is located in Akcakale-Harran Plain, in the South east Anatolian Region of Turkey (38°56′ N 36°41′ E) and has Mediterranean climatic conditions. In Harran Plain, yearly average temperature is 17.1°C, average rainfall is 369.3 mm and average relative humidity is 49% according to a long time data. Generally, average temperature exceeds 30°C in July and August and the highest temperature is 46.8°C in July in this plain. Rainfall is experienced mostly in winter and it does not rain between June and September. Relative humidity reduces in summer and, in July, the average relative humidity is 32% while total evaporisation is 2061.7 mm (Anonymous, 2005). Some of the climatic data of the experiment location (Akcakale-Harran Plain, Sanliurfa) in 2001 to 2004 cotton seasons are given in Table 3.

Previously, maize and wheat were grown in 2001 to 2002 and 2003 to 2004, respectively. After harvesting of previous crops, experimental field was plowed deeply and cultivated in winter twice using cultivator shallow depending on weeds. Before sowing, the experimental field was again tilled using cultivator shallow, and it was harrowed two times. And then, seed bed was prepared.

The seeds were sown by experimental mechanical planter in four-row plots of 12 m length at a row space of 70 cm in 27 April, 2 May, 8 May and 11 May in 2001, 2002, 2003 and 2004, respectively. Later, germinated plants were thinned to 20 cm within row twice. During the growing season, plants were hoed three times

Table 1. Some agronomical and technological characteristics of the varieties tested.

Variety	Sowing-first boll opening duration (day)	Plant height (cm)	Seed cotton yield (kg ha^{-1})	Seed cotton weight per boll (g)	100 Seed weight (g)	Ginning outturn (%)	Fiber length (mm)	Fiber fineness (micro-naire)	Fiber strength (g tex^{-1})
Agdas-3	112	95-100	3410	6.0	11.2	37.5	33.5	4.9	29.3
Agdas-6	119	90-95	3870	5.8	11.0	37.0	32.6	4.8	28.8
Agdas-7	118	85-95	3930	6.9	11.6	39.3	33.5	4.5	29.0
Agdas-17	124	90-100	3650	6.3	12.7	38.6	34.8	4.8	27.0
Maras-92 (STD)	-	90-100	3500-4000	6.0	9-10	39	29-30	3.6	31.0
Sayar-314 (STD)	-	95-100	3500-4000	5.8-6.0	10	41-42	29.5-31.5	4.4-4.5	25.7-26.7
Stoneville-453 (std)	120	-	-	5.6	9.4	41.9	29.3	3.84	28-30

STD: Local standard variety (Anonymous, 1978; 1979; 1982; 1983; 2001; Harem, 2000; Calhoun et al., 1997). (Each of the references (Anonymous (1978, 1979, 1982 and 1983) is registration certificate for each variety. If you look at reference list you can see it. The other two references in this table are also related to varieties).

Table 2. Soil physical and chemical properties of the experiment location (Akcakale-Harran Plain, Sanliurfa) (Dinc et al., 1988).

Horizon	Soil depth (cm)	pH	CEC (Meq/100 g)	Salt (%)	Loam (%)	Silt (%)	Clay (%)	Class of structure
Ap	0 - 18	7.5	52.1	0.090	2.6	36.6	60.8	c
A_{12}	18 - 48	7.5	56.1	0.080	3.0	27.8	69.2	c
B_2	48 - 87	7.4	64.4	0.080	4.0	22.9	73.1	c
B_3	87 - 112	7.6	58.0	0.075	2.4	25.8	71.8	c
C	112 - 150	7.6	54.4	0.078	8.9	23.6	67.5	c

Table 3. The some climatic data of the experiment location in 2001 to 2004 cotton seasons.

Month	Average temperature (°C)				Average relative humidity (%)				Total rainfall (mm)			
	2001	2002	2003	2004	2001	2002	2003	2004	2001	2002	2003	2004
April	16.7	14.8	15.6	15.7	61.1	69.1	61.4	53.4	25.1	53.9	35.5	27.0
May	20.5	20.8	23.5	21.9	50.3	44.9	41.7	46.5	52.1	4.0	13.4	14.0
June	29.0	27.6	27.9	28.7	30.0	32.2	31.4	40.7	-	-	-	-
July	32.2	30.6	30.9	31.8	37.3	35.4	34.5	50.7	-	-	-	-
August	30.6	29.0	31.1	29.7	44.4	41.5	37.2	55.5	-	-	-	-
September	25.8	25.3	25.0	25.7	48.4	47.6	42.2	50.9	2.5	6.3	2.2	-
October	19.2	21.0	20.4	20.2	50.0	47.2	55.0	56.2	29.7	9.5	42.5	2.5

Anonymous (2005).

Table 4. Combined analysis of variance of the varieties tested for yield,yield components, earliness and fiber quality characters for four years (2001 to 2004).

Source of variation	df	Plant height	Sympo-dial number	Boll number per plant	Seed cotton weight per boll	Seed cotton yield	Ginning outturn	100-seed weight	First harvest ratio	Fiber length	Fiber fineness	Fiber strength	Fiber uniformity
Block/Year	12	226.33	4.98	1.00	0.37	177.70	3.89	0.19	12.64	0.98	0.18	1.81	1.38
Years (Y)	3	8661.89**	91.25**	91.05**	13.49**	136375.02**	7.24	3.18**	2920.98**	13.16**	7.36**	195.36**	10395.76**
Varieties (V)	6	764.81**	4.63**	31.06**	0.56*	6431.91**	9.42**	0.90**	107.09**	1.65*	0.17	0.86	2.57
YxV Interaction	18	158.89**	1.85	4.47**	0.14	7516.93**	3.98	0.41	36.26**	1.12	0.11	1.16	1.24
Error	72	68.36	1.09	1.43	0.19	114.18	2.86	0.27	6.34	0.66	0.14	1.73	1.86
Coefficient of variation (CV) (%)		8.21	8.07	8.51	8.09	3.09	4.28	5.05	3.20	2.79	8.08	4.92	2.04

(*)P<0.05; (**)P<0.01; df: Degree of freedom.

and four times with harrow.

In each year, compost fertilizer of 20:20:0 was applied at presowing at a rate of 8 kg da^{-1} N, and P_2O_5. 26% of ammonium nitrate was treated at a rate of 5 kg da^{-1} N by using fertilizer spreader in inter-rows prior to first and second irrigation. After treatment, plants were irrigated by furrow method 8 to 9 times until bolls opened up to 60%. Plants were harvested on 27 September and 12 October in 2001; on 28 September and 26 October in 2002; on 26 September and 24 October in 2003 two times by hand. In 2004, harvesting was done on 19 October only once due to environmental conditions. At harvesting time, samples of 20 bolls were taken from each plot.

Morphological and agronomical characteristics of the varieties were determined according to Gencer et al. (1992). Boll samples of each plot were ginned in experimental gin roller and fiber technological traits were determined using lint samples in high volume instruments (HVI) analyser. Obtained data were analysed for each year separately and over four years according to randomized block design, using MSTAT C statistical package. The means were compared by LSD test.

RESULTS AND DISCUSSION

The results of combined analysis of variance for four years (2001-2004) of varieties tested for yield, yield components, earliness and fiber quality characters are given in Table 4.

According to four years' combined analysis of variance, it was determined that the differences among varieties were statistically significant for plant height, sympodial number, boll number per plant, cotton seed yield, ginning outturn, 100 seed weight, first harvest ratio (P<0.01), cotton seed weight per boll and fiber length (P<0.05). Differences among years were statistically significant (P<0.01) for investigated traits, except ginning outturn. Furthermore, it was determined that year x variety interactions were statistically significant (P<0.01) for plant height, boll number per plant, cotton seed yield and first harvest ratio (Table 4).

Plant height

In Table 5, it is seen that the plant height means for years varied between 86.59 and 126.00 cm; the shortest plants were obtained in 2002 (86.59 cm), and the longest plants were obtained in 2003 (126.00 cm). The plant height of varieties among years varied between 68.25 cm (Stoneville-453) and 102.97 cm (Agdas-3) in 2001; 77.17 cm (Stoneville-453) and 91.42 cm (Maras-92) in 2002; 108.20 cm (Stoneville-453) and 139.05 cm (Agdas-6) in 2003; 90.90 cm (Agdas-3) and 107.00 cm (Agdas-6) in 2004.

According to mean results of over four years, Agdas-6 was the longest variety with 109.15 cm. Maras-92 (103.78 cm), Agdas-7 (103.61 cm) and Agdas-3 (103.21 cm) followed this. The shorter variety was Stoneville-453 (standard variety) with 87.12 cm (Table 5). Basbag and Temiz (2004) reported similar results (89.46 to 90.44 cm). Hanif et al. (2005) reported that the highest plant height was 159 cm. This finding is higher than that of our results. Plant height is a cultivar trait but it can be affected by environmental and cultivation conditions.

Year × variety interaction was statistically significant for plant height. It can be explained that varieties with genotypic differences respond to ecological variablesdifferently (Figure 1). The highest plant height values were seen in Agdas-6 in 2003, while the lowest plant height values were seen in Stoneville-453 in 2001.

Sympodial number

It has been relized that sympodial numbers of years varied between 10.39 number/plant and 14.61 number/plant. While the least sympodial number was obtained in 2002 (10.39 number/plant), the highest sympodial number was

Table 5. Means for plant height and sympodia number and arised groups.

Variety	Plant height (cm)					Sympodia number				
	2001	2002	2003	2004	Means of varieties	2001	2002	2003	2004	Means of varieties
Agdas-3	102.97^{a}	88.62	130.35^{ab}	90.90	103.21^{ab}	14.65^{a}	10.72	15.55	14.25	13.79^{a}
Agdas-6	101.07^{a}	89.47	139.05^{a}	107.00	109.15^{a}	13.37^{abc}	10.92	14.15	12.80	12.81^{abc}
Agdas-7	96.35^{ab}	86.80	130.60^{ab}	100.70	103.61^{ab}	14.90 a	10.22	14.25	14.15	13.38^{ab}
Agdas-17	95.07^{ab}	88.25	115.60^{bc}	99.85	99.69^{b}	14.30^{ab}	10.07	14.70	13.85	13.23^{ab}
Maras-92 (std)	94.72^{ab}	91.42	131.90^{ab}	97.10	103.78^{ab}	12.07^{c}	9.80	14.45	13.60	12.48^{bc}
Sayar-314 (std)	81.87^{bc}	84.40	126.30^{abc}	100.75	98.33^{b}	12.40^{bc}	10.55	14.70	13.40	12.76^{bc}
Stoneville-453 (std)	68.25^{c}	77.17	108.20^{c}	94.85	87.12^{c}	11.74^{c}	10.42	14.45	12.40	12.25^{c}
LSD (1%)	17.24	ns	8.47	ns	7.73	1.93	ns	ns	ns	0.97
Means of years	91.47^{c}	86.59^{c}	126.00^{a}	98.74^{b}		13.37^{b}	10.39^{c}	14.61^{a}	13.49^{b}	
LSD (1%)			5.85							
CV (%)	9.26	8.09	7.20	8.49	8.21	7.11	6.29	8.88	7.11	8.07

ns: Not significant; std: local standard variety; * means followed by the same letter are not significantly different at 5% level of probability.

obtained in 2003 (14.61 number/plant) (Table 5). Sympodial number of varieties varied between 11.74 number/plant (Stoneville-453) and 14.90 number/plant (Agdas-7) in 2001; 9.80 number/plant (Maras-92) and 10.92 number/plant (Agdas-6) in 2002; 14.15 number/plant (Agdas-6) and 15.55 number/plant (Agdas-3) in 2003; 12.40 number/plant (Stoneville-453) and 14.25 number/plant (Agdas-3) in 2004.

According to mean results of over four years, the highest sympodial number was taken from Agdas-3 (13.79 number/plant); Agdas-7 and Agdas-17 varieties followed it respectivelly. The lowest sympodial number was taken from Stoneville-453 (12.25 number/plant) (Table 5). Singh et al. (2008) noticed that the highest sympodial number was 29.5 plant^{-1}. Ashokkumar and Ravikesavan ((2011) also reported higher sympodial number (24.17 and 28.97). These findings are very higher than that of our results. Sympodial number is a trait of cultivar. It can occur due to genotypic differences.

Boll number per plant

The highest number of boll per plant was obtained in 2003 (16.11), while the lowest boll number per plant was obtained in 2004 (12.30). Boll number per plant of varieties was between 16.30 (Agdas-3) and 13.25 (Stoneville-453) in 2001; 14.27 (Agdas-6) and 10.90 (Maras-92) in 2002; 18.85 (Agdas-6) and 12.75 (Maras-92) in 2003; 14.20 (Agdas-17) and 10.17 (Maras-92) in 2004 (Table 6). All Agdas varieties had more boll number than standard varieties in all the experiment years except 2002.

According to mean results of over four years, Agdas-17 had the highest boll number per plant with 15.41; Agdas-6 (15.39), Agdas-7 (14.97) and Agdas-3 (14.56) followed this. Maras-92 had the lowest boll number per plant with 11.79 (Table 6). Singh et al. (2008), Hanif et al. (2005), Hassan et al. (2007) and Arshad et al. (2003) reported that the highest boll number per plant was 33.2, 54 to 59, 40.2 to 52.1 and 36 respectively. Boll number per plant is related to plant height and sympodial number. Researchers noticed that the cultivar having the highest boll number was the cultivar having the highest sympodial number. Soomro et al. (2005) noticed that boll number per plant varried between 11.22 and 25.85. Basbag and Temiz (2004) and Ashokkumar and Ravikesavan (2011) reported that boll number per plant was 22 and 25.17 to 26.47, respectively.

Moreover, year x variety interaction was found statistically significant for boll number per plant (Figure 2). This could be due to different responses of varieties to ecological conditions of years and to cultural treatments. While the highest boll numbers per plant were taken from Agdas-6, Agdas-17 and Agdas-7 in 2003, the lowest boll number per plant was taken from Maras-92 in 2004. Being statistically significant, year x variety interaction showed that different varieties can be different in different years for boll number. Similar results were also reported by Bridge et al. (1969), Gill and Singh (1982), Gill et al. (1992) and

Table 6. Means for boll number per plant and seed cotton weight per boll and arised groups.

Variety	Boll number per plant					Seed cotton weight per boll (g)				
	2001	2002	2003	2004	Means of varieties	2001	2002	2003	2004	Means of varieties
Agdas-3	16.30^{a}	12.97^{bc}	16.65^{ab}	12.35 ^{abc}	14.56 ^{a}	4.63	4.92	5.17 ^{a}	5.94	5.16 ^{bc}
Agdas-6	15.87^{ab}	14.27^{a}	18.85^{a}	12.57 ^{abc}	15.39 ^{a}	4.75	4.82	5.80 ^{ab}	6.14	5.38 ^{abc}
Agdas-7	16.15^{ab}	13.37^{abc}	17.85^{a}	12.52 ^{abc}	14.97 ^{a}	4.65	5.38	6.12 ^{a}	6.00	5.54 ^{a}
Agdas-17	16.02 ^{ab}	12.82^{c}	18.60^{a}	14.20 ^{a}	15.41 ^{a}	4.58	5.22	5.94 ^{a}	6.08	5.45 ^{ab}
Maras-92 (std)	13.35^{c}	10.90^{d}	12.75 ^{c}	10.17 ^{c}	11.79 ^{c}	4.63	5.10	5.93 ^{a}	5.98	5.41 ^{a}
Sayar-314 (std)	13.97^{bc}	11.40^{d}	14.40 ^{bc}	12.65 ^{ab}	13.10 ^{b}	4.82	4.92	6.16 ^{a}	6.02	5.48 ^{a}
Stoneville-453 (std)	13.25^{c}	13.85^{ab}	13.70 ^{bc}	11.65 ^{bc}	13.11 ^{b}	4.14	4.80	5.57 ^{ab}	5.68	5.05 ^{c}
LSD (1%)	2.31	1.01	3.39	2.42	1.12	ns	ns	(5%) 0.67	ns	(5%) 0.30
Means of years	14.98^{b}	12.80^{c}	16.11 ^{a}	12.30 ^{c}		4.49 c	5.02 b	5.81 a	5.98 a	
LSD (1%)	0.84									
CV (%)	7.57	3.89	10.34	9.65	8.51	7.02	8.76	7.84	4.98	8.09

ns: Not significant; std: local standard variety; *means followed by the same letter are not significantly different at 5% level of probability.

Gencer et al. (1992).

Cotton seed weight per boll

In Table 6, it can be seen that the highest cotton seed weight per boll was obtained in 2003 and 2004 (5.98 and 5.81 g), followed by 2002 (5.02 g). The lowest cotton seed weight per boll was determined in 2001 (4.49 g). Over the years, cotton seed weight per boll of varieties varied between 4.82 g (Sayar-314) and 4.14 g (Stoneville-453) in 2001; between 5.38 g (Agdas-17) and 4.80 g (Stoneville-453) in 2002; 6.12 g (Agdas-7) and 5.57 g (Stoneville-453) in 2003; 6.14 g (Agdas-6) and 5.68 g (Stoneville-453) in 2004. In all the years of experiment, variety Stoneville-453 had the lowest cotton seed weight per boll (Table 6). Begum and Hossain (2011), Begum et al. (2005), Hanif et al. (2005), Arshad et al. (2003) and Ashokkumar and Ravikesavan (2011) reported similar results (5.6 to 5.7 g, 4.9 to 5.7 g, 4.6, 4.34, 4.20 g respectivelly). Although Soomro et al. (2005) noticed that cotton seed weight per boll was 3.07 to 3.30 g. These findings are lower than that of our findings. However boll weight and cotton seed weight per boll are characters depending on varieties. Some varieties form smaller bolls than others. From the same table, when the four years's mean results are looked at, it is seen that there is also a similar case; Stoneville-453 had the lowest cotton seed weight per boll. Among Agdas varieties only Agdas-7 had more cotton seed weight per boll than standard varieties.

Ginning outturn

From Table 7, it can be seen that in years, there were no significant differences for ginning outturn, and that it varied between 38.83 and 39.97%. Also, from Table 7, when years of trial are investigated separately, it is seen that there were no significant differences among ginning outturn of varieties in the first three years. In 2004, the highest ginning outturn was obtained from standard varieties (Maras-92, Sayar-314 and Stoneville-453, respectively). Ginning outturns of Azerbaijan varieties were between 37.87% (Agdas-7) and 39.60 % (Ağdaş-6).

According to four years' mean results, there were two different groups among varieties for ginning outturn (Table 7). The highest ginning outturns were taken from standard varieties (Sayar-314 and Stoneville-453), followed by Agdas-17 (39.96%), Agdas-3 (39.27%) and Agdas-6 (38.68%), respectively. Agdas-7 had the lowest ginning outturn (38.39%). Agdas varieties having lower ginning outturn is due to their higher 100 seed weights. Arshad et al. (2003) and Basbag and Temiz (2004) reported similar results (37.98 and 39.45%). Ashokkumar and Ravikesavan (2011) reported lower ginning outturn (35.43 and 36.24%).

100 seed weight

The highest 100 seed weight was taken in 2004 (10.81 g) and there were no significant differences

Table 7. Means for ginning outturn and 100 seed weight and arised groups.

Variety	Ginning outturn (%)					100 Seed weight (g)				
	2001	2002	2003	2004	Means of varieties	2001	2002	2003	2004	Means of varieties
Agdas-3	38.33	39.97	40.70	38.10[c]	39.27[ab]	10.12	10.22	10.12[ab]	11.09	10.38[a]
Agdas-6	39.10	38.35	37.67	39.60[abc]	38.68[ab]	10.71	10.04	10.18[ab]	10.98	10.48[a]
Agdas-7	37.62	38.96	39.12	37.87[c]	38.39[b]	9.82	10.06	10.71[a]	11.12	10.43[a]
Agdas-17	41.25	38.84	40.47	39.27[bc]	39.96[ab]	10.25	10.09	10.87[a]	10.94	10.54[a]
Maras-92 (std)	38.92	39.14	40.02	41.87[a]	39.99[ab]	9.87	9.97	10.58[a]	10.63	10.26[ab]
Sayar-314 (std)	40.22	38.82	40.75	41.13[ab]	40.23[a]	10.38	10.16	11.07[a]	10.64	10.56[a]
Stoneville-453 (std)	40.13	38.75	41.05	40.72[ab]	40.16[a]	9.78	10.01	9.42[b]	10.29	9.87[b]
LSD	ns	ns	ns	(1%)2.58	(1%)1.58	ns	ns	(5%) 1.00	ns	(1%) 0.48
Means of years	39.37	38.83	39.97	39.79		10.13 [b]	10.08 [b]	10.42[b]	10.81[a]	
LSD (5%)	ns					(1%) 0.37				
CV (%)	5.78	1.02	5.29	3.19	4.28	5.42	3.97	6.55	3.80	5.05

ns: Not significan; std: local standard variety; *means followed by the same letter are not significantly different at 5% level of probability.

among average 100 seed weight values obtained in the other years of trial. In 2003, the highest 100 seed weights were obtained from Sayar-314 (11.07 g), Agdas-17 (10.87 g), Agdas-7 (10.71 g) and Maras-92 (10.58 g). Stoneville-453 had the lowest 100 seed weight (9.42 g) (Table 7). Ashokkumar and Ravikesavan ((2011) reported similar results (10.44 g).

In the same table, looking at the four years' mean results, it is seen that the highest 100 seed weight was taken from Sayar-314, Agdas-17, Agdas-6, Agdas-7 and Agdas-3 and that the lowest 100 seed weight was taken from Stoneville-453. Varieties having higher 100 seed weights except Sayar-314 (especially Agdas varieties) had lower ginning outturn than the others.

First harvest ratio

In Table 8, it can be seen that the highest first harvest ratio was obtained in 2001 (90.52%), followed by 2003 (73.95%) and that the lowest first harvest ratio was obtained in 2002 (71.89%). In 2001 there was not any difference among varieties for first harvest ratio. In 2002, Stoneville-453 had the highest first harvest ratio (80.41%), followed by Agdas-17 (73.02%). The lowest first harvest ratio was taken from Agdas-7 (66.28%). In 2003, the highest first harvest ratios were obtained from Stoneville-453 (83.87%) and Agdas-3 (78.57%), followed by Sayar-314 (76.68%) and Agdas-17 (73.02%). Agdas-7 had the lowest first harvest ratio (68.78%).

According to three years' mean results, Stoneville-453 had the highest first harvest ratio (83.87%), followed by Agdas-3 (81.30%). For this trait among Agdas varieties, Agdas-3 attracted attention because of higher first harvest ratio (Table 8).

Furthermore, first harvest ratio year x variety interaction was found statistically significant (Figure 3).

While the highest first harvest ratio was taken from Stoneville-453 in 2003, the lowest first harvest ratio was taken from Agdas-7 in 2002. Earliness is an important factor in cotton cultivation. In hand harvest, as long as numbers of opened boll are more in first picking, the variety has been defined as "early maturing". In machine harvesting, it is desired that most bolls open at harvesting time. From this point of view, Agdas-3 can be defined as an early maturing variety because approximately 81% of its bolls open at harvest time.

Cotton seed yield

The highest seed cotton yield was taken in 2003 (4490.5 kg ha^{-1}) followed by 2001 (3271.3 kg ha^{-1}) and 2004 (3052.5 kg ha^{-1}) (Table 8). The lowest seed cotton yield was obtained in 2002 (3015.5 kg ha^{-1}). In 2003, cotton seed yield was more (120 kg) than the other years of trial. This could be

Table 8. Means for first harvest ratio and seed cotton yield and arised groups.

Variety	First harvest ratio (%)					Seed cotton yield (kg ha^{-1})				
	2001	2002	2003	2004	Means of varieties	2001	2002	2003	2004	Means of varieties
Agdas-3	93.11	72.23^{bc}	78.57^{a}	-	81.30^{ab}	3050.8^{b}	3233.9^{b}	5026.3^{b}	3064.1^{b}	3593.8^{ab}
Agdas-6	89.86	69.40^{bc}	69.40^{c}	-	76.22^{d}	3194.6^{b}	3226.7^{b}	4702.2^{c}	2899.3^{b}	3505.7^{bc}
Agdas-7	91.75	66.28^{c}	68.78^{c}	-	75.60^{d}	3200.8^{b}	3087.5^{bc}	4697.4^{c}	2925.5^{b}	3477.8^{c}
Agdas-17	87.88	73.02^{b}	73.02^{bc}	-	77.97^{cd}	2852.2^{c}	3512.5^{a}	5346.3^{a}	2905.7^{b}	3654.2^{a}
Maras-92 (std)	90.42	70.25^{bc}	70.25^{c}	-	76.97^{cd}	3859.8^{a}	2408.9^{e}	4117.8^{d}	3420.5^{a}	3451.7^{c}
Sayar-314 (std)	90.33	71.68^{bc}	76.68^{ab}	-	79.56^{bc}	3694.6^{a}	2933.7^{c}	3980.9^{d}	3321.1^{a}	3482.6^{c}
Stoneville-453 (std)	90.31	80.41^{a}	80.91^{a}	-	83.87^{a}	3045.9^{b}	2705.3^{d}	3552.4^{e}	2831.4^{b}	3033.7^{d}
LSD (1%)	ns	6.36	4.63		2.72	17.13	21.36	22.82	24.93	9.99
Means of years	90.52^{a}	71.89^{c}	73.95^{b}	-		3271.3^{b}	3015.5^{c}	4490.5^{a}	3052.5^{c}	
LSD (1%)	2.05					7.55				
CV (%)	2.23	4.35	3.08	-	3.20	2.57	3.48	2.50	4.01	3.09

ns: Not significant ; std: local standard variety; * means followed by the same letter are not significantly different at 5% level of probability.

because farm manure was applied to experiment field in the previous year.

In 2001, Maras-92 (3859.8 kg ha^{-1}) and Sayar-314 (3694.6 kg ha^{-1}) have the highest cotton seed yield. Agdas-7, Agdas-6, Agdas-3 and Stoneville-453 followed them. The lowest cotton seed yield was obtained from Agdas-17 (2852.2 kg ha^{-1}). In 2002, the highest cotton seed yield was taken from Agdas-17 (3512.5 kg ha^{-1}), followed by Agdas-3 (3233.9 kg ha^{-1}) and Agdas-6 (3226.7 kg ha^{-1}). Maras-92 was the variety with the lowest cotton seed yield (2408.9 kg ha^{-1}). In 2003, the highest cotton seed yield was taken from Agdas-17 (5346.3 kg ha^{-1}), followed by Agdas-3 (5026.3 kg ha^{-1}). Stoneville-453 had the lowest cotton seed yield (3552.4 kg ha^{-1}). In 2004, local standard varieties of Maras-92 (3420.5 kg ha^{-1}) and Sayar-314 (3321.1 kg ha^{-1}) had the highest cotton seed yields and the other varieties followed in the same group (Table 8).

According to four years' mean results, significant differences among the varieties were observed for cotton seed yield. The highest cotton seed yield was taken from Agdas-17 (3654.2 kg ha^{-1}), followed by Agdas-3 (3593.8 kg ha^{-1}); and Stoneville-453 (3033.7 kg ha^{-1}) had the lowest cotton seed yield. Begum and Hossain (2001) reported lower cotton seed yields than that of our results (2634 and 2632 kg ha^{-1}). Copur (2006) reported that cotton seed yields were found between 1884 and 4322 kg ha^{-1}. Basbağ and Temiz (2004) and Arshad et al (2003) reported similar results to that of ours for cotton seed yield (3907.5 kg ha^{-1} and 2896-3026 kg ha^{-1}) . Hanif et al (2005) reported higher cotton seed yield (4294-4629 kg ha^{-1}), Hassan et al (2007) also reported higher cotton seed yield (5095 to 5417 kg ha^{-1}) and Karademir et al (2010) reported higher cotton seed yield (4087 kg ha^{-1}) more than that of our four years' mean results. The different ecological conditions and varieties with different genotypes can give different results because these traits are quantitative. All the Agdas varieties except Agdas-seven gave higher cotton seed yield more than standard varieties. The reason might be that these varieties had more sympodial and boll numbers than standards. It can be said that among Agdas varieties the most hopeful varieties for this region are Agdas-17 and Agdas-3. Agdas-17 gave more cotton seed yield of 170 kg ha^{-1} than standard Sayar-314 with higher cotton seed yield; Agdas-3 gave more cotton seed yield of 110 kg ha^{-1}. The other varieties (Agdas-6 and Agdas-7) could not pass standard varieties.

Cotton seed yield year x variety interaction was found statistically significant (Figure 4). This could be due to ecological and cultural treatment differences in different years. Essentially, cotton seed yield has quantitative heredity and is controlled by more than one gene pair. Therefore, it can be affected by environmental conditions. This finding is similar to the results of Killi (1994), Kaynak et al. (1997), Copur and Oglakci (1997) and Mert and Caliskan (1999).

In Figure 4, it is seen that the highest cotton seed yield was taken from Agdas-17 in 2003 and

Table 9. Means for fiber length and fiber fineness and arised groups.

Variety	Fiber length (mm)					Fiber fineness (micronaire)				
	2001	2002	2003	2004	Means of varieties	2001	2002	2003	2004	Means of varieties
Agdas-3	28.40	28.85	29.12	29.55^{ab}	28.98^{b}	5.17	4.70	4.00	5.07	4.73
Agdas-6	29.12	29.62	29.92	28.40^{c}	29.26^{b}	4.90	4.57	3.77	5.12	4.59
Agdas-7	28.27	28.25	30.52	28.80^{bc}	28.96^{b}	4.92	4.95	4.17	5.10	4.78
Agdas-17	28.35	28.67	30.37	28.85^{bc}	29.06^{b}	5.10	4.75	4.17	5.15	4.79
Maras-92 (std)	28.85	28.12	30.47	29.52^{ab}	29.24^{b}	4.97	4.37	3.65	5.12	4.53
Sayar-314 (std)	29.37	29.15	31.25	29.85^{a}	29.90^{a}	4.72	4.62	4.20	5.02	4.64
Stoneville-453 (std)	29.15	29.25	30.07	28.45^{c}	29.23^{b}	4.92	4.87	3.62	5.02	4.61
LSD (1%)	ns	ns	ns	(5%) 0.96	(5%) 0.57	ns	ns	ns	ns	ns
Means of years	28.79^{b}	28.85^{b}	30.25^{a}	29.06^{b}		4.96^{a}	4.69^{b}	3.94^{c}	5.08^{a}	
LSD (1%)	0.58					0.26				
CV (%)	2.45	3.12	3.17	2.24	2.79	4.40	5.04	15.27	6.31	8.08

ns: Not significant ; std: local standard variety; * means followed by the same letter are not significantly different at 5% level of probability.

that the lowest cotton seed yield was taken from Maras-92 in 2002. All the varieties had the highest yield performance in 2003.

Fiber length

As seen in Table 9, the longest fiber was obtained in 2003 (30.25 mm) and fiber length values obtained in the other years took place at the same group. In the same table, years of trial were investigated separately. It can be seen that there was no statistically difference among varieties for fiber length in the first three years and that fiber length values of varieties varied between 28.12 mm and 31.25 mm in this years. In 2004, the longest fiber was taken from Sayar-314 (29.85 mm), followed by Agdas-3 and Maras-92. Stoneville-453 and Agdas-6 had the lowest fiber.

According to mean results of over four years, Sayar-314 had the longest fiber (29.90 mm) and the other varieties had similar fiber length. Arshad et al. (2003) reported that the longest fiber was 28.13 mm followed by 26.43 mm. Basbağ and Temiz (2004) reported that the longest fiber was 30.02 mm. But Karademir et al. (2010) reported that longer fibers were obtained (32.33 mm). However, researchers also reported that advanced cotton strains derived from a cotton breeding program were used in the study. Ashokkumar and Ravikesavan (2011) also reported they obtained long fibers (32.90 mm).

Fiber fineness

Over the years, fiber fineness varied between 5.08 and 3.94 micronaire (Table 9). The thickest fibers were obtained in 2004, while the thinest fibers were obtained in 2003. Fiber fineness of varieties varied between 4.72 and 5.17 micronaire in 2001; between 4.37 and 4.95 micronaire in 2002; between 3.62 and 4.20 micronaire in 2003; and between 5.02 and 5.15 micronaire in 2004. The thickest fibers were taken from all the varieties in 2004.

According to four years' mean results, there was no statistical signifacant difference among fiber fineness of varieties and these values varied between 4.53 (Maras-92) and 4.79 (Agdas-17) micronaire (Table 9). Basbağ and Temiz (2004) reported that the thickest fibers were 2.87 micronaire. Otherwise, Ashokkumar and Ravikesavan ((2011) noticed that the thickest fibers were 4.6 micronaire. This finding is similar to that of our results.

Fiber strength

As seen in Table 10, the strongest fibers were obtained in 2003 (30.53 g tex^{-1}), followed by 2004 (26.45 g tex^{-1}). The weakest fiber values were obtained in 2001 and 2002. No statistical differences were found among fiber strength of varieties in individual years of trial and combined

Table 10. Means for fiber strength and fiber uniformity and arised groups.

Variety	Fiber strength (g tex^{-1})					Fiber uniformity (%)				
	2001	2002	2003	2004	Means of varieties	2001	2002	2003	2004	Means of varieties
Agdas-3	25.32	25.17	30.20	26.52	26.80	50.97	49.50	83.12	82.85	66.61
Agdas-6	25.02	25.37	30.60	25.85	26.71	49.65	49.90	83.82	83.20	66.64
Agdas-7	24.60	24.70	30.52	27.02	26.71	49.80	50.92	84.20	84.10	67.25
Agdas-17	24.65	25.02	30.62	25.60	26.47	50.52	49.90	84.40	83.65	67.12
Maras-92 (std)	24.52	24.87	31.50	27.27	27.04	51.65	50.20	83.52	83.82	67.30
Sayar-314 (std)	25.00	25.35	29.87	27.35	26.89	51.65	49.82	84.47	84.05	67.50
Stoneville-453 (std)	24.27	25.36	30.37	25.50	26.37	49.55	49.80	83.17	83.15	66.42
LSD (%1)	ns	ns	ns	ns	ns	ns	ns	ns	ns	ns
Means of years	24.77^{c}	25.2^{c}	30.53^{a}	26.45^{b}		50.54^{b}	50.01^{b}	83.82^{a}	83.47^{a}	
LSD (1%)	0.93					0.96				
CV (%)	2.87	4.43	6.60	3.99	4.92	2.79	2.26	2.23	0.99	2.04

ns: Not significant; std: local standard variety; * means followed by the same letter are not significantly different at 5% level of probability.

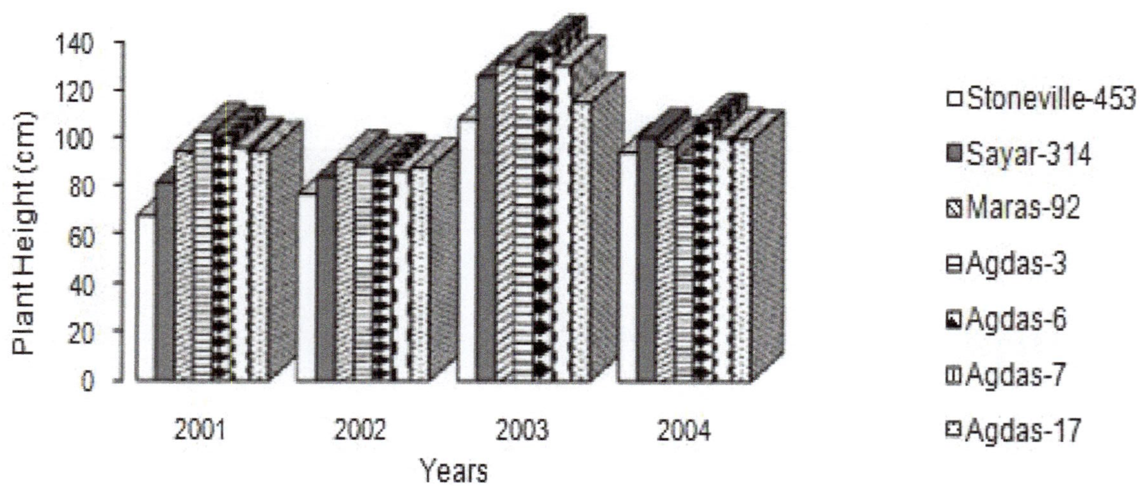

Figure 1. Year × Variety interaction for plant height.

anlysis. Averaged fiber strength values of varieties were in the range of 26.37 to 27.04 g tex^{-1}. According to these results, varieties were found in the strong class (25 to 27 g tex^{-1}): [medium class: 22-24 g tex^{-1}; strong class: 25 to 27 g tex^{-1}; very strong class: 28 to 35 g tex^{-1}]. Basbağ and Temiz (2004) reported that the strongest fibers had 33.41 g tex^{-1}. However, Ashokkumar and Ravikesavan (2011) noticed that the strongest

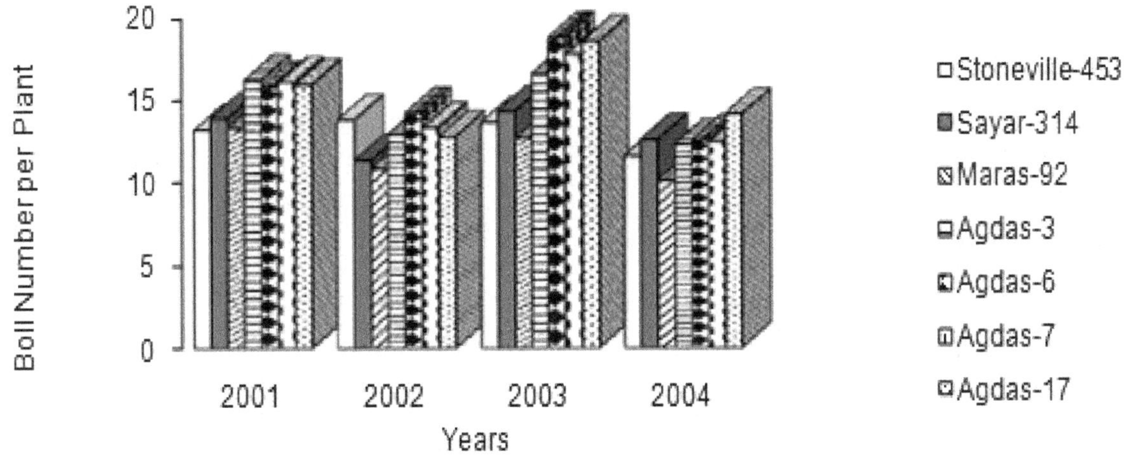

Figure 2. Year × Variety interaction for boll number per plant.

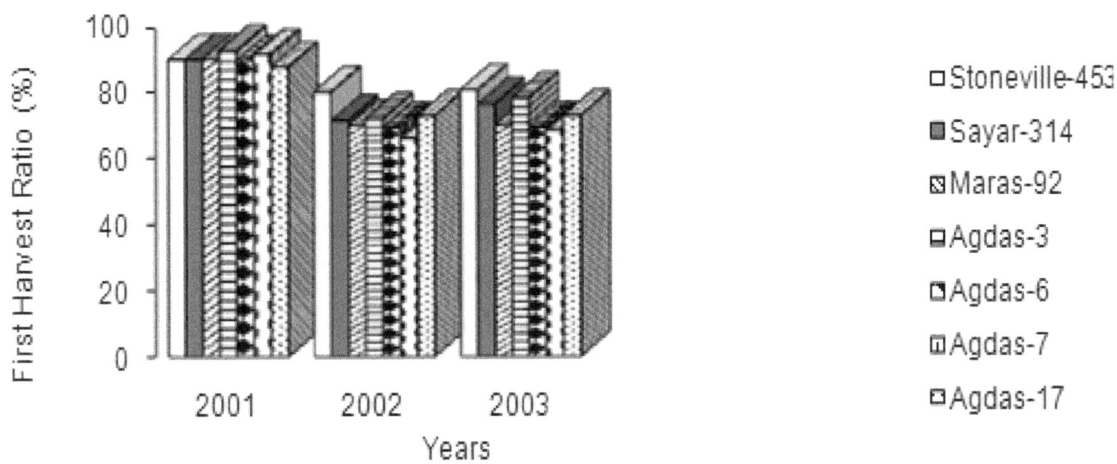

Figure 3. Year × Variety interaction for first harvest ratio.

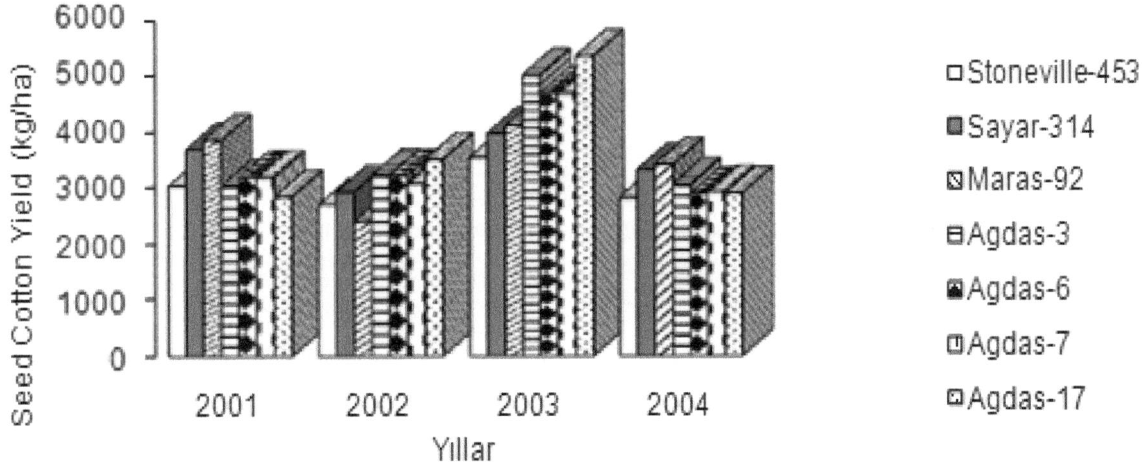

Figure 4. Year × Variety interaction for seed cotton yield.

fibers had 22.9 g tex^{-1}.

Fiber uniformity

In Table 10, it can be seen that the highest fiber uniformity was got in 2003 and 2004. In the individual trial years, statistical differences were not found among varieties for fiber uniformity. In 2003 and 2004, fiber uniformity of varieties increased greatly compared to the two previous years. According to four years' mean results, fiber uniformity of varieties varied in the range of 66.42 to 67.50%.

Conclusion

According to the results of over four years, differences among varieties were statistically significant for investigated traits except for fiber fineness, fiber strength and fiber uniformity; and differences among years were statistically significant for investigated traits except for ginning outturn. Furthermore, it was determined that year x variety interactions were statistically significant for plant height, boll number per plant, cotton seed yield and first harvest ratio. Agdas-3 variety had the highest sympodial number (13.79). Agdas-17 was the variety with the highest boll number per plant (15.41). The highest cotton seed weight per boll was taken from Agdas-7 (5.54 g). All the Agdas varieties had less ginning outturn than local standard varieties. The highest cotton seed yield was obtained from Agdas-17 (3654.2 kg ha^{-1}), followed by Agdas-3 (3593.8 kg ha^{-1}). All the Agdas varieties had more cotton seed yield than standards except Agdas-7. Among Azerbaijan varieties, Agdas-3 variety was the earliest maturing variety with 81.3% of first harvest ratio. All investigated varieties were similar to one another in terms of fiber technological properties.

In conclusion, it can be said that among Azerbaijan varieties, Agdas-17 and Agdas-3 had higher cotton seed yield than local standard varieties, and that they are the most hopeful varieties for South east Anatolian Region.

REFERENCES

Anonymous (1978). The Soviet Union, Azerbaijan National Academy, Decision of The Senate of Institute of Genetics and Selection for Agdas-6 (*G. hirsutum* L.) Variety.

Anonymous (1979). The Soviet Union, Azerbaijan National Academy, Decision of The Senate of Institute of Genetics and Selection for Agdas-7 (*G. hirsutum* L.) Variety.

Anonymous (1982). Ministry of Agriculture of The Soviet Union. Patent for Agdas-3 (*G. hirsutum* L.) Variety. p. 3515.

Anonymous (1983). The Soviet Union, Azerbaijan National Academy, Decision of The Senate of Institute of Genetics and Selection for Agdas-17 (*G. hirsutum* L.) Variety.

Anonymous (2005). Turkish State Meteorological Sevice, Meteorological Directorate of Şanliurfa, Monthly Weather Reports, Şanliurfa. http://tumas.mgm.gov.tr/wps/portal/ (Accessed 15 January 2012).

Anonymous (2010). World Cotton Supply and Use, USDA. http://www.fas.usda.gov/psdonline/psdHome.aspx (Accessed 15 January 2012).

Anonymous (2011). http://www.tuik.gov.tr/bitkiselapp/bitkisel.zul (20.07.2012).

Arshad M, Afzal M, Khan MI, Mahmood R (2003). Performances of newly developed cotton strains for economic and fibre traits in national coordinated varietal trials. Pakistan J. Scientific Ind. Res. 46(5):373-375.

Ashokkumar K, Ravikesavan R (2011). Morphological diversity and *per se* performance in upland cotton (*Gossypium hirsutum* L.). J. Agric. Sci. 3(2):107-113.

Basbag S, Temiz MG (2004). Determinations of some agronomical and technological properties on cotton having different colors fiber. J. Agron. 3(4):301-304.

Begum S, Hossain MS (2011). Characterization and evaluation of advanced cotton genotypes in Bangladesh. SAARC J. Agric. 9(1):1-12.

Begum S, Kashem MA, Miah MAE, Ali MMA, Hossain S, Akter N (2005). Multilocation yield trial of American cotton (*Gossypium hirsutum* L.). SAARC J. Agric. 3:29-36.

Bridge RR, Meredith WR, Chism JF (1969). Variety x environment interactions in cotton variety tests in Delta of Mississippi. Crop Sci. 9:837-839.

Brown JS (1991). Principal component and cluster analyses of cotton cultivar variability across the U.S. cotton belt. Crop Sci. 13:307-312.

Calhoun DS, Bowman DT, May OL (1997). Pedigrees of Upland and Pima cotton cultivars released between 1970 and 1995. Division of Agriculture, Forestry and Veterinary, Medicine Communications, Mississippi State University, USA. http://msucares.com/pubs/bulletins/b1069.pdf (Accessed 1 January 2012).

Copur O (2006). Determination of yield and yield components of some cotton cultivars in semi arid conditions. Pak. J. Biol. Sci. 9(14):2572-2578.

Copur O, Oglakci M (1997). A research on determining the yield and yield components of 12 cotton cultivars relating to *Gossypium hirsutum* L. at the Harran plain conditions. In Proceedings of the 2nd National Field Crops Congress, University of 19 May, Agricultural Faculty, Field Crops Department, 22-25 September, 1997, Samsun, Turkey. pp. 310-314.

Dinc U, Senol S, Sayin, M, Kapur S, Guzel N (1988). The soils of South East Anatolian region. 1. Harran plain. The Scientific and Technological Research Council of Turkey (TUBITAK), Agriculture and Forestry Group, Final Report of Research Project, Project No: TOAG-534, Adana.

Gencer O, Sinan S, Yelin D, Kaynak MA, Gormuş O (1992). Determination of cotton varieties high yielding and having superior fiber technological properties in GAP region. Cukurova University, Agricultural Faculty, The Final Reports of The Package of Project of GAP Agricultural Research, Investigation and Development. Project No: 5.2.1, Publication No: 31, Publications of GAP, No: 60, Adana, Turkey.

Gill KS, Singh TH, Chanal GS (1992). Analysis of genotype environment interactions in upland cotton. Field Crop Abst. 45:8.

Gill SS, Singh TH (1982). Stability for fibre and morphological characters in upland cotton. Indian J. Gen. Plant Breed. 41(2):292-296.

Hanif CM, Hassan SW, Mehdi SS, Iqbal K, Hayat F (2005). Performance of candidate cotton genotypes in respect of yield and yield components. Indus Cottons 2(2):151-154.

Harem E (2000). Domestic and foreign cotton varieties registered in Turkey and their characteristics. Ministry of Agriculture and Rural Affairs, Directorate of Nazilli Cotton Research Institute, Nazilli-Aydin-Turkey. p. 55.

Hassan M, Haq MA, Aslam M, Maqsood M, Alam SS (2007). Evaluation of new cotton (*Gossypium hirsutum* L.) lines for seed cotton yield and leaf curl virus infection. Int. J. Biol. Biotechnol. 4(2/3):223-226.

Karademir E, Karademir C, Ekinci R, Gencer O (2010). Relationship between yield, fiber length and other fiber-related traits in advanced cotton strains. Not. Bot. Hort. Agrobot. Cluj 38(3):111-116.

Kaynak MA, Unay A, Acarturk E, Ozkan I (1997). Detemination of cotton (*Gossypium hirsutum* L.) varieties which have high yield and high

fiber technological properties in the Buyuk Menderes basin. In Proceedings of the 2nd National Field Crops Congress, University of 19 May, Agricultural Faculty, Field Crops Department, 22-25 September, 1997, Samsun, Turkey. pp. 315-319.
Killi F (1994). İnvestigations on genotype x environment interactions, estimates of heritability and adaptability to environment related to yield, yield components and fiber technological properties of 8 cotton cultivars belonging to *Gossypium hirsutum* L. under east Mediterranean region and GAP (south east Anatolian project) region conditions. Disertation Thesis, Cukurova University, Institute of Natural Sciences, Department of Field Crops.
Long RL, Bange MP, Gordon SG, Sluijs MHJ, Naylor GRS, Constable GA (2010). Fiber quality and textile performance of some Australian cotton genotypes. Crop Sci. 50(4):1509-1518.
Meredith WR (1990). Yield and fiber quality potential for second-generation cotton hybrids. Crop Sci. 31:1045-1048.
Meredith WR, Bridge RR (1973). Yield, yield component and fiber property variation of cotton within and among environments. Crop Sci. 13:307-312.
Mert M, Caliskan ME (1999). Determination of agronomical and technological traits on 16 cotton varieties belong to *Gossypium hirsutum* L. under conditions of Amik Plain. In Proceedings of the 3th national congress of field crops, (Eds.: A. C. Ulger and R. Hatipoglu), 15-18 November, 1999, Adana, Turkey. Industrial Plants 2:259-263.
Mahmood-ul-Hassan, Haq MA, Maqsood M, Alam SS (2007). Evaluation of new cotton (Gossypium hirsutum L.) lines for seed cotton yield and leaf curl virus infection. Int. J. Biol. Biotechnol. 4(2-3):223-226.
Murray JC, Verhalen LM (1970). Genotype by environment interaction study of cotton in Oklahoma. Crop Sci. 10:197-200.
Singh SP, Paramjit S, Buttar GS, Bains GS (2008). Evaluation of cotton genotypes for their performance during spring summer season. Journal of Agrometeorology 10(1):82-85.
Soomro AR, Kakar RG, Hakoomat A, Abid SA (2005). Comparison of yield and its components in some commercial cotton varieties. Indus J. Plant Sci. 4(4):545-552.
Wondimu A (2000). Adaptation of cotton cultivars. Afr. Crop Sci. J. 8(2):153-157.

Analysis of diversity in rice (*Oryza sativa* L.) using random amplified polymorphic DNA (RAPD) and simple sequence repeats (SSR) markers

Harsh Bansal[1], Ravindra Kumar[1] , Vivek[2] and Sanjay[3]

[1]Department of Fingerprinting,Sardar Vallabhbhai Patel University of Agriculture And Technology,Modipuram, Meerut, (U.P), India.
[2]Department of Agronomy, Sardar Vallabhbhai Patel University of Agriculture And Technology,Modipuram, Meerut, (U.P), India.
[3]Division of Genetics, Indian Agriculture Research Institute, New Delhi, India.

Molecular markers are useful tool for assessing genetic variations and resolving genotype identity. In the current study, genetic diversity among 20 rice genotypes was assessed using the random amplified polymorphic DNA (RAPD) and simple sequence repeat (SSR). In RAPD analysis, 20 primers generated a total of 116 bands of which 114 were polymorphic. The number of amplification products produced by each primer varied from 4 to 7 with an average of 5.8 bands per primer. Twenty (20) SSR primers generated a total of 65 alleles with an average 3.2 alleles per primer. Genetic diversity of 20 genotypes estimated by polymorphic information content (PIC) value ranged from 0.62 to 0.97 in SSR and 0.33 to 0.88 with RAPD analysis. The cluster dendrogram by SSR revealed two major clusters. Rajeshwari was the only genotype in cluster I. The cluster II further divided into two sub clusters IIA and IIB. II A consisted of 17 genotypes while II B consisted of two genotypes (Apo and Kalakeni). The information generated from this study can be used to maximize selection of diverse parents and broaden the germplasm base for the future rice breeding programs.

Key words: Genetic diversity, molecular markers, rice, markers based estimation of genetic diversity in rice.

INTRODUCTION

Rice (*Oryza* spp.) is one of the most important food crops in the world, being planted on almost 11% of the Earth's cultivated land area over a wide number of ecosystems (Cuevas and Fitzgerald, 2012). Rice belongs to the genus *Oryza* and has two cultivated and 22 wild species.

Rice is an ideal model plant for the study of grass genetics and genome organization due to its diploid genetics, relatively small genome size of 430 Mb (Causse et al., 1994), significant level of genetic polymorphism (McCouch et al., 1998), large amount of well conserved genetically diverse material (about 100,000 accessions world wide) and the availability of widely collected compatible wild species. The success of breeding program also depends upon the amount of genetic variability present in the population and extent to which the desirable traits are heritable. The assessment of phenotype may not be a reliable measure of genetic differences. The rapid development of biotechnology allows easy analysis of a large number of loci distributed throughout the genome of plants. Molecular markers have proven to be powerful tool in the assessment of genetic variation and in the elucidation of genetic relationships within and among species (Matin et al., 2012). Molecular markers originate from different parts of the genome including coding and non-coding regions and can cover either the full genome or large genomic segments. Morphological traits are controlled by a relatively small number of loci (Onaga et al., 2013) however; they serve as a valuable guide for effective collection and use of genetic resources. Molecular markers (RAPD) without prior knowledge of DNA sequences are especially useful for unzipping the variations in species with low genetic variability. RAPD markers are considered to be unbiased and neutral markers for genetic mapping applications, in population genetics, taxonomy as well as for genetic diagnostics. Microsatellites are the popular molecular markers in rice for various applications in genetics and breeding. SSR markers are important tool for estimation of genetic variation and identification of germplasm. These markers have some merits like quickness, simplicity, rich polymorphism and stability, thus being widely used in molecular map construction and gene mapping, construction of fingerprints and genetic purity test (Ma et al., 2011), analysis of germ plasm diversity (Jin et al., 2010), utilization of heterosis, especially in identification of species with closer genetic relationship. SSR markers are more popular in rice because they are highly informative, mostly monolocus, co dominant, easily analyzed and cost effective (Prabakaran et al., 2010). Using PCR rapid amplification and gel electrophoresis of high resolving power, we can test SSR length polymorphism rapidly and economically. In the current studies, RAPD, SSR and combined analysis was performed to assess the genetic diversity among 20 genotypes of rice.

MATERIALS AND METHODS

Plant material

Leaf samples of 20 rice genotypes (25 days after transplanting.) were collected from the Genetics Division, Indian Agriculture research institute, Delhi, India from wet season 2010. The details of rice genotypes are presented in Table 1.

DNA extraction

Total DNA was extracted from fresh leaves by the cetyl tri-methyl ammonium bromide (CTAB) method (Murray and Thompson, 1980). The quality and concentration of extracted DNA were estimated by using a UV-Vis spectrophotometer. The DNA was spooled out, washed twice with 70% ethanol and dissolved in TE (10 mM Tris, 0.1 mM EDTA, pH 8.0) containing 25 µg/ml RNase-A, incubated at 37°C for 30 min and extracted with chloroform:*iso*-amyl alcohol (24:1 v/v). DNA was re-precipitated and dissolved in TE buffer. DNA was checked for its quality and quantity by 0.8% agarose gel electrophoresis.

PCR analysis and gel electrophoresis

In case of SSR, a set of 20 primers (Chakravarthi et al., 2006) were used. The details of primers used for SSR analysis are presented in Table 2. The PCR reaction was carried out using Taq polymerase (Genei) in 20 ml reaction volume containing 1.5X PCR buffer, 2 mM $MgCl_2$, 0.02 mM of each dNTPs, 1 mM of forward and reverse primers, 0.5 µl (3 unit) Taq polymerase and 50 ng genomic DNA. Profile used was as follows: an initial hot start and denaturing step at 95°C for 5 min followed by 35 cycles at 94°C for 1 min, appropriate annealing temperature 55°C for 1 min, and primer elongation at 72°C for 2 min. Final extension step at 72°C for 7 min was performed.

RAPD analysis was performed according to Williams et al. (1990) with minor modifications. The primers used for RAPD are given in Table 3. PCR reactions were carried out in 20 µl volume containing 50 ng of total genomic DNA, 10 pmol primer, 200 µM dNTPs, 2 mM MgCl2, 2.5X PCR buffer and 0.4 µl (3 units) AmpliTaq Polymerase. Twenty (20) 10-mer oligonucleotide random primers were selected for analysis. These primers were obtained from Banglore Genei, India. Amplification was performed in a Astec thermal cycler with the following profile: 94°C for 4 min (initial denaturation), 94°C for 1 min, 32°C for 1 min, 72°C for 2 min for 42 cycles with a final extension at 72°C for 7 min.

The RAPD-PCR products were analyzed directly on 1.5% agarose gels in TAE buffer while SSR-PCR products were analyzed on 4% agarose gel, visualized by staining with ethidium bromide under short-wave UV light. DNA ladder used in the electrophoresis was of 100 bp. Agrose gel electrophoresis for PCR products of RM-222, SSR marker and OPA-07, RAPD marker is shown in Figure 4 and 5 respectively.

Data analysis

Pair wise comparison of genotypes, based on the presence (1) or absence (0) of unique and shared polymorphic products was used to generate similarity coefficients of Jaccord's coefficient by NTSYS-pc version 2.1 software (Rohlf, 2000). The similarity coefficient was used to construct a dendrogram by the unweighted pair group method with arithmetic averages (UPGMA) according to Rohlf (1993). The polymorphism information content (PIC) value described by Botstein et al. (1980) and modified by Anderson et al. (1993) for self-pollinated species was calculated as follows:

$$\mathrm{PIC}_i = 1 - \sum_{j=1}^{n} P_{ij}^2$$

Where, p_i equals to the frequency of the *i*th allele and p_j the frequency of the allele. Only data from polymorphic loci were used for analysis.

Table 1. Characteristics of rice genotypes used in investigation.

Name of Variety	Sources/origin	Characteristic features
Apo	IRRI, Philippines	Suitable for aerobic cultivation, larger and fuller grains, tall
Kalakeni	Orissa, India	Semi dwarf, Grain, long slender, tolerant to alkalinity.
Satabdi	West Bengal, India	Semi-dwarf, grains: LS, white, resistant to Sh.B, BB & Sh.R;
Ratna	Orissa, India	Dwarf, long slender and white grain. Moderately resistant to blast, leaf hopper and tolerant to stem borer.
Khandagini	Orrisa , India	Semi dwarf grains- LS, white tolerant to blast, BPH & GM;
IR-36	IRRI, Philippines	Medium duration, irrigated, rain fed upland or lowland areas. semi dwarfs, grain long slender and white.
Swarna	Andhra Pradesh, India	Semi dwarf, grains: LS, resistant to blast, WBPH, moderately resistant to SB
IR-64	IRRI, Philippines	Semi dwarfs, mid/long duration
Safri -17	Chattisgarh , India	Tall, grains –LB , resistant to blast , BPH & BLB
Lalat	Orrisa, India	Semi dwarfs grains – LS, white, resistant to blast, Suitable for aerobic and boro conditions
Khitish	West Bengal, India	Dwarf , grains: LS, moderately resistant to blast & BS,
Naveen	Orissa, India	Medium duration
Suskhasmanat	Bihar , Jharkhand, India	Tall, grains: short bold, white, susceptible to blast, BLB and resistant to drought
Sambhamahsuri	Andhra Pradesh, India	Semi dwarf grains -MS, tolerant to SB,GM & RTV; resistant to blast
Samanta	Andra Pradesh, India	Semi dwarf short duration lodging & disease resistance
Dandi	Gujrat, India	Semi dwarf, grains-MB, white, moderately resistant to Sh.B, BS,BPH,& BPH
Rajeshwari	Orissa, India	Semi dwarf, grains- MB, white,. resistant to blast & BLB
Shravani	Jharkhand, India	Grains-LS, white; resistant to blast, Helminthosporium, tolerant to BLB, susceptible to RTV
Abhishek	Jharkhand, India	Medium duration
Udaya	Orissa, India	Semi dwarf grains: long bold, resistant to BPH, blast, root rot and tolerant to RTV;

RESULTS AND DISCUSSION

SSR analysis

In present study, 20 SSR primers distributed from chromosome 7 to 12 were used to estimate genetic diversity among 20 genotypes. A total of 65 alleles were detected among all genotypes. The number of alleles per locus varied from 2 to 7. The mean allele in current studies (3.25 alleles) was comparable with the results of Etemad et al. (2012) as 3.57 in rice. However, it was reported lower by Prathepha et al. (2012) and Rahman et al. (2012) with an average of 11.85 and 4.18 alleles per locus, respectively.

The overall size of amplified products ranged from 100 (RM264) to 250 bp (RM286). In the current studies, 15 out of 20 SSR primer pairs generated polymorphic bands. PIC values for SSR ranged from 0.62 to 0.97 with mean value of 0.81 (Table 4). The highest PIC value was observed with primer RM-222 and RM-206 (0.97). However, it varied from 0.19 to 0.90 with an average of 0.75 as reported by Borba et al. (2009). About similar average PIC values (0.78) was reported by Upadhyay et al. (2011). The cluster analysis showed a significant genetic variation among the genotypes with similarity coefficients ranging from 0.63 to 0.85 (Figure 1). The dendrogram revealed two distinct clusters at a similarity coefficient level of0.63. Cluster II was the largest and included 19 genotypes while, clusters I comprised only one genotype, that is, Rajeshwari. The cluster II further divided into two sub clusters IIA and IIB. II A consisted of 17 genotypes while II B consisted of only two genotypes (Apo and Kalakeni). Apo and

Table 2. List of RAPD Primers used for the analysis of twenty rice genotypes.

Code	Primer sequences 5'-3'	Code	Primer sequences 5'-3'
OPA-04	AATCGGGCTG	OPA-16	AGCCAGCGAA
OPA-07	GAAACGGGTG	OPA-17	GACCGCTTGT
OPA-08	GTGACGTAGG	OPA-18	AGGTGACCGT
OPA-09	GGGTAACGCC	OPA-19	CAAACGTCGG
OPA-10	GTGATCGCAG	OPA-20	GTTGCGATCC
OPA-11	CAATCGCCGT	OPB-06	TGCTCTGCCC
OPA-12	TCGGCGATAG	OPB-07	GGTGACGCAG
OPA-13	CAGCACCCAC	OPB-08	GTCCACACGG
OPB-11	GTAGACCCGT	OPB-17	AGGGAACGAG
OPB-13	TTCCCCCGCT	OPB-18	CCACAGCAGT

Table 3. List of SSR Primers and their sequences used for the analysis of twenty rice genotypes.

Primer code	Primer sequences 5'-3'	
	Forward	Reverse
RM 21	ACAGT AT TCCGTAGACGG	GCTCCATGAGGG TGGTA AG
RM 536	TCTCTCCTCTTGTTTGGCTC	ACACACCAACACGACCACAC
RM 206	CCCATGCGTTTAACATTC T	CGTTCCATCGATCCGTATGG
RM 566	ACCCAACTACGATCATCG	CTCCAGGAACACGCTCTTTC
RM 320	CAACGTGATCGAGGAGAT C	GGATTTGCTTACCACAGCTC
RM 247	TAGTGCCGATCGATGTA ACG	CATATGGTTTTGACAAAGCG
RM 544	TGTGAGCCTGAGCAAT AACG	GAAGCGTGTGATATCGCATG
RM 346	CGAGAGAGCCCATAACTACG	ACAAGACGACGAGGAGGGAC
RM 547	TAGGTTGGCAGACCTTTTCG	GTCAAGATCATTCTCGTAGCG
RM 519	AGAGAGCCCCTAAATTT CCG	AGGTACGCTCACCTGTGGAC
RM 286	GGCTTCATCTTTGGCGAC	CCGGATTCACGAGATAAACTC
RM 581	GAGCTGTTTTGGACTACGGC	GAGTAGCTTTCTCCCACCCC
RM 222	CTTAAATGGGCCACATGCG	CAAAGCTTCCGGCCAAAAG
RM 47	ACTCCACTCCACTCCCCA	GTCAGCAGGTCGGACGTC
RM 10	TTGTCAAGAGGAGGCATCG	CAGAATGGGAAATGGTCC
RM 19	CA AAAACAGAGCAGATGAC	CTCAAGATGGACGCCAAGA
RM 20	ATCTTGTCCCTGCAGGTCAT	GAAACAGAGGCACATTTCATTG
RMR	ACGAGCTCTCGATCAGCCTA	TCGGTCTCCATGTCCCAC
RM 167	GATCCAGCGTGAGGAACACGT	AGTCCGACCACAAGGTGCGTTGTC
RM 264	GTTGCGTCCTACTGCTACTTC	GATCCGTGTCGATGATTAGC

Rajeshwari were distantly related at a similarity coefficient of 0.63. Cluster analysis showed that Satabdi and Ratna showed maximum similarity with the similarity coefficient of 0.85. Based on dendrogram, the genotypes Apo, Samantha and Rajeshwari belong to different clusters and are genetically diverse.

RAPD analysis

Using 20 RAPD primers, a total of 116 alleles were detected among the 20 rice genotypes. The number of alleles per locus varied from 4 to 7. The average number of alleles per locus was 5.8. The overall size of amplified products ranged from 100 (OPA-10) to 1200 bp (OPA-11). All the RAPD primers used for analysis of genetic diversity and relationship generated polymorphic bands among the genotypes. The PIC value for the RAPD ranged from 0.33 to 0.88 with an average of 0.65 (Table 5). Similar results were also reported by Shiva et.al. (2013) in rice. The present investigation revealed the effectiveness of RAPD in detecting polymorphism among different genotypes of rice. Nair et al. (2013) also reported

Table 4. Polymorphism Information Content (PIC) of SSR Loci across various genotypes.

Primer	Chromosomal location	Motifs	Molecular weight range (bp)	Total number of alleles	Number of Polymorphic alleles	Number of monomorphic alleles	% Polymorphism	Diversity in value of PIC
RM21	11	(GA)18	130-190	5	5	0	100.00	0.90
RM536	11	(CT)16	150-240	4	4	0	100.00	0.88
RM206	11	(CT)21	200-140	7	7	0	100.00	0.97
RM566	9	(AG)15	250-200	4	4	0	100.00	0.74
RM320	7	(AT)11GTAT(GT13)	250-150	6	6	0	100.00	0.94
RM247	12	(CT)16	250-210	4	4	0	100.00	0.93
RM544	8	(TC)9	180-130	2	2	0	100.00	0.68
RM346	7	(CTT)18	200-100	5	5	0	100.00	0.94
RM547	8	(ATT)19	100-280	5	5	0	100.00	0.90
RM519	12	(AAG)8	150	1	0	1	0.00	0.00
RM286	11	(GA)16	100-150	5	5	0	100.00	0.96
RM581	-	(GA)12	200-150	3	3	0	100.00	0.90
RM222	10	(CT)18	150-250	4	4	0	100.00	0.97
RM47	7	(AG)7(AG)11	200-250	6	6	0	100.00	0.96
RM10	7	(GA)15	160	1	0	1	100.00	0.00
RM19	12	(ATC)10	250	1	1	0	100.00	0.00
RM20	12	(ATT)14	-	0	0	0	0.00	0.00
RMR	-	-	-	0	0	0	0.00	0.00
RM167	11	(GA)16	-	0	0	0	0.00	0.00
RM264	8	(GA)27	100-150	2	2	0	100.00	0.62

that RAPD analysis in *O. sativa* accessions can be successfully used to estimate the genetic diversity in rice.

The highest PIC value (0.88) was observed for primers OPA-17. Cluster analysis revealed two major clusters at a cut-off similarity coefficient of 0.57 (Figure 2). Cluster I was largest and included 14 genotypes while cluster II consisted of six genotypes. Jaccard's coefficient of similarity revealed that 83.8% exist between genotypes Shravani and Abhishek. Whereas, Apo keeps very low level of similarity with Rajeshwari at similarity coefficient of 0.57%. The major cluster I further divided into two sub clusters IA and IB. The sub cluster IA consist of 10 genotypes namely Khitish, Naveen, Udaya, Sukhasamat, Shravani, Abhishek, Sambhamasuri, Dandi,Samanta and Rajeshwari with the similarity coefficient ranged between 0.62 to 0.83. While sub cluster IB consist of 4 genotypes namely Swarna, IR-64, Safri and Lolut with similarity coefficient from 0.67 to 0.78.

Combined analysis

The cluster dendrogram with combined analysis revealed two major clusters. Cluster I was the largest and included 14 genotypes, namely Swarna. IR64, Lolat, Safri, Suskhasamanat, Khitis, Naveen, Sambhamasuri, Dandi, Samanta, Rajaewari, Shravani, Abhishek and Udaya with the similarity

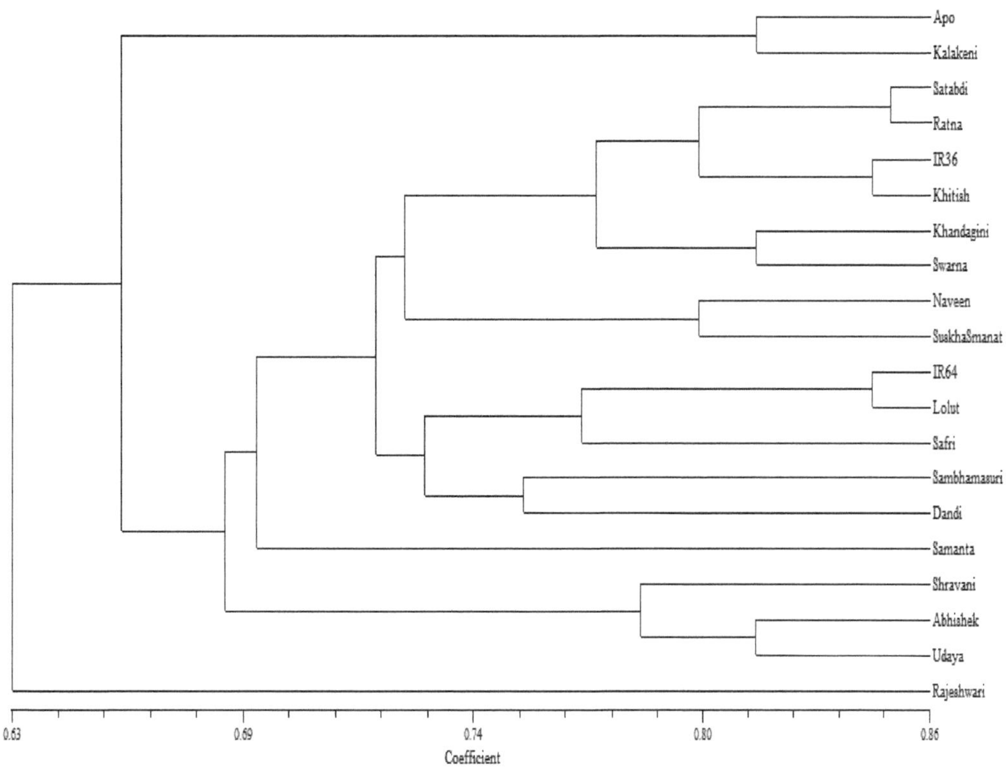

Figure 1. Dendrogram showing clustering of 20 genotypes of rice genotypes based on SSR data.

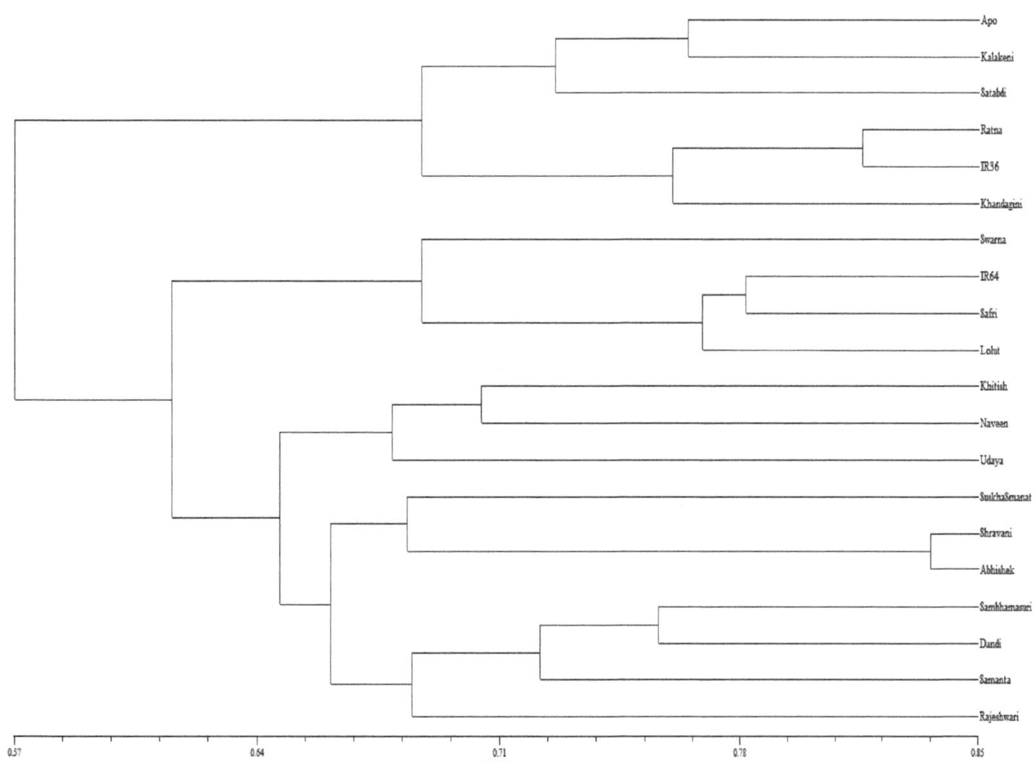

Figure 2. Dendrogram showing clustering of 20 genotypes of rice genotypes based on RAPD data.

Table 5. Polymorphism information content (PIC) of RAPD loci across various genotypes.

Primer	Annealing temperature	Molecular weight range (bp)	Total number of alleles	Number of polymorphic alleles	Number of Monomorphic alleles	% of polymorphism	PIC
OPA-04	30	350-800	5	5	0	100.00	0.64
OPA-07	30	200-700	7	7	0	100.00	0.68
OPA-08	30	200-1100	5	4	1	80.00	0.33
OPA-09	32	100-1000	5	4	1	80.00	0.46
OPA-10	30	100-800	6	6	0	100.00	0.64
OPA-11	30	100-1200	5	5	0	100.00	0.66
OPA-12	30	200-800	7	7	0	100.00	0.72
OPA-13	32	150-700	5	5	0	100.00	0.62
OPA-16	30	300-700	5	5	0	100.00	0.77
OPA-17	30	200-1000	6	6	0	100.00	0.88
OPA-18	30	250-600	5	5	0	100.00	0.82
OPA-19	30	400-1000	7	7	0	100.00	0.66
OPA-20	30	400-900	7	7	0	100.00	0.64
OPB-06	32	300-800	4	4	0	100.00	0.67
OPB-07	32	200-800	5	5	0	100.00	0.69
OPB-08	32	200-550	6	6	0	100.00	0.79
OPB-11	30	300-800	6	6	0	100.00	0.76
OPB-13	32	300-700	6	6	0	100.00	0.78
OPB-17	30	200-900	7	7	0	100.00	0.86
OPB-18	30	300-1000	7	7	0	100.00	0.59

coefficient that ranged between 0.65 to 0.82 while clusters II comprised six genotypes namely Apo, Kalakeni, Satabdi, Ratna, IR36 and Khandagini with the similarity coefficient ranging from 0.69 to 0.82 (Figure 3). Jaccard's coefficient of similarity revealed that high degree of similarity up to the extent of 82.1% existed between IR36 and Ratna. Dendrogram revealed that Apo was distantly located with Udaya with similarity coefficient of 0.62. Dendrogram revealed that genotypes Swarna, Satabdi, and Sukhasmanat, Samanta, Rajeshwari, Udaya and Apo belong to different clusters and are genetically diverse. SSR markers showed higher PIC value compared to RAPD which indicates that SSR markers are highly informative and are reliable. In the current study, larger range of distinct values for genotypes revealed by microsatellite markers provides greater confidence for the assessments of genetic diversity and relationships, which can be used in future breeding programs. With the aid of microsatellite makers and clustering data, different distantly related rice genotypes may be combined by intercrossing genotypes, for instance, aromatic rice genotypes with non-aromatic rice genotypes from different clusters to get hybrid varieties with highest heterosis (Sajib et al., 2012). Markers with PIC values of 0.5 or higher are highly informative for genetic diversity studies and can be successfully used to distinguish the polymorphism at a specific locus. The joint use of primer is excellent way of identification of genotypes. A combination of RAPD and SSR help to provide whole genome coverage and

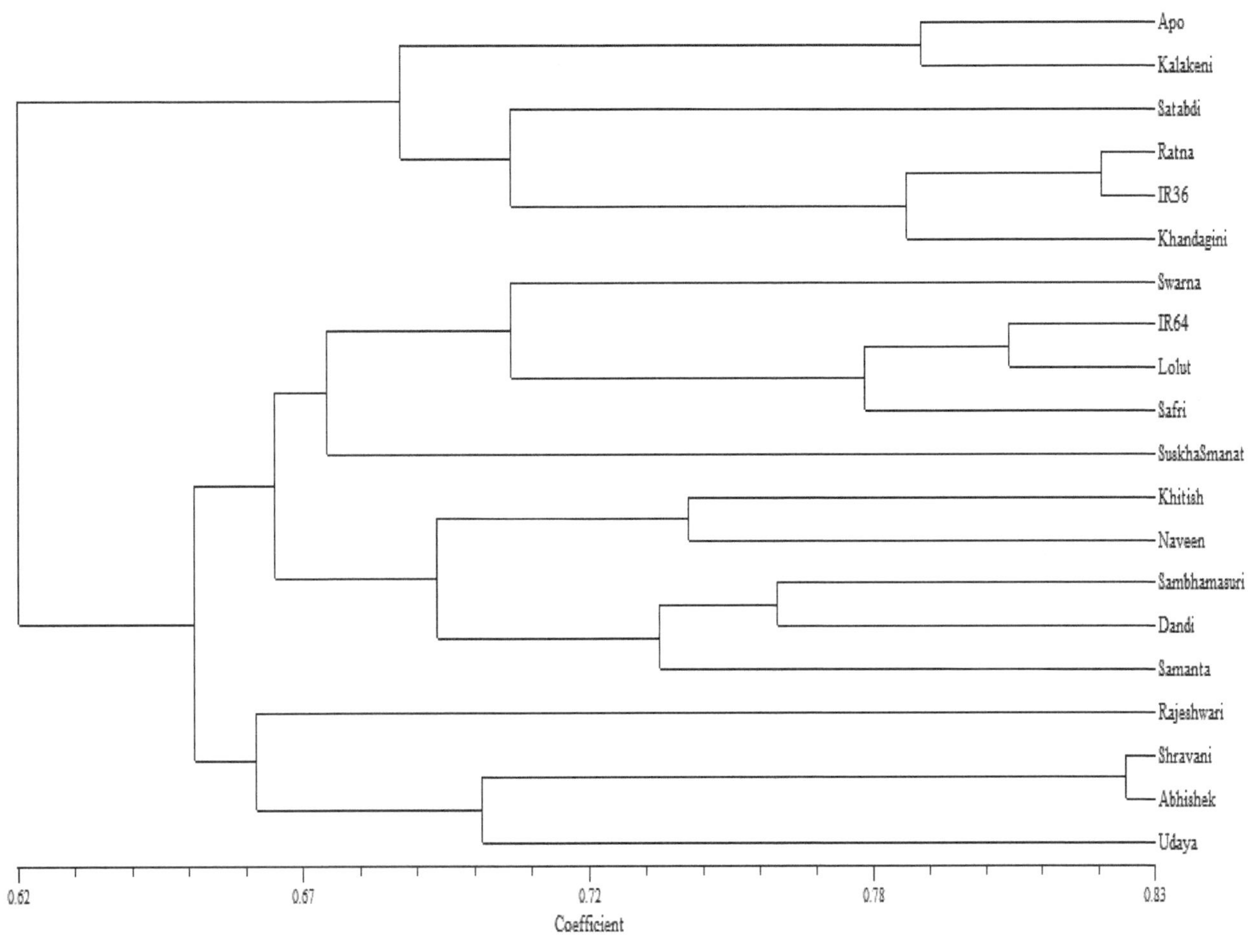

Figure 3. Dendrogram showing clustering of 20 genotypes of rice based on combined (SSR and RAPD) data.

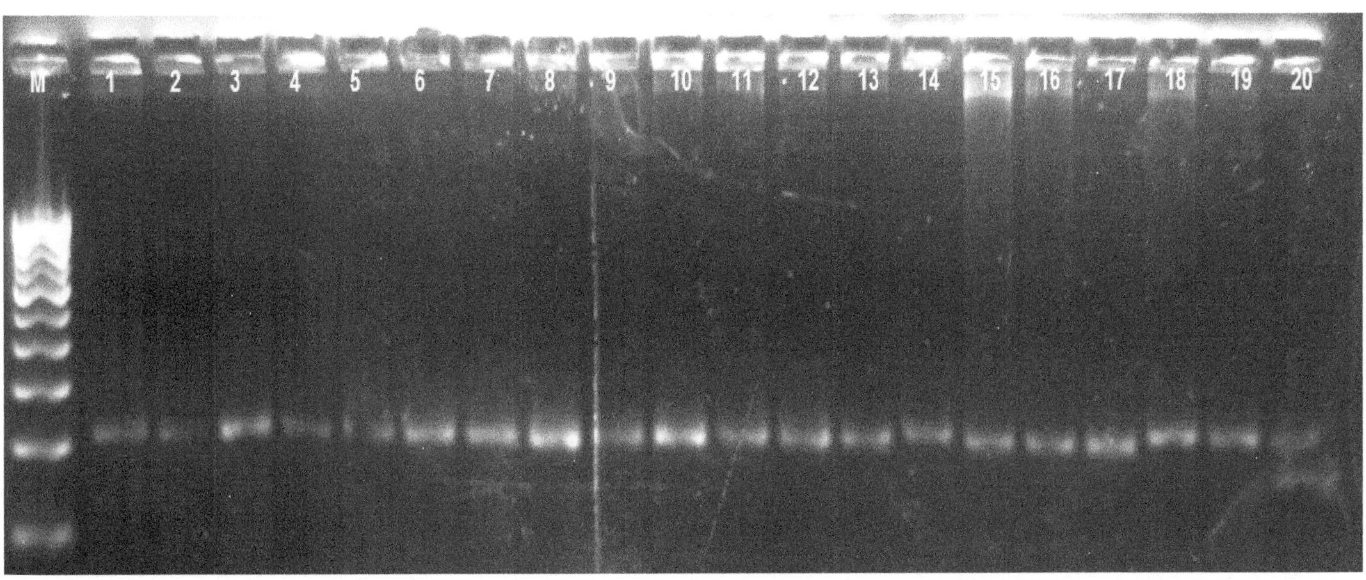

Figure 4. Agrose gel electrophoresis for PCR products of SSR marker (RM-222) of 1 - 20 rice genotypes (As described in Table 1). M=DNA ladder (100 bp).

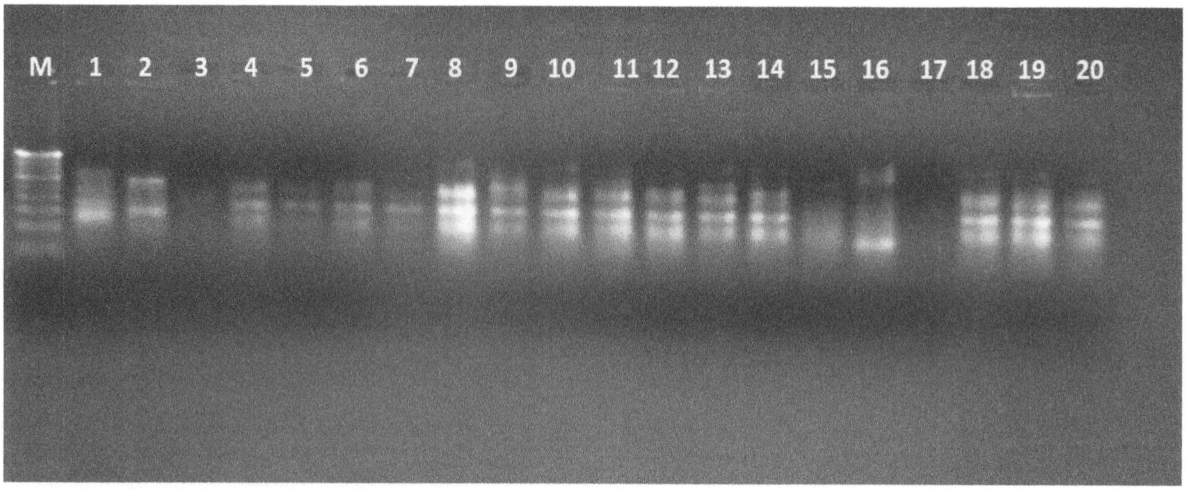

Figure 5. Agrose gel electrophoresis for PCR products of RAPD marker (OPA-07) of 1 - 20 rice genotypes (as described in Table 1). M=DNA ladder (100 bp).

and reduce the error in estimation of genetic similarity based on any one marker system alone.

ACKNOWLEDGEMENTS

The technical help provided by Mrs. Rekha Satpuri, Technician and Mr. Ashwani Kumar Singh, Senior Research Fellow during the course of investigation is gratefully acknowledged. Financial help provided by SVPUA&T, Meerut is also gratefully acknowledged.

REFERENCES

Anderson JA, Churchill GA, Autrique JE, Tanksley SD, Sorrells ME (1993). Optimizing parental selection for genetic linkage maps. Genom. 36:181-6.

Botstein D, White R, Skolnick M, Davis R (1980). Construction of genetic linkage map in man using restriction fragment length polymorphism. Am. J. Genet. 32:314–331.

Causse M, Fulton TM, Cho YG, Ahn SN, Chunwonge J, Wu J, Xiao ZY, Ronald PC (1994). Saturated molecular map of rice genome based on inter specific back cross. Phyto. pathol. 85:1251-1274.

Chakravarthi KB, Rambabu N (2006). SSR marker based DNA fingerprinting and diversity study in rice (*Oryza sativa* L.). Afr. J. Biotechnol. 5:684–688.

Cuevas RP, Fitzgerald MA (2012). Genetic Diversity of Rice Grain Quality Grain Quality and Nutrition Centre, Int. Rice Res. Inst. Philippines 15:285-310.

Etemad A, Maziah M, Daud SK (2012). Determination of genetic relatedness among selected rice (Oryza sativa, L.) cultivars using microsatellite markers. Afr. J. Biotechnol. 11(28):7158-7165.

Jin L, Lu Y, Xiao P, Sun M, Corke H, Bao J (2010). Genetic diversity and population structure of a diverse set of rice germplasm for association mapping. Theor. Appl. Genet. 121(3):475-487.

Ma H, Yin Y, Guo ZF, Cheng LJ, Zhang L, Zhong M, Shao GJ (2011). Establishment of DNA fingerprinting of Liaojing series of japonica rice. Middle-East J. Sci. Res. 8(2):384-392.

Matin S, Mohmmad A, Islam MM, Sikdar SU, Zobayer N (2012). Molecular marker based (SSR) genetic diversity analysis in deep water rice germplasms of Bangladesh. Int. J. Biosci. 10(2):64-72

McCouch SR, Kochert G, Yu ZH, Wang ZY, Khush GS, Coffman WR, Tanksley SD (1998). Molecular mapping of rice chromosomes. Theor. Appl. Genet. 76:815-829.

Murray MG, Thompson WF (1980). Rapid isolation of high molecular weigh plant DNA. Nucleic Acids Res. 8:4321-4325.

Nair GM, Nair AJ, Rajani J, Deepu V (2013). Molecular characterization of selected cultivars of rice, *Oryza sativa* L. using Random Amplified Polymorphic DNA (RAPD) markers. Int. Food Res. J. 20(2):919-923.

Onaga G, Egdane J, Edema R, Abdelbagi I (2013). Morphological and Genetic Diversity Analysis of Rice Accessions (*Oryza sativa* L.) Differing in Iron Toxicity Tolerance. J. Crop Sci. Biotech. 16(1):53-62.

Prabakaran A, Paramasivam K, Rajesh T, Rajarajan D (2010). Molecular characterization of rice land races using SSR markers. Electron. J. Plant Breed. 1(4):512-516.

Prathepha P(2012). Genetic diversity and population structure of wild rice, *Oryza rufipogon* from Northeastern Thailand and Laos. Australian J. Crop Sci. 6(4):717-723.

Rahman MM, Rasaul MG, Hossain MA, Iftekharuddaula KM, Hasegawa H (2012). Molecular Characterization and Genetic Diversity Analysis of Rice (*Oryza sativa* L.) Using SSR Markers. J. Crop Improv. 26(2):244-257.

Rohlf FJ (2000). NTSYS-pc: Numerical taxonomy and multivariate analysis system. Version 2.1. Exeter Publications, New York, USA.

Rohlf FJ (1993).Numerical taxonomy and multivarieties analysis system NTSys-PC version 1.80 Exeter software, New York.

Sajib AM, Hossain MM, Mosnaz ATMJ, Hossain H, Islam MM, Md. Shamsher Ali MS, Prodhan SH (2012). SSR marker-based molecular characterization and genetic diversity analysis of aromatic landreces of rice (*Oryza sativa* L.). J. BioSci. Biotech. 1(2):107-116.

Siva R, Kumar K, Rajasekaran C (2013). Genetic diversity study of important Indian rice genotypes using biochemical and molecular markers. Afr. J. Biotechnol. 12(10):1004-1009.

Williams K, Kubelik A, Livak K, Rafalski J, Tingey V (1990). DNA polymorphisms amplified by arbitrary primers are useful as genetic markers. Nucleic Acid Res. 18:6531–6535.

Permissions

All chapters in this book were first published in AJB, by Academic Journals; hereby published with permission under the Creative Commons Attribution License or equivalent. Every chapter published in this book has been scrutinized by our experts. Their significance has been extensively debated. The topics covered herein carry significant findings which will fuel the growth of the discipline. They may even be implemented as practical applications or may be referred to as a beginning point for another development.

The contributors of this book come from diverse backgrounds, making this book a truly international effort. This book will bring forth new frontiers with its revolutionizing research information and detailed analysis of the nascent developments around the world.

We would like to thank all the contributing authors for lending their expertise to make the book truly unique. They have played a crucial role in the development of this book. Without their invaluable contributions this book wouldn't have been possible. They have made vital efforts to compile up to date information on the varied aspects of this subject to make this book a valuable addition to the collection of many professionals and students.

This book was conceptualized with the vision of imparting up-to-date information and advanced data in this field. To ensure the same, a matchless editorial board was set up. Every individual on the board went through rigorous rounds of assessment to prove their worth. After which they invested a large part of their time researching and compiling the most relevant data for our readers.

The editorial board has been involved in producing this book since its inception. They have spent rigorous hours researching and exploring the diverse topics which have resulted in the successful publishing of this book. They have passed on their knowledge of decades through this book. To expedite this challenging task, the publisher supported the team at every step. A small team of assistant editors was also appointed to further simplify the editing procedure and attain best results for the readers.

Apart from the editorial board, the designing team has also invested a significant amount of their time in understanding the subject and creating the most relevant covers. They scrutinized every image to scout for the most suitable representation of the subject and create an appropriate cover for the book.

The publishing team has been an ardent support to the editorial, designing and production team. Their endless efforts to recruit the best for this project, has resulted in the accomplishment of this book. They are a veteran in the field of academics and their pool of knowledge is as vast as their experience in printing. Their expertise and guidance has proved useful at every step. Their uncompromising quality standards have made this book an exceptional effort. Their encouragement from time to time has been an inspiration for everyone.

The publisher and the editorial board hope that this book will prove to be a valuable piece of knowledge for researchers, students, practitioners and scholars across the globe.

List of Contributors

Manisha P. Shinde
National Research Centre for Grapes, Solapur Road, Pune -412307, Maharashtra, India

Anuradha Upadhyay
National Research Centre for Grapes, Solapur Road, Pune -412307, Maharashtra, India

Lalitkumar B. Aher
National Research Centre for Grapes, Solapur Road, Pune -412307, Maharashtra, India

G. S. Karibasappa
National Research Centre for Grapes, Solapur Road, Pune -412307, Maharashtra, India

Saïd Ben Hamed
Laboratory of Microbiology, Bioprocess Group, Institute Pasteur of Tunisia (IPT), Bp 74, 13 place Pasteur, Belvédère-1002 Tunis, Tunisia
Laboratoire de Microbiologie IRD, Biotechnologie Microbienne des Environnements Chauds, UMR D180, IRD ESIL MIO, Aix-Marseille Université, 163 avenue de Luminy, 13288 Marseille cedex 9, France

Abderrazak Maaroufi
Laboratory of Microbiology, Bioprocess Group, Institute Pasteur of Tunisia (IPT), Bp 74, 13 place Pasteur, Belvédère-1002 Tunis, Tunisia
Abdeljelil GhramLaboratory of Microbiology, Bioprocess Group, Institute Pasteur of Tunisia (IPT), Bp 74, 13 place Pasteur, Belvédère-1002 Tunis, Tunisia

Ben Ali Gam Zouhaier
Laboratoire de Microbiologie IRD, Biotechnologie Microbienne des Environnements Chauds, UMR D180, IRD ESIL MIO, Aix-Marseille Université, 163 avenue de Luminy, 13288 Marseille cedex 9, France

Marc Labat
Laboratoire de Microbiologie IRD, Biotechnologie Microbienne des Environnements Chauds, UMR D180, IRD ESIL MIO, Aix-Marseille Université, 163 avenue de Luminy, 13288 Marseille cedex 9, France

Anita Rawat
Division of Genetics and Tree Propagation, Forest Research Institute, P.O.I.P.E. Kaulagarh Road, Dehradun 248195, Uttarakhand, India

Santan Barthwal
Division of Genetics and Tree Propagation, Forest Research Institute, P.O.I.P.E. Kaulagarh Road, Dehradun 248195, Uttarakhand, India

H. S. Ginwal
Division of Genetics and Tree Propagation, Forest Research Institute, P.O.I.P.E. Kaulagarh Road, Dehradun 248195, Uttarakhand, India

K. M. Alagappan
Centre of advanced study in Marine Biology, Faculty of Marine Sciences, Annamalai University, Parangipettai -608502, Tamil Nadu South India
Department of Biochemistry and Biotechnology, Pondicherry University, Puducherry - 605014, India

B. Deivasigamani
Department of Biochemistry and Biotechnology, Pondicherry University, Puducherry - 605014, India

S. Balamurugan
Department of Biochemistry and Biotechnology, Pondicherry University, Puducherry - 605014, India

Ahmed M. El-Shehawi
Department of Biotechnology, Faculty of Science, Taif University, Taif 21974, Saudi Arabia
Department of Genetics, Faculty of Agriculture, Alexandria University, Alexandria, Egypt

Mona M. Elseehy
Department of Genetics, Faculty of Agriculture, Alexandria University, Alexandria, Egypt

R. Pushpa
Centre for Plant Breeding and Genetics, Tamil Nadu Agricultural University, Coimbatore-641 003, India

T. S. Raveenderan
Centre for Plant Breeding and Genetics, Tamil Nadu Agricultural University, Coimbatore-641 003, India

S. Rajeswari
Centre for Plant Breeding and Genetics, Tamil Nadu Agricultural University, Coimbatore-641 003, India

P. Amalabalu
Centre for Plant Breeding and Genetics, Tamil Nadu Agricultural University, Coimbatore-641 003, India

D. Punitha
Centre for Plant Breeding and Genetics, Tamil Nadu Agricultural University, Coimbatore-641 003, India

Akin-Idowu Pamela Eloho
Biotechnology Unit, National Horticultural Research Institute, P.M.B. 5432 Idi-Ishin, Ibadan, Oyo State, Nigeria
Cancer Research and Molecular Biology Laboratories, Department of Biochemistry, University of Ibadan, Oyo State, Nigeria

Oyeronke Adunni Odunola
Cancer Research and Molecular Biology Laboratories, Department of Biochemistry, University of Ibadan, Oyo State, Nigeria

Michael Adedapo Gbadegesin
Cancer Research and Molecular Biology Laboratories, Department of Biochemistry, University of Ibadan, Oyo State, Nigeria

Abiola Oke
Vegetable Program, National Horticultural Research Institute, P.M.B. 5432 Idi-Ishin, Ibadan, Oyo State, Nigeria

Uterdzua Orkpeh
Vegetable Program, National Horticultural Research Institute, P.M.B. 5432 Idi-Ishin, Ibadan, Oyo State, Nigeria

F. Bnejdi
Laboratoire de Génétique et Biométrie, Département de Biologie, Faculté des Sciences de Tunis, Université Tunis El Manar 2092, Tunisia

M. Saadoun
Laboratoire de Protection des Végétaux, Institut National de la Recherche Agronomique de Tunisie, Tunisia

N. Rassaa
Laboratoire de physiologie des céréales, Ecole supérieure d'agriculture du Kef. Tunisie

M. El Gazzah
Laboratoire de Génétique et Biométrie, Département de Biologie, Faculté des Sciences de Tunis, Université Tunis El Manar 2092, Tunisia

Akbar Azimi
Department of Plant Breeding and Biotechnology, Ferdowsi University of Mashhad, Iran

Farajollah Shahriari
Department of Plant Breeding and Biotechnology, Ferdowsi University of Mashhad, Iran

Amir Fotovat
Department of Soil Science, Ferdowsi University of Mashhad, Iran

Reza Kazemi Qale
Department of Horticultural Science, Islamic Azad University of Abhar, Iran

Khodaverd Hajizade Agje
Department of Chemistry Engineering, Sahand University of Tabriz, Iran

Kamal Sharma
International Institute of Tropical Agriculture, Oyo Road, PMB 5320, Ibadan, Nigeria

R. Bhattacharjee
International Institute of Tropical Agriculture, Oyo Road, PMB 5320, Ibadan, Nigeria

Alieu Sartie
International Institute of Tropical Agriculture, Oyo Road, PMB 5320, Ibadan, Nigeria

P. Lava Kumar
International Institute of Tropical Agriculture, Oyo Road, PMB 5320, Ibadan, Nigeria

S. H. Mohamed
Department of Biology, Faculty of Science, Taif University, P.O. Box 888, Taif, Kingdom of Saudi Arabia
Soil, Water and Environmental Research Institute, Agricultural Research Center, 9 Gamaa St., P.O. Box 12619, Giza, Egypt

S. D. El-Helafiy Seham
Department of Biology, Faculty of Science, Taif University, P.O. Box 888, Taif, Kingdom of Saudi Arabia

A. Ismail Mona
Department of Biology, Faculty of Science, Taif University, P.O. Box 888, Taif, Kingdom of Saudi Arabia

A. S. Sadik
Department of Biology, Faculty of Science, Taif University, P.O. Box 888, Taif, Kingdom of Saudi Arabia
Department of Agricultural Microbiology, Faculty of Agriculture, Ain Shams University, P.O. Box 68 Hadayek Shubra, 11241, Cairo, Egypt

Zhao Hong Weng
Key Laboratory of Healthy Mariculture for the East China Sea, Ministry of Agriculture, Fisheries College, Jimei University, Xiamen 361021, China
State Key Laboratory of Freshwater Ecology and Biotechnology, Institute of Hydrobiology, Chinese Academy of Sciences, Wuhan 430072, China
Graduate School of Chinese Academy of Sciences, Beijing 100039, China

Yang Jie Xie
Key Laboratory of Healthy Mariculture for the East China Sea, Ministry of Agriculture, Fisheries College, Jimei University, Xiamen 361021, China

Zhi Qun Xiao
Key Laboratory of Healthy Mariculture for the East China Sea, Ministry of Agriculture, Fisheries College, Jimei University, Xiamen 361021, China

Zhi Yong Wang
Key Laboratory of Healthy Mariculture for the East China Sea, Ministry of Agriculture, Fisheries College, Jimei University, Xiamen 361021, China

Jian Fang Gui
State Key Laboratory of Freshwater Ecology and Biotechnology, Institute of Hydrobiology, Chinese Academy of Sciences, Wuhan 430072, China

Thiago Alexandre Mota
Programa de Pós-graduação em Entomologia e Conservação da Biodiversidade, Faculdade Ciências Biológicas e Ambientais, Universidade Federal da Grande Dourados (UFGD), 79804-970, Dourados, MS, Brasil

Maria Freire de Souza
Programa de Pós-graduação em Entomologia e Conservação da Biodiversidade, Faculdade Ciências Biológicas e Ambientais, Universidade Federal da Grande Dourados (UFGD), 79804-970, Dourados, MS, Brasil

Marcos Gino Fernandes
Programa de Pós-graduação em Entomologia e Conservação da Biodiversidade, Faculdade Ciências Biológicas e Ambientais, Universidade Federal da Grande Dourados (UFGD), 79804-970, Dourados, MS, Brasil

Paulo Rogério Beltramin da Fonseca
Programa de Pós-graduação em Entomologia e Conservação da Biodiversidade, Faculdade Ciências Biológicas e Ambientais, Universidade Federal da Grande Dourados (UFGD), 79804-970, Dourados, MS, Brasil.

Samir Oliveira Kassab
Programa de Pós-graduação em Entomologia e Conservação da Biodiversidade, Faculdade Ciências Biológicas e Ambientais, Universidade Federal da Grande Dourados (UFGD), 79804-970, Dourados, MS, Brasil

Jean Carlo de Quadros
Programa de Pós-graduação em Entomologia e Conservação da Biodiversidade, Faculdade Ciências Biológicas e Ambientais, Universidade Federal da Grande Dourados (UFGD), 79804-970, Dourados, MS, Brasil

Huaqiang Tan
College of Horticulture, Sichuan Agricultural University, Ya'an 625014, Sichuan, China

Manman Tie
College of Horticulture, Sichuan Agricultural University, Ya'an 625014, Sichuan, China

Li Zhang
College of Horticulture, Sichuan Agricultural University, Ya'an 625014, Sichuan, China

Yanxia Zhu
College of Horticulture, Sichuan Agricultural University, Ya'an 625014, Sichuan, China

Huanxiu Li
College of Horticulture, Sichuan Agricultural University, Ya'an 625014, Sichuan, China

Viviane Costa Silva Zaffani
Department of Food and Nutrition at the University of Campinas (UNICAMP), Campinas, Brazil

Carolina Cauduro Bensabath Carneiro-Leão
Department of Food and Nutrition at the University of Campinas (UNICAMP), Campinas, Brazil

Giovana Ermetice de Almeida Costa
Department of Food and Nutrition at the University of Campinas (UNICAMP), Campinas, Brazil

Pablo Christiano Barboza Lollo
Department of Food and Nutrition at the University of Campinas (UNICAMP), Campinas, Brazil

Emilianne Miguel Salomão
Department of Anatomy, Cell Biology and Physiology, at the University of Campinas (UNICAMP), Campinas, Brazil

Maria Cristina Cintra Gomes-Marcondes
Department of Anatomy, Cell Biology and Physiology, at the University of Campinas (UNICAMP), Campinas, Brazil

Jaime Amaya-Farfán
Department of Food and Nutrition at the University of Campinas (UNICAMP), Campinas, Brazil

N. Du Rand
Discipline of Plant Pathology, School of Agriculture, Earth and Environmental Sciences, College of Agriculture and Environmental Sciences, University of KwaZulu-Natal, Private Bag X01, Scottsville 3201, Pietermaritzburg, South Africa

M. D. Laing
Discipline of Plant Pathology, School of Agriculture, Earth and Environmental Sciences, College of Agriculture and Environmental Sciences, University of KwaZulu-Natal, Private Bag X01, Scottsville 3201, Pietermaritzburg, South Africa

A.H. Al-Shehri
Ministry of Interior, Public Security, Criminal Evidence Administration, P. O. Box 271198, Riyadh 11352, Saudi Arabia

A.A. Al-Saleh
Department of Zoology, College of Science, King Saudi University, P. O. Box 2455, Riyadh 11451, Saudi Arabia.

Maria do Carmo Catanho Pereira de Lyra
Instituto Agronômico de Pernambuco (IPA), Laboratório de Genômica. Av. Gal. San Martin, 1371, Bongi. Recife, PE, CEP 50761-000, Brasil

Ana Dolores Santiago de Freitas
Universidade Federal Rural de Pernambuco (UFRPE), Departamento de Agronomia, Laboratório de Microbiologia do Solo. Rua Dom Manoel de Medeiros, s/n, Dois Irmãos, Recife, PE, CEP 52171-900, Brasil

Taciana Amorim Silva
Universidade Federal do Amazonas (UFAM), Instituto de Ciências Biológicas. Rua Gen. Rodrigo Octávio Jordão Ramos, 3000, setor sul, Bloco D, Sala 08 Coroado I, Manaus, AM, CEP:69077-000, Brasil

Carolina Etienne de Rosália e Silva Santos
Universidade Federal Rural de Pernambuco (UFRPE), Departamento de Agronomia, Laboratório de Microbiologia do Solo. Rua Dom Manoel de Medeiros, s/n, Dois Irmãos, Recife, PE, CEP 52171-900, Brasil

K. Kumar Marmath
Molecular Biology and Genetic Engineering, College of Basic Science and Humanities, G.B. Pant University of Agriculture and Technology, Pantnagar 263145, Dist. US Nagar, Uttarakhand, India

P. Giri
Molecular Biology and Genetic Engineering, College of Basic Science and Humanities, G.B. Pant University of Agriculture and Technology, Pantnagar 263145, Dist. US Nagar, Uttarakhand, India

G. Taj
Molecular Biology and Genetic Engineering, College of Basic Science and Humanities, G.B. Pant University of Agriculture and Technology, Pantnagar 263145, Dist. US Nagar, Uttarakhand, India

D. Pandey and A. Kumar
Molecular Biology and Genetic Engineering, College of Basic Science and Humanities, G.B. Pant University of Agriculture and Technology, Pantnagar 263145, Dist. US Nagar, Uttarakhand, India

Cristina Helena Maria Moreira Vernetti
Laboratório de Engenharia Genética Animal - LEGA/ Departamento de Zoologia e Genética/Instituto de Biologia/Universidade Federal de Pelotas, Campus Universitário, s/n°. 96010-900 - Pelotas, RS, Brazil

Marília Danyelle Nunes Rodrigues
Programa de Pós Graduação em Zootecnia, Faculdade de Agronomia Eliseu Maciel; Laboratório de Engenharia Genética Animal -LEGA/DZ/IB/UFPel, Brazil

Harold Julian Perez Gutierrez
Programa de Pós Graduação em Zootecnia, Faculdade de Agronomia Eliseu Maciel; Laboratório de Engenharia Genética Animal -LEGA/DZ/IB/UFPel, Brazil

Cecilia Perez Calabuig
Departamento de Ecologia - Universidade Federal do Rio Grande do Sul, Porto Alegre, RS, Brazil

Carla Giovane Ávila Moreira
Programa Pós-graduação em Genética e Biologia Molecular - Universidade Federal do Rio Grande do Sul, Porto Alegre, RS, Brazil

Anthony A. Nlewadim
Fisheries and Aquatic Resources Management, College of Natural Resources and Environmental Management, Michael Okpara University of Agriculture, Umudike, Abia State, Nigeria

Heden Luiz Marques Moreira
Laboratório de Engenharia Genética Animal - LEGA/ Departamento de Zoologia e Genética/Instituto de Biologia/Universidade Federal de Pelotas, Campus Universitário, s/n°. 96010-900 - Pelotas, RS, Brazil

B. Okon
Department of Animal Science, University of Calabar, Calabar, Nigeria

L. A. Ibom
Department of Animal Science, University of Calabar, Calabar, Nigeria

G. N. Njume
Department of Animal Science, University of Calabar, Calabar, Nigeria

Vikram Singh
Department of Genetics and Plant Breeding, Chaudhary Charan Singh Haryana Agricultural University, Hisar (Haryana) 125004, India

R. K. Jain
Department of Molecular Biology and Biotechnology, Chaudhary Charan Singh Haryana Agricultural University, Hisar (Haryana) 125004, India

Mukesh Kumar
Department of Genetics and Plant Breeding, Chaudhary Charan Singh Haryana Agricultural University, Hisar (Haryana) 125004, India

F. W. Jansen van Rijssen
Phytomedicine Programme, Department of Paraclinical Sciences, Faculty of Veterinary Science,University of Pretoria, Private Bag X04, Onderstepoort, 0110, South Africa

E. Jane Morris
Department of Biochemistry, Faculty of Natural and Agricultural Sciences, University of Pretoria and African Centre for Gene Technologies, PO Box 75011, Lynnwood Ridge 0040, South Africa

J. N. Eloff
Phytomedicine Programme, Department of Paraclinical Sciences, Faculty of Veterinary Science,University of Pretoria, Private Bag X04, Onderstepoort, 0110, South Africa

J. A. Anejo-Okopi
AIDS Prevention Initiative in Nigeria (APIN), Jos University Teaching Hospital, Nigeria

O. O. Agbaji
AIDS Prevention Initiative in Nigeria (APIN), Jos University Teaching Hospital, Nigeria
Department of Medicine, University of Jos/Jos University Teaching Hospital, Nigeria

P. A. Agaba
AIDS Prevention Initiative in Nigeria (APIN), Jos University Teaching Hospital, Nigeria
Department of Family Medicine, University of Jos/Jos University Teaching Hospital, Nigeria

P. O. Ugoagwu
AIDS Prevention Initiative in Nigeria (APIN), Jos University Teaching Hospital, Nigeria

K. Were
HIV-R Laboratory, Kenya Medical Research Institute, Kisian Kisumu, Kenya

H. Onywera
HIV-R Laboratory, Kenya Medical Research Institute, Kisian Kisumu, Kenya

P. Owiti
HIV-R Laboratory, Kenya Medical Research Institute, Kisian Kisumu, Kenya

S. E Isa
AIDS Prevention Initiative in Nigeria (APIN), Jos University Teaching Hospital, Nigeria
Department of Medicine, University of Jos/Jos University Teaching Hospital, Nigeria

N. Otecko
HIV-R Laboratory, Kenya Medical Research Institute, Kisian Kisumu, Kenya

A. E. J Okwori
National Veterinary Research Institute, Vom, Nigeria

J. Musa
AIDS Prevention Initiative in Nigeria (APIN), Jos University Teaching Hospital, Nigeria
Department of Obstetrics and Gynaecology, University of Jos/Jos University Teaching Hospital, Nigeria

S. Oguche
AIDS Prevention Initiative in Nigeria (APIN), Jos University Teaching Hospital, Nigeria
Department of Paediatrics, University of Jos/Jos University Teaching Hospital, Nigeria

A. S. Sagay
AIDS Prevention Initiative in Nigeria (APIN), Jos University Teaching Hospital, Nigeria
Department of Obstetrics and Gynaecology, University of Jos/Jos University Teaching Hospital, Nigeria

J. A Idoko
AIDS Prevention Initiative in Nigeria (APIN), Jos University Teaching Hospital, Nigeria
Department of Medicine, University of Jos/Jos University Teaching Hospital, Nigeria

L. Nimzing
AIDS Prevention Initiative in Nigeria (APIN), Jos University Teaching Hospital, Nigeria
Department of Medical Microbiology, University of Jos, Nigeria

E. D. Jatau
Department of Microbiology, Ahmadu Bello University Zaria, Nigeria

O. S. Olonitola
Department of Microbiology, Ahmadu Bello University Zaria, Nigeria

A Mahalingam
Centre for Plant Breeding and Genetics, Tamil Nadu Agricultural University, Coimbatore, India

R. Saraswathi
Tamil Nadu Rice Research Institute, Tamil Nadu Agricultural University, Coimbatore, India

J. Ramalingam
Centre for Plant Molecular Biology, Tamil Nadu Agricultural University, Coimbatore, India

Chenguang Zhu
Shanghai Key Laboratory of Bio-energy Crops, School of Life Sciences, Shanghai University, Shanghai 200444, P. R. China

Peipei Wang
Shanghai Key Laboratory of Bio-energy Crops, School of Life Sciences, Shanghai University, Shanghai 200444, P. R. China

Chunlei Shen
Shanghai Key Laboratory of Bio-energy Crops, School of Life Sciences, Shanghai University, Shanghai 200444, P. R. China

Wei Wang
Shanghai Key Laboratory of Bio-energy Crops, School of Life Sciences, Shanghai University, Shanghai 200444, P. R. China

Yuanping Tang
Shanghai Key Laboratory of Bio-energy Crops, School of Life Sciences, Shanghai University, Shanghai 200444, P. R. China

Bing Mei
Shanghai Key Laboratory of Bio-energy Crops, School of Life Sciences, Shanghai University, Shanghai 200444, P. R. China

Zhengkai Xu
Shanghai Key Laboratory of Bio-energy Crops, School of Life Sciences, Shanghai University, Shanghai 200444, P. R. China

Rentao Song
Shanghai Key Laboratory of Bio-energy Crops, School of Life Sciences, Shanghai University, Shanghai 200444, P. R. China

Conrad A. Omonhinmin
Applied Biology and Biotechnology Unit, Department of Biological Sciences, College of Science and Technology, Covenant University, Ota, Ogun State, Nigeria

Oluwaseun O. Ogunbodede
Applied Biology and Biotechnology Unit, Department of Biological Sciences, College of Science and Technology, Covenant University, Ota, Ogun State, Nigeria

Đorđe Gatarić
Agriculture Faculty, University of Banja Luka, Banja Luka, Republic of Srpska

Vojo Radić
Agriculture Faculty, University of Banja Luka, Banja Luka, Republic of Srpska

Branko Đurić
Agriculture Faculty, University of Banja Luka, Banja Luka, Republic of Srpska

Milenko Šarić
Center for the Advancement and Rural Development, Banja Luka, Republic of Srpska

Zora Čolović
Center for the Advancement and Rural Development, Banja Luka, Republic of Srpska

Boro Petković
Center for the Advancement and Rural Development, Banja Luka, Republic of Srpska

Peter M Waziri
Department of Biochemistry, Kaduna State University, Main Campus, P.M.B. 2339, Kaduna, Nigeria

Festo J Massawe
Department of Crop biotechnology, School of Biosciences, The University of Nottingham, Malaysia Campus, Jalan Broga, 43500 Semenyih, Selangor, Malaysia

Samson B Wayah
Department of Biochemistry, Kaduna State University, Main Campus, P.M.B. 2339, Kaduna, Nigeria

Jaafar M Sani
Department of Biochemistry, Kaduna State University, Main Campus, P.M.B. 2339, Kaduna, Nigeria

Lale EFE
Kahramanmaras Sutcu Imam University, Faculty of Agriculture, Field Crops Department, Kahramanmaras, Turkey

Fatih KILLI
Kahramanmaras Sutcu Imam University, Faculty of Agriculture, Field Crops Department, Kahramanmaras, Turkey

A. Sefer MUSTAFAYEV
Kahramanmaras Sutcu Imam University, Faculty of Agriculture, Field Crops Department, Kahramanmaras, Turkey

Harsh Bansal
Department of Fingerprinting,Sardar Vallabhbhai Patel University of Agriculture And Technology,Modipuram, Meerut, (U.P), India

Ravindra Kumar
Department of Fingerprinting,Sardar Vallabhbhai Patel University of Agriculture And Technology,Modipuram, Meerut, (U.P), India

Vivek
Department of Agronomy, Sardar Vallabhbhai Patel University of Agriculture And Technology,Modipuram, Meerut, (U.P), India

Sanjay
Division of Genetics, Indian Agriculture Research Institute, New Delhi, India